Accounts in Drug Discovery
Case Studies in Medicinal Chemistry

RSC Drug Discovery Series

Editor-in-Chief
Professor David Thurston, *London School of Pharmacy, UK*

Series Editors:
Dr David Fox, *Pfizer Global Research and Development, Sandwich, UK*
Professor Salvatore Guccione, *University of Catania, Italy*
Professor Ana Martinez, *Instituto de Quimica Medica-CSIC, Spain*
Dr David Rotella, *Wyeth Research, USA*

Advisor to the Board:
Professor Robin Ganellin, *University College London, UK*

Titles in the Series:
1: Metabolism, Pharmacokinetics and Toxicity of Functional Groups: Impact of Chemical Building Blocks on ADMET
2: Emerging Drugs and Targets for Alzheimer's Disease; Volume 1: Beta-Amyloid, Tau Protein and Glucose Metabolism
3: Emerging Drugs and Targets for Alzheimer's Disease; Volume 2: Neuronal Plasticity, Neuronal Protection and Other Miscellaneous Strategies
4: Accounts in Drug Discovery: Case Studies in Medicinal Chemistry

How to obtain future titles on publication:
A standing order plan is available for this series. A standing order will bring delivery of each new volume immediately on publication.

For further information please contact:
Book Sales Department, Royal Society of Chemistry, Thomas Graham House, Science Park, Milton Road, Cambridge, CB4 0WF, UK
Telephone: +44 (0)1223 420066, Fax: +44 (0)1223 420247,
Email: books@ rsc.org
Visit our website at http://www.rsc.org/Shop/Books/

Accounts in Drug Discovery
Case Studies in Medicinal Chemistry

Edited by

Joel C. Barrish
Bristol-Myers Squibb, Princeton, NJ , USA

Percy H. Carter
Bristol-Myers Squibb, Princeton, NJ , USA

Peter T. W. Cheng
Bristol-Myers Squibb, Pennington, NJ, USA

Robert Zahler
PharmD Consulting, LLC, Pennington, NJ, USA

RSCPublishing

RSC Drug Discovery Series No. 4

ISBN: 978-1-84973-126-3
ISSN: 2041-3203

A catalogue record for this book is available from the British Library

Published by The Royal Society of Chemistry,
Thomas Graham House, Science Park, Milton Road,
Cambridge CB4 0WF, UK

Registered Charity Number 207890

For further information see our web site at www.rsc.org

Foreword

The great challenge of medicinal chemistry is to connect a preclinical hypothesis about how a disease might be treated to clinical validation of the concept with a new medicine. The history of medicinal chemistry tells a remarkable story of success in achieving this challenging goal. For example, new medicines for treating psychosis, depression, infectious diseases, high blood pressure, hypercholesterolemia and inflammation transformed the practice of medicine and improved the quality of life for countless people around the world. Despite these highly significant accomplishments, the need for new or better medicines to help manage unmet or unsatisfied clinical conditions continues. In *Accounts in Drug Discovery*, Joel Barrish, Percy Carter, Peter Cheng and Robert Zahler have assembled an interesting collection of more recent case histories in drug discovery that will be of value to the drug discovery and development community.

As the pharmaceutical industry evolved during the 20th century, R & D organizations focused on the mechanism of drug action as a key driver for the discovery of new medicines. This point is clearly visible in *Accounts in Drug Discovery*, as each case history describes the pursuit of a drug discovery opportunity based on a defined mechanism for drug action. Some of the targets presented here, such as DPP4 and CCR5, already have been validated through a clinical proof-of-concept backed by a successful product launch, while others like PAR1 and TRPV1 are still faced with this challenge. Those who have proclaimed the end of drug discovery based on a natural product lead should take note of the fact that the PAR1 antagonist described in this book was derived from himbacine. Since only a small fraction of targets that have been studied for drug discovery purposes in the past have actually achieved the goal of creating a new medicine, it is apparent that this is a difficult task. It is interesting to note that information gleaned from the Human Genome Project has been used to rapidly expand the number of preclinical hypotheses for new medicines that await clinical validation. Yet progress in the use of this type of information to bridge the gap between hypothesis and clinical validation has been much slower. Thus, target selection, the first step in rational drug discovery, remains a challenge despite great advances in molecular biology, biochemistry, cell biology and pharmacology. Perhaps siRNA science and technology will improve this situation.

In addition to target selection, lead identification and lead optimization are critical components of the drug discovery process. The tools that support lead identification and lead optimization have improved over time. High-through-put technology has facilitated lead identification by providing better access to libraries of compounds appropriate for screening in the search for activity at a selected target. Advances in molecular and cellular biology and biochemistry have greatly facilitated the design of appropriate assays that can be used in high-capacity mode to define mechanism and function in both the screening and optimization processes. The data from these operations can in turn be more efficiently stored and managed with the use of bioinformatics. In terms of drug design, crystal structures of target proteins with inhibitors bound to them have provided insight for the optimization of lead compounds for some targets. Fragment-based drug design often builds on this type of information. High-throughput technology also has facilitated the evaluation of ADMET properties for compounds that are proceeding through the lead optimization process. Since it is necessary to analyze what the body does to a drug as well as what the drug does to the body, this is a very important use of new technology. Case histories as set forth in *Accounts in Drug Discovery* are a valuable source of first-hand information on how these various factors have contributed to the drug discovery process.

While the new science and technology mentioned above have produced valuable information that is highly relevant to drug discovery, it has yet to increase productivity as measured by new product launches. Since the rate of new product launches has been flat for a long period of time, perhaps it is too soon to be able to evaluate whether or not new science and technology can change it. Be that as it may, it is reassuring to see in *Accounts of Drug Discovery* that creative medicinal chemistry is alive and well and medicinal chemists are still engaged in life-long learning as they use new science and technology to solve challenging problems on the road to new medicines.

Paul S. Anderson, Ph.D.

Preface

At its heart, the practice of medicinal chemistry is an art form influenced by data, style, intuition, and experience. Three decades after the discovery of captopril ushered in a new era of "rational drug design", empiricism remains an important element of most drug discovery projects. As such, the knowledge gleaned from our successes and failures, and those of others, provides the best opportunity to avoid the pitfalls and overcome the vagaries of drug discovery. In *Accounts in Drug Discovery*, we focus on case studies in medicinal chemistry with a particular emphasis on how the inevitable problems that arise during any project can be surmounted.

Our intent is to cover a wide range of therapeutic areas and medicinal chemistry strategies, including lead optimization starting from high-throughput screening "hits" as well as rational, structure-based design. These chapters include "follow-ons" and "next generation" compounds that aim to improve upon first-generation agents. This volume surveys the range of challenges commonly faced by medicinal chemistry researchers, including the optimization of metabolism and pharmacokinetics, toxicology, pharmaceutics, and pharmacology, including proof-of-concept in the clinic for novel biological targets.

These case studies include medicinal chemistry stories on recently approved and marketed drugs, but also chronicle "near-misses", *i.e.* exemplary compounds that may have proceeded well into the clinic but for various reasons did not result in a successful registration. As the vast majority of projects fail prior to registration, much can be learned from such narratives.

By sharing a wide range of drug discovery experiences and information across the community of medicinal chemists in both industry and academia, we believe that these accounts will provide insights into the art of medicinal chemistry as it is currently practiced and will help to serve the needs of active

RSC Drug Discovery Series No. 4
Accounts in Drug Discovery: Case Studies in Medicinal Chemistry
Edited by Joel C. Barrish, Percy H. Carter, Peter T. W. Cheng and Robert Zahler
© Royal Society of Chemistry 2011
Published by the Royal Society of Chemistry, www.rsc.org

medicinal chemists. Our sincere appreciation goes to the authors for sharing their stories and for their contributions to science and patients in need.

Joel C. Barrish, Ph.D.
Percy H. Carter, Ph.D.
Peter T. W. Cheng, Ph.D.
Robert Zahler, Ph.D.

Contents

RSC Drug Discovery Series No. 4
Accounts in Drug Discovery: Case Studies in Medicinal Chemistry
Edited by Joel C. Barrish, Percy H. Carter, Peter T. W. Cheng and Robert Zahler
© Royal Society of Chemistry 2011
Published by the Royal Society of Chemistry, www.rsc.org

CHAPTER 1

The Discovery of the Dipeptidyl Peptidase-4 (DPP4) Inhibitor Onglyza™: From Concept to Market

JEFFREY A. ROBL AND LAWRENCE G. HAMANN

Bristol-Myers Squibb Research & Development, Department of Discovery Chemistry – Metabolic Diseases, P.O. Box 5400, Princeton, NJ 08543, USA

1.1 Introduction

The prevalence of diabetes in developed and now emerging countries represents a significant health burden to a large portion of the world's population. Type-2 diabetic patients, characterized in part by elevated fasting plasma glucose of $>125\,mg\,dL^{-1}$ ($7.0\,mmol\,L^{-1}$) and glycosylated hemoglobin (HbA1c) $\geq 6\%$, are at increased risk for the development of both microvascular (retinopathy, neuropathy, nephropathy) as well as macrovascular complications (myocardial infarction, stroke). As such, diabetes is the leading cause of blindness, kidney failure, and limb amputation worldwide.[1] Diabetes is a progressive disease, with morbidity and mortality risk increasing with both duration and severity of hyperglycemia. Additionally, diabetes is also now impacting different population sectors (adolescents, developing countries) not typically associated with the disease 30 years ago. Consequently, the continually increasing diabetes prevalence is placing greater strain on both health care systems and economies on a global scale. In 2007 alone, studies have shown that diabetes cost the US

RSC Drug Discovery Series No. 4
Accounts in Drug Discovery: Case Studies in Medicinal Chemistry
Edited by Joel C. Barrish, Percy H. Carter, Peter T. W. Cheng and Robert Zahler
© Royal Society of Chemistry 2011
Published by the Royal Society of Chemistry, www.rsc.org

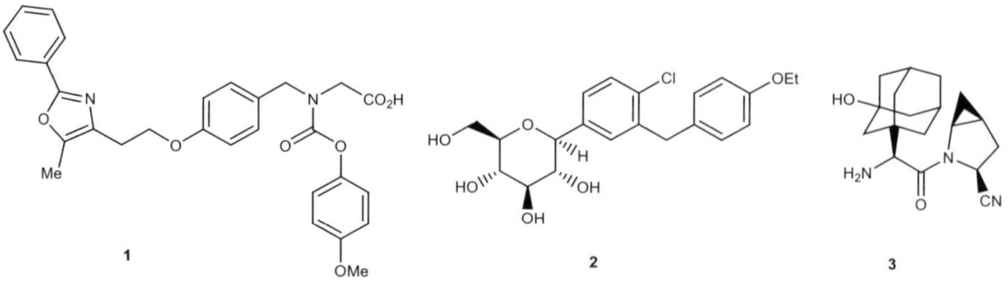

Figure 1.1   Late-stage development candidates from the Bristol-Myers Squibb diabetes research portfolio.

economy \$174 billion in medical expenses and lost productivity.[2] While death rates related to heart disease, stroke, and cancer have all decreased since 1987, the death rate due to diabetes has increased by 45% during this same period.[3] Thus, the discovery and development of new therapies for treating and preventing diabetes continue to be a major emphasis of health care companies.

In response to this landscape, the Discovery organization at Bristol-Myers Squibb (BMS) made the strategic decision to refocus efforts in the late 1990's towards identifying and progressing novel targets for the treatment of diabetes. This was in part aimed at building upon BMS's already established presence in the anti-diabetes market through the Glucophage™ franchise and in recognition of the significant unmet medical need for novel, more efficacious, and well tolerated treatments for the disease. It was from these efforts that advanced clinical candidates such as muraglitazar (**1**, Pargluva™, dual PPAR agonist),[4] dapagliflozin (**2**, SGLT2 inhibitor),[5] and saxagliptin (**3**, Onglyza™, DPP4 inhibitor)[6] were discovered within the BMS Discovery organization (Figure 1.1).

1.2   Modulation of GLP-1 in the Treatment of Diabetes

At the start of this effort, several oral anti-diabetic agents (OADs) were available to patients suffering from type-2 diabetes. These included hepatic glucose suppressors (*e.g.* metformin), insulin secretagogues (*e.g.* sulfonylureas), glucose absorption inhibitors (*e.g.* acarbose), and insulin sensitizers (*e.g.* thiazolidinediones or TZDs such as rosiglitazone and pioglitazone). While all have shown utility in lowering HbA1c levels in diabetic patients, current OADs come with a variety of safety and/or tolerability issues. The biguanindes such as metformin, currently the most widely prescribed therapy for diabetes, have issues related to gastrointestinal (GI) tolerability and lactic acidosis.[7] Sulfonylurea treatment is often accompanied by higher incidences of hypoglycemia and weight gain,[8] while glucose absorption inhibitors exhibit modest efficacy and GI disturbance.[9] Finally, TZDs have been associated with edema, worsening of congestive heart failure, negative effects on bone fracture rate, and, in recent studies, mixed results regarding cardiovascular (CV) safety profiles.[10]

With this background in mind, we sought to identify new targets which would not only provide an efficacious alternative mechanism for lowering blood glucose and HbA1c levels, but would also present an opportunity for achieving a superior safety and tolerability profile when compared to current standards of care. Ideally, such a drug would be suitable for combination with existing agents, as poly-pharmacology with multiple OADs is emerging as the standard treatment paradigm for type-2 diabetes therapy.

Glucagon like peptide-1 (GLP-1) is a 30-amino acid peptide incretin hormone derived from processing of pro-glucagon and is secreted by the L-cells of the intestinal mucosa in response to glucose stimulation. Since the early 1990's, GLP-1 had been known to be a potent insulin secretagogue and glucagon suppressor, with robust anti-diabetic and pro-satiety effects in diabetic humans,[11,12] but efforts to advance GLP-1 itself as a pharmaceutical agent were hampered by its extremely short pharmacokinetic half-life *in vivo* (plasma $t_{1/2} \approx 2\,\text{min}$). As a result, considerable effort in the drug discovery community was expended toward the identification of small-molecule GLP-1 receptor agonists that would capture the beneficial effects of GLP-1 while exhibiting oral bioavailability and a superior pharmacokinetic duration of action. Unfortunately, efforts to identify such small-molecule agonists have to date been unsuccessful, due in part to a dearth of viable *bona fide* screening hits.[13] In light of this shortcoming, a number of pharmaceutical and biotech companies have advanced subcutaneously administered, peptide GLP-1 receptor agonists with superior duration of action *in vivo*. Among the most advanced agents are exenatide (Byetta™)[14] and liraglutide (Victoza™),[15] both of which have been approved by regulatory agencies for the treatment of type-2 diabetes. While these drugs are effective in lowering HbA1c and demonstrate a net beneficial effect on weight gain and other CV risk factors, they require parenteral administration (once or twice daily dosing), and patient uptake of these agents has been limited despite their robust efficacy and promising safety profile.

1.3 Dipeptidyl Peptidase-4 as a Target for Diabetes Treatment

While the advancement of orally active, small-molecule GLP-1 receptor agonists remains elusive, another opportunity to modulate GLP-1 receptor activity *in vivo* focused on preventing the degradation of endogenous GLP-1 with small-molecule inhibitors of the primary peptidase responsible for the *in vivo* degradation of GLP-1, dipeptidyl peptidase-4 (DPP4), a non-classical serine protease.[16] Our initial interest in DPP4 inhibitors was piqued by a report from Holst and Deacon, wherein the authors outlined a compelling argument for the utility of DPP4 inhibition in the treatment of diabetes, primarily *via* the potentiation of endogenous GLP-1.[17] DPP4 belongs to a family of aminodipeptidases and is both a cell surface and circulating enzyme. Historically, it had also been identified as the lymphocyte cell surface marker CD26, and as such DPP4/CD26 exhibits pharmacology related to cell membrane-associated activation of intracellular signal transduction pathways and cell–cell interaction in addition to its peptidase enzymatic activity.[18] It is widely believed that the signaling function of DPP4/CD26 is distinct from its enzymatic function.

The concept of generating targeted protease/peptidase inhibitors as therapeutic agents is well documented in the literature.[19] In the majority of cases, selective enzyme inhibitors have been used to prevent the conversion of an endogenous "non-functional" peptide/protein precursor (*e.g.* angiotensin I) to a physiologically active peptide/protein (*e.g.* angiotensin II), thereby attenuating formation of the protagonist bio-active enzyme product to effect amelioration of the disease state. Use of such approaches has led to marketed drugs for numerous indications, including ACE and renin inhibitors for hypertension,[20] HIV protease inhibitors for AIDS,[21] and thrombin inhibitors for DVT.[22] Less prevalent are approaches targeting proteases/peptidases that degrade endogenous substrates which are known to exert a beneficial effect. For example, neutral endopeptidase (NEP) proteolytically degrades the endogenous vasodilator atrial natriuretic peptide (ANP) to inactive fragments. By retarding this degradation, NEP inhibitors have found use in the treatment of hypertension.[23] In common with NEP, where inhibition of protease mediated protein degradation was the pharmacological objective, BMS and several other research groups engaged in the search for DPP4 inhibitors to maximize the beneficial effects of endogenously released GLP-1.

1.4 Early Inhibitors of DPP4

To jump-start the BMS DPP4 chemistry program, the group was able to capitalize on groundwork laid in the mid-late 1990's when several potent inhibitors of DPP4/CD26 were reported that could be classified as either "irreversible" or "reversible", depending on the mechanism of inhibition (Figures 1.2 and 1.3). Hydroxamates such as **4** were proposed to be both substrates and inhibitors of DPP4, presumably *via* direct covalent modification of the enzyme through the active-site serine residue (Ser630).[24] Phosphate-based inhibitors such as **5** were also reported to undergo covalent addition to DPP4 but exhibited weak potency.[25] The interesting boronate-based inhibitors (*e.g.* **6**), originally advanced by the Tufts University and Boehringer Ingelheim groups, exhibited exceptionally high inhibitory activity *in vitro*, presenting

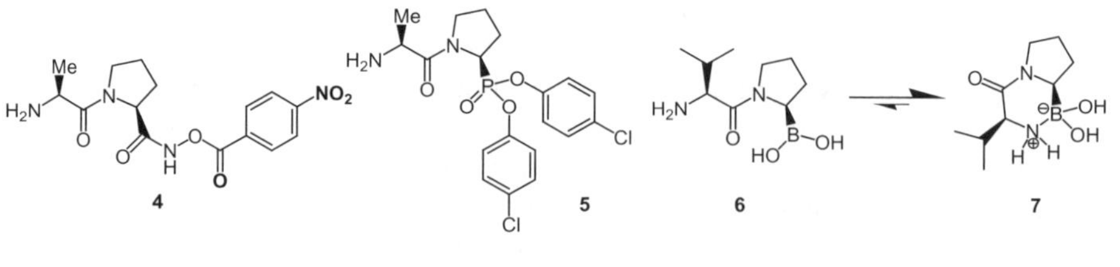

Figure 1.2   Early examples of "irreversible" DPP4 inhibitors.

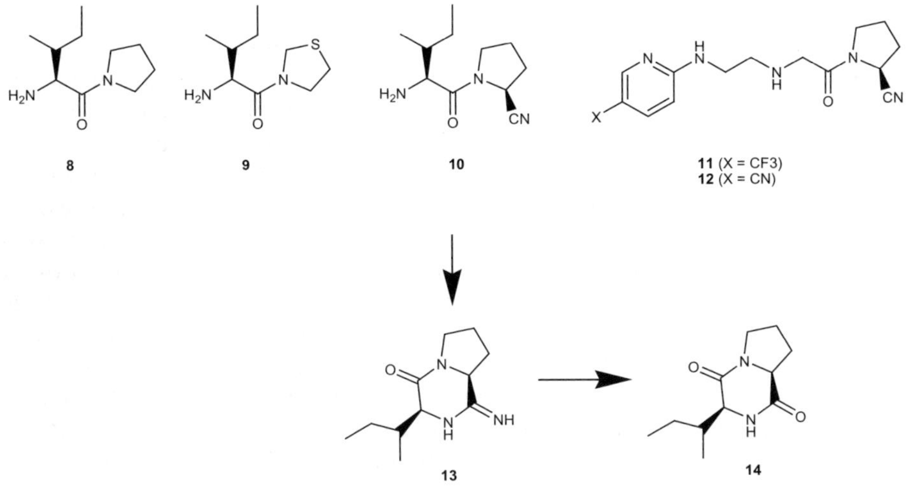

Figure 1.3   Early examples of "reversible" DPP4 inhibitors

themselves as "transition-state" inhibitors which presumably form tetrahedral boronate esters involving the Ser630 hydroxyl group.[26–28] As such, many of these compounds exhibited slow tight-binding kinetics with K_{off} rates of several days (*versus* seconds/minutes with non-covalently bound inhibitors). However, these compounds also suffered from poor solution stability arising from intramolecular cyclization of the terminal primary amine with the boronate, affording an inactive product (*e.g.* **7**). The propensity of compounds of this chemotype to undergo internal cyclization limited their viability as drug candidates.

Due to the uncertain risks associated with advancing irreversible inhibitors as drug candidates, the team viewed the reversible inhibitors exploited by the Mount Sinai, Probiodrug, Ferring Research, and Novartis scientific teams as more attractive starting points for a lead finding effort (Figure 1.3). Probiodrug had described simple Ile-pyrrolidides (**8**) and Ile-thiazolidides (**9**, later advanced to the clinic by Merck and Probiodrug as P32/98) which exhibited *in vitro* potency in the nanomolar range, were chemically stable, and, in the case of **9**, demonstrated glucose area under the curve (AUC) reductions in Zucker[fa/fa] rats in an oral glucose tolerance test (oGTT).[29–31] Reports from Li *et al.* at Mount Sinai highlighted early examples of nitrilo-pyrrolidines specifically designed as inhibitors of DPP4.[32] Equally intriguing was the work described by Ferring in which nitrilo-pyrrolidines such as **10** were identified as exceptionally potent inhibitors of the enzyme.[33,34] Initially targeted as agents for immunomodulation (via CD26 inhibition), these compounds represented "drug-like" scaffolds and exhibited exceptional inhibitory potency. Additionally, a contemporaneous patent application from Novartis[35] described the structure of compound **11** (a related analogue **12** would later be disclosed as Novartis' first clinical compound, DPP-728)[36] and its ability to increase plasma insulin in fasted, high-fat fed rats in an oGTT. The inhibitors represented by compounds **8–12** provided useful insights for the design of DPP4 inhibitors at BMS.

At the initiation of the program, there were still many unknown factors related to the pharmacology and safety of DPP4 inhibition. It was clear from earlier work that Fischer 344 rats possessing a naturally occurring loss-of-function mutation of the DPP4 gene were healthy, viable, and free of serious immunological complications.[37] It was later shown that these rats also exhibited a favorable metabolic phenotype on a high-fat diet and demonstrated improved glucose tolerance and GLP-1 secretion.[38] Thus, complete ablation of DPP4 did not appear to represent a serious safety concern in rats, but rather the Fischer rat provided support for the concept that inhibition of DPP4 could be both safe and efficacious. Others questions still remained. Was high selectivity for DPP4 *versus* other related peptidases (*e.g.* DPP8, DPP9, FAP, *etc.*) an absolute requirement for this target? Would inhibition of DPP4 potentiate other endogenous peptides, leading to unintended deleterious (or beneficial) consequences? Would potentiating endogenous GLP-1 (*versus* exogenous administration) be sufficient to affect a robust anti-diabetic response in humans? Finally, what potential mechanism-based toxicological effects, if any, would be seen upon chronic administration of a DPP4 inhibitor?

In light of limited literature in the field, and with no reports of a compound having advanced to clinical trials, these questions would ultimately need to be addressed during the execution of our discovery and subsequent clinical development programs. Despite the unknowns, the positive aspects of this target were numerous. Potent small-molecule inhibitors with systemic exposure upon oral dosing were known. Although limited, DPP4 inhibitors had demonstrated pharmacodynamic efficacy in genetic animal models of type 2 diabetes in preliminary pharmacological studies. Preclinical proof-of-concept for the anti-diabetic actions of GLP-1 was already established and suggested a low potential for hypoglycemia. *In vitro* assays and several *in vivo* models were already described in the literature, enabling rapid program initiation. It was against this backdrop that significant medicinal chemistry and biology resources at BMS were deployed on this newly emerging target in the early months of 1999.

1.5 Design of BMS's DPP4 Medicinal Chemistry Program

Given the attractiveness of DPP4 as a therapeutic target, it was anticipated that this field would soon become highly competitive, and we therefore sought to accelerate the program. High-throughput screening (HTS), routinely a key component of drug discovery programs, was deemed as too time consuming to rapidly afford a chemical starting point. Thus, we decided to initially adopt a design optimization approach, improving upon the leads reported by the Probiodrug, Ferring, and Novartis research groups; HTS would later be used to provide leads for a second-generation effort. From a potency perspective, the

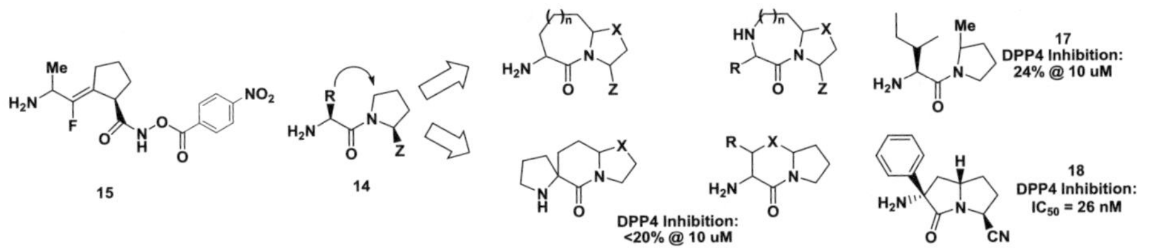

Figure 1.4   Rationale for generation of conformationally restricted DPP4 inhibitors.

nitrilo-pyrrolidines were deemed to be highly attractive, but were reported to exhibit modest pharmacokinetic duration of action and suffered from chemical instability.[33,34] In solution, the proximal amino group attacks the nitrile functionality (see Figure 1.3), eventually leading to the intermediate cyclic imidate **13** and ultimately the diketopiperzine **14**, both of which are inactive *versus* DPP4. Addressing this issue was viewed as a critical component of the medicinal chemistry effort due to considerations regarding both the half-life of the compound *in vivo*, and for high purity processing of the active pharmaceutical ingredient (API) on large scale in a drug manufacturing setting.

From earlier work by Lin *et al.*,[39] it was demonstrated that replacement of the prolyl amide bond with a fluoroalkene isostere resulted in the generation of potent DPP4 inhibitor **15** (Figure 1.4). This finding was significant in that it suggested that the critical prolyl and amino pharmacophores in **16** may be conformationally locked in an extended arrangement which is favorable for enzyme inhibition. In addition, because incorporation of the alkene prohibits intramolecular attack of the amine onto the acyl hydroxamate, the finding suggested novel paths for inhibitor design to retard intramolecular cyclization. Taking a cue from earlier work performed at BMS in the design of dual ACE/NEP inhibitors,[40] we applied the concept of conformationally restricted dipetide mimetics in our search for novel inhibitor chemotypes. Many of these cores seemed to possess the critical elements required for DPP4 inhibition, including a prolyl amide group and a charged amino functionality at the P2 position. It was hoped that locking the inhibitor conformation by this approach would not only enhance binding affinity, but also prevent inactivating cyclization in compounds possessing an electrophilic pharmacophore (*e.g.* nitrile, phosphate, *etc.*) on the proline ring. Unguided by the availability of a DPP4 X-ray crystal structure at that time, our design efforts led to a variety of different bi- and monocyclic dipetide mimetics, generically represented in Figure 1.4. Unfortunately, all of the compounds generated in this series were inactive against DPP4, which we attributed to either incorrect conformational geometry required for inhibition, or to steric intolerance for substitution on the proline ring. The latter hypothesis was supported by the poor activity exhibited by the simple methyl-substituted prolyl derivative **17** as compared to its unsubstituted counterpart **8**. Interestingly, a recent report from Phenomix disclosing 5,5-fused bicyclic lactams such as **18** as potent DPP4 inhibitors provides validation for this initial approach.[41]

1.6   Design of Cyclopropyl-fused Nitrilo-pyrrolidines

In concert with the effort described above, the discovery team examined alternative approaches to the generation of novel inhibitors that might minimize or obviate the undesired cyclization reaction. Cognizant of the impact of pyrrolidine substitution on DPP4 activity (*e.g.* **17**), we proposed that cyclopropyl fusion to the prolyl ring, represented by **19**,[42] might represent a new and viable approach (Figure 1.5). Incorporation of a fused cyclopropane ring constituted a minimal steric burden at P1, would impact the planarity of the prolyl ring, and importantly have the potential to conformationally retard intramolecular attack of the amine to the nitrile. Hence, a series of cyclopropane-fused nitrilo-pyrrolidines **20–23** were synthesized and assessed for both DPP4 inhibition and solution stability (pH 7.2 phosphate buffer, 39.5 °C).[43] As compared to the unsubstituted prototype **10**, the *trans*-4,5- and *trans*-2,3-isomers **22** and **23**, respectively, were not well tolerated by the enzyme, negating further evaluation of these as viable program leads. In contrast, both the *cis*-4,5- (**20**) and the *cis*-3,4- (**21**) analogues exhibited acceptable, though slightly diminished, activity as compared to **10**. More interestingly, **20** exhibited a significant enhancement in solution stability as compared to either **10** or **21**, confirming that *cis*-4,5-cyclopropyl fusion did indeed retard the intramolecular cyclization process. Expanding upon this finding, compounds which incorporated a corresponding *tert*-leucine at P2 as in **24–26** minimized potency differences between the respective cores, though solution stability was still

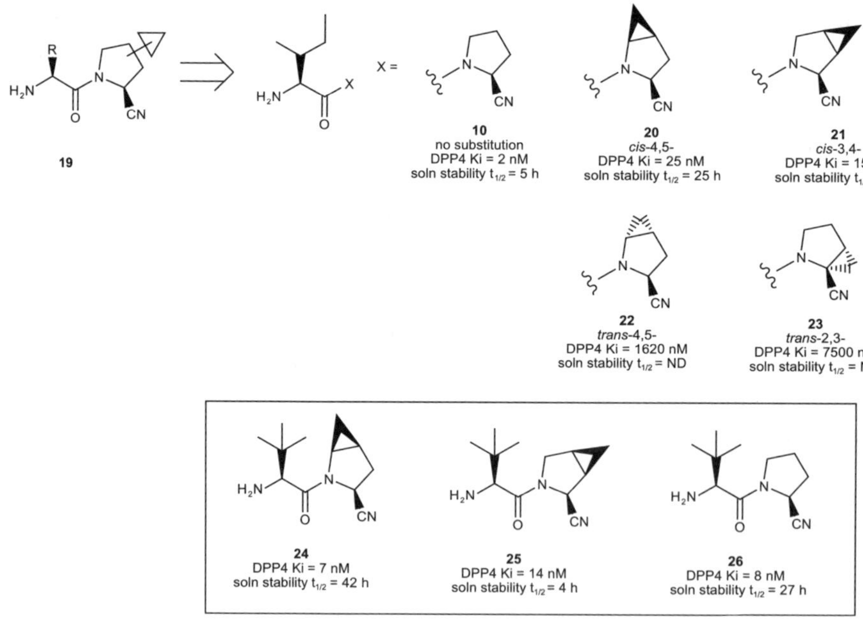

Figure 1.5   Genesis of cyclopropyl-fused nitrilo-pyrrolidine chemotype.

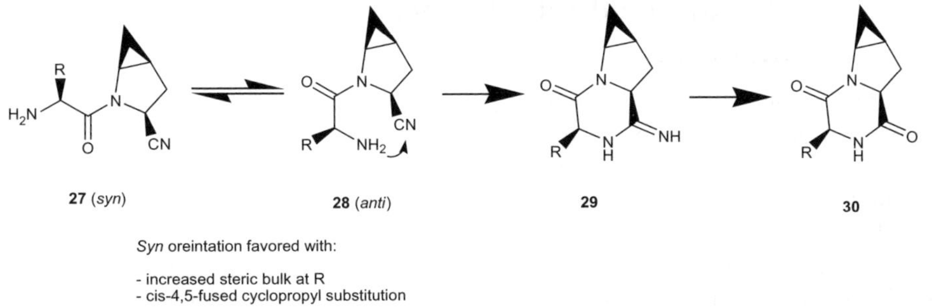

27 (*syn*)   **28** (*anti*)   **29**   **30**

Syn oreintation favored with:

- increased steric bulk at R
- cis-4,5-fused cyclopropyl substitution

Figure 1.6 Factors determining *syn versus anti* orientation of the methano-nitrilo-pyrrolidine chemotype.

favorably preserved in the *cis*-4,5-methano core (*e.g.* **24**). These and other data suggested that increased steric and lipophilic bulk at the P2 position of the inhibitor (*e.g.* ethyl *vs.* isobutyl *vs. tert*-butyl, *etc.*) enhanced both *in vitro* potency as well as solution stability. A similar trend was also observed by the Ferring Research group in the simple nitrilo-pyrrolidine series.[34]

Clearly both the steric bulk of the P2 side-chain and the cyclopropyl fusion on the pyrrolidine ring cooperatively enhanced comformational stability. Computational analyses later demonstrated that, primarily because of van der Waals interactions, increasing the size of the side-chain R (Figure 1.6) disfavored the *anti* orientation **28** required for irreversible intramolecular cyclization to the inactive products **29** and **30**. For example, where R is Me *versus tert*-butyl, the $\Delta\Delta H$ between the *anti* (**28**) and *syn* (**27**) orientations were calculated to be $0.8\,\text{kcal}\,\text{mol}^{-1}$. Furthermore, addition of the *cis*-4,5-cyclopropyl moiety also disfavors adoption of the *anti* orientation by an additional $0.6\,\text{kcal}\,\text{mol}^{-1}$ as compared to the unsubstituted nitrilo-pyrrolidine ring. Thus, by combining these conformational elements, we were able to identify novel inhibitors of DPP4 with high *in vitro* potency and enhanced solution stability.

Compound **24** represented a major breakthrough in the chemistry program. Profiling of this lead demonstrated a low potential for off-target liabilities [hERG inhibition, cytochrome P450 (CYP450) inhibition, broad receptor screening, *etc.*], as well as favorable pharmacokinetic properties in the rat ($F = 77\%$, $t_{1/2} = 2.8\,\text{h}$). Additionally, compound **24** was efficacious in Zucker[fa/fa] rats, reducing glucose AUC in an acute oGTT ($ED_{50} = 3.3\,\text{mg}\,\text{kg}^{-1}$, glucose challenge 30 min post-dose) when compared to vehicle control. While the duration of action of **24** was not particularly long ($ED_{50} = 92\,\text{mg}\,\text{kg}^{-1}$, glucose challenge 5 h post-dose), the *in vivo* potency was still significantly greater than that of the literature thiazolidine lead **9** ($ED_{50} = 38\,\text{mg}\,\text{kg}^{-1}$, glucose challenge 30 min post-dose). On the basis of these findings, additional staff were assigned to the program to permit a detailed and robust exploration of the structure–activity relationship (SAR), primarily focusing on further modifications at P2.

1.7 SAR Optimization Leading to the Discovery of Saxagliptin

Building upon the *cis*-4,5-methano-2-nitrilopyrrolidine core, a wide variety of substituents, represented generically in **31**, were explored (Figure 1.7). As highlighted earlier, greater potency and stability were realized via introduction of highly β-branched P2 side-chains. Due to the very limited commercial availability of β-quaternary α-amino acids, the team devised an efficient strategy to generate novel P2 units by several complementary paths, each deriving from cyclic and/or symmetrical ketones to avoid unnecessary introduction of additional stereocenters.[6,43] In one arm of this approach, ketones underwent Knoevenagel condensation with a malonate diester, and the resulting Michael acceptor was subjected to conjugate addition to introduce the alkyl-substituted quaternary center. Mono-hydrolysis and subsequent Curtius rearrangement yielded the desired β-branched amino acids. A parallel approach began with Horner–Emmons condensation of the ketone to give an α,β-unsaturated ester. Reduction of the ester to the primary allylic alcohol and esterification with Boc-Gly set up a zinc-mediated ester enolate Claisen rearrangement, which provided the desired amino acids with a vinyl functional handle at the β-position. Alternatively, a standard Strecker synthesis could be used on available aldehydes to ultimately afford the desired amino acid building blocks. In all cases, the racemic amino acids were then coupled to the

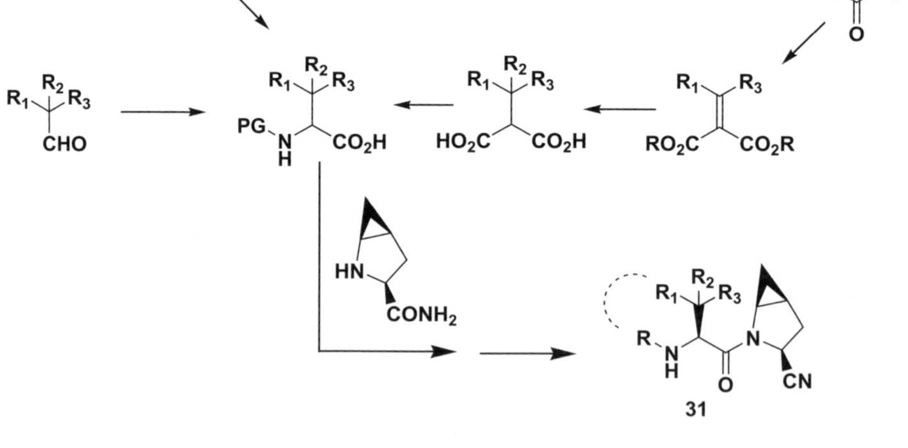

Figure 1.7 Various synthetic routes for the generation of highly β-branched amino acids.

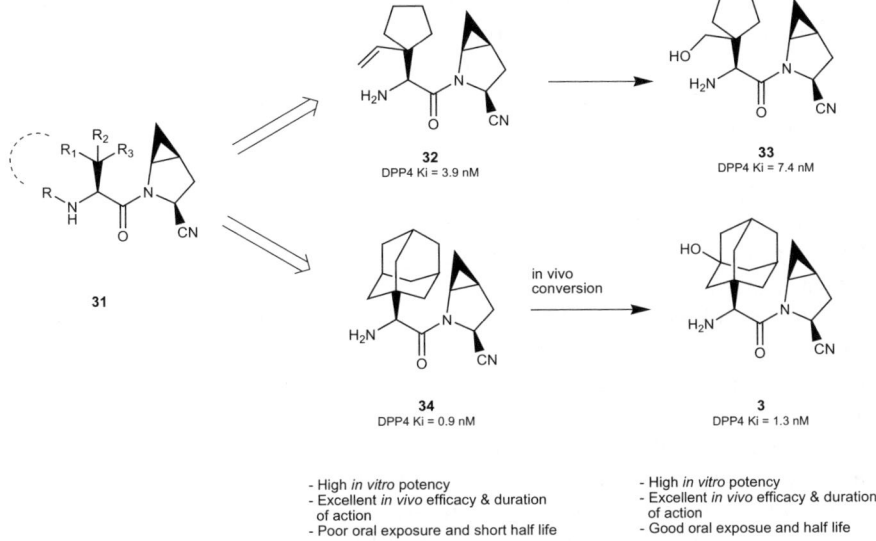

Figure 1.8 Classical PK/PD disconnect in rats leads to hypothesis of active metabolite generation *in vivo*, resulting in the discovery of saxagliptin.

enantiomerically pure P1 nitrilo-methanopyrrolidine fragment and the resulting diastereomers were chromatographically resolved.

From this effort emerged two compounds of particular interest, inhibitors **32** and **34** (Figure 1.8). In the case of **32**, this compound demonstrated superior *in vivo* efficacy and a sustained duration of response for inhibition of plasma DPP4 in normal fasted SD rats at a dose of $4 \, \mu mol \, kg^{-1}$ (71% @ 30 min post-dose and 64% @ 4 h post-dose). Duration of the pharmacodynamic (PD) response was particularly long when compared to compounds generated earlier in the program. Intriguingly, this extended PD response was most often observed in analogues containing a vinyl substituent (*e.g.* **32**) at P2. Despite its robust *in vivo* activity, compound **32** demonstrated low oral bioavailability (5%) and a short half-life ($t_{1/2} = 1.2$ h) in rats. In contrast, the related saturated analogue of **32** (vinyl replaced with ethyl) exhibited significantly higher bioavailability (31%) and greater *in vitro* stability in rat and human liver microsomes, but a weaker response in the rat plasma DPP4 inhibition assay (despite equivalent *in vitro* potency). The disconnect between the pharmacokinetic and pharmacodynamic profile of **32** immediately suggested *in vivo* conversion to an active metabolite possessing superior target potency and/or reduced clearance profile. Based on this hypothesis, a variety of putative or surrogate hydroxylated metabolites related to **32** were prepared. While unequivocal characterization of **33** as an active metabolite of **32** was never established, the activity of **33** mirrored that of **32** in the aforementioned PD model. While the rat half-life of **33** ($t_{1/2} = 1.3$ h) was not significantly longer when compared to that of **32**, the absolute bioavailability increased 10-fold to 53%. A more striking finding was observed with compound **34**. In concordance with established SAR, incorporation of the bulkier adamantyl hydrophobic group in the P2 side-chain enhanced *in vitro* potency ($K_i = 0.9$ nM) as compared to compounds with smaller fragments in this position. In the normal rat, this compound (4 µmol kg^{-1}, po) afforded robust plasma DPP4 inhibition (84% @ 0.5 h post-dose and 83% @ 4 h post-dose) despite exceptionally low bioavilability (2%) and modest half-life ($t_{1/2} = 1.4$ h). Capitalizing on the learnings from compound **33**, and considering the propensity of adamantyl groups to undergo CYP-mediated hydroxylation, the corresponding metabolite **3** (saxaglipitin) was synthesized. The compound proved to be highly potent *in vitro*, exhibited good pharmacokinetic properties in the rat ($F = 75\%$, $t_{1/2} = 2.1$ h) and effected near maximal ($\sim 87\%$) plasma DPP4 inhibition in the rat at both 0.5 and 4 h post-dose when administered at 4 µmol kg^{-1}.

In addition to saxagliptin, a number of other inhibitors incorporating P2 side-chains with various substituted adamantanes were prepared and evaluated, including other positional isomers of hydroxyadamantane, dihydroxyadamantane, and fluoroadamantane. While these other adamantane-derived compounds also exhibited potent DPP4 inhibition *in vitro*, saxagliptin provided a comparatively superior PK and PD profile, and was ultimately chosen for development.

1.8 Binding of Saxagliptin to Human DPP4: A Slow Tight-binding Inhibitor

Through more extensive characterization of the binding kinetics of saxagliptin (and select analogs) to DPP4, it was noted that several compounds showed evidence of slow tight-binding. Further analysis of SAR patterns relating to this slow on/off-rate feature revealed correlations with both steric bulk in the P2 side-chain (specifically β-quaternary substitution) and the presence of a nitrile functionality. Prior to our undertaking of detailed kinetic studies, it had been speculated by several groups that a transient covalent bond was formed between the hydroxyl group of the active-site Ser630 of the DPP4 catalytic triad and the nitrile carbon of nitrile-based inhibitors, yet early X-ray co-crystal structures with such compounds lacked adequate resolution to confirm appropriate electron density where such a bond would exist.[44] Our findings were consistent with a hypothesis whereby the bulky P2 side-chain's displacement of water in the S2 pocket drove entropic aspects governing on-rate (for **3**, $K_{on} = 4.6 \times 10^5$ M^{-1} s^{-1}), and, once anchoring the inhibitor in the active site, covalent interaction with the nitrile drove enthalpic aspects governing off-rate (for **3**, $K_{off} = 23 \times 10^{-5}$ s^{-1}), resulting in overall K_i enhancement for certain inhibitors possessing these features.[45] As a result, the $t_{1/2}$ for dissociation of saxagliptin from DPP4 was determined to be ~ 50 min at 37 °C (~ 250 min at room temperature). In comparison to the rapid off rates of non-nitrilo-pyrrolidines such as sitagliptin, the slow off-rates exhibited by saxagliptin and, to a lesser extent, other nitrilo-pyrrolidines such as **12**,[45b,46] proved to be a unique attribute of saxagliptin among the clinically advanced inhibitors. The slow off-rate kinetics exhibited by saxagliptin likely serves to enhance saxagliptin's pharmacodynamic potency and duration *in vivo*, and may also contribute to its enzymatic selectivity.[47,48]

Further understanding of this finding was obtained through: (1) a high-resolution X-ray co-crystal structure of saxagliptin bound to DPP4, definitively revealing the presence of a covalent O–C bond; (2) a series of biophysical studies performed using wild-type and mutant DPP4 proteins S630A and H740Q characterizing interactions with both saxagliptin and its de-nitrilo analogue; and (3) comparative ¹H NMR studies of DPP4 in apo form and in complex with saxagliptin,[45,49] revealing the characteristic downfield shift indicative of a short, strong H-bond upon complex formation.[50,51] The X-ray co-crystal structure of saxagliptin bound to DPP4 (Figure 1.9) reveals several key features: (1) the ionic interaction of the primary amine with the Glu205 and Glu206 residues; (2) hydrogen bonding of the adamantyl hydroxyl with Tyr547; (3) efficient hydrophobic space-filling by the adamantyl and methano-pyrrolidine groups; and (4) the aforementioned covalent bonding of the catalytic Ser630 hydroxyl with the pendant nitrilo functional group. Despite the clear covalent nature of the bond, this interaction was shown to be fully reversible (as demonstrated by complete recovery of enzyme activity upon dialysis) and enzymatically unproductive (since no saxagliptin hydrolysis products were

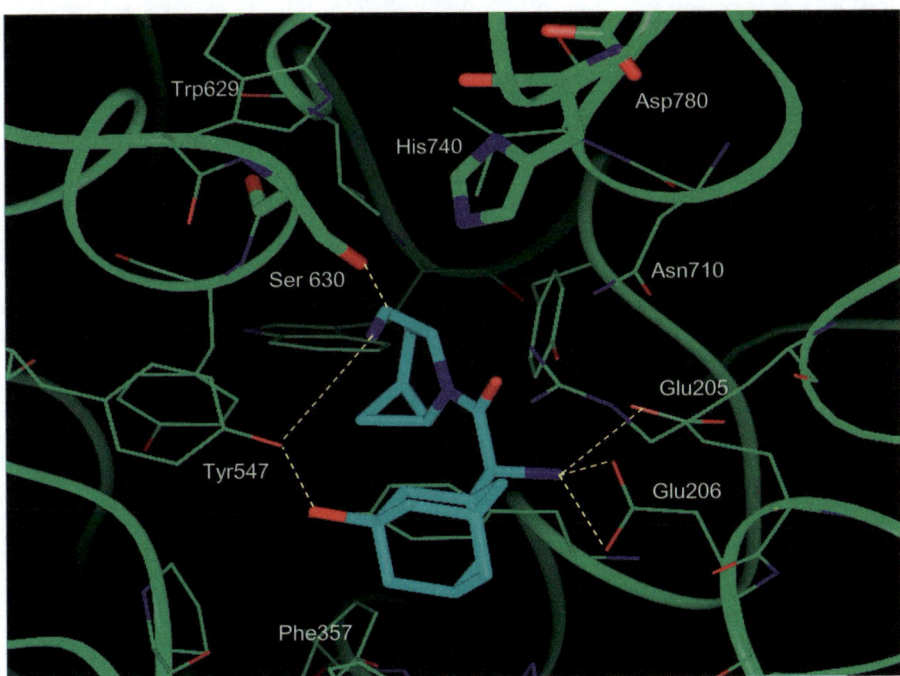

Figure 1.9 X-Ray co-crystal structure of saxagliptin bound to human DPP4.

formed). It is speculated that the methano ring of saxagliptin displaces any active-site water molecules required for hydrolysis, preventing the inhibitor from being a substrate for the enzyme. A unique constellation of features exhibited by both saxagliptin and its biologically active, circulating metabolite (5-hydroxy saxagliptin), including: (1) slow tight-binding kinetics; (2) low nM *in vitro* potency; (3) good pharmacokinetics; and (4) high tissue distribution at the primary site of action (GI tract),[52] all contribute to saxagliptin's low clinically efficacious human dose.

1.9   Chemical Stability of Saxagliptin and Analogs

As highlighted earlier, a property shared by many of the nitrilo-pyrrolidine based inhibitors bearing an amine at the P2 N-terminus is the propensity for intramolecular cyclization at or above neutral pH (see Figures 1.3 & 1.6). Such instability could complicate manufacturing and formulation, and potentially shorten pharmacokinetic half-life. As a result, the stability of the BMS nitrilo-pyrrolidine analogs were routinely assessed in pH 7.4 buffer. While introduction of bulky alkyl substitution on the terminal amine of Novartis' DPP-728 (**12**)[36] and LAF-237 (**35**, vildagliptin)[53] greatly retarded intramolecular cyclization, incorporating a secondary amine structural element into the BMS methano-pyrrolidine series routinely attenuated potency (Figure 1.10). While the solution stability of saxagliptin ($t_{1/2} \approx 4$ days @ pH 7.4/37 °C) bode well for manufacturing and drug pharmacokinetics, lingering formulation concerns spurred later discovery efforts to obviate any unforeseen chemical stability issues associated with the nitrile functionality. Interestingly, wholesale removal of the nitrile moiety from saxagliptin afforded BMS-538305 (**36**), a highly potent ($K_i = 10$ nM) inhibitor of the enzyme with a favorable *in vitro* and *in vivo* development profile.[54] SAR in this non-nitrile series was narrow, with a precipitous drop-off of activity accompanying even modest structural changes. Because BMS-538305 was completely stable in solution, it was nominated as a back-up development candidate and went on to exhibit robust pharmacodynamic efficacy in humans, although it was later suspended from further development, in part due to a clear lack of superiority *versus* saxagliptin.

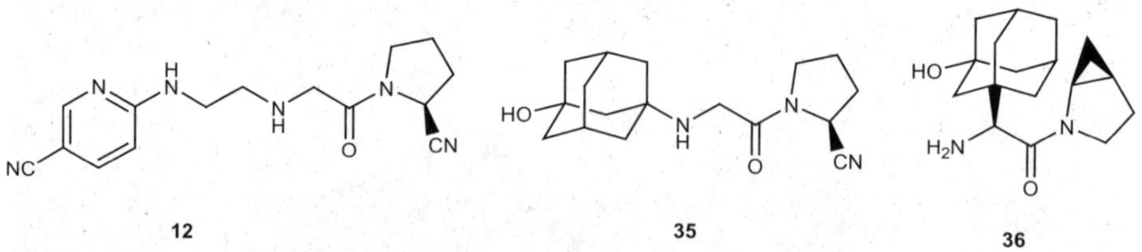

Figure 1.10   Structure of nitrilo-pyrrolidine-based DPP4 inhibitors with enhanced solution stability.

The potential for formulation-related challenges with saxagliptin also spurred the application of a unique formulation strategy which may ultimately translate into a significant advantage for the development of future single-tablet saxagliptin drug combinations. It was discovered that in the process of compressing the saxagliptin API salt together with excipients to develop a standard tablet formulation, unacceptably high levels (2%) of the undesired cyclization products were produced. A pharmaceutics solution to the problem, made possible by the low dose (2.5–10 mg) of saxagliptin required to effect maximal HbA1c lowering in humans, was realized by spray coating the drug substance (sandwiched between two polymeric vapor barrier layers) onto an inert tablet core. This formulation is currently used in the marketed brand of saxagliptin, Onglyza™. It is anticipated that the spray coating technique, a procedurally simple technology, will greatly facilitate the development of future fixed-dose combinations of saxagliptin with other OADs.[55] As proof, saxagliptin has been readily formulated with extended-release metformin, leading to the most clinically advanced once-daily DPP4/metformin single-tablet combination. This innovative combination was submitted for FDA regulatory review in December 2009.

1.10 *In vivo* Efficacy of Saxagliptin

As SAR within the series developed, a simple *ex vivo* plasma DPP4 inhibition assay was used to efficiently differentiate among the growing number of potent compounds. To further understand particularly promising compounds, a clinically relevant pharmacodynamic assay (an oGTT in Zucker$^{fa/fa}$ rats or ob/ob mice) was also utilized. These models served as an efficient means to phenotypically stratify compounds without the need for prior, time-consuming PK assessments, thereby reducing cycle times. In these models, saxagliptin afforded quite robust effects on glucose excursion and insulin release at doses as low as 0.3–1 µmol kg^{-1}.[6] However, preclinical demonstration of reduced fasting plasma glucose and/or HbA1c in subacute or chronic disease models proved to be elusive. In-house experience gained in earlier discovery programs targeting PPARα agonists led the team to evaluate the chronic dosing of saxagliptin in a variety of genetic models of type-2 diabetes, such as in ZDF and Zucker$^{fa/fa}$ rats. Despite numerous attempts to optimize these models, the group was unable to demonstrate statistically significant, dose-dependent efficacy (fasting plasma glucose lowering), irrespective of the model or dosing duration. Despite this lack of validation in pre-clinical disease models, timely disclosures of clinical data by Novartis helped to bolster internal confidence in the mechanism. Novartis scientists had recently reported promising reductions in HbA1c for DPP-728 (12) in a 12-week Phase IIa trial in diabetic subjects.[56] Based on our belief that saxagliptin exhibited superior potency and pharmacokinetic profiles compared to DPP-728, the decision was made to forego further efforts to develop chronic *in vivo* efficacy models and to move forward into clinical development. Despite more than 10 years of pre-clinical testing of DPP4 inhibitors *in vivo*, there are scant reports in the literature regarding the efficacy of chronic DPP4 inhibition in rodent models. These observations highlight the limitations of these pre-clinical disease-pertinent models in fully or accurately mirroring the pathophysiology of human disease.

1.11 Peptidase Selectivity of Saxagliptin

At program initiation, the precise requirements for DPP4 selectivity *versus* inhibition of related proteases were not entirely clear. SAR progression and compound optimization outpaced the full implementation of our protease selectivity panel, and it was not until identification of saxagliptin as a development candidate that all elements of the selectivity panel were in place. Saxagliptin (similar to other analogs within the series) was found to be highly selective (> 4000-fold) for inhibition of DPP4 as compared to a variety of other proteases, including FAP, PEP, and DPP2. Against the most closely related enzymes DPP8 and DPP9, selectivity was attenuated but still high (400- and 75-fold, respectively).[45,57] Contributing to saxagliptin's enhanced selectivity is the observation that, unlike its slow tight-binding kinetics for the inhibition of DPP4, saxagliptin did not display slow tight-binding when inhibiting DPP8 or DPP9. Additionally, since DPP8 and DPP9 are intracellular enzymes, the poor cellular permeability of saxagliptin attenuates access to these off-target proteases. As a result, saxagliptin behaves as a highly selective inhibitor of DPP4 with little or no propensity for off-target protease inhibition.

In 2005, scientists at Merck published results of two-week rat (10, 30, and 100 mg kg^{-1}) and single-dose dog (10 mg kg^{-1}) toxicity studies with a cohort of compounds which included a selective DPP4 inhibitor closely related to sitagliptin, a selective inhibitor of DPP8 and DPP9 (L-*allo*-isoleucyl isoindolide), and a somewhat weak and non-selective inhibitor of all three enzymes (L-*allo*-isoleucyl thiazolidide), as well as several other controls.[58] Both the DPP8/9 selective compound and the weak non-selective compound (which share a common structural element in the L-*allo*-isoleucine moiety) caused alopecia, thrombocytopenia, anemia, reticulocytopenia, splenomegaly, and mortality in rats, and bloody diarrhea in dogs. None of these effects were observed in animals dosed with the DPP4 selective compound. Based on these empirical findings, the Merck team concluded that the observed toxicities were a direct result of inhibition of DPP8 and/or DPP9. In formal toxicology studies, neither saxagliptin nor vildagliptin produced the spectrum of pathology changes ascribed by Merck to DPP8/9 inhibition, even at doses predicted to give plasma levels much higher than the K_i's required for inhibition of DPP8/9. Thus, the findings suggested both saxagliptin and vildagliptin are highly selective DPP4 inhibitors *in vivo*, and/or the theory regarding the consequences of DPP8/9 inhibition proposed by the Merck group may be incorrect. An alternate hypothesis, that the observed toxicology profiles of the Merck tool compounds were structure (compound) related, rather than based on their DPP8/9 inhibition profiles, was not initially embraced as a possible cause of the observed toxicities. The Merck findings undoubtedly influenced the inhibitor design efforts of nearly every DPP4 discovery program in the industry, resulting in a proliferation of varied chemotypes with "atypical" protease inhibitor structural features as compared to the more dipeptidic motifs shared by saxagliptin and vildagliptin.

Several groups attempting to better understand the consequences of DPP8/9 inhibition in animals have subsequently published their findings using more structurally diverse inhibitors of DPP8/9 as tool molecules. Rosenblum *et al.* recently showed that treatment of dogs with high doses of a selective, cell-permeable DPP4/8/9 inhibitor (producing extensive inhibition of DPP4/8/9 in GI tissues) did not afford toxicological responses similar to those observed by Merck with their tool compounds.[59] In a separate study, Chen *et al.* demonstrated that repeated exposure (iv, QD for 14 days) of IG244, a potent and cell-permeable DPP8/9 inhibitor, did not result in severe toxicity in rats, despite complete DPP8/9 inhibition achieved with this inhibitor.[60] Additionally, Burkey *et al.* demonstrated that chronic dosing of vildagliptin (achieving plasma levels well above its K_i for DPP8 and DPP9 inhibition) failed to recapitulate the toxicity profile observed with the Merck DPP8/9 inhibitor.[61] Collectively, these empirical studies offer little to mechanistically connect the inhibition of DPP8 or DPP9 to a phenotype, and as such, the functional consequences of DPP8 and DPP9 inhibition in animals and humans remains unknown.[62] Thus, while DPP4 selectivity has become cemented as an intrinsic component of most DPP4 inhibitor programs industry-wide, the need for exquisitely high selectivity against the DPP8 and DPP9 proteases remains an unproven concept in DPP4 inhibitor design.

1.12   Synthesis of Saxagliptin

The structural complexity of saxagliptin posed some synthesis development challenges, in that the molecule possesses four asymmetric centers, including a neopentyl carbon center, where only one of the stereochemical centers was available through a commercially available chiral building block. However, the team's willingness to employ challenging synthetic chemistry to solve medicinal chemistry problems was key to the discovery of highly potent, more stable, and patentable lead compounds. The discovery synthesis of saxagliptin, used to prepare ~ 100–200 g of material for pre-development work, was in need of substantial optimization for large-scale production of this rather complex molecule. In particular, the need to perform a zinc-mediated cyclopropanation at low temperature with high diastereoselectivity, and the development of a robust large-scale synthesis of the chiral adamantyl fragment, proved to be a daunting task. The process chemistry group at BMS tackled each portion of this molecule with creative solutions to provide an efficient and reproducible manufacturing route. Key synthetic transformations are depicted in Figure 1.11, and include: (1) a Reformatsky reaction of bromoadamantane with a silyl-protected dichloroketene acetal, followed by oxidation and hydrolysis to give keto

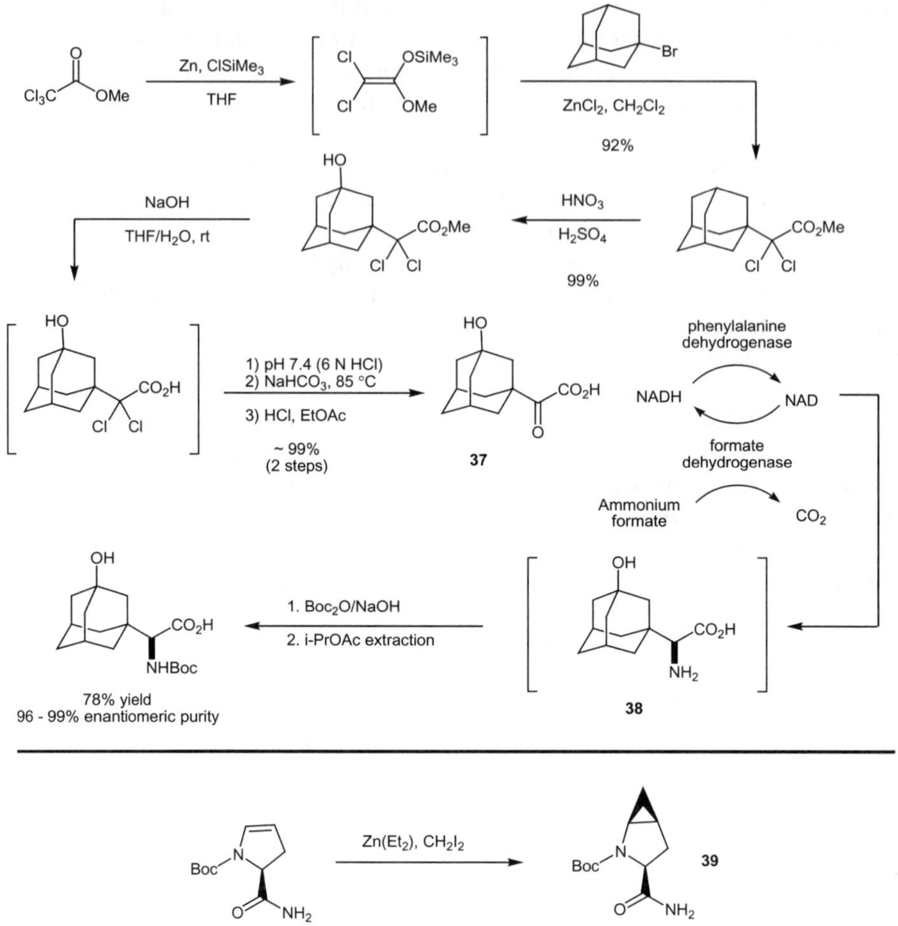

Figure 1.11 Development synthesis of the hydroxyadamantylglycine and methano-proline fragments of saxagliptin.

acid **37**;[63] (2) an enzymatic asymmetric reductive amination with ammonia to give the requisite unnatural amino acid **38**;[64] and (3) a continuous flow process for large-scale diethylzinc-mediated diastereoselective cyclopropanation of a protected dehydroprolinamide **39**.[65]

1.13   Saxagliptin Development

Saxagliptin formally entered development at BMS in April of 2001, and the first human clinical dose was administered to normal healthy volunteers in December of that same year. As highlighted earlier, genetic rodent models of diabetes proved unsuitable for accurately assessing the chronic efficacy of DPP4 inhibitors, and so the translational leap from acute plasma enzyme inhibition and post-prandial antihyperglycemic effects to a registrational Phase IIb clinical trial would require more detailed dose-ranging efforts early on. Because the majority of the preclinical efficacy data generated for saxagliptin was in an acute setting, accurately projecting an efficacious human clinical dose was somewhat challenging. Saxagliptin was very well tolerated in Phase I/II single and multiple ascending dose studies ranging up to 100 mg and exhibited 24 h maximal inhibition of plasma DPP4 activity at steady state, beginning with the 5 mg dose. Additional higher dose panels (up to 600 mg) were added at a later date to examine the possibility of achieving greater efficacy. As the top marketed dose of saxagliptin is only 5 mg once-daily, the high doses tested in the Phase I clinical program provided the development team with a deeper understanding of and confidence in the compound's safety margin. Dose ranging proof-of-concept studies with saxagliptin followed a standard randomized, double-blind, placebo-controlled Phase IIb protocol in drug naïve type-2 diabetic patients measuring HbA1c reductions.[66] Five doses of saxagliptin were studied (2.5, 5, 10, 20, and 40 mg, q.d.) and compared with placebo in an active 12-week treatment period. All doses of saxagliptin significantly lowered HbA1c from baseline (mean baseline HbA1c 7.9%) compared with placebo, and placebo-subtracted adjusted mean changes from baseline ranged from − 0.45 to − 0.63%, with no apparent statistically significant dose-relationship in this study. These results supported saxagliptin's advancement to Phase III, where additional efficacy and safety data were generated in > 3000 drug-treated patients to support the registrational package.[67] In January of 2007, BMS entered into a co-development and co-marketing agreement with Astra-Zeneca encompassing saxagliptin and the SGLT2 inhibitor dapagliflozin, as well as the back-ups for each. This partnership was established in order to share in the late-stage development costs/risks and to broaden the sales and marketing capacity required to effectively reach the target clinical population, generally served by primary care physicians. Ultimately, the BMS/Astra-Zeneca collaboration was successful in obtaining regulatory approval from both the FDA (July 2009) and the EMEA (October 2009) for the use of saxagliptin, either as monotherapy or in combination with metformin,[68,69] sulfonylureas,[70] or TZDs, in the treatment of type-2 diabetes. The retrospective CV safety package generated to support the filing indicated an acceptable CV safety risk in the patient population studied, though further prospective post-marketing studies will be conducted in other patient populations to more comprehensively assess the impact of chronic saxagliptin treatment on CV outcomes. Interestingly, trends towards reduced events in this analysis have prompted BMS and Astra-Zeneca to initiate larger, more extensive CV safety trials, with the hope of demonstrating a statistically significant and meaningful reduction in CV morbidity and mortality in diabetic patients.

1.14 Summary

In the late 1960's, the pioneering team of Miguel Ondetti and David Cushman gave birth to the concept of rational drug design with the invention of captopril, a small-molecule, peptidomemtic inhibitor of angiotensin-converting enzyme (ACE). Using a novel approach based on knowledge of a target's relevance to a disease state, a mechanistic understanding of biological function of the target, and an ability to design small molecules to interact with critical elements of the target's active site (transition-state pharmacophores, hydrophobic, hydrogen binding, and ionic interactions), they demonstrated that rational drug design (*versus* an empirical approach) can lead to new medicines. As a result, modern drug discovery was born, and continues to evolve, from this approach. Nearly 40 years after the discovery of captopril, these same principles are still being applied in medicinal chemistry, as exemplified in part with the discovery and development of saxagliptin.

Our interest in DPP4 as a target coincided with the completion of several other metabolic disease discovery programs at BMS, thereby permitting a near immediate influx of staffing for the nascent program. Saxagliptin was discovered and advanced for development by utilizing an efficient and timely expenditure of resources (approximately 320 analogs, 26 months, 6–7 full-time chemists, and 20 total full-time scientist-years across chemistry, biology, ADME and pharmaceutics). Prior to submission of saxagliptin for regulatory approval, the development team also contended with several hurdles, including synthetic and formulation challenges and a continually evolving competitive and regulatory landscape. Now that compounds such as saxagliptin, sitagliptin (Januvia™, Merck),[71] and vildagliptin (Galvus™, Novartis, marketed ex-US) have entered the market, with others likely to follow, DPP4 inhibitors have robustly entered the pharmacopeia of diabetes treatments. The efficacy and safety profiles to date for agents acting by this mechanism appear promising, and future studies may elucidate differentiation of the clinical profiles of these agents, as well as establish their potential for long-term benefit in the treatment of diabetes.

References

1. *Diabetes Facts and Figures*, International Diabetes Federation, Brussels, October 6 2009; <http://www.idf.org/Facts_and_Figures>.
2. "Undiagnosed diabetes cost US $18 billion a year" May 6 2009; <www.reuters.com/articles/healthNews 5/6/2009>.
3. B. Tauzin, *Medicines in Development for Diabetes*, Pharmaceutical Research and Manufacturers of America, Washington, 2009; <http://www.phrma.org/news_room/press_releases/report_shows_record_ number_of_medicines_in_development_to_treat_diabetes>.
4. P. V. Devasthale, S. Chen, Y. Jeon, F. Qu, C. Shao, W. Wang, H. Zhang, D. Farrelly, R. Golla, G. Grover, T. Harrity, Z. Ma, L. Moore, J. Ren, R. Seethala, L. Cheng, P. Sleph, W. Sun, A. Tieman, J. R. Wetterau, A. Doweyko, G. Chandrasena, S. Y. Chang, W. G. Humphreys, V. G. Sasseville, S. A. Biller, D. E. Ryono, F. Selan, N. Hariharan and P. T. W. Cheng, *J. Med. Chem.*, 2005, **48**, 2248.

5. W. Meng, B. A. Ellsworth, A. A. Nirschl, P. J. McCann, M. Patel, R. N. Girotra, G. Wu, P. M. Sher, E. P. Morrison, S. A. Biller, R. Zahler, P. P. Deshpande, A. Pullockaran, D. L. Hagan, N. Morgan, J. R. Taylor, M. T. Obermeier, W. G. Humphreys, A. Khanna, L. Discenza, J. G. Robertson, A. Wang, S. Han, J. R. Wetterau, E. B. Janovitz, O. P. Flint, J. M. Whaley and W. W. Washburn, *J. Med. Chem.*, 2008, **51**, 1145.

6. D. J. Augeri, J. A. Robl, D. A. Betebenner, D. R. Magnin, A. Khanna, J. G. Robertson, L. M. Simpkins, P. C. Taunk, Q. Huang, S.-P. Han, B. Abboa-Offei, A. Wang, M. Cap, L. Xin, L. Tao, E. Tozzo, G. E. Welzel, D. M. Egan, J. Marcinkeviciene, S. Y. Chang, S. A. Biller, M. S. Kirby, R. A. Parker and L. G. Hamann, *J. Med. Chem.*, 2005, **48**, 5025.

7. T. Strack, *Drugs Today*, 2008, **44**, 303.

8. J. B. Green and M. N. Feinglos, *Curr. Diabetes Rep.*, 2006, **6**, 373.

9. A. J. Scheen, *Drugs*, 2003, **63**, 933.

10. R. E. Buckingham and A. Hanna, *Diabetes Obes. Metab.*, 2008, **10**, 312.

11. M. Zander, S. Madsbad, J. L. Madsen and J. J. Holst, *Lancet*, 2002, **359**, 824.

12. M. A. Nauck, D. Wollschlaeger, J. Werner, J. J. Holst, C. Orskov, W. Creutzfeldt and B. Williams, *Diabetologia*, 1996, **39**, 1546.

13. (a) D. Chen, J. Liao, N. Li, C. Zhou, Q. Liu, G. Wang, R. Zhang, S. Zhang, L. Lin, K. Chen, X. Xie, F. Nan, A. A. Young and M.-W. Wang, *Proc. Natl. Acad. Sci. U. S. A.*, 2007, **104**, 943; (b) A. M. M. Mjalli, D. R. Polisetti, T. S. Yokum, S. Kalpathy, M. Guzel, C. Behme and S. T. Davis, *US Pat. Appl.* 2009/111700-A2, 2009.

14. R. Gentilella, C. Bianchi, A. Rossi and C. M. Rotella, *Diabetes Obes. Metab.*, 2009, **11**, 544.

15. T. Vilsboell, *Therapy*, 2009, **6**, 199.

16. S. H. Havale and M. Pal, *Bioorg. Med. Chem.*, 2009, **17**, 1783.

17. J. J. Holst and C. J. Deacon, *Diabetes*, 1998, **47**, 1663.

18. E. Matteucci and O. Giampietro, *Curr. Med. Chem.*, 2009, **16**, 2943.

19. S. L. Johnson and M. Pellecchia, *Curr. Top. Med. Chem.*, 2006, **6**, 317.

20. H. Krum and R. E. Gilbert, *J. Hyperten.*, 2007, **25**, 25.

21. A. Mastrolorenzo, S. Rusconi, A. Scozzafava, G. Barbaro and C. T. Supuran, *Curr. Med. Chem.*, 2007, **14**, 2734.

22. M. Di Nisio, S. Middeldorp and H. R. Bueller, *New Engl. J. Med.*, 2005, **353**, 1028.

23. M. Wegner, D. Ganten and J.-P. Stasch, *Hyperten. Res.*, 1996, **19**, 229.

24. H. U. Demuth, R. Baumgrass, C. Schaper, G. Fischer and A. Barth, *J. Enzyme Inhib.*, 1988, **2**, 129.

25. B. Boduszek, J. Oleksyszyn, C.-H. Kam, J. Selzler, R. E. Smith and J. C. Powers, *J. Med. Chem.*, 1994, **37**, 3969.

26. G. R. Flentke, E. Munoz, B. T. Huber, A. G. Plaut, C. A. Kettner and W. W. Bachovchin, *Proc. Natl. Acad. Sci. U. S. A.*, 1991, **88**, 1556.

27. R. J. Snow and W. W. Bachovchin, *Adv. Med. Chem.*, 1995, **3**, 149.

28. S. J. Coutts, T. A. Kelly, R. J. Snow, C. A. Kennedy, R. W. Barton, J. Adams, D. A. Krolikowski, D. M. Freeman, S. J. Campbell, J. F. Ksiazek and W. W. Bachovchin, *J. Med. Chem.*, 1996, **39**, 2087.
29. L. A. Sorbera, L. Revel and J. Castaner, *Drugs Future*, 2001, **26**, 859.
30. R. P. Pauly, H.-U. Demuth, F. Rosche, J. Schmidt, H. A. White, F. Lynn, C. H. S. McIntosh and R. A. Pederson, *Metab. Clin. Exp.*, 1999, **48**, 385.
31. R. A. Pederson, H. A. White, D. Schlenzig, R. P. Pauly, C. H. S. McIntosh and H.-U. Demuth, *Diabetes*, 1998, **47**, 1253.
32. J. Li, E. Wilk and S. Wilk, *Arch. Biochem. Biophys.*, 1995, **323**, 148.
33. D. M. Ashworth, B. Atrash, G. R. Baker, A. J. Baxter, P. D. Jenkins, D. M. Jones and M. Szelke, *Bioorg. Med. Chem. Lett.*, 1996, **6**, 2745.
34. D. M. Ashworth, B. Atrash, G. R. Baker, A. J. Baxter, P. D. Jenkins, D. M. Jones and M. Szelke, *Bioorg. Med. Chem. Lett.*, 1996, **6**, 1163.
35. E. B. Villhauer, *Pat. Appl.* WO 9819998 A2, 1998.
36. E. B. Villhauer, J. A. Brinkman, G. B. Naderi, B. E. Dunning, B. L. Mangold, M. D. Mone, M. E. Russell, S. C. Weldon and T. E. Hughes, *J. Med. Chem.*, 2002, **45**, 2362.
37. N. L. Thompson, D. C. Hixson, H. Callanan, M. Panzica, D. Flanagan, R. A. Faris, W. J. Hong, S. Hartel-Schenk and D. Doyle, *Biochem. J.*, 1991, **273**, 497.
38. N. Yasuda, T. Nagakura, K. Yamazaki, T. Inoue and I. Tanaka, *Life Sci.*, 2002, **71**, 227.
39. J. Lin, P. J. Toscano and J. T. Welch, *Proc. Natl. Acad. Sci. U. S. A.*, 1998, **95**, 14020.
40. J. A. Robl, C.-Q. Sun, J. Stevenson, D. E. Ryono, L. M. Simpkins, M. P. Cimarusti, T. Dejneka, W. A. Slusarchyk, S. Chao, L. Stratton, R. N. Misra, M. S. Bednarz, M. M. Asaad, H. S. Cheung, B. E. Abboa-Offei, P. L. Smith, P. D. Mathers, M. Fox, T. R. Schaeffer, A. A. Seymour and N. C. Trippodo, *J. Med. Chem.*, 1997, **40**, 1570.
41. J. M. Betancort, D. T. Winn, R. Liu, Q. Xu, J. Liu, W. Liao, S.-H. Chen, D. Carney, D. Hanway, J. Schmeits, X. Li, E. Gordon and D. A. Campbell, *Bioorg. Med. Chem. Lett.*, 2009, **19**, 4437.
42. J. A. Robl, R. B. Sulsky, D. J. Augeri, D. R. Magnin, L. G. Hamann and D. A. Betebenner, *U.S. Pat.* 6 395 767, 2002.
43. D. R. Magnin, J. A. Robl, R. B. Sulsky, D. J. Augeri, Y. Huang, L. M. Simpkins, P. Taunk, D. A. Betebenner, J. G. Robertson, B. Abboa-Offei, A. Wang, M. Cap, L. Xing, L. Tao, D. F. Sitkoff, M. F. Malley, J. Z. Gougoutas, A. Khanna, Q. Huang, S.-P. Han, R. A. Parker and L. G. Hamann, *J. Med. Chem.*, 2004, **47**, 2587.
44. H. B. Rasmussen, S. Branner, F. C. Wiberg and N. Wagtmann, *Nat. Struct. Biol.*, 2003, **10**, 19.
45. (a) Y. B. Kim, L. M. Kopcho, M. S. Kirby, L. G. Hamann, C. A. Weigelt, W. J. Metzler and J. Marcinkeviciene, *Arch. Biochem. Biophys.*, 2006, **445**, 9; (b) M. S. Kirby, C. Dorso, A. Wang, C. Weigelt, L. Kopcho, L. Hamann and J. Marcinkeviciene, *Clin. Chem. Lab. Med.*, 2008, **46**, A29.

46. T. E. Hughes, M. D. Mone, M. E. Russell, S. C. Weldon and E. B. Villhauer, *Biochemistry*, 1999, **38**, 11597.
47. R. Zhang and F. Monsma, *Curr. Opin. Drug Discovery Dev.*, 2009, **12**, 488.
48. R. A. Copeland, D. L. Pompliano and T. D. Meek, *Nat. Rev. Drug Discovery*, 2006, **5**, 730.
49. W. J. Metzler, J. Yanchunas, C. Weigelt, K. Kish, H. E. Klei, D. Xie, Y. Zhang, M. Corbett, J. K. Tamura, B. He, L. G. Hamann, M. S. Kirby and J. Marcinkeviciene, *Protein Sci.*, 2008, **17**, 240.
50. E. L. Ash, J. L. Sudmeier, E. C. De Fabo and W. W. Bachovchin, *Science*, 1997, **278**, 1128.
51. K. S. Kim, K. S. Oh and J. Y. Lee, *Proc. Natl. Acad. Sci. U. S. A.*, 2000, **97**, 6373.
52. A. Fura, A. Khanna, V. Vyas, B. Koplowitz, S.-Y. Chang, C. Caporuscio, D. W. Boulton, L. J. Christopher, K. Chadwick, L. G. Hamann, W. G. Humphreys and M. S. Kirby, *Drug Metab. Dispos.*, 2009, **37**, 1164.
53. E. B. Villhauer, J. A. Brinkman, G. B. Naderi, B. F. Burkey, B. E. Dunning, K. Prasad, B. L. Mangold, M. E. Russell and T. E. Hughes, *J. Med. Chem.*, 2003, **46**, 2774.
54. L. M. Simpkins, S. Bolton, Z. Pi, J. C. Sutton, C. Kwon, G. Zhao, D. R. Magnin, D. J. Augeri, T. Gungor, D. P. Rotella, Z. Sun, Y. Liu, W. S. Slusarchyk, J. Marcinkeviciene, J. G. Robertson, A. Wang, J. A. Robl, K. S. Atwal, R. L. Zahler, R. A. Parker, M. S. Kirby and L. G. Hamann, *Bioorg. Med. Chem. Lett.*, 2007, **17**, 6476.
55. C. J. Bailey and C. Day, *Diabetes Obes. Metab.*, 2009, **11**, 527.
56. Investor relations presentations, Morgan Stanley Dean Witter Pan-European Healthcare Conference, Novartis AG Company Presentation, September 2001.
57. M. Kirby, D. M. T. Yu, S. P. O'Connor and M. D. Gorrell, *Clin. Sci.*, 2010, **118**, 31.
58. G. R. Lankas, B. Leiting, R. S. Roy, G. J. Eiermann, M. G. Beconi, T. Biftu, C. C. Chan, S. Edmondson, W. P. Feeney, H. He, D. E. Ippolito, D. Kim, K. A. Lyons, H. O. Ok, R. A. Patel, A. N. Petrov, K. A. Pryor, X. Qian, L. Reigle, A. Woods, J. K. Wu, D. Zaller, X. Zhang, L. Zhu, A. E. Weber and N. A. Thornberry, *Diabetes*, 2005, **54**, 2988.
59. J. S. Rosenblum, L. C. Minimo, Y. Liu, J. Wu and J. W. Kozarich, presented at the 2007 American Diabetes Association Meeting, Chicago, abstr. 0519-P.
60. J.-J. Wu, H.-K. Tang, T.-K. Yeh, C.-M. Chen, H.-S. Shy, Y.-R. Chu, C.-H. Chien, T.-Y. Tsai, Y.-C. Huang, Y.-L. Huang, C.-H. Huang, H.-Y. Tseng, W.-T. Jianng, Y.-S. Chao and X. Chen, *Biochem. Pharmacol.*, 2009, **78**, 203.
61. B. F. Burkey, P. K. Hoffmann, U. Hassiepen, J. Trappe, M. Juedes and J. E. Foley, *Diabetes Obes. Metab.*, 2008, **10**, 1.
62. B. Ahren, *Best Pract. Res. Clin. Endocrinol. Metab.*, 2009, **23**, 487.
63. J. D. Godfrey, Jr., R. T. Fox, F. G. Buono, J. Z. Gougoutas and M. F. Malley, *J. Org. Chem.*, 2006, **71**, 8647.

64. R. L. Hanson, S. L. Goldberg, D. B. Brzozowski, T. P. Tully, D. Cazzulino, W. L. Parker, O. K. Lyngberg, T. C. Vu, M. K. Wong and R. N. Patel, *Adv. Synth. Catal.*, 2007, **349**, 1369.
65. I. Gill and R. Patel, *Bioorg. Med. Chem. Lett.*, 2006, **16**, 705.
66. J. Rosenstock, S. Sankoh and J. F. List, *Diabetes Obes. Metab.*, 2008, **10**, 376.
67. FDA Endocrinologic and Metabolic Drugs Advisory Committee, April 1 2009; < http://www.fda.gov/AdvisoryCommittees/CommitteesMeeting-Materials/Drugs/EndocrinologicandMetabolicDrugsAdvisoryCommittee/ucm149225.htm > .
68. M. Jadzinsky, A. Pfützner, E. Paz-Pacheco, Z. Xu, E. Allen and R. Chen, *Diabetes Obes. Metab.*, 2009, **11**, 611.
69. R. A. DeFronzo, M. N. Hissa, A. J. Garber, J. L. Gross, R. Y. Duan, S. Ravichandran and R. S. Chen, *Diabetes Care*, 2009, **32**, 1649.
70. A. R. Chacra, G. H. Tan, A. Apanovitch, S. Ravichandran, J. List and R. Chen, *Int. J. Clin. Pract.*, 2009, **63**, 1395.
71. A. E. Weber and N. Thornberry, *Annu. Rep. Med. Chem.*, 2007, **42**, 95.

CHAPTER 2

The Discovery of Vorapaxar (SCH 530348), a Thrombin Receptor (Protease Activated Receptor-1) Antagonist with Potent Antiplatelet Effects

SAMUEL CHACKALAMANNIL

Merck, 2015 Galloping Hill Rd, Kenilworth, NJ 07033, USA

2.1 Introduction

Great strides have been made in the prevention and treatment of cardiovascular disease (CVD), thereby slowing down the upsurge that existed during the early second half of the 20th century.[1] Nevertheless, CVD continues to be the major cause of mortality and morbidity in the industrialized world and its incidence is on fast ascendancy in the developing world. According to the World Health Organization, CVD will be the major cause of mortality worldwide by 2020, with fatalities projected to rise above 20 million.[2] About 40% of deaths in the United States are attributed to diseases of cardiovascular origin, with similar statistics existing in the European countries. Nearly 80 million US adults are affected by CVD, causing a staggering economic burden of about 500 billion dollars annually.[1]

   Coronary artery disease (CAD) accounts for nearly half of the cardiovascular deaths.[1] Over half a million people die annually of CAD in the US alone,

RSC Drug Discovery Series No. 4
Accounts in Drug Discovery: Case Studies in Medicinal Chemistry
Edited by Joel C. Barrish, Percy H. Carter, Peter T. W. Cheng and Robert Zahler
© Royal Society of Chemistry 2011
Published by the Royal Society of Chemistry, www.rsc.org

with global deaths estimated over 7 million.[3] The underlying etiology for CAD is atherosclerosis, a chronic, progressive, inflammatory, and proliferative response to cholesterol infiltration into arterial wall, leading to arterial plaques. Arterial plaques grow in size with time, thereby causing luminal narrowing that results in restricted blood flow to the myocardium, which often gives rise to symptoms of stable angina, such as exercise-induced chest pain.[4,5] The abrupt clinical manifestations of CAD, however, are triggered by the rupture of an atherosclerotic plaque, which marks the sudden transition from a stable, clinically-silent disease to a symptomatic, life-threatening condition.[6] When a plaque ruptures in a coronary artery, blood comes into contact with the vasoactive components of the endothelial matrix, initiating two principal mechanisms of thrombosis—activation of platelets and activation of blood coagulation—thereby leading to an occlusive thrombus formation.[7,8] The obstruction of blood flow and the myocardial underperfusion that results leads to ischemic disorders, presenting a spectrum of clinical conditions known as acute coronary syndrome (ACS) that includes Q-wave myocardial infarction, non-Q-wave myocardial infarction, and unstable angina.[9–11]

Thrombin plays a very important dual role in thrombosis (Figure 2.1). In its role as a procoagulant, thrombin cleaves fibrinogen to fibrin, which polymerizes to form a meshwork. Thrombin, being the most potent activator of platelets, also causes platelets to aggregate. The fibrin meshwork traps aggregated platelets and other plasma particles such as red blood cells to give rise to a fast growing thrombus, which gains further rigidity and mechanical strength by factor XIIIa-mediated cross-linking reactions.[12]

Antithrombotic agents are the mainstay of pharmacological therapy for thrombotic disorders.[13–15] Thrombosis, a localized clotting of blood, can occur either in the arterial or venous segment of the circulation system, and, depending on the local milieu, thrombus composition and pathophysiology may vary and need to be treated with different antithrombotic agents.[16,17] Mechanistically, the function of an antithrombotic agent is to prevent the formation of new thrombi in the blood vessels or to disrupt the existing thrombi and restore blood flow.[18] In general, arterial thrombi are abundant in aggregated platelets and venous thrombi have polymerized fibrin as the major component. Depending on their functional mechanism, the antithrombotic agents can be divided into anticoagulants, antiplatelet agents, and fibrinolytic agents (Table 2.1). The anticoagulants either modulate the endogenous levels of thrombin or inactivate the enzymatic function of thrombin.[19,20] Antiplatelet agents inhibit platelet activation and aggregation, integral processes of hemostasis and thrombosis.[21] Fibrinolytic agents, which are enzyme or protein preparations, are intravenously administered under clinical emergency to cause lysis of an existing thrombus.[22]

2.2   Mechanism of Platelet Activation

Normally, platelets circulate freely in blood vessels without interacting with each other or the vascular endothelium. In the context of endothelial damage, a

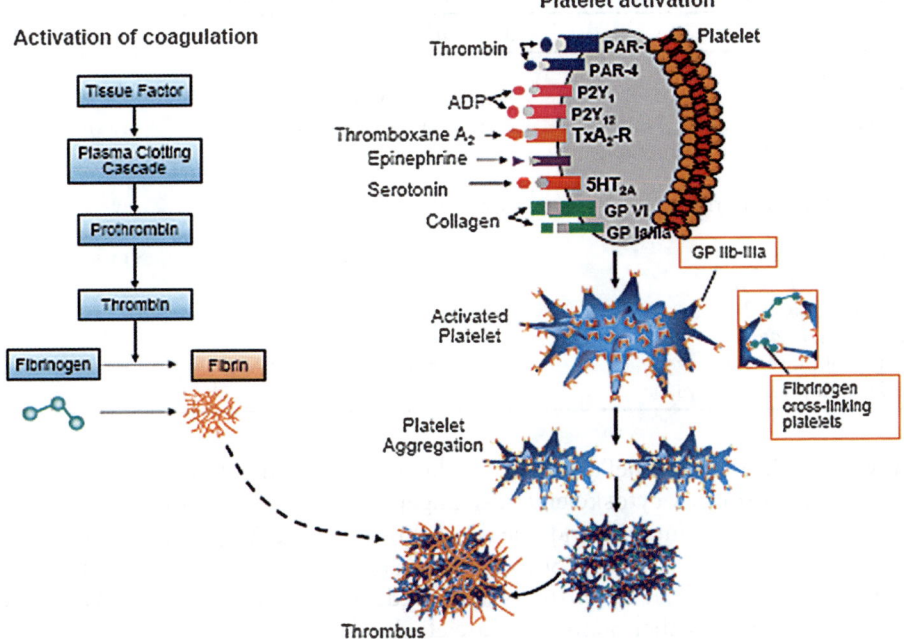

Figure 2.1　Mechanism of thrombosis. The platelet activation mechanisms and activation of the coagulation system synergize in thrombus formation. Platelets have different cell-surface receptors that are activated by specific agonists. Upon activation, these receptors trigger intracellular signal transduction mechanisms, resulting in the activation of integrin receptors GP IIb/IIIa. Activated GP IIb/IIIa receptors bind to the RGD motif of fibrinogen, thereby cross-linking activated platelets which leads to platelet aggregation. Among the various platelet surface receptors, PAR-1 (thrombin receptor) is by far the most potent inducer of platelet activation. Concurrent with platelet activation, the coagulation process is also triggered by the exposure of tissue factor to blood, resulting in the production of thrombin. In its procoagulant role, thrombin cleaves soluble fibrinogen to fibrin which cross-links to form an insoluble fibrin meshwork that traps aggregated platelets and other plasma particles, leading to a growing thrombus.

chain of events leads to the aggregation of platelets. Depending on the nature of the vascular injury, this may develop into either a normal hemostasis or a pathologic condition (the latter resulting in vascular thrombosis, ischemic stroke, *etc.*). The underlying platelet events constitute a complex series of biochemical and cellular processes that can be classified as adhesion of platelets to damaged vessel wall, activation of platelets, secretion of granular contents from activated platelets, and aggregation of platelets.[23,24]

The adhesion of platelets to denuded endothelium represents the primary hemostatic response to vessel wall injury.[25] When endothelial layer disruption occurs as a result of a vascular trauma, platelets adhere to the exposed endothelium to form a discontinuous platelet monolayer. The adhered platelets

Table 2.1 Classification of major marketed antithrombotic agents.

Name	Mechanism of action	Formulation
Anticoagulants		
Warfarin	Vitamin K antagonist	Oral
Heparin, Danaparoid, Dalteparin, Tinzaparin, Enoxaparin, Fondaparinux	Antithrombin cofactor	iv or sc
Lepirudin, Bivaluridin, Argatroban	Direct thrombin inhibitors	iv
Rivaroxaban	Factor Xa inhibitor	Oral
Antiplatelet agents		
See Table 2.2		
Fibrinolytic agents		
Streptokinase, Alteplase, Tenectaplase, Reteplase	Promote plasmin activity	iv

interact with sub-endothelial collagen, which causes their activation. Activation of platelets prompts cytoskeletal rearrangements, membrane fusion, exterization, and internal synthesis and release of thromboxane A2 (TxA_2), which itself is a potent platelet activator. Secretion of platelet granular contents releases a variety of biochemical agonists—such as adenosine diphosphate (ADP) and serotonin—that further activate platelets by interacting with their specific platelet surface receptors (Figure 2.1).[26,27] Additionally, thrombin that is locally generated at the site of injury activates platelets via protease-activated receptors (PARs).[28]

Among the various platelet stimulants, thrombin is the most potent activator of platelets.[29] Vascular injury and inflammation expose tissue factor, resulting in the formation tissue factor/factor VIIa complex that leads to the local generation of thrombin from prothrombin. Platelets also facilitate thrombin generation by providing procoagulant phospholipid surfaces that anchor various coagulation factors. Thrombin activates platelets at very low concentrations by interacting with PARs.

Activation of platelets and the ensuing intracellular biochemical events lead to the activation of surface GP IIb/IIIa receptors—the final, convergent pathway in the platelet activation mechanism. GP IIb/IIIa is a member of the integrin family of receptors, composed of α- and β-subunits (α_{IIb}, β_3). In the resting state, platelet GP IIb/IIIa receptors do not bind to fibrinogen (or bind with a very low affinity).[30] Activation alters the conformation of GP IIb/IIIa, rendering itself capable of binding to extracellular macromolecular ligands, including fibrinogen and von Willebrand Factor (vWF). The arginine-glycine-aspartic acid (RGD) sequence of the adhesive proteins binds to the GP IIb/IIIa receptor. Fibrinogen contains two RGD sequences on its α-chain, one in the N-terminal region and the other in the C-terminal region, and is therefore bivalent in its binding to GP IIb/IIIa receptors, which allows efficient cross-linking of platelets.[31] Although vWF also binds to the GP IIb/IIIa receptor at its various RGD sites, studies in fibrinogen knockout mice have shown that vWF alone is not sufficient to achieve stable platelet aggregation.[32]

Table 2.2 Major marketed antiplatelet agents and candidates in advanced stage of development.

Generic name or code name	Brand name	Formulation	Status
Cyclooxygenase inhibitor			
Acetylsalicylic acid	Aspirin	Oral	launched
ADP antagonists			
Ticlopidine	Ticlid	Oral	launched
Clopidogrel	Plavix	Oral	launched
Prasugrel	Effient	Oral	launched
Ticagrelor (AZD6140)	Brilinta	Oral	P-III
Glycoprotein IIb/IIIa antagonists			
Abciximab	Reopro	iv	launched
Eptifibatide	Integrilin	Iv	launched
Tirofiban	Aggrastat	iv	launched
Thrombin receptor (PAR-1) antagonists			
Vorapaxar (SCH 530348)	–	Oral	P-III
E5555	–	Oral	P-II
Phosphodiesterase-III Inhibitors			
Cilostazol	Pletal	Oral	launched
Adenosine Reuptake Inhibitors			
Dipyridamole	Persantine,	Oral	launched
	Aggrenox (in combination with aspirin)	Oral	launched

2.3 Antiplatelet Agents

Antiplatelet agents have been the mainstay of treatment for ACS over the past 40 years, and exhibit well-established benefits in preventing coronary thrombosis and myocardial infarction in patients with ACS. Results of a meta-analysis of 145 clinical studies reported a 25% reduction of cardiovascular events in high-risk patients treated with antiplatelet therapy.[33] Another meta-analysis of 109 clinical trials has concluded that antiplatelet medications were able to reduce transient ischemic attacks and strokes by 22%, reduce coronary artery disease by 29%, and reduce peripheral artery disease by 23%.[34] Table 2.2 summarizes the major antiplatelet agents available commercially and promising compounds in late stage development.

2.4 Thrombin Receptor (Protease-activated Receptor-1) Antagonists as Promising Antiplatelet Agents with Improved Safety Margin

As mentioned before, thrombin plays a dual procoagulant and proaggregatory role in the context of hemostasis (see Figure 2.1). The platelet activation of thrombin is mediated by specific cell-surface receptors known as protease-activated receptors (PARs).[35,36] The prototype of these receptors is PAR-1, which is also known as the thrombin receptor.

Subsequent to the identification of the initial protease-activated receptor (PAR-1), three additional protease-activated receptors have been identified.[37] The current family of PARs comprises PAR-1, PAR-2,[38] PAR-3,[39] and PAR-4.[40] Of these, PAR-1, PAR-3, and PAR-4 are activated by thrombin. PAR-2 is activated by trypsin, tryptase, and coagulation factors VIIa and Xa, but not by thrombin. It has been established that the "tethered ligand" activation paradigm described below (see Section 2.4.1) applies to all the PARs. There is considerable species specificity to the nature of PARs on platelets. In primates, PAR-1 is the main platelet protease-activated receptor. It is also present on other cells such as endothelial cells, smooth muscle cells, monocytes, and fibroblasts. PAR-4 is the second PAR on human platelets. Since PAR-4 has only weak affinity for thrombin, it is activated only at high thrombin concentrations. It is believed to be a "rescue receptor" that is activated in the event of a serious vascular lesion and the resultant high thrombin concentration. PAR-3 is found in mouse platelets where it is the major regulator of thrombin response, along with PAR-4.[41] Since PAR-1 is not present in mouse and rat platelets, but is present in monkey platelets, non-human primate models have been used to study the *in vivo* antithrombotic effects of thrombin receptor antagonists. However, due to the presence of PAR-1 on rodent coronary artery smooth muscle cells, rodent models have been used to study the effect of thrombin receptor antagonists on restenosis and neointimal proliferation.[42]

2.4.1   Mechanism of Thrombin Receptor Activation

Although it was known for some time that a thrombin-specific receptor on platelets mediates platelet activation, the exact mechanism of thrombin-specific cellular activation was unknown until 1991,[43] when expression cloning of a functional thrombin receptor was achieved by two groups. These studies also unveiled the intriguing mechanistic details of the cellular activation of thrombin.[44–46] The amino acid sequence deduced from the mRNA encoding the thrombin receptor revealed a new G-protein-coupled seven-transmembrane domain receptor with a large extracellular domain.[47] The authors postulated that thrombin binds to the cellular receptor through its anion binding exo-site and subsequently cleaves the extracellular domain at Arg41-Ser42 (Figure 2.2). The newly unmasked amino terminus acts as a "tethered ligand," binding intramolecularly to the proximal heptahelical segment and eliciting G-protein-coupled transmembrane signaling.[48,49] Peptides having the sequence Ser-Phe-Leu-Leu-Arg-Asn, which mimics the new amino terminus of the activated receptor, function as agonists of PAR-1 and produce functional responses such as platelet aggregation and mitogenesis.[50] These peptides and their functional analogs are known as thrombin receptor activating peptides (TRAPs). Uncleavable mutant thrombin receptors failed to respond to thrombin, but were responsive to TRAPs.

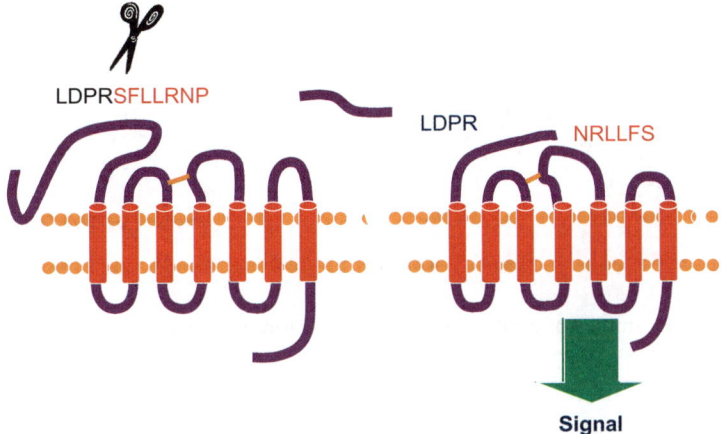

Figure 2.2 The tethered ligand mechanism of thrombin receptor activation. Thrombin binds to the extracellular domain of the receptor and cleaves it at Arg41-Ser42. The newly generated serine N-terminus binds intramolecularly to the second loop of the receptor, triggering cellular activation.

2.4.2 Early Preclinical Evidence Suggesting the Antiplatelet Effect of a Thrombin Receptor Antagonist with Improved Safety Margin with Regard to Bleeding

A number of early preclinical experiments pointed to the fact that PAR-1 antagonism could indeed engender strong antithrombotic effects without the attendant bleeding that is common to most anticoagulants and antiplatelet agents. Early PAR-1 antagonists were peptides and peptidomimetics modeled after the tethered ligand of PAR-1. Several of these compounds showed potent PAR-1 affinity in radioligand binding assays against PAR-1 isolated from platelet membrane. In platelet aggregation functional assays, these compounds exhibited potent inhibition of aggregation caused by either TRAP or α-thrombin.[51,52] These compounds also blocked thrombin-induced calcium mobilization and smooth muscle cell proliferation mediated by the activation of PAR-1. Two salient examples from these early studies are illustrated in Figure 2.3.[53,54]

Due to the absence of PAR-1 in mouse and rat platelets, non-human primate models have been often used to test the antiplatelet effect of PAR-1 antagonism. *In vivo* studies carried out by PAR-1 antagonists of peptide and non-peptide origin have shown inhibition of platelet-rich thrombus formation in both baboon and guinea pig thrombosis models, suggesting the promise of a PAR-1 antagonist for inhibiting arterial thrombosis.[55,56] In another important study, an antibody to the PAR-1 N-terminus was reported to inhibit mechanical injury-induced thrombosis in a baboon carotid artery model without affecting bleeding time and coagulation parameters.[57] The peptidomimetic PAR-1 antagonist **2** (RWJ-58259) was tested in a cynomolgus monkey

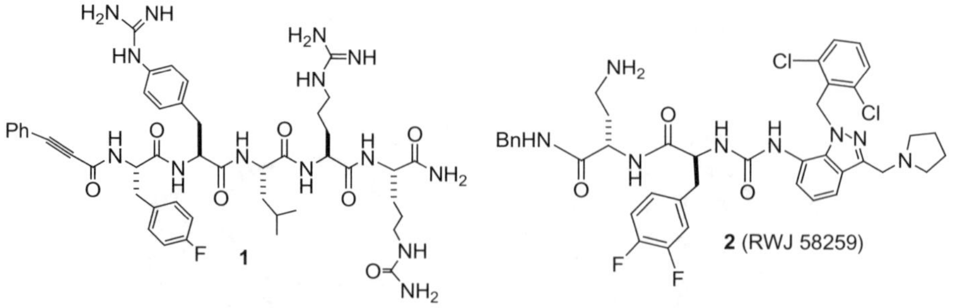

Figure 2.3   Examples of early peptide and peptidomimetic thrombin receptor antago-
                nists.

vascular injury model.[58] The compound was administered intravenously and
the degree of vessel occlusion caused by electrolytic injury-induced thrombus in
each carotid artery was characterized. Compound **2** significantly reduced
occlusion in the vessels of all animals; of note that these vessels were completely
occluded under experimental conditions in the absence of drug. In the drug-
treated group, not only was the thrombus size reduced, but the composition of
the thrombus indicated a switch from platelet-rich thrombi to platelet-depleted
thrombi, demonstrating the antiplatelet property of a PAR-1 antagonist. In a
related study in rat, compound **2** showed significant reduction of neointima
thickness in a rat restenosis model after perivascular administration, suggesting
the potential utility of a thrombin receptor antagonist to treat vascular proliferative disorders such as restenosis.[42]

As described below, more recent studies using potent PAR-1 antagonists in
non-human primate thrombosis models have corroborated the outcome of
these earlier studies and established the therapeutic potential of PAR-1
antagonists as promising antithrombotic agents.[59] By selectively inhibiting
PAR-1-mediated platelet activation, a PAR-1 antagonist should exhibit strong
antiplatelet action under conditions in which thrombin-stimulated platelet
activation is important. A PAR-1 antagonist is specific for thrombin-mediated
platelet aggregation and it would not interfere with platelet activation by other
platelet receptors needed for normal hemostasis. Additionally, fibrin generation
by thrombin would not be affected, keeping the normal coagulation processes
intact. For these reasons, a PAR-1 antagonist is expected to give an improved
safety margin with regard to any hemorrhagic side effect, which is a complicating factor for the current antithrombotic therapy.

2.4.3   Small-molecule Thrombin Receptor Antagonists

There has been considerable progress in the discovery of orally active small-
molecule thrombin receptor antagonists.[60] Some examples of early small-
molecule thrombin receptor antagonists are shown in Figure 2.4. In the radioligand binding assay using [³H]-TRAP, these compounds showed promising

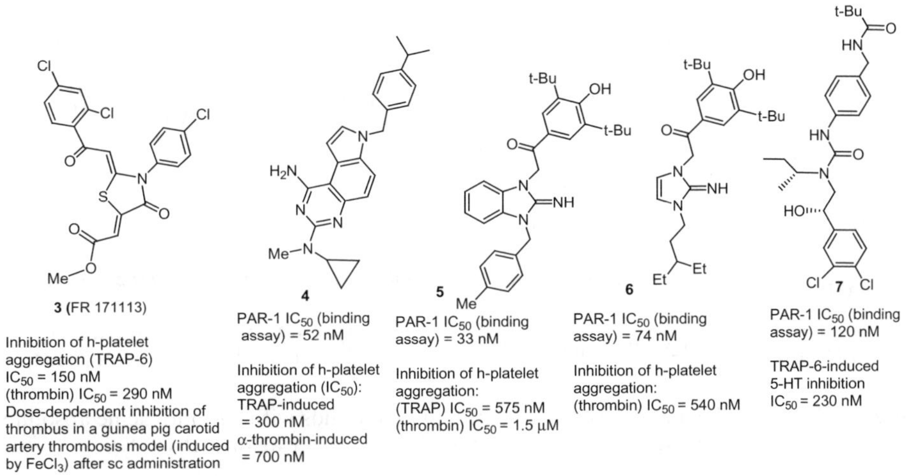

Figure 2.4 Examples of early small-molecule thrombin receptor (PAR-1) antagonists.[60,61]

inhibitory values. Modest inhibition was noted in platelet aggregation functional assays using TRAP as the agonist, while the inhibitory effect on aggregation induced by α-thrombin was weak. However, compound **4** was reported to show a strong inhibitory effect in functional assays in human coronary artery smooth muscle cells that measure calcium transients induced by α-thrombin (3 nM) or a peptide agonist (TFLLRNPNDK-NH$_2$, 30 μM), with IC$_{50}$ values of 82 and 55 nM, respectively.[61] In a proliferation functional assay in human coronary artery smooth muscle cells (hCASMC) that measured the incorporation of [^{3}H]thymidine induced by α-thrombin or TRAP, compound **4** showed apparent K_i values of 88 and 32 nM, respectively. In contrast to platelets, where the inhibition of any thrombin-induced effect was transient, inhibition of calcium transients and thymidine incorporation in hCASMC mediated by thrombin was sustained over the time course of the assay.

2.4.4　Himbacine-derived Thrombin Receptor Antagonists

While the early peptidomimetic and small-molecule thrombin receptor antagonists provided useful tools for the study of the pharmacology of the thrombin receptor, they did not advance into clinical development. A more promising series emerged from the natural product himbacine **8** (Figure 2.5). Himbacine is a complex alkaloid isolated from the bark of *Galbulimima belgraveana*, a tall growing species of white magnolia found in the northern Queensland regions of Australia and nearby Pacific islands. In the early 1990's, the antimuscarinic properties of himbacine were reported, which stimulated synthetic efforts in several laboratories to identify analogs with selective antimuscarinic activity for the M$_2$ subtype as a cholinergic approach to treat the symptoms of Alzheimer's disease. Our own efforts in this area culminated in a

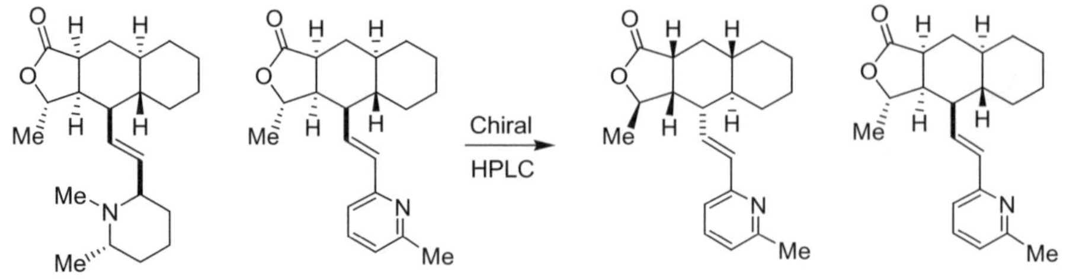

8 (+)-Himbacine	Racemic lead	*ent*-Himbacine series	Himbacine series
M_2 Ki = 4.8 nM PAR-1 inactive	**(±)-9** R = Me IC_{50} = 300 nM	**(+)-10a** R = Me IC_{50} = 150 nM	**(-)-10b** R = Me IC_{50} = 1500 nM

Figure 2.5 Himbacine-derived PAR-1 lead. Compounds with *ent*-himbacine absolute chirality showed consistently superior activity against PAR-1.

total synthesis of himbacine and several of its analogs.[62,63] Our detailed structure–activity study of himbacine-based muscarinic antagonists has also been published.[63] One of these racemic synthetic analogs (**9**), which replaces the 6-methylpiperidine motif with the corresponding methylpyridine, was totally devoid of any antimuscarinic activity. However, this compound was identified as a hit against PAR-1 in a later high-throughput PAR-1 assay.

2.4.4.1 *Lead Optimization*

In the initial phase of lead optimization, the racemic himbacine derivative **9** was resolved using chiral HPLC. The enantiomer **10a** bearing absolute chirality opposite to that of himbacine (*ent*-himbacine) was found to have 10-fold more affinity toward PAR-1 in the binding assay than the enantiomer with natural absolute chirality (**10b**). The absolute chirality was confirmed by enantiospecific synthesis starting from (*R*)-but-3-yn-2-ol (Scheme 2.1).[59]

The synthesis of himbacine-derived PAR-1 antagonists outlined in Scheme 2.1 employs a highly diastereoselective intramolecular Diels–Alder (IMDA) reaction to generate the tricyclic intermediate **18**. The protocol for the IMDA reaction was adapted from the total synthesis of himbacine.[59,62] The synthesis commences with commercially available (*R*)-but-3-yn-2-ol, which was *O*-protected to give **11**, and subsequently elaborated to the pentynoic acid benzyl ester **12** as shown (Scheme 2.1). Esterification of **12** with dienoic acid **13** and subsequent Lindlar reduction of the triple bond gave the IMDA precursor **15**. Thermal cyclization of **15** gave 66% of the required tricyclic carboxylic ester **17** after a brief *in situ* treatment with DBU to epimerize the stereogenic center α to the lactone carbonyl group. Catalytic hydrogenation of **17** over platinum oxide effected diastereospecific reduction of the double bond as well *O*-debenzylation to produce the tricyclic carboxylic acid **18**. The acid chloride generated from **18** was reduced to the aldehyde **19** by treatment with tributyltin hydride in the

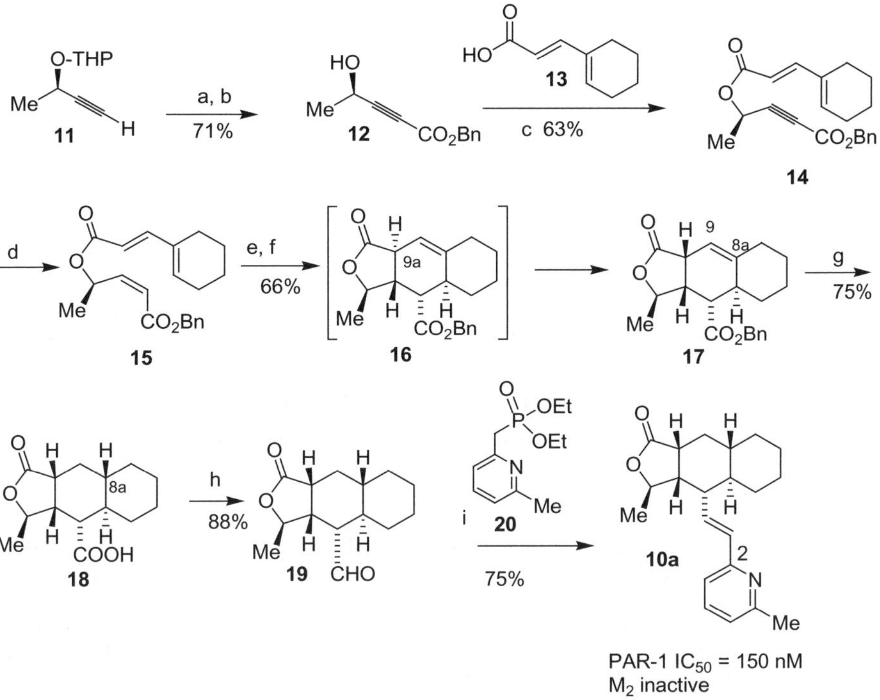

Scheme 2.1 A representative synthesis of himbacine-derived thrombin receptor antagonists.[59] The key *exo*-selective intramolecular Diels–Alder reaction proceeds with high diastereoselectivity. Reagents and conditions: (a) Bu^{n}Li/THF, BnOCOCl, $-78\,^\circ$C; (b) DOWEX/MeOH; (c) **13**, 4-pyrrolidinylpyridine, DCC/CH$_2$Cl$_2$; (d) H$_2$, Lindlar/THF; (e) *o*-xylene, $185\,^\circ$C, (f) DBU; (g) H$_2$, PtO$_2$/MeOH; (h) SOCl$_2$/toluene, $80\,^\circ$C, then Bu$_3$SnH, (Ph$_3$P)$_4$Pd; (i) **20**, Bu^{n}Li, hexane/THF, $0\,^\circ$C.

presence of palladium(0). Horner–Wadsworth–Emmons reaction of aldehyde **19** with the appropriate phosphonate **20** gave the vinylpyridine derivative **10a** in excellent yield.

The diastereoselectivity of the IMDA reaction can be rationalized as shown in Figure 2.6. Within the ordered transition state of the thermal cyclization, conformation A is disfavored due to allylic A^{1,3} strain. The favored conformation B facilitates the facial selectivity of C$_{3a}$–C$_{9a}$ bond formation, engendering a *cis* relationship between the C$_3$ methyl group and C$_{3a}$ hydrogen in **16**. The relative stereochemistry between C$_{3a}$ and C$_{9a}$ is a function of the *exo*-selective nature of the IMDA reaction, and the relative stereochemistry between C$_{3a}$ and C$_4$ is controlled by the choice of *cis*-dienophile. The C$_{9a}$ stereogenic center was readily epimerized by treatment with DBU. Catalytic reduction of the internal double bond incorporated the stereogenic center at C$_{8a}$ in the required relative stereochemistry. In short, the intramolecular Diels–Alder reaction lends itself to the construction the tricyclic motif by incorporating five new stereogenic centers with

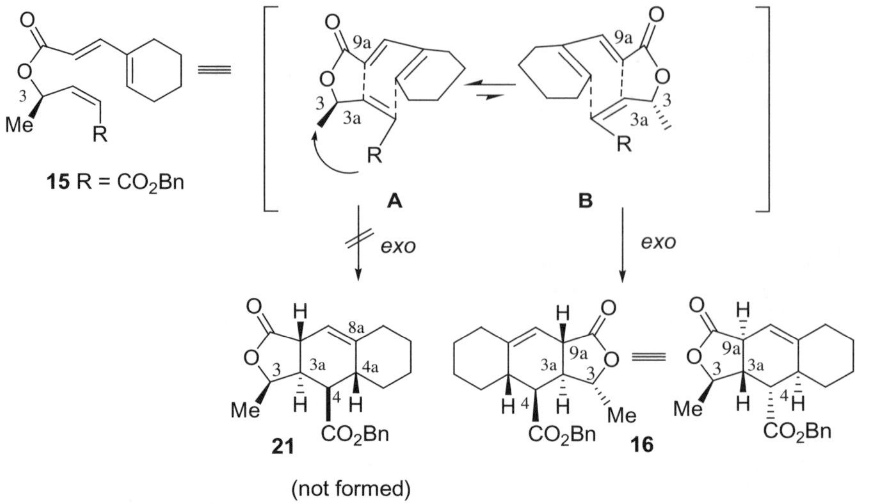

Figure 2.6 The high diastereoselectivity of the IMDA can be explained by an $A^{1,3}$ strain-induced conformational preference and the *exo*-selective nature of the cyclization. Conformation **B** is preferred due to $A^{1,3}$ strain. The preferred conformation dictates the facial selectivity of C3a–C9a bond formation, rendering a *cis* relationship between the C3 methyl and C3a hydrogen. The relative stereochemistry between C3a and C9a is controlled by the *exo*-selective transition state.

high diastereoselectivity. By employing optically pure IMDA precursor **11**, this synthesis translates to an enantiospecific construction of the tricyclic ring system.

2.4.4.1.1 Structure–Activity Relationship (SAR) Studies. Key points of the SAR studies of himbacine-derived PAR-1 antagonists are shown in Table 2.3.[59,64] A small alkyl group such as methyl or ethyl is preferred at the $C_{6'}$ position of the pyridine (**9** and **22**). However, alkyl substitution at $C_{5'}$ is not favored (**23** versus **9** or **10**). The substituted pyridine can be replaced with a 2-vinylisoquinoline (*e.g.*, **28**). Although phenyl substitution is not tolerated at $C_{6'}$ (*e.g.*, **24**), it is highly desirable at $C_{5'}$ (*e.g.*, **25**). More importantly, the $C_{5'}$ aryl derivatives showed markedly improved rat oral absorption relative to the $C_{6'}$ alkyl-substituted derivatives.

The lactone ring is essential for PAR-1 activity. The corresponding ring-opened hydroxy acid (not shown) is unstable and undergoes spontaneous ring closure to the lactone. Reduced versions of the lactone, such as lactol (**26**) and the fused tetrahydrofuran (**27**), are far less active. The C_3 methyl group is optimal: while C_3-unsubstituted lactone (**29**) and C_3 *gem*-dimethyl substituents are tolerated, larger substituents at C_3 are not tolerated (*e.g.*, **31**). The internal double bond within the tricyclic unit does not affect the PAR-1 affinity (structure type **B**). However, the *trans*-disubstituted vinyl group that links the tricyclic unit to the heteroaryl group is essential for activity. Reduction of this double bond resulted in greater than 10-fold loss in potency (data not shown).

Table 2.3 Structure–activity relationship of himbacine-derived PAR-1 antagonists

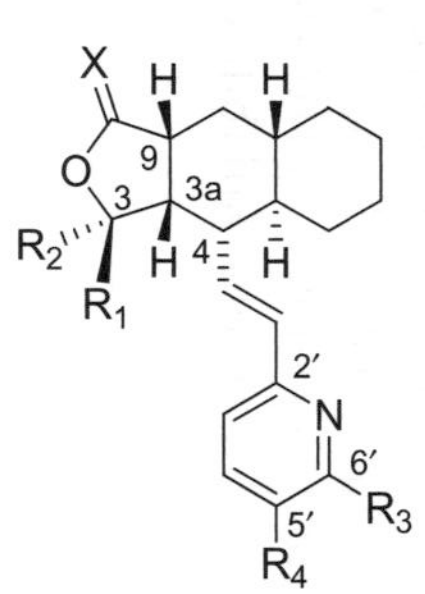

Structure type A

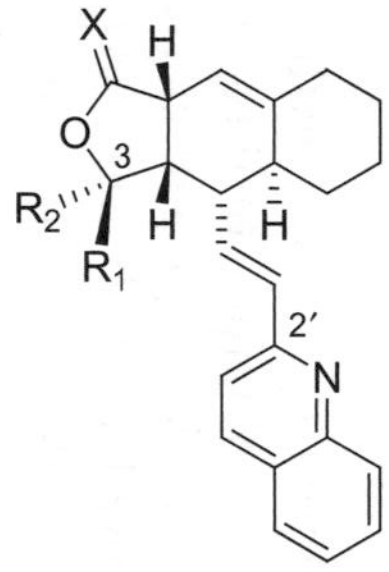

Structure type B

No.	Str. type	X	R1	R2	R3	R4	PAR-1 IC50 (nM)	Rat oral absorption
(±)-9	A	O	Me	H	Me	H	300 nM	low
(+)-10a	A	O	Me	H	Me	H	150	low
(+)-22	A	O	Me	H	Et	H	20	low
(±)-23	A	O	Me	H	H	Me	1100	ND[a]
(±)-24	A	O	Me	H	Ph	H	>10 000	ND
(+)-25	A	O	Me	H	H	Ph	27	high
(±)-26	A	H, OH	Me	H	Et	H	3500	ND
(±)-27	A	H, H	Me	H	fused phenyl (quinolyl)		8000	ND
(±)-28	B	O	Me	H	quinolyl		150	low
(±)-29	B	O	H	H	quinolyl		200	ND
(±)-30	B	O	Me	Me	quinolyl		150	ND
(±)-31	B	O	Et	H	quinolyl		1500	ND

[a]ND = not determined.

2.4.4.1.2 First-generation Thrombin Receptor Antagonists. The $C_{6'}$-ethyl-pyridine derivative **22** was the first benchmark compound which showed a promising spectrum of activity in the binding assay, functional assays, and a monkey *ex vivo* platelet aggregation assay (Figure 2.7). It showed a K_i value of 12 nM against PAR-1 in the binding assay, and it inhibited aggregation of washed human platelets induced by TRAP with an IC_{50} of 70 nM. In a cyno-molgus monkey *ex vivo* platelet aggregation model, compound **22** caused complete inhibition of TRAP-induced platelet aggregation at $10\,mg\,kg^{-1}$ upon intravenous administration (over 30 min).

However, compound **22** showed poor oral bioavailability ($F = 3\%$ in rat). Our efforts to identify compounds with good oral bioavailability led to $C_{5'}$-aryl derivatives (*e.g.*, **25**), which, in general, showed good potency and oral bioa-vailability. The synthesis of $C_{5'}$-aryl derivatives was carried out by Suzuki reaction of the corresponding triflate or bromide (Scheme 2.2).[59] From this sub-series, compound **33** was identified as our first recommended candidate for development (Table 2.4).

The $C_{5'}$-aryl derivative **33** was a competitive antagonist of PAR-1 with a K_i value of 2.7 nM in the binding assay. It showed robust inhibition of thrombin-induced platelet aggregation and TRAP-induced platelet aggregation (IC_{50} values of 44 and 24 nM, respectively), whereas it showed no inhibition of ADP- and collagen-induced aggregation of platelets, thereby demonstrating the spe-cificity of the compound within the platelet activation mechanism.[21] In a thrombin-induced calcium transient assay[32] in human coronary artery smooth muscle cells (HCASM), this compound showed a K_d value of 2.6 nM. The compound inhibited thrombin-induced thymidine incorporation[32] in HCASM with a K_i of 13.0 nM. In rat, compound **33** showed an oral bioavailability of

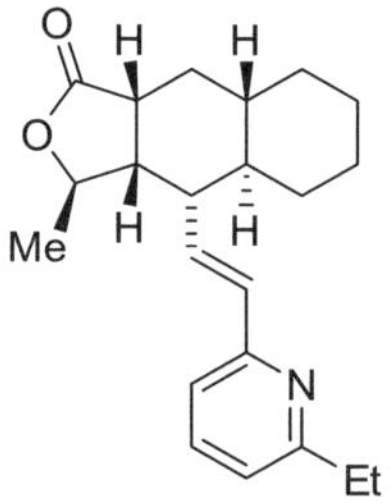

22

PAR-1 Ki = 20 nM [59]
Inhibition of human
platelet aggregation, IC_{50} = 70 nM

Figure 2.7   An early benchmark PAR-1 antagonist, himbacine derivative **22** showed potent PAR-1 inhibition in the binding assay and in functional assays. It also showed complete inhibition of TRAP-induced platelet aggregation in an ex vivo assay in cynomolgus monkeys following intravenous admin-istration ($10\,mg\,kg^{-1}$, infusion over 30 min).

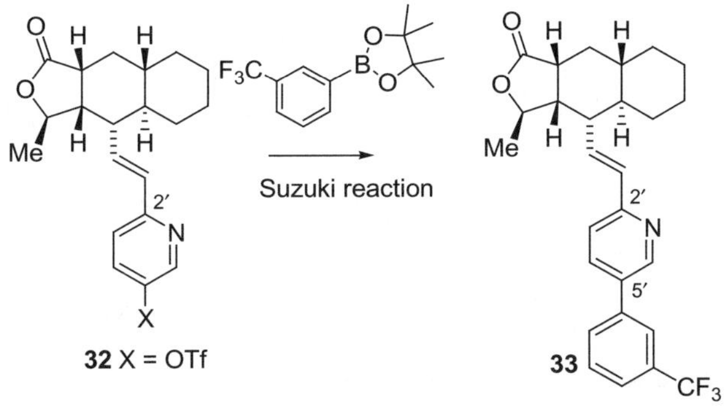

Scheme 2.2 Representative synthesis of orally bioavailable C5'-aryl derivatives.[59]

Table 2.4 Summary of *in vitro* and pharmacokinetic data for compound **33**.

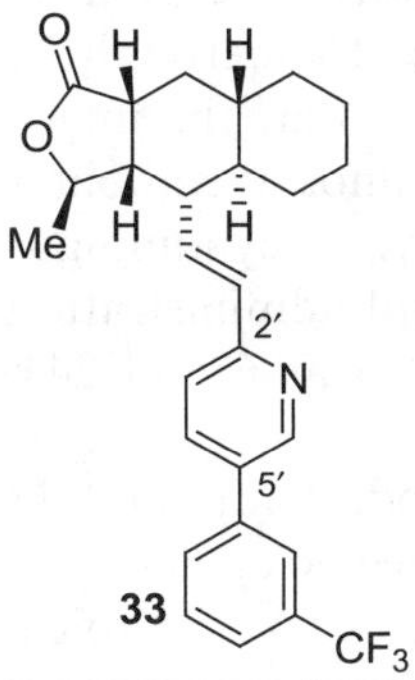

PAR-1 K_i	2.7 nM
Inhibition of human platelet aggregation (IC$_{50}$)	44 nM,[a] 24 nM[b]
Ca^{2+} transient assay $(K_j)^c$	2.6 nM
Proliferation assay $(K_i)^d$	13 nM
Rat PK (oral): *F*	30%
Monkey PK (oral): *F*	50%

[a]Induced by 10 nM thrombin.
[b]Induced by 15 μM haTRAP.
[c]Inhibition of thrombin-stimulated Ca^{2+} transient in HCASM.
[d]Inhibition of [^{3}H]thymidine incorporation in HCASM.

30%. The oral bioavailability of **33** in cynomolgus monkey was 50%; the half life after intravenous administration in e. monkey was 12.4 h.

The effect of **33** on TRAP-induced *ex vivo* platelet aggregation in whole blood in conscious, fasted cynomolgus monkeys was determined. Following oral administration, the compound showed dose-dependent inhibition of platelet aggregation for the duration of the experiment (6 h), with complete suppression of platelet aggregation at or above a dose of 3 mg kg^{-1}. This compound did not affect clotting parameters (PT, prothrombin time, and APTT, activated partial thromboplastin time), confirming that its mechanism of action is not by active site inhibition of thrombin or other coagulation proteases. The compound was inactive in the PAR-2 and PAR-4 functional assays, and was selective against a number of GPCRs, ion-channels, and receptors that it was tested against.

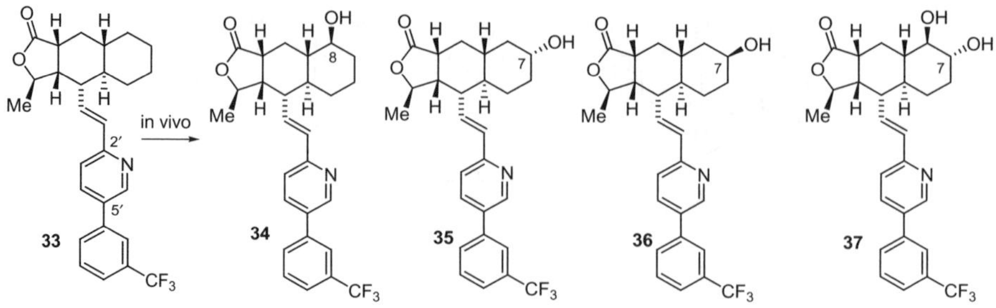

Figure 2.8   Major metabolites of compound **33**.

2.4.4.1.3   Identification of a Metabolism-based Second-generation Thrombin Receptor Antagonist.

Metabolite profiling of [^{3}H]-**33** done in rat hepatocytes indicated three monohydroxylated and multiple dihydroxylated metabolites, according to mass spectral characterization (Figure 2.8).[65] Quantities of metabolites needed for NMR characterization were initially generated by incubation of the parent compound with pregnenolone 16a-carbonitrile (PCN)-induced rat liver microsomes. NMR studies indicated that the major metabolite was the 8β-OH compound **34**, and that the minor metabolites were 7α- and 7β-hydroxy derivatives **35** and **36**, respectively. In cynomolgus monkeys, a similar pattern of metabolites was seen after oral administration of the compound and analysis of plasma samples withdrawn at 6 and 24 h time points.

In a multiple rising-dose study in rat over a 10-day period, compound **33** (300 mg kg^{-1}) induced a very high elevation (21-fold) of rat-specific cytochrome P4502B (CYP2B) isozyme and considerable elevation of CYP1A (3.6-fold). Additionally, there was a concomitant reduction of the plasma concentration of the parent drug from day one to day eight, suggesting an auto-induction/ metabolism pattern. Since this profile would make it difficult to sustain high levels of the parent in the plasma in the rodent species required for long-term toxicological studies, this compound was dropped from further consideration for development.

The mono-oxygenated metabolites of **33** were evaluated as potential replacement candidates.[65] The syntheses of these compounds were carried out in a fashion similar to the original synthesis, but using preinstalled ketal groups at the carbons that would eventually bear the hydroxyl groups (Scheme 2.3). The 7α- and 7β-hydroxy derivatives showed comparable IC$_{50}$ values. The 8-hydroxy analogs were slightly less active.

In the *ex vivo* platelet aggregation inhibition assay model in cynomolgus monkeys, compounds **35** (α-OH) and **36** (β-OH) showed equivalent potency: both exhibited robust inhibition of platelet aggregation after an oral administration of 1 mg kg^{-1}, with complete suppression of platelet aggregation at earlier time points and ~60% inhibition at the 24-h time point.[65] However, the β-OH isomer showed interconversion to the α-OH isomer in cynomolgus monkeys. This interconversion was also observed in monkey and human

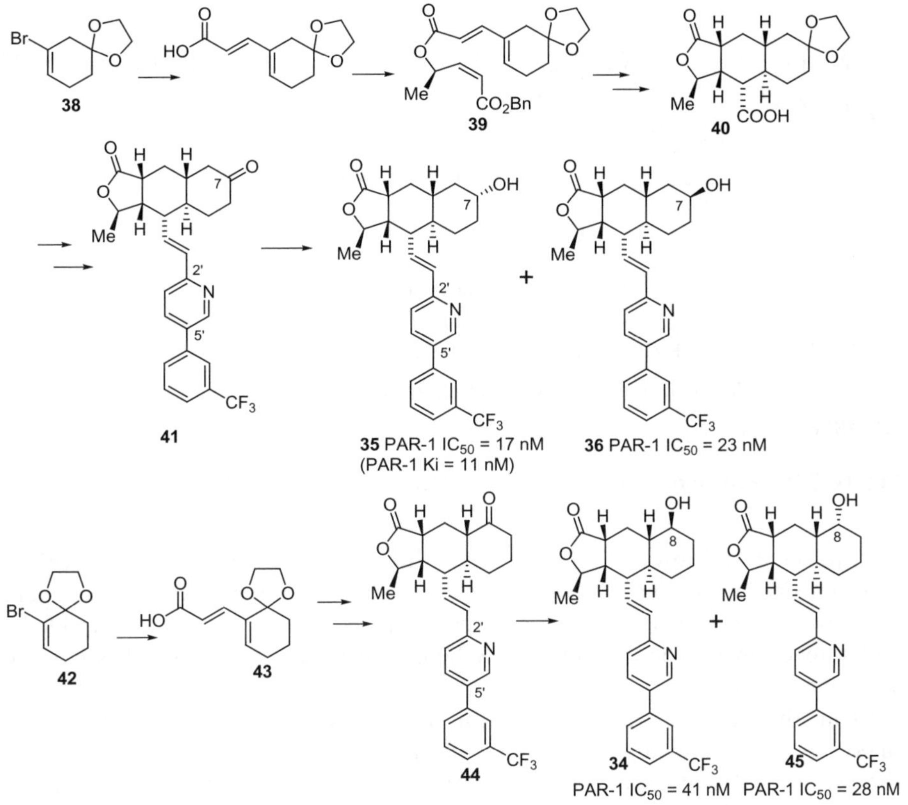

Scheme 2.3 Synthesis of 7- and 8-hydroxyl derivatives.[65]

hepatocytes. On the contrary, the α-OH isomer **35** did not undergo conversion to the β-OH isomer under the same conditions. This interconversion profile was reversed in rat (Figure 2.9). The favorable metabolic stability of **35** in cyno-molgus monkeys and in human liver hepatocytes prompted us to select this compound as a replacement for **33**.

In a multiple rising-dose enzyme-induction study in rat for eight days, compound **35** showed an excellent therapeutic window with no sign of auto-induction. However, in a later mass balance study in cynomolgus monkeys, the targeted recovery of radioactivity could not be achieved within the required 10 days after intravenous administration of [^{3}H]-**35**. This raised concerns about insufficient clearance and the potential for the compound to persist within the body for a prolonged period of time, resulting in the discontinuation of the development of the second-generation development candidate **35**.

2.4.4.1.4 Third-generation Thrombin Receptor Antagonists. The next phase of lead optimization efforts was directed to incorporating heteroatoms into the c-ring of the tricyclic skeleton in order to increase the overall polarity of

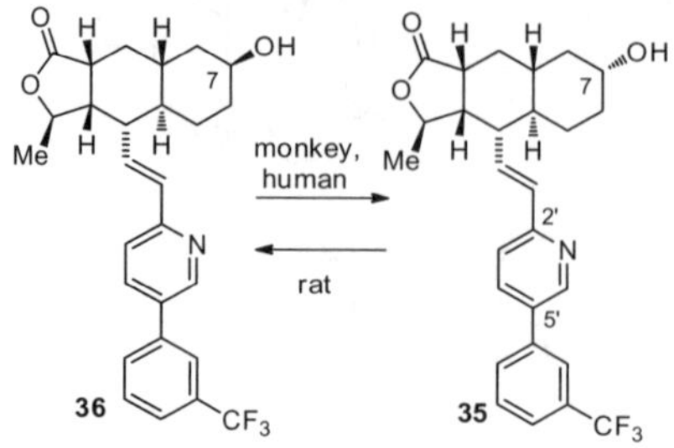

PAR-1 Ki = 11 nM

Inhibition of human platelet aggregation $(IC_{50})^a$ = 60 nM

Ca^{+2} transient assay $(Ki)^c$ = 85 nM

Proliferation assay $(Ki)^d$ = 22 nM

Rat PK (oral): F = 80%

Monkey PK (oral): F = 89%

[a] Induced by 15 µM haTRAP . [c] Inhibition of thrombin-stimulated Ca^{+2} transient in HCASM. [d] Inhibition of ^{3}H thymidine incorporation in human coronary artery smooth muscle cells (HCASM).

Figure 2.9   Profile of second-generation PAR-1 antagonist **35**.[65]

the molecule and lower its log P. Since the C_7 carbon was the "hot-spot" of metabolism, we replaced it with oxygen, sulfur, and nitrogen atoms. The syntheses of the derivatives were carried out using the intramolecular Diels–Alder reaction protocol as before; Scheme 2.4 illustrates the synthesis of the aza derivatives.[66]

From this structure–activity relation study, the ethylurethane **51** was identified as a promising thrombin receptor antagonist. This compound showed a PAR-1 affinity of 4.7 nM in the binding assay. In the *ex vivo* platelet aggregation inhibition assay, compound **51** was quite potent, completely inhibiting platelet aggregation at earlier time points and showing a robust 75% inhibition of platelet aggregation at the 24-h time point, after oral administration of 1 mg kg^{-1}. The compound showed excellent pharmacokinetics in rat and monkey species, and it was clean in an eight-day enzyme induction assay. More importantly, this compound showed good recovery in the mass balance study in cynomolgus monkey. However, compound **51** showed solubility issues that could not be satisfactorily addressed using formulation techniques.

2.4.4.1.5   Fourth-generation Thrombin Receptor Antagonists: Discovery of Vorapaxar (SCH 530348).   Encouraged by the clean profile of ethylurethane **51**, we decided to further pursue the modified c-ring amine derivatives. In this context, the exocyclic C_7-amine derivatives were pursued. Since the earlier compound **35** that bears a hydroxyl group (a hydrogen bond donor) had excellent solubility and pharmacokinetic properties, and the secondary urethane **51** (a hydrogen bond acceptor, but not a donor) retained the potency and showed good clearance pattern, we thought that an exocyclic primary urethane that provides both a hydrogen bond donor and multiple hydrogen

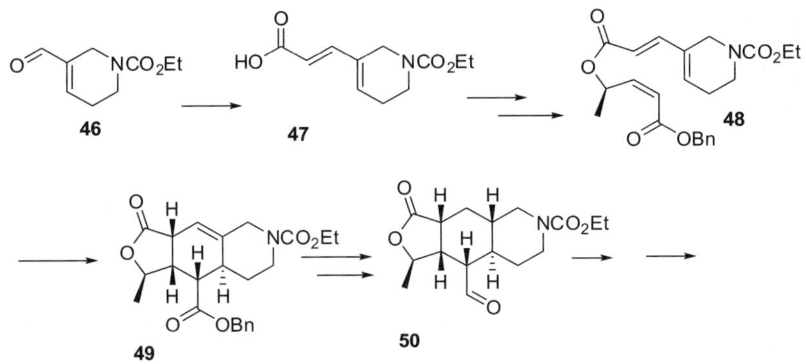

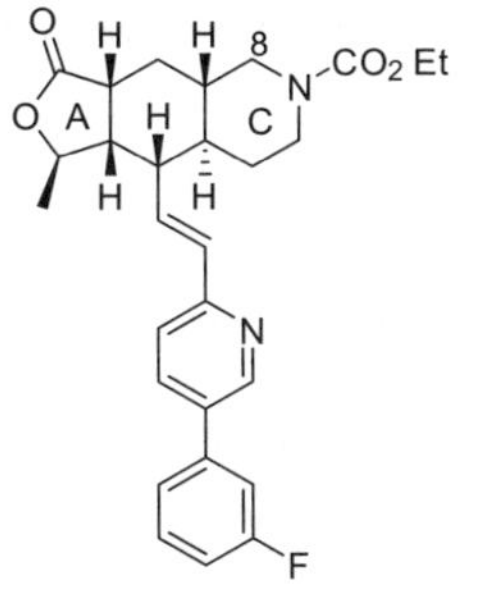

PAR-1 K_i = 4.7 nM

Ex vivo platelet aggregation inhibition @ 24 h = 75%[a]

Monkey PK (oral): $F = 62\%$

Enzyme induction study (8 days, mice): clean

Mass balance study (1 mg kg^{-1}, iv in cynomolgus monkey): clean

[a]After an oral dose of 1 mg kg^{-1} in cynomolgus monkey; agonist: TRAP.

Scheme 2.4 Synthesis of heterocyclic analog **51** and its biological profile.[66]

bond acceptors would be interesting. However, at the outset, we had some concerns about the hydrolytic stability of such a primary urethane. The initial synthesis of 7-amino derivatives was carried out from an available supply of **35** by sequential oxidation, reductive amination, and *N*-derivatization (Scheme 2.5).

The primary urethane **52** had a PAR-1 K_i of 13 nM. In the *ex vivo* platelet aggregation inhibition assay, this compound showed complete ablation of platelet aggregation for >24 h after a single oral dose of 1 mg kg^{-1}. This result promoted us to study the 7-amino series more thoroughly, and we investigated amides, ureas, sulfonamides, and carbamates. Following the previously established SAR for the pendant phenyl group, we also modified this substitution pattern. Together, these efforts lead to the identification of SCH 530348 (**53**) as our current development candidate.[67]

Compound **53** showed a K_i of 8.1 nM in the binding assay (Table 2.5). Detailed binding studies were consistent with a competitive binding of **53** to the PAR-1 receptor. Furthermore, compound **53** showed a long dissociation $t_{1/2}$ of about 20 h, which may explain the superb *ex vivo* potency and long duration of action for this compound.

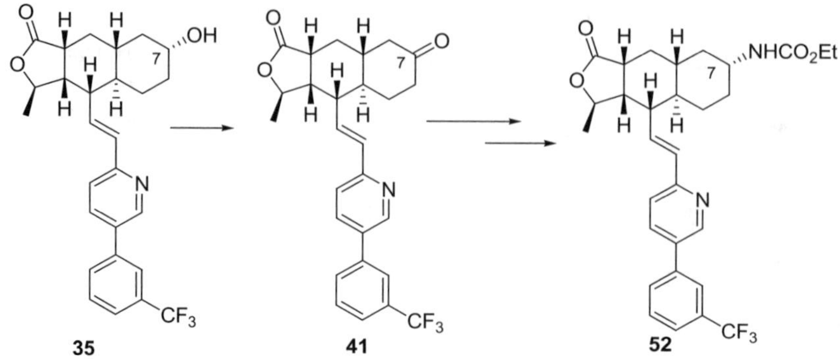

Scheme 2.5   Initial synthesis of analogs that incorporate primary urethane at C7 proceeded from **35**.[67]

Table 2.5   Summary of *in vitro* and pharmacokinetic data for compound **53**.[67]

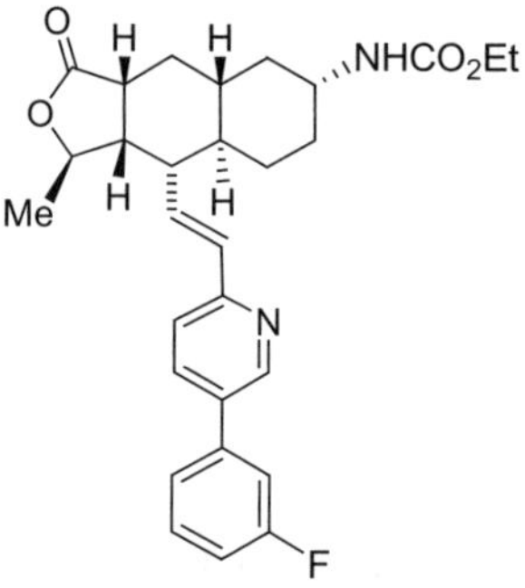

PAR-1 K_i	8.1 nM
Inhibition of human platelet aggregation (IC$_{50}$)	47 nM,[a] 25 nM[b]
Ca^{2+} transient assay (K_i)[c]	1.1 nM
Proliferation assay (K_i)[d]	13 nM
Rat PK (oral):[e] AUC$_{(0-24 h)}$, C_{max}, F; iv $t_{1/2}$	5.3 (µM · h), 0.67 (µM), 33%; 5.1 h
Monkey PK (oral):[f] AUC$_{(0-24 h)}$, C_{max}, F; iv $t_{1/2}$	10 (µM · h), 1.3 (µM), 86%; 13 h

53 (absolute chirality)
SCH 530348

[a] Induced by 10 nM thrombin.
[b] Induced by 15 µM haTRAP.
[c] Inhibition of thrombin-stimulated Ca^{2+} transient in HCASM.
[d] Inhibition of [³H]thymidine incorporation in HCASM.
[e] Oral and intravenous (iv) dosing of HCl salt at 10 mg kg^{-1}.
[f] Oral and iv dosing of HCl salt at 1 mg kg^{-1}. Vehicle for oral dosing: 0.4% methylcellulose; vehicle for iv dosing: 20% hydroxypropyl-β-cyclodextrin.

Compound **53** showed potent inhibition of platelet aggregation induced by either thrombin (IC$_{50}$ = 47 nM) or haTRAP (IC$_{50}$ = 25 nM). However, the compound did not inhibit platelet aggregation induced by ADP, collagen, the thromboxane mimetic 9,11-dideoxy-11α,9α-epoxymethanoprostaglandin F2α (U46619), or a PAR-4 agonist peptide. In additional functional assays, compound **53** inhibited thrombin-induced calcium transients and thymidine incorporation in human coronary artery smooth muscle cells (HCASMC) with K_i values of 1.1 and 13 nM, respectively. Compound **53** did not affect clotting parameters (PT and APTT), suggesting that its mechanism of action was not through direct inhibition of thrombin or other coagulation proteinases. The compound was selective over a number of GPCRs, ion channels, and receptors. Furthermore, it was inactive in the PAR-2, PAR-3 binding, and PAR-4 functional assays.

The pharmacokinetics of **53** were studied in both rat and monkey models. In rats, compound **53** displayed an elimination half-life of 5.1 h and an oral bioavailability of 33%. In monkey, the compound exhibited a half-life of 13 h and an oral bioavailability of 86%. Even at high concentrations (90 μM), compound **53** did not show CYP450 inhibition potential, including metabolism- and mechanism-based inhibition against various isozymes (CYP 1A2, 2C9, 2C19, 2D16, and 3A4) in human liver microsomes. Furthermore, the compound showed a clean profile in a 14-day enzyme induction study in rats, and mass balance studies using tritiated **53** gave full recovery of radioactivity within the targeted period of seven days in both rat and monkey models. Due to its excellent safety margin and superior potency, carbamate **53** (SCH 530348) was advanced to clinical development.

2.4.4.2 Clinical Studies of SCH 530348

SCH 530348 demonstrated excellent safety and tolerability in Phase I clinical studies. In pharmacodynamic platelet aggregation studies, the compound showed a robust >90% inhibition of platelet aggregation at all tested doses for a sustained period of time.[68] The target level of platelet aggregation throughout the 28-day treatment period was maintained by a 2.5 mg once-daily maintenance dose.

In a Phase II randomized double-blind placebo-controlled clinical study, SCH 530348 was tested in patients who underwent non-urgent percutaneous coronary intervention (PCI) in a trial known as Thrombin Receptor Antagonist for Cardiovascular Event Reduction in Percutaneous Coronary Interventions (TRA-PCI).[69–73] The primary endpoint was the bleeding risk associated with SCH 530348 using the thrombolysis in myocardial infarction (TIMI) scale, and the secondary endpoint was a composite of death or major adverse cardiac events (myocardial infarction, urgent coronary revascularization, and ischemia requiring hospitalization) in PCI patients who received a loading dose followed by a 60-day maintenance dose of the drug. Patients also received other standard therapies such as aspirin, clopidogrel, and anticoagulants.

For the primary safety endpoint, SCH 530348 was not associated with increased TIMI major plus minor bleeding when compared with placebo. For the secondary outcome endpoint, SCH 530348 was associated with a numerical reduction in periprocedural myocardial infarction, although the study was not powered to detect statistical significance. Overall, treatment with SCH 530348 reduced arterial thrombotic events without an increase in bleeding risk.[69] A key sub-study to evaluate the inhibition of platelet aggregation showed that the drug achieved a sustained, potent (>80%) inhibition of TRAP-induced platelet aggregation. Two additional Phase-II studies conducted in Japanese patients (one in patients with acute coronary syndromes and the other in patients with ischemic stroke) have confirmed the favorable safety profile of SCH 530348. In addition, the study performed in Japanese patients with acute coronary syndromes demonstrated a significant reduction in periprocedural myocardial infarctions, similar to the findings reported in the TRA-PCI study. SCH 530348 is currently undergoing two large Phase III studies in patients with acute coronary syndrome (TRA*CER, with 10,000 patients) and for secondary prevention of ischemic events (TRA*2PTIMI 50, with 25,000 patients).[74,75]

2.5   Conclusion

Thrombin receptor antagonists are a new class of antiplatelet agents. Unlike direct thrombin inhibitor anticoagulants, thrombin receptor antagonist antiplatelet agents do not inhibit the enzymatic activity of thrombin, and thereby spare its role in fibrin generation. Rather, these antiplatelet agents inhibit thrombin-mediated activation of protease activated receptor-1 (PAR-1), which is the most potent cell surface inducer of human platelet activation. SCH 530348 (**53**) is a potent, competitive antagonist of platelet thrombin receptor ($K_i = 8.1$ nM), with excellent activity in multiple functional assays. As described herein, this compound was discovered through a research effort focused on modulating "metabolic hot spots" identified in a series of extensively modified himbacine derivatives. Of note, the active series was enantiomeric to the natural product, and thus would not have been discovered through straight derivatization of himbacine itself. In addition to SCH 530348, another thrombin receptor antagonist, E5555, has also advanced to clinical trials.[76]

SCH 530348 has demonstrated potent *ex vivo* antiplatelet effect in non-human primates and similar compounds from the same structural series have demonstrated dose-dependent inhibition of thrombus formation in non-human primate thrombosis models. As discussed above, Phase II clinical trials of SCH 530348 showed a promising safety margin with regard to TIMI-major or TIMI-minor bleeding episodes, and a numerical reduction in thrombotic events in ACS patients during a two-month follow-up. Due to these encouraging results, SCH 530348 has been advanced to Phase III clinical trails to evaluate its efficacy and safety in patients with acute coronary syndrome (TRA*CER) and for secondary prevention of ischemic events (TRA*2P TIMI 50).[74,75]

Bleeding is the main side effect for all antithrombotic therapies, including antiplatelet therapy.[77] In fact, the current view of antithrombotic therapy is "no bleeding, no efficacy." Therefore, clinicians have to strike a balance between bleeding and outcome by careful selection of dose and adjunct therapies. This scenario is often made more complex by virtue of patient frailty, advanced age, and other underlying pathologic conditions. Therefore, it will be very desirable to achieve improved safety margins in a new antiplatelet drug. The thrombin receptor antagonists hold some promise in this area. Based on the published data for SCH 530348 and E5555, these compounds confer potent antiplatelet effect without attendant bleeding in the preclinical models and, in the case of SCH 530348, in the published Phase II clinical trial results.

There are plausible explanations for the absence of significant bleeding that is noted for the thrombin receptor antagonists in preclinical animal models and in the TRA-PCI study. Although proteolytic activation of the thrombin receptor is the most potent mechanism of platelet stimulation, other platelet activation mechanisms that remain intact are perhaps sufficient for normal hemostasis. Thrombin-mediated platelet activation therefore might represent a more pathologically-relevant platelet activation mechanism in the context of a major vascular injury, such as rupture of an atherosclerotic plaque or PCI procedures. In a recent study in PAR4$^{-/-}$ mice, Furie and Coughlin have demonstrated, using real-time digital fluorescence microscopy, that platelet activation by thrombin is necessary for thrombus growth, but not for primary hemostasis.[78] During laser-induced thrombosis, juxtamural platelet accumulation immediately after laser injury was not affected in PAR4$^{-/-}$ mice. However, subsequent growth of platelet thrombi was markedly diminished in PAR4$^{-/-}$ mice. Also, fibrin generation was preserved in the knockout mice. The conclusion from these studies is that a thrombin receptor antagonist could produce potent antiplatelet effect without increased bleeding. The ongoing clinical studies will be needed to validate the long-term sustainability of the anticipated safety margins and efficacy for thrombin receptor antagonists.

Acknowledgement

The author acknowledges the support of Dr. William J. Greenlee. Also acknowledged are contributions from Drs. Mariappan Chelliah, Martin Clasby, Yan Xia, Yuguang Wang, and Mr. Keith Eagen. Their individual contributions have been cited in the reference section.

References

1. *Heart Disease and Stroke Statistics – 2009 Update*, American Heart Association, Dallas, Texas, 2009.
2. *The Atlas of Heart Disease and Stroke*, ed. G. Mackay and G. Mensah, World Health Organization, Geneva, 2004; available also at http://www.who.int/cardiovascular_diseases/resources/atlas/en/; accessed Sept 4 2009.
3. E. S. Ford, U. A. Ajani, J. B. Croft, J. A. Critchley, D. R. Labarthe, T. E. Kottke, W. H. Giles and S. Capewell, *New Engl. J. Med.*, 2007, **356**, 2388.
4. J. J. Boyle, *Curr. Vasc. Pharmacol.*, 2005, **3**, 63.
5. G. K. Hansson, *New Engl. J. Med.*, 2005, **352**, 1685.
6. L. W. Klein, P. R. Liebson and A. P. Selwyn, *Curr. Cardiol. Rev.*, 2005, **1**, 171.
7. M. C. Fishbein and R. J. Siegel, *Circulation*, 1996, **94**, 2662.
8. Y. Yeghiazarians, J. B. Braunstein, A. Askari and P. H. Stone, *New Engl. J. Med.*, 2000, **342**, 101.
9. M. R. Tamberella 3rd and J. G. Warner Jr., *Postgrad. Med.*, 2000, **107**, 87.

10. J. L. Vacek, *Postgrad. Med.*, 2002, **112**, 71.
11. P. Libby, *Circulation*, 2001, **104**, 365.
12. Z. Bereczky, E. Katona and L. Muszbek, *Pathophysiol. Haemost. Thromb.*, 2003, **33**, 430.
13. M. P. Bonaca, P. G. Steg, L. J. Feldman, J. F. Canales, J. J. Ferguson, L. Wallentin, R. M. Califf, R. A. Harrington and R. P. Giugliano, *J. Am. Coll. Cardiol.*, 2009, **54**, 969.
14. G. S. Bisacchi, in *Burger's Medicinal Chemistry and Drug Discovery*, 6th edn, ed. D. J. Abraham, Wiley, Hoboken, NJ, 2003, **vol. 3**, p. 283.
15. K. K. Wu and N. Matijevic-Aleksic, *Crit. Rev. Clin. Lab. Sci.*, 2005, **42**, 249.
16. S. Moll and I. I. White, in *Principles of Molecular Medicine*, 2nd edn, ed. M. S. Runge and C. Patterson, Humana Press, Totowa NJ, 2006, p. 871.
17. D. Feinbloom and K. A. Bauer, *Arterioscl. Thromb. Vasc. Biol.*, 2005, **25**, 2043.
18. J. Harenberg, *Semin. Thromb. Hemost.*, 2008, **34**, 779.
19. J. A. Cairns, P. Theroux, H. D. Lewis Jr, M. Ezekowitz and T. W. Meade, *Chest*, 2001, **119**(1S), 228S.
20. J. I. Weitz and L. A. Linkins, *Expert Opin. Invest. Drugs*, 2007, **16**, 271.
21. S. Goto, *Curr. Vasc. Pharmacol.*, 2004, **2**, 23.
22. H. R. Lijnen and D. Collen, *Thromb. Haemost.*, 1995, **74**, 387.
23. H. J. Weiss, *New Engl. J. Med.*, 1975, **293**, 531.
24. H. J. Weiss, *New Engl. J. Med.*, 1975, **293**, 580.
25. Z. M. Ruggeri, J. A. Dent and E. Saldivar, *Blood*, 1999, **94**, 172.
26. R. K. Andrews and M. C. Berndt, *Histol. Histopathol.*, 1998, **13**, 837.
27. H. Shankar, B. Kahner and S. P. Kunapuli, *Curr. Drug Targets*, 2006, **7**, 1253.
28. C. S. Abrams and L. F. Brass, in *Hemostasis and Thrombosis*, 5th edn, ed. R. W. Colman, V. J. Marder, A. W. Clowes, J. N. George and S. Z. Goldhaber, Lippincott Williams & Wilkins, Philadelphia, 2006, p. 617.
29. S. R. Coughlin, *Thromb. Haemost.*, 1999, **82**, 353.
30. S. Offermanns, *Circ. Res.*, 2006, **99**, 1293.
31. H. Mohri and T. Ohkubo, *Peptides*, 1993, **14**, 353.
32. H. Ni, C. V. Denis, S. Subbarao, J. L. Degen, T. N. Sato, R. O. Hynes and D. D. Wagner, *J. Clin. Invest.*, 2000, **106**, 385.
33. Antithrombotic Trialists' Collaboration, *Br. Med. J.*, 2002, **324**, 71.
34. H. Tran and S. S. Anand, *J. Am. Med. Assoc.*, 2004, **292**, 1867.
35. S. R. Coughlin, *Proc. Natl. Acad. Sc. U. S. A.*, 1999, **96**, 11023.
36. S. R. Coughlin, *J. Thromb. Haemost.*, 2005, **3**, 1800.
37. P. J. O'Brien, M. Molino, M. Kahn and L. Brass, *Oncogene*, 2001, **20**, 1570.
38. B. Al-Ani, M. Saifeddine and M. D. Hollenberg, *Can. J. Physiol. Pharmacol.*, 1995, **73**, 1203.
39. H. Ishihara, A. J. Connolly, D. Zeng, M. L. Kahn, Y. W. Zheng, C. Timmons, T. Tram and S. R. Coughlin, *Nature*, 1997, **386**, 502.

40. W. F. Xu, H. Anderson, T. E. Whitmore, S. R. Presnell, D. P. Yee, A. Ching, T. Gilbert, E. W. Davie and D. C. Foster, *Proc. Natl. Acad. Sci. U. S. A.*, 1998, **95**, 6642.
41. G. R. Sambrano, E. J. Weiss, Y. W. Zheng, W. Huang and S. R. Coughlin, *Nature*, 2001, **413**, 74.
42. H. C. Zhang, C. K. Derian, P. Andrade-Gordon, W. J. Hoekstra, D. F. McComsey, K. B. White, B. L. Poulter, M. F. Addo, W. M. Cheung, B. P. Damaino, D. Oksenberg, E. E. Reynolds, A. Pandey, R. M. Scarborough and B. Maryanoff, *J. Med. Chem.*, 2001, **44**, 1021.
43. C. Tapparelli, R. Metternich and N. S. Cook, *Trends Pharm. Sci.*, 1993, **14**, 426.
44. T.-K. H. Vu, D. T. Hung, V. I. Wheaton and S. R. Coughlin, *Cell*, 1991, **64**, 1057.
45. U. B. Rasmussen, V. Vouret-Craviari, S. Jallat, Y. Schlesinger, G. Pages, A. Pavirani, J. P. Lecocq, J. Pouyssegur and E. Van Obberghen-Schilling, *FEBS Lett.*, 1991, **288**, 123.
46. D. T. Hung, T.-H. Vu, N. A. Nelken and S. R. Coughlin, *J. Cell. Biol.*, 1992, **116**, 827.
47. S. R. Coughlin, T.-K. H. Vu, D. T. Hung and V. I. Wheaton, *J. Clin. Invest.*, 1992, **89**, 351.
48. T.-K. H. Vu, V. I. Wheaton, D. T. Hung, I. Charo and S. R. Coughlin, *Nature*, 1991, **353**, 674.
49. M. D. Hollenberg, *Trends Pharmacol. Sci.*, 1996, **17**, 3.
50. T. Nanevicz, M. Ishii, L. Wang, M. Chen, J. Chen, C. W. Turck, F. E. Cohen and S. R. Coughlin, *J. Biol. Chem.*, 1995, **270**, 21619.
51. H. S. Ahn and S. Chackalamannil, *Drugs Future*, 2001, **26**, 1065.
52. S. M. Seiler and M. S. Bernatowicz, *Curr. Med. Chem. – Cardiovasc. Hematol. Agents*, 2003, **1**, 1.
53. G. D. Barry, G. T. Le and D. P. Fairlie, *Curr. Med. Chem.*, 2006, **13**, 243.
54. M. S. Bernatowicz, C. E. Klimas, K. S. Hartl, M. Peluso, N. J. Allegreto and S. M. Seiler, *J. Med. Chem.*, 1996, **39**, 4879.
55. L. A. Harker, S. R. Hanson and M. Runge, *Am. J. Cardiol.*, 1995, **75**, 12B.
56. Y. Kato, Y. Kita, Y. Hirasawa-Taniyama, M. Nishio, K. Mihara, K. Ito, T. Yamanaka, J. Seki, S. Miyata and S. Mutoh, *Eur. J. Pharmacol.*, 2003, **473**, 163.
57. J. J. Cook, G. R. Sitko, B. Bednar, C. Condra, M. J. Mellott, D. M. Feng, R. F. Nutt, J. A. Shafer, R. J. Gould and T. M. Connolly, *Circulation*, 1995, **91**, 2961.
58. C. K. Derian, B. P. Damiano, M. F. Addo, A. L. Darrow, M. R. D'andrea, M. Nedelman, H. C. Zhang, B. E. Maryanoff and P. Andrade-Gordon, *J. Pharmacol. Exp. Ther.*, 2003, **304**, 855.
59. S. Chackalamannil, Y. Xia, W. J. Greenlee, M. Clasby, D. Doller, H. Tsai, T. Asberom, M. Czarniecki, H. S. Ahn, G. Boykow, C. Foster, J. Agans-Fantuzzi, M. Bryant, J. Lau and M. Chintala, *J. Med. Chem.*, 2005, **48**, 5884.
60. S. Chackalamannil, *J. Med. Chem.*, 2006, **49**, 5389.

61. H. S. Ahn, C. Foster, G. Boykow, A. Stamford, M. Manna and M. Graziano, *Biochem. Pharmacol.*, 2000, **60**, 1425.
62. S. Chackalamannil, R. J. Davies, T. Asberom, D. Doller and D. Leone, *J. Am. Chem. Soc.*, 1996, **118**, 9812.
63. D. Doller, S. Chackalamannil, M. Czarniecki, R. McQuade and V. Ruperto, *Bioorg. Med. Chem. Lett.*, 1999, **9**, 901.
64. Y. Xia, S. Chackalamannil, T. M. Chan, M. Czarniecki, D. Doller, K. Eagen, W. J. Greenlee, H. Tsai, Y. Wang, H. S. Ahn, G. Boykow and A. T. McPhail, *Bioorg. Med. Chem. Lett.*, 2006, **16**, 4969.
65. M. C. Clasby, S. Chackalamannil, M. Czarniecki, D. Doller, K. Eagen, W. Greenlee, G. Kao, Y. Lin, H. Tsai, Y. Xia, H. S. Ahn, J. Agans-Fantuzzi, G. Boykow, M. Chintala, C. Foster, A. Smith-Torhan, K. Alton, M. Bryant, Y. Hsieh, J. Lau and J. Palamanda, *J. Med. Chem.*, 2007, **50**, 129.
66. M. V. Chelliah, S. Chackalamannil, Y. Xia, K. Eagen, M. C. Clasby, X. Gao, W. Greenlee, H. S. Ahn, J. Agans-Fantuzzi, G. Boykow, Y. Hsieh, M. Bryant, J. Palamanda, T. M. Chan, D. Hesk and M. Chintala, *J. Med. Chem.*, 2007, **50**, 5147.
67. S. Chackalamannil, Y. Wang, W. J. Greenlee, Z. Hu, Y. Xia, H. S. Ahn, G. Boykow, Y. Hsieh, J. Palamanda, J. Agans-Fantuzzi, S. Kurowski, M. Graziano and M. Chintala, *J. Med. Chem.*, 2008, **51**, 3061.
68. M. Chintala, K. Shimizu, M. Ogawa, H. Yamaguchi, M. Doi and P. Jensen, *J. Pharmacol. Sci.*, 2008, **108**, 433.
69. R. C. Becker, D. J. Moliterno, L. K. Jennings, K. S. Pieper, J. Pei, A. Niederman, K. M. Ziada, G. Berman, J. Strony, D. Joseph, K. W. Mahaffey, F. Van de Werf, E. Veltri and R. A. Harrington, *Lancet*, 2009, **373**, 919.
70. E. M. Tuzcu, R. C. Starling and J. D. Thomas, *J. Am. Coll. Cardiol.*, 2007, **50**, C2.
71. M. Chintala, M. P. Graziano, S. Chackalamannil, W. J. Greenlee, T. Kosoglou, J. Strony, G. Berman and E. Veltri, *Nature*, 2008, **451**; Schering-Plough-sponsored feature.
72. M. P. Bonaca and D. A. Morrow, *Future Cardiol.*, 2009, **5**, 435.
73. J. Oestreich, *Curr. Opin. Invest. Drugs*, 2009, **10**, 988.
74. The TRACER Executive and Steering Committees, *Am. Heart J.*, 2009, **158**, 327.
75. The TRACER Executive and Steering Committees, *Am. Heart J.*, 2009, **158**, 335.
76. M. Kogushi, H. Yokohoma, S. Kitamura and I. Hishinuma, *J. Thromb. Haemost.*, 2007, **5**(S2), P–M-059.
77. W. Alvarez Jr, *Am. J. Health-System Pharm.*, 2008, **65**, 1017.
78. E. R. Vandendries, J. A. Hamilton, S. R. Coughlin, B. C. Furie and B. Furie, *Proc. Natl. Acad. Sci. U. S. A.*, 2007, **104**, 288.

The Discovery of Piragliatin, a Glucokinase Activator

RAMAKANTH SARABU,[a],* JEFFERSON W. TILLEY[a] AND JOSEPH GRIMSBY[b]

[a] Departments of Discovery Chemistry, Hoffmann-La Roche, 340 Kingsland Street, Nutley, NJ 07110, USA; [b] Metabolic and Vascular Diseases, Hoffmann-La Roche, 340 Kingsland Street, Nutley, NJ 07110, USA

3.1 Introduction and Target Background

Type 2 diabetes (T2D) is a metabolic disorder characterized by elevated blood glucose levels resulting from a combination of pancreatic β-cell dysfunction and consequent deficiencies in insulin secretion, insulin resistance, and increased hepatic glucose production. Defects in the ability of the pancreatic β-cell to secrete insulin in response to a meal and the diminished capacity of the liver and skeletal muscle to utilize glucose in response to insulin represent core patho-physiological defects associated with T2D. Other tissues such as adipose, the gastrointestinal tract, kidney, and brain also contribute to dysglycemia in patients with T2D.[1] As T2D progresses, patients rely on higher doses of oral antidiabetic agents, addition of additional drugs to the treatment regimen, and often require insulin to gain adequate glycemic control. Therefore, more effective therapeutic agents targeting the multiple pathogenic abnormalities of T2D along with sustained efficacy represents a major unmet medical need.

Type 2 diabetes has been an area of focus for drug discovery research at Roche for nearly two decades. As a part of this effort, a Roche research team in the mid 1990s recognized the important role of glucokinase (GK) in regulating

RSC Drug Discovery Series No. 4
Accounts in Drug Discovery: Case Studies in Medicinal Chemistry
Edited by Joel C. Barrish, Percy H. Carter, Peter T. W. Cheng and Robert Zahler
© Royal Society of Chemistry 2011
Published by the Royal Society of Chemistry, www.rsc.org

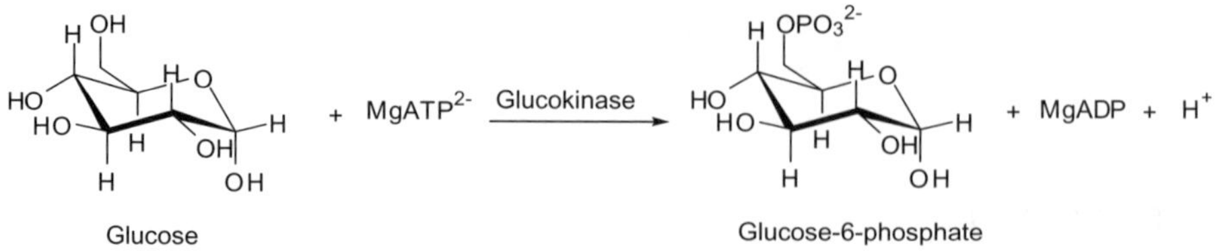

Figure 3.1   Glucokinase enzymatic reaction.

glucose homeostasis and initiated a drug discovery program to identify small molecules to increase its enzymatic activity.[2] Glucokinase is primarily expressed in pancreatic β-cells and liver. In pancreatic β-cells, GK has a control strength approaching unity, allowing for a tight coupling of GK enzymatic activity and pathway changes that lead to glucose-stimulated insulin release. Glucokinase, also known as hexokinase IV or D, is one of four hexokinase isozymes that metabolize glucose. It utilizes ATP for the phosphorylation of glucose to glucose-6-phosphate (G-6-P), and plays an important role in regulating glucose-stimulated insulin release in β-cells, and in hepatic glucose utilization as discussed below (Figure 3.1).

Matschinsky and Ellerman[3] first proposed the "gluco-stat" concept, today known as the GK glucose-sensor concept, in which the GK rate sets the threshold concentration of glucose ($\sim 5\,mM$) required to initiate the signaling cascade leading to insulin release. A key feature of this concept is the enzyme's low affinity ($S_{0.5} = 7\,mM$, glucose concentration at half-maximal enzyme velocity) for glucose and positive cooperativity (nH = 1.7). This translates into a sigmoidal velocity versus [glucose] curve whereby GK is essentially inactive at low glucose concentrations (*i.e.* during a fasting period). As glucose levels approach and exceed its $S_{0.5}$ (*i.e.* during a post-prandial state), GK enzymatic activity rapidly increases. Thus glucose metabolism is tightly coupled to glucose levels. Following β-cell GK-mediated phosphorylation of glucose, glycolysis occurs leading to an increase in the ATP : ADP ratio. This in turn results in the closing of ATP-sensitive K^+ channels, membrane depolarization, opening of the voltage-gated Ca^{2+} channel, and an influx of Ca^{2+} ions which then triggers insulin release (Figure 3.2).

Sulfonylureas, which are a popular class of oral antidiabetic drugs, cause release of insulin *via* their direct action on ATP-sensitive K^+ channels, but act independently of plasma glucose levels. This mechanism is uncoupled from glucose levels and can lead to episodes of hypoglycemia. In contrast, insulin release *via* GK activation is coupled to plasma glucose levels and thus should be safer as it is a natural physiological mechanism. The decreased stress on the β-cells by coupling insulin release with demand could also lead to improved preservation of β-cell mass and a decrease in the rate of progression of T2D. Thus GK plays a critical role in the regulation of insulin secretion and has been termed a pancreatic β-cell glucose sensor on account of its kinetics which allow β-cells to change the glucose phosphorylation rate over a range of physiological glucose concentrations.

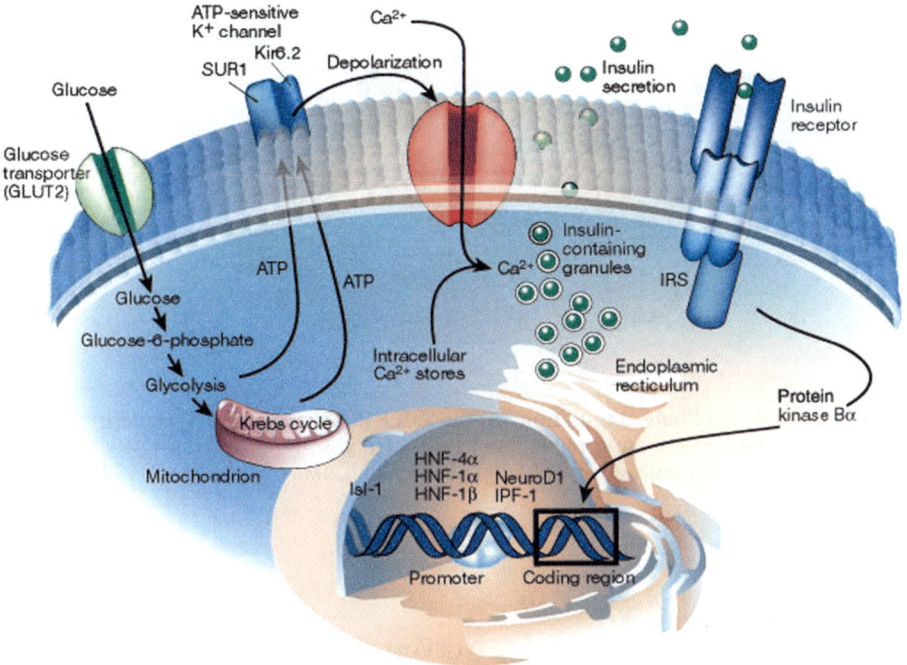

Figure 3.2 Role of glucokinase in the pancreatic β-cell. Reproduced with permission from *Nature*, 2001, **414**, 788.

In 1989 a Belgian research group discovered a protein in the liver called the glucokinase regulatory protein (GKRP) that functions as a competitive GK inhibitor.[4] When bound to GKRP/fructose-6-phosphate, hepatic GK is sequestered into the nucleus in an inactive state. Under hyperglycemic conditions or when fructose-6-phosphate (F-6-P) is displaced by fructose-1-phosphate (F-1-P), a conformational change in GKRP takes place which dissociates the GK/GKRP complex, releasing GK in its active form (Figure 3.3).[4] This observation provided a rational basis to seek low molecular weight compounds which might modulate the GK/GKRP association, *via* either by design of F-1-P mimetics or discovery of other small molecules through a high-throughput screen (HTS).

The biological rationale for this approach was derived from mice with a genetic disruption of the GK gene[5] and data from humans[6] with loss and gain of function GK mutations.[7] Loss of function GK mutations causes a type of diabetes known as maturity onset of diabetes of the young type 2 (MODY-2)[8] or a more severe phenotype presenting at birth referred to as permanent neonatal diabetes mellitus (PNDM),[9] while rare activating or gain of function GK mutations cause persistent hyperinsulinemic hypoglycemia of infancy (PHHI-GK).[7] At the time this research was initiated, in the mid-1990s, there were no reports of small-molecule or protein GK activators.

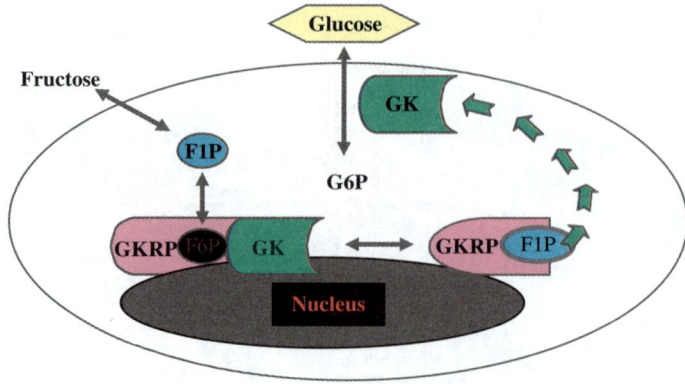

Figure 3.3   Glucokinase regulation by the GKRP (GK regulatory protein) in liver.
G6P = glucose-6-phosphate; F6P = fructose-6-phosphate; F1P = fructose-
1-phosphate. Adapted from R. Nordlie, *Annu. Rev. Nutr.*, 1999, **19**, 379–
406.

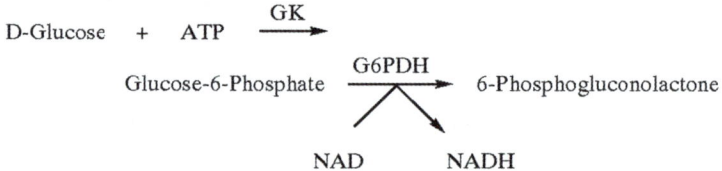

Figure 3.4   Glucokinase activators: high-throughput screening (HTS) strategy. G6PDH =
glucose-6-phosphate dehydrogenase.

3.2   Hit Generation and Hit-to-lead Strategies[10]

A high-throughput screening campaign was conducted based on the reaction
scheme shown in Figure 3.4. Briefly, the GK activity was assayed by coupling
production of G-6-P to the generation of NADH, with glucose-6-phosphate
dehydrogenase (G6PDH) as the coupling enzyme. Screening of 120 000 small
molecules from the Roche compound library identified a single weak activator
1 (see Figure 3.5 below), a racemate, which was shown to increase GK enzymatic activity by 1.5-fold at a concentration of $24 \mu M$ ($SC_{1.5}$ value). This
unique activity, and a combination of low molecular weight and mix of polar
and non-polar features associated with **1**, prompted us to investigate this hit
further. The molecule was conceptually divided into three regions, each of
which was optimized independently as described in detail below. This effort led
to discovery of the prototypical GK activators, **2a** (RO0274375) and **9a**
(RO0281675), which displayed robust glucose lowering effects in preclinical
models of T2D.[11] These findings have stimulated intense interest within the
pharmaceutical industry, leading to identification of several structural
mimics of **9a**.[12,13]

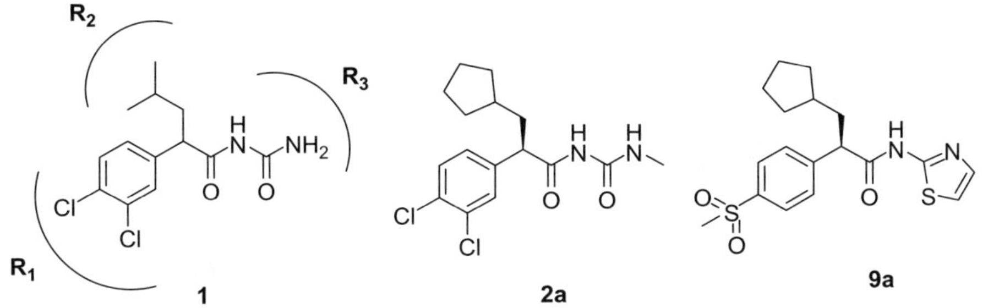

Figure 3.5 Glucokinase activator hit molecule **1** and optimized molecules **2a** and **9a**.

For the purpose of further exploration of the acyl urea hit molecule **1**, each of the three regions around the chiral center (R_1, the substituted aromatic ring; R_2, the alkyl group; and R_3, the *N*-acylurea group; Figure 3.5) was considered independently. Our initial studies centered on a systematic evaluation of several R_2 analogues that varied in size and polarity. These studies indicated a clear preference for a non-polar, five- or six-membered alicyclic group. The structure–activity relationship (SAR) of the substituted aromatic ring of the R_1 group of **1** indicated the importance of an aromatic ring substituted with one or two electron-withdrawing groups for potency, with tolerance for a wide variety of substituents at the *para* position. Attempts to replace the aromatic ring with alicyclic moieties led to a loss of activity. In general, *meta* or *para* substitutions were well tolerated, while *ortho* substitution dramatically reduced or completely eliminated GK activator activity. Investigation of the R_3 modifications of **1**, which included small alkyl groups, hydroxyalkyl groups, or alkyl ester groups, resulted in improvements in potency, identified areas to introduce functionality, and provided clues regarding the active conformation required for GK activation.

These efforts culminated in the discovery of the racemic *N*-acylurea analog **2**. The optically pure (*R*)- and (*S*)-stereoisomers, **2a** and **2b**, were prepared from the corresponding chiral phenylacetic acids and their absolute stereochemistry was assigned based on a single-crystal X-ray structure of **2a**. Thus, the (*R*)-isomer **2a** was found to have an $SC_{1.5}$ value of 0.45 µM, while the corresponding (*S*)-isomer **2b** had no effect on GK activity at concentrations up to 10 µM. The crystal structure of **2a** further revealed that the *N*-acylurea moiety exists in a six-membered cyclic conformation, characterized by an intramolecular H-bond, between the carbonyl of the acyl oxygen atom and the terminal NH of the urea, leaving the amide bond in a *cisoid* conformation as shown in Figure 3.6.

A variety of compounds (Figure 3.7) were prepared to explore the significance of this *cisoid* conformation of the acylurea. The cyclic urea **3**, the *N*-methyl acylurea **4**, and the terminal *N,N'*-dimethyl analog **5**, all of which lack the ability to make an intramolecular H-bond, failed to activate GK. The guanidine analog **6** potentially could have an intramolecular H-bond, but

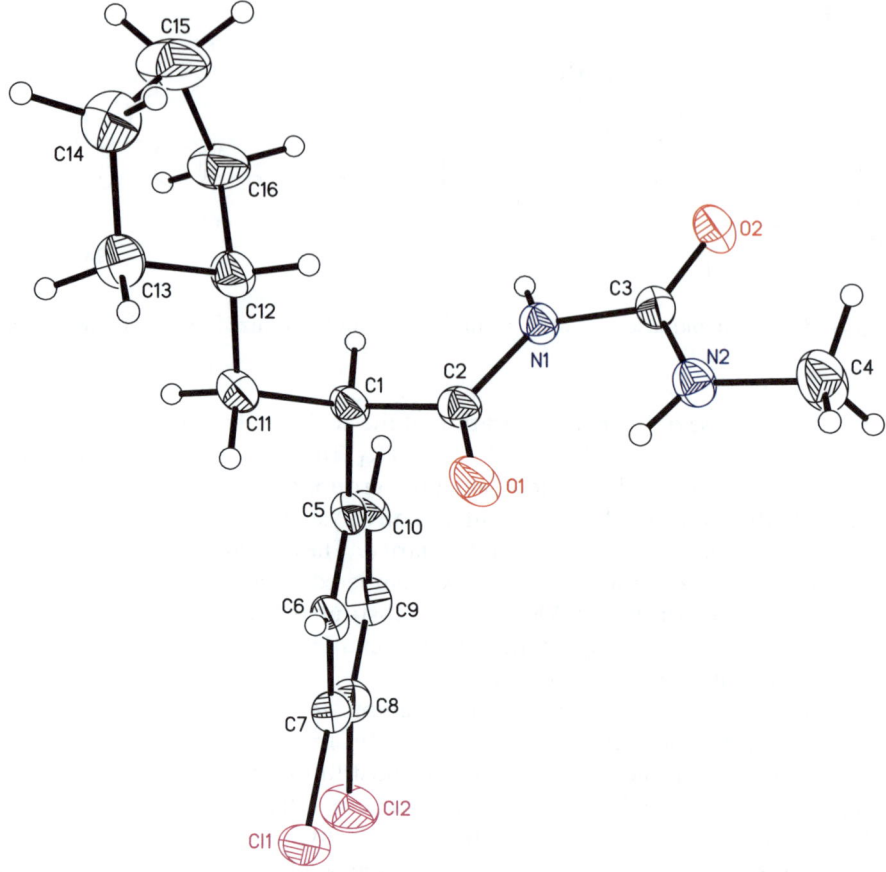

Figure 3.6 Single-crystal X-ray structure of **2a**.

lacking the urea carbonyl that we surmised may accept a hydrogen bond from glucokinase, also failed to activate GK. The thiourea analog **7** ($SC_{1.5} = 3.8\,\mu M$) lost about four-fold in potency relative to the urea **2**, possibly as a reflection of the reduced capacity of the sulfur atom to accept a hydrogen bond. The C_α-methylated analog **8** was also inactive, suggesting a steric effect between the C_α-methyl and the *N*-acylurea moiety, preventing it from making a productive interaction with GK. Collectively, these observations suggested that the cyclic, intramolecularly hydrogen bonded conformation adopted by the *N*-acylureas was necessary to support optimal donor–acceptor hydrogen bonding with GK.

Later, modeling experiments with the *N*-acylurea GK activators using the X-ray co-crystal structures of other GK-GK activator complexes identified key donor–acceptor H-bond interactions with the backbone NH and carbonyl of the Arg63 residue in GK, as shown in Figure 3.8.[14] These observations led to a hypothesis that amides derived from heterocyclic amines of formula I

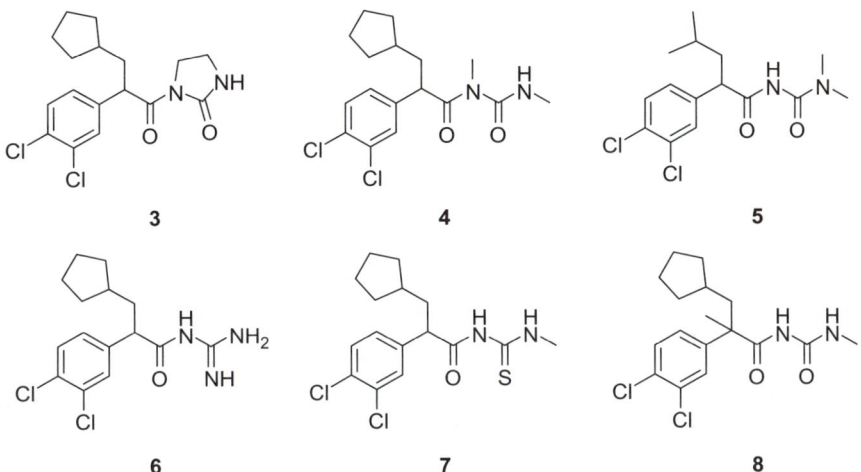

Figure 3.7   Active conformation probe compounds.

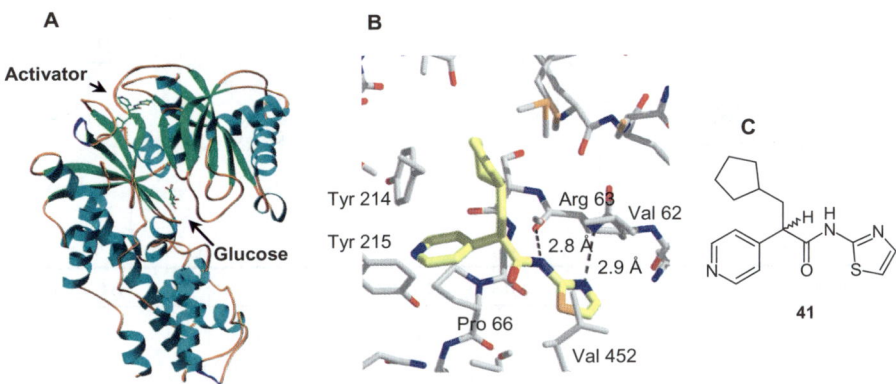

Figure 3.8   GK–GK activator complex crystal structure. A, the GK–GKA complex; B, GKA binding site interactions; C, GK activator structure.

(Figure 3.9) could mimic the H-bond donor–acceptor interactions of the *cis*-amide bond of the *N*-acylurea.

Thus, several phenylacetamides of various 2-amino *N*-containing hetero-aromatic systems (azoles and azines) were evaluated as GK activators and found to be active. We discovered a number of heterocyclic ring systems that were capable of activating GK, the main requirement being the presence of a lone pair-bearing nitrogen atom adjacent to the acylated heteroaromatic amino group. The data for several of these are shown in Table 3.1 (compounds **10–18**). Since 2-aminothiazoles consistently proved to be highly potent, they were employed in further optimization experiments. In addition, there was a wide tolerance for substitution at the 4- and 5-positions of the thiazole rings, as

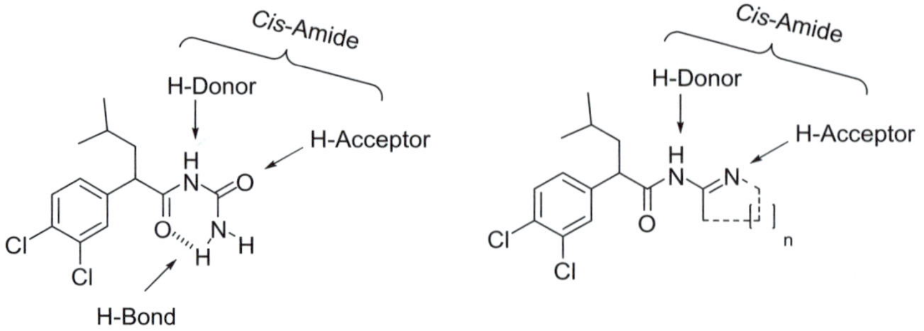

Figure 3.9 Intramolecular H-bond and *cis*-amide characteristics of **1** and the resulting 2-amino heteroaromatic ring hypothesis.

Table 3.1 Key aromatic ring SAR findings related to **9a**.

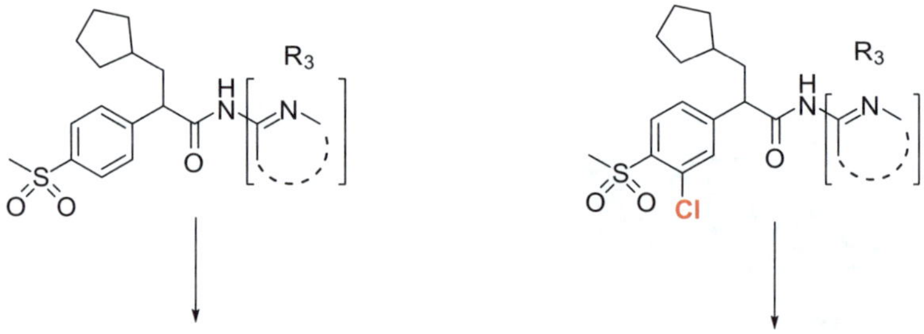

Compound	R_3	$SC_{1.5}$ (μM)	Compound	R_3	$SC_{1.5}$ (μM)
9	thiazolyl	0.34	14	thiazolyl	0.12
10	2-pyridyl	1.21	15	2-pyridyl	0.083
11	2-pyrazinyl	1.78	16	2-pyrazinyl	0.18
12	2-quinolyl	1.38	17	2-quinolyl	0.12
13	3-pyrimidinyl	1.74	18	3-pyrimidinyl	0.19

summarized in Table 3.2 (compounds **19–26**). In general, the SAR at R_1 and R_2 of phenylacetamidothiazoles was parallel to that of the *N*-acylureas. Similar to *N*-acylureas, the (*R*)-stereoisomer **9a** potently activated GK ($SC_{1.5} = 0.35\,\mu M$), while the corresponding (*S*)-isomer **9b** did not activate GK at up to $10\,\mu M$.

The structural elements required in a phenylacetamide-based GK activator provided by SAR studies can be summarized as follows: (i) an electron-deficient aromatic ring is necessary at R_1; (ii) a five- or six-membered alicyclic ring is optimal at R_2; (iii) an H-bond donor–acceptor pair, either as part of an *N*-acylurea or an amide of a 2-amino heterocycle with H-bond acceptor capability, is required at R_3; and (iv) the (*R*) stereochemistry is preferred at the chiral center.

Table 3.2 Thiazole analogs of **9a** and relative thiourea concentration in HLM incubations.

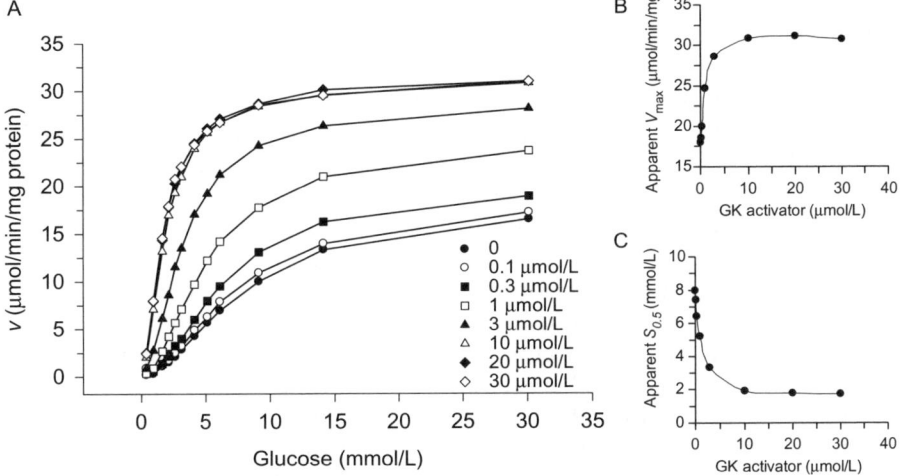

Compound	Substitution	$SC_{1.5}$ (μM)	Thiourea ($ng\,mL^{-1}$)
9a (R-isomer)	X = Y = H	0.18	506
19	X = CH$_2$CH$_2$OH, Y = H	0.67	167
20	X = CH$_2$OH, Y = H	0.48	89
21	X = H, Y = Cl	0.15	48
22	X = H, Y = Br	0.14	43
23	X = CONHCH$_3$, Y = H	> 30	13
24	X = H, Y = CONH$_2$	0.73	< 10
25	X = H, Y = CONHCH$_3$	1.40	< 10
26	benzothiazolyl	0.50	Not detectable

Figure 3.10 Effects of **2a** (RO0274375) on GK kinetics. A, the rate *versus* glucose plot in the absence and presence of **2a**; B, apparent V_{max} *versus* [**2a**] plot; C, apparent $S_{0.5}$ *versus* [**2a**] plot.

Compound **9a** increased the enzymatic activity of GK across a wide range of glucose concentrations (Figure 3.10A), whereas the corresponding (*S*)-enantiomer was inactive. Activation of GK was driven by dual effects of increasing the enzyme's V_{max} (Figure 3.10B) and lowering its $S_{0.5}$ (concentration at

Table 3.3 *In vitro* GK enzyme kinetics data for compounds **2a** and **9a**.

Parameter	2a	9a
V_{max} (mmol min^{-1} mg protein^{-1})		
Control	18.0	19.1
1 μM (fold change *versus* control)	24.7 (1.4)	23.7 (1.2)
20 μM (fold change *versus* control)	31.2 (1.7)	26.1 (1.4)
$S_{0.5}$ (mmol L^{-1})		
Control	8.0	8.7
1 μM (fold change *versus* control)	5.2 (0.66)	3.3 (0.38)
20 μM (fold change *versus* control)	1.8 (0.22)	0.84 (0.10)

half-maximal velocity) for glucose (Figure 3.10C). The magnitude of these effects at 1 μM and 20 μM activator concentrations was determined by fitting the data to the Hill equation (Table 3.3). Compound **2a** exerted a relatively larger increase in V_{max} (a 1.7-fold change for **2a** *versus* a 1.4-fold change for **9a** at 20 μM concentration of each), whereas **9a** showed a greater effect on reducing the enzyme's $S_{0.5}$ for glucose (1.8 mmol L^{-1} for **2a** *versus* 0.84 mmol L^{-1} for **9a** at 20 μM concentration of each). Consistent with the role of pancreatic β-cell GK, treatment of rodent islets with either **2a** or **9a** shifted the glucose-stimulated insulin release curve to the left, as demonstrated in perifusion studies using isolated pancreatic islets.[16]

In order to better discriminate between the two GK activators, head-to-head *in vivo* efficacy studies were performed. Acute oral administration of **2a** and **9a** (50 mg kg^{-1}) to male normal C57Bl/6J mice caused a statistically significant reduction in fasting glucose levels and improvement in glucose tolerance relative to the vehicle-treated animals (Figure 3.11A). Compound **9a** showed a statistically significant superior effect during an oral glucose tolerance test (OGTT) relative to **2a** (Figure 3.11B). Furthermore, comparison of rat pharmacokinetic (PK) parameters indicated that **9a** displayed lower clearance and higher oral bioavailability compared to **2a** (Table 3.4).

Based on these and other safety related properties, **9a** was advanced to a single ascending dose (25, 100, 200, and 400 mg) study in healthy volunteers. Following an oral dose, **9a** reduced fasting and postprandial glucose levels (minimal effective dose was between 25 and 100 mg) following an OGTT, was well tolerated, and displayed no adverse effects related to drug administration other than hypoglycemia at a single dose of 400 mg, which defined the maximum tolerated dose.[15] Thus, **9a** was the first GK activator that provided the proof-of-concept for this mechanism in humans.

3.3   Lead Optimization and Discovery of Piragliatin[16]

Compound **9a**, however, caused reversible hepatic lipidosis in rats and dogs in sub-chronic toxicology studies and thus was unsuitable for full clinical development. With this observation, we were faced with a key question: is this finding related to the compound structure or due to drug-induced GK

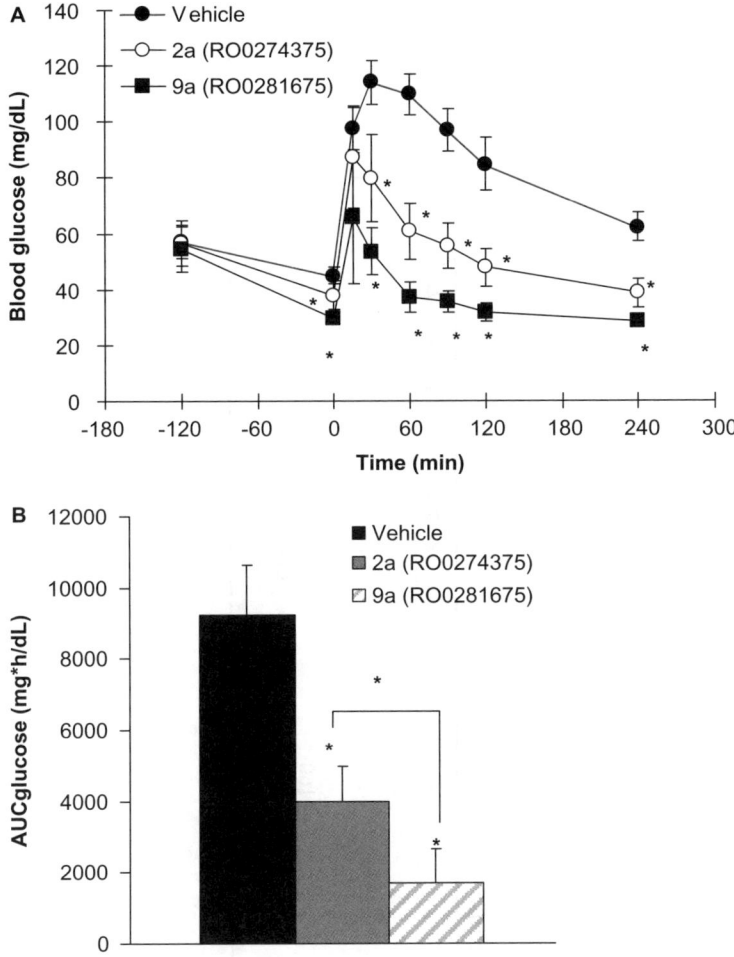

Figure 3.11 Acute effects of orally administered **2a** (RO0274375) and **9a** (RO-0281675) at 50 mg kg^{-1} on an oral glucose tolerance test in male normal C57Bl/6J mice. All results are reported as the mean ± STDEV ($n = 6$/time point). A Student's t-test was used to test for statistical significance (*, $P < 0.05$).

activation? First, we investigated whether the hepatic lipidosis was due to either the parent molecule or its metabolites. The metabolite profile of **9a** indicated that: (1) products of oxidation at the cyclopentyl ring led to the formation of the corresponding hydroxylated and ketone products; and (2) oxidative ring opening of the thiazole ring led to the formation of the corresponding thiourea (Figures 3.12 and 3.13). The thiourea metabolites were suspected to be the causative agents of hepatic lipidosis, based on literature precedent.[17] As described earlier, the (*S*)-isomer of **9b** does not activate GK or lower glucose levels. Therefore, we used the inactive thiazole **9b** and the thiourea **10**

Table 3.4 Rat pharmacokinetics data for compounds **2a** and **9a**.

Cpd	Dose $(mg\,kg^{-1})$	Route	$Cl\,(mL\,kg^{-1}\,h^{-1})$	$V_{ss}\,(mL\,kg^{-1})$	$AUC_{0-inf}\,(ng\,h\,mL^{-1})$	$C_{max}\,(\mu g\,mL^{-1})$	$T_{max}\,(h)$	$F_{po}\,(\%)$
2a	5	iv	2609	4211	1877	–	–	–
	10	po	–	–	455	133	1.4	12
	30	po	–	–	1464	471	3.3	13
9a	5	iv	1800	2160	2774	–	–	–
	10	po	–	–	5150	1140	3.3	92.8
	30	po	–	–	51748	4302	4.7	311

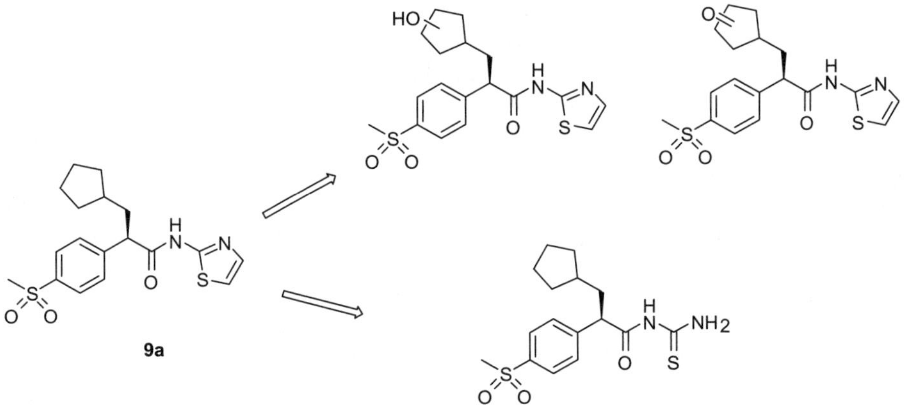

Figure 3.12 Primary metabolite profile of **9a**.

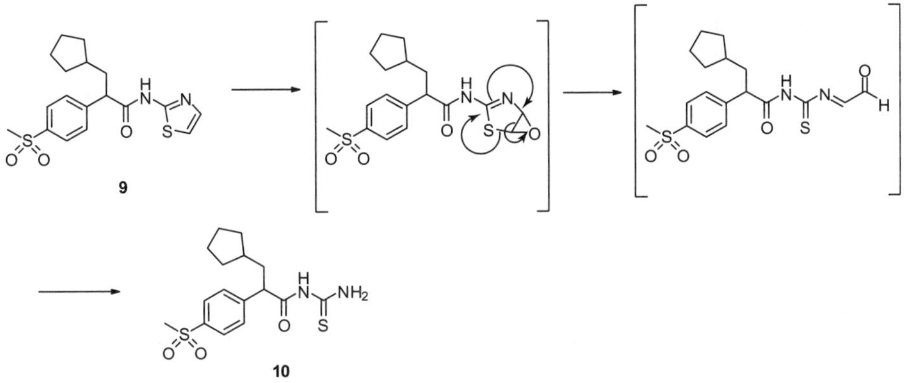

Figure 3.13 Mechanism of thiourea formation.

($SC_{1.5} = 45 \, \mu M$) as structural probes for **9a** and its key metabolite. In a five-day rat toxicology study, either a dose of $250 \, mg \, kg^{-1}$ of **9b** or doses from 10 to $300 \, mg \, kg^{-1}$ of **10** caused hepatic lipidosis in rats in the *absence* of glucose lowering. These findings strongly supported the hypothesis that the thiourea metabolite formed *via* oxidative thiazole ring opening was the primary cause for the observed hepatic lipidosis associated with **9a**, rather than the GK activity of **9a**, and set the stage for the discovery of a safer compound incapable of this transformation.[18]

As a first step, several substituted thiazole analogs (Table 3.2) of **9a** that were either monosubstituted (compounds **19–25**) or the benzothiazole **26** were investigated for their propensity to form the thiourea metabolite in human liver microsomes. These studies demonstrated a marked reduction in their propensity to undergo oxidation of the thiazole ring, particularly when the thiazole substituent was an electron-withdrawing group, such as chloro or nitro.

However, several of these analogs were also less potent than the parent unsubstituted thiazole *in vitro* and *in vivo*, had poor pharmacokinetic properties, and were not considered further. Hence, we reinvestigated alternative heterocyclic ring systems as replacements to the thiazole moiety.

We observed earlier during SAR studies that introduction of a small electron-withdrawing group such as a halogen or cyano group next to the methyl sulfone moiety on R_3 generally led to a modest improvement in potency (Table 3.1). More interestingly, when the thiazole of **9a** was replaced by a six-membered heterocycle, such as a pyridine, pyrazine, pyrimidine, or quinoline, there was a significant loss of activity, whereas incorporation of a 3-chloro-4-(methylsulfonyl)phenyl ring was found to enhance potency at least 10-fold over the corresponding 4-(methylsulfonyl)phenyl derivative of these six-membered heterocycles. From these observations, and extensive comparison of *in vitro* profiles, *in vivo* potency, rodent pharmacokinetics, and five-day high-dose rat exploratory toxicity profiles of several compounds, we selected **27** (RO0505082) as a development candidate. This compound was found to be as efficacious as compound **9a** in reducing fasting and post-prandial glucose levels and was profiled in a rat toxicity study. In a rodent 14-day safety range finding study at doses up to 600 mg kg^{-1}, there was no evidence for hepatic lipidosis, supporting our hypothesis that the lipidosis observed with **9a** was related to metabolic thiourea formation rather than GK activation.

However, **27** was found to be a potent inhibitor of the hERG potassium channel (IC$_{20}$ ≈ 3 µM) and showed activity in a rabbit Purkinje fiber assay (EC$_{50}$ @ 1 Hz = 8.2 µM), suggesting a high potential for cardiovascular risk based on the low margin between the plasma efficacious exposure levels and hERG IC$_{20}$ values. In addition, **27** caused a time-dependent inhibition of CYP3A4 (52% inh @ 24 min, 10 µM incubation), indicating a potential for causing drug–drug interactions and undesired non-specific covalent modification of proteins, which could lead to unpredictable idiosyncratic toxicity. These findings were considered to pose an unacceptable risk for further development and consequently **27** was abandoned.

Detailed characterization of the metabolites of **27** from liver microsomal studies, *in vivo* rat PK, and safety studies showed a significant formation of alcohol and ketone metabolites resulting from oxidation of the cyclopentyl ring at C2 and C3 (Figure 3.14). The extent of metabolism in liver microsomal studies was found to be 76% in human and rat, while in dog it was 56%. Initially, the authentic diastereomeric mixtures of metabolites were synthesized for potency and *in vitro* safety evaluations.[20] The diastereomeric mixtures of the C2-hydroxyl **28** and C3-hydroxyl **29** metabolites were less potent in GK activation compared to the parent **27** (Figure 3.15). However, the diastereomeric mixtures of C2- (**30**) and C3- (**31**) ketone metabolites had comparable potency to **27**. Both the hydroxyl and keto metabolites had reduced hERG inhibition and CYP 3A4 time-dependent inhibition compared to the parent **27** (Figure 3.15). We chose to further evaluate the C3-keto metabolite **31** over the C2-keto metabolite **30** in view of the potential of the C2-ketone to promote epimerization of the adjacent chiral center.

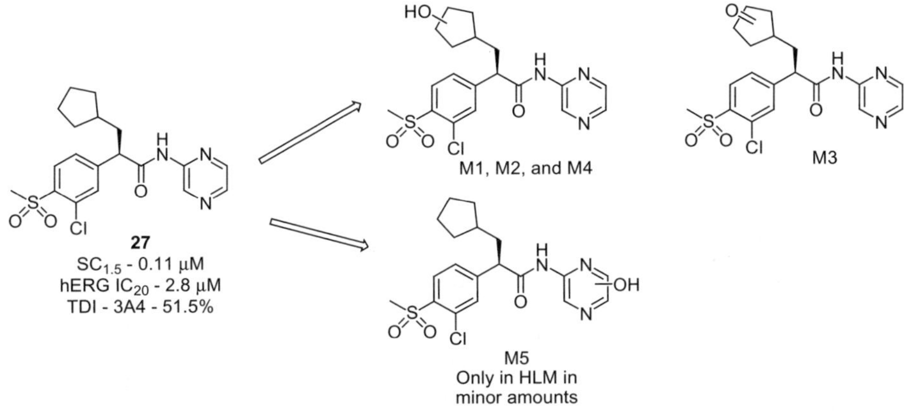

Figure 3.14 Compound **27** and its metabolites. TDI – Time-dependent inhibition of CYP3A4 at 10 μM concentration and 30 min incubation.

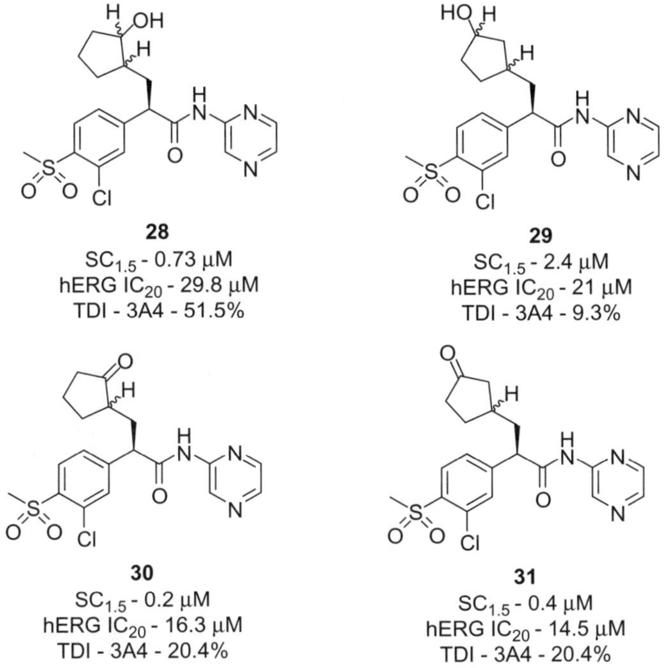

Figure 3.15 The *in vitro* activity profiles of the metabolites of compound **27**.

A stereoselective synthesis starting from either the (*S,S*)- or (*R,R*)-hydro-benzoin[19] ketal of cyclopent-2-enone **34** to prepare either the (*R,S*)- (**32**) or (*R,R*)- (**33**) diastereomer was developed (Scheme 3.1). In this synthesis, the key transformations were: (a) the stereoselective Simmons–Smith cyclopropanation

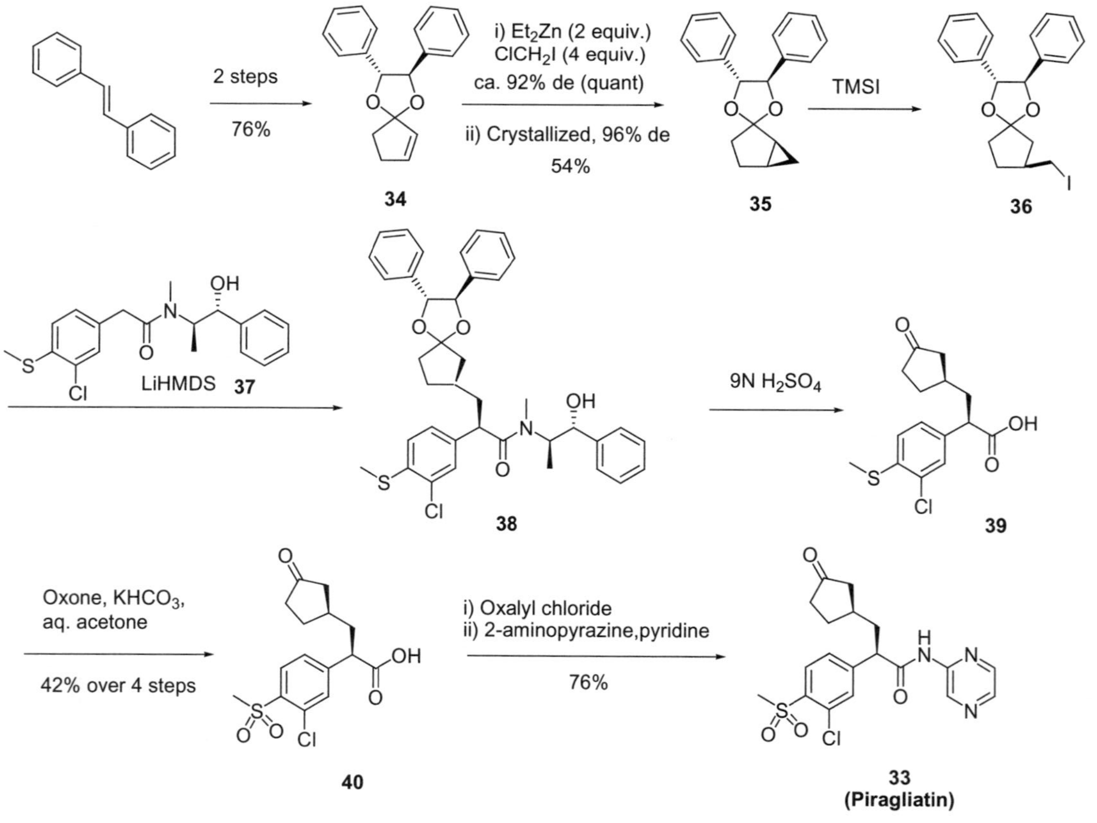
2 steps
76%
34
i) Et_2Zn (2 equiv.)
$ClCH_2I$ (4 equiv.)
ca. 92% de (quant)
ii) Crystallized, 96% de
54%
35
TMSI
36
LiHMDS 37
38
9N H_2SO_4
39
Oxone, $KHCO_3$,
aq. acetone
42% over 4 steps
40
i) Oxalyl chloride
ii) 2-aminopyrazine,pyridine
76%
33
(Piragliatin)

TMSI = Trimethylsilyliodide

Scheme 3.1 Stereoselective synthesis of piragliatin (**33**)

reaction of the chiral ketal **34** to the bicyclic intermediate **35**; (b) opening of the cyclopropane of **35** to alkyl iodide **36** using iodotrimethylsilane; and (c) the diastereoselective alkylation of chiral amide **37** to yield the critical phenylacetamide **38**. Thus, both the (*R,S*)- (**32**) and (*R,R*)- (**33**) diastereomers, corresponding to the C3-keto metabolites of **27**[20] (Scheme 3.1), were available. *In vitro* and *in vivo* evaluation and comparison of both diastereomers (Figures 3.16 and 3.17)

32
$SC_{1.5}$ - 5.6 μM
hERG IC_{20} - 10 μM

33
(Piragliatin)
$SC_{1.5}$ - 0.18 μM
hERG IC_{20} - 6μM
TDI - 3A4 - 12%

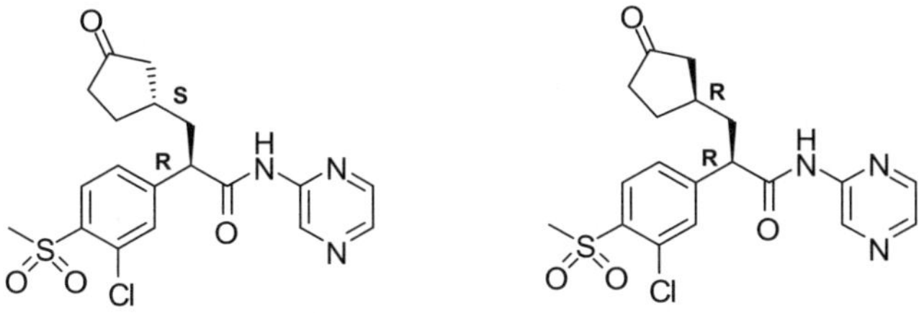

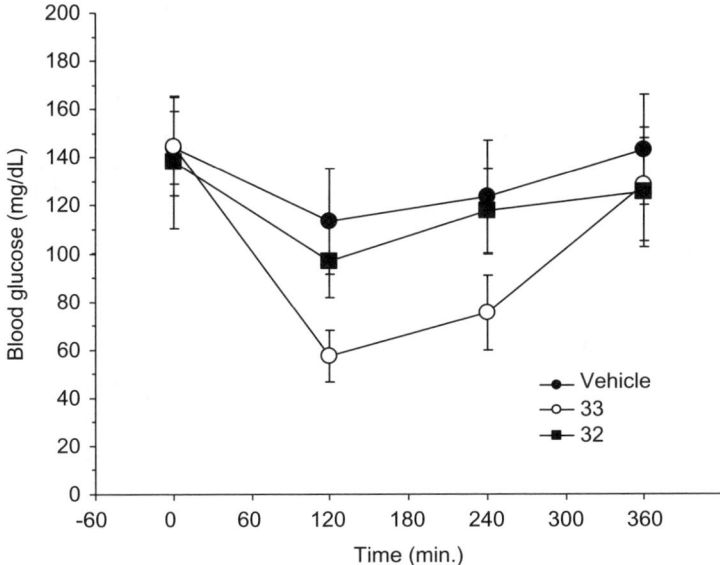

Figure 3.16 *In vitro* potency of C3 ketone metabolites of **27**.

Figure 3.17 *In vivo* potency of **32** and **33**, C3 ketone metabolites of **27**, in diabetic diet-induced obese mice ($n = 8$/group). C57Bl/6J mice were fed a high-fat diet for 16.7 weeks and fasted for 2 h prior to oral administration of vehicle or test compound (15 mg kg^{-1}). Animals were fasted for the duration of the study.

clearly indicated the superior potency of **33** over **32**, and led to the selection of the (*R*,*R*)- diastereomer **33**, piragliatin,[20] as the clinical lead.

3.4 Clinical Proof-of-concept

Piragliatin, in a battery of *in vitro* and *in vivo* efficacy tests, PK, and toxicology profiling in rats and dogs, was superior to the earlier clinical lead **9a**, and was advanced to clinical testing. In a Phase I proof-of-concept study,[21b] piragliatin was evaluated in 15 T2D patients at 25 and 100 mg (po, qd) and in a subsequent multiple ascending dose study at doses ranging from 25 mg to 400 mg either qd or bid in T2D patients.[21a] In these trials, piragliatin decreased fasting and post-OGTT glucose levels,[21] improved the insulin secretory profile, increased β-cell sensitivity to glucose, and decreased hepatic glucose output. Following completion of Phase II clinical studies, a decision was made not to develop piragliatin further. Based on the experience from piragliatin, Roche is continuing development of other GK activators.

3.5 Outlook and Conclusion

GK's key role in glucose homeostasis has long been recognized. Developments over the past decade, including the discovery of MODY-2, PNDM, and PHHI-related GK mutations, and discovery of novel GK activators and their co-crystal structures with GK, have significantly enhanced the understanding of GK structure and function. Nearly 100 patent applications and 10 papers reporting the characterization of lead molecules have been published to date. GK activators from several companies, including Merck-Banyu, AstraZeneca, Lilly, and Array Biopharma, are reported to be in Phase I or II stages of clinical development. In the next few years, we look forward to understanding the impact of GK activator therapy on glycemic control and safety in T2D patients, and specifically their differentiation from the widely prescribed sulfonylureas.

Acknowledgements

The authors thank Stephen To for editing this article and to Christophe Arbet-Engels for helpful discussions on the clinical data.

References

1. R. A. DeFronzo, *Diabetes*, 2009, **58**, 773–795.
2. J. Grimsby, F. M. Matschinsky and J. F. Grippo, in *Glucokinase and Glycemic Disease: From Basics to Novel Therapeutics*, ed. F. M. Matschinsky and M. A. Magnuson, Karger, Basel, Switzerland, 2004, pp. 360–378.

3. F. M. Matschinsky and J. E. Ellerman, *J. Biol. Chem.*, 1968, **243**, 2730–2736.

4. E. Van Schaftingen, *Eur J. Biochem.*, 1989, **179**, 179–184.

5. T. Aizawa, N. Asanuma, Y. Terauchi, N. Suzuki, M. Komatsu, N. Itoh, T. Nakabayashi, H. Hidaka, H. Ohnota, K. Yamauchi, K. Yasuda, Y. Yazaki, T. Kadowaki and K. Hashizume, *Biochem. Biophys. Res. Commun.*, 1996, **229**, 460–5.

6. P. Froguel, H. Zouali, N. Vionnet, G. Velho, M. Vaxillaire, F. Sun, S. Lesage, M. Stoffel, J. Takeda, P. Passa, M. A. Permutt, J. S. Beckmann, G. I. Bell and D. Cohen, *New Engl. J. Med.*, 1993, **328**, 697–702.

7. B. Glaser, P. Kesavan, M. Heyman, E. Davis, A. Cuesta, A. Buchs, C. A. Stanley, P. S. Thornton, M. A. Permutt, F. M. Matschinsky and K. C. Herold, *New Engl. J. Med.*, 1998, **338**, 226–230.

8. S. S. Fajans, G. I. Bell and K. S. Polonsky, *New Engl. J. Med.*, 2001, **345**, 971–980.

9. P. R. Njolstad, O. Sovik, A. Cuesta-Munoz, L. Bjorkhaug, O. Massa, F. Barbetti, D. E. Undlien, C. Shiota, M. A. Magnuson, A. Molven, F. M. Matschinsky and G. I. Bell, *New Engl. J. Med.*, 2001, **344**, 1588–1592.

10. N.-E. Haynes, W. L. Corbett, F. T. Bizzarro, K. G. Guertin, W. D. Hilliard, H. Holland, R. F. Kester, P. E. Mahaney, L. Qi, C. L. Spence, J. Tengi, M. Dvorozniak, A. Railkar, F. Matschinskky, J. F. Grippo, J. Grimsby and R. Sarabu, *J. Med. Chem.*, 2010, **53**, 3618–3625.

11. J. Grimsby, R. Sarabu, W. L. Corbett, N. E. Haynes, F. T. Bizzaro, J. W. Coffey, K. R. Guertin, D. H. Hilliard, R. F. Kester, P. E. Mahaney, L. Marcus, L. Qi, C. L. Spence, J. Tengi, M. A. Magnuson, C. A. Chu, M. T. Dvorozniak, F. M. Matschinsky and J. F Grippo, *Science*, 2003, **301**, 370–373.

12. See, for example: (a) L. S. Bertram, D. Black, P. H. Biner, R. Chatfield, A. Cooke, M. C. T. Fyfe, P. J. Murray, F. Naud, M. Nawano, M. J. Procter, G. Rakipovski, C. M. Rasamison, C. Reynet, K. L. Schofield, V. K. Shah, F. Spindler, A. Taylor, R. Turton, G. M. Williams, P. Wong-Kai-In and K. Yasuda, *J. Med. Chem.*, 2008, **51**, 4340–4345; (b) A. M. Efanov, D. G. Barrett, M. B. Brenner, S. L. Briggs, A. Delaunois, J. D. Durbin, U. Giese, H. Guo, M. Radloff, G. S. Gil, S. Sewing, Y. Wang, A. Weichert, A. Zaliani and J. Gromada, *Endocrinology*, 2005, **146**, 4696; (c) A. L. Castelhano, H. Dong, M. C. T. Fyfe, L. S. Gardner, Y. Kamikozawa, S. Kurabayashi, M. Nawano, R. Ohashi, M. J. Procter, L. Qiu, C. M. Rasamison, K. L. Schofield, V. K. Shah, K. Ueta, G. M. Williams, D. Witter and K. Yasuda, *Bioorg. Med. Chem. Lett.*, 2005, **15**, 1501–1504.

13. R. Sarabu, S. J. Berthel, R. F. Kester and J. W. Tilley, *Expert Opin. Ther. Pat.*, 2008, **18**, 759–768.

14. P. Dunten, A. Swain, U. Kammlott, R. Crowther, C. M. Lukacs, W. Levin, L. Reik, J. Grimsby, W. L. Corbett, M. A. Magnuson and F. M. Matschinsky, in *Glucokinase and Glycemic Disease – From Basics to Therapeutics*, ed. F. M. Matschinsky and M. A. Magnuson, Karger, Basel, Switzerland, 2004, pp. 145–154.

15. J. Grimsby, J. Zhi, M. E. Mulligan, C. Arbet-Engels, R. Taub and R. Balena, presented at the Keystone Symposia: Diabetes Mellitus, Insulin Action and Resistance, January 2008, Breckenridge, CO, USA, poster 151.

16. R. F. Kester, W. L. Corbett, R. Sarabu, P. E. Mahaney, N.-E. Haynes, K. R. Guertin, F. T. Bizzarro, D. W. Hilliard, L. Qi, J. Tengi, J. F. Grippo, J. Grimsby, L. Marcus, C. Spence, M. Dvorozniak, J. Racha and K. Wang, presented at MEDI-005, 238th ACS National Meeting, Washington, August 2009.

17. See, for example: (a) T. Mizutani, K. Yoshida and S. Kawazoe, *Drug Metab. Dispos.*, 1994, **22**, 750; (b) G. J. Stevens, K. Hitchcock, Y. K. Wang, G. M. Coppola, R. W. Versace, J. A. Chin, M. Shapiro, S. Suwanrumpha and B. L. Mangold, *Chem. Res. Toxicol.*, 1997, **10**, 733.

18. J. Racha, W. Geng, M. Pignatello, Z. Liang, R. Sarabu, J. Grimsby and D. M. Moore, *Drug Metab. Rev.*, 2009, **41**(S3), 127 (abstract 259, ISSX Meeting, Baltimore, MD, October 2009).

19. Z.-M. Wang and K. B. Sharpless, *J. Org. Chem.*, 1994, **59**, 8302.

20. W. L. Corbett, J. Grimsby, N.-E. Haynes, R. F. Kester, P. E. Mahaney, J. K. Racha, R. Sarabu and K. Wang, *US Pat.* 7 105 671, 2006.

21. (a) J. Zhi, S. Zhai, M. -E. Mulligan, J. Grimsby, C. Arbet-Engels, M. Boldrin and R. Balena, *Diabetologia*, 2008, **51**(suppl. 1, abstr. 42), S23; (b) R. C. Bonadonna, C. Kapitza, T. Heise, A. Avogaro, M. Boldrin, J. Grimsby, M.-E. Mulligan, C. Abert-Engels and R. Balena, *Diabetologia*, 2008, **51**(suppl. 1, abstr. 927), S371; (c) S. Zhai, M. -E. Mulligan, J. Grimsby, C. Arbet-Engels, M. Boldrin, R. Balena and J. Zhi, *Diabetologia*, 2008, **51**(suppl. 1, abstr. 928), S372.

CHAPTER 4

The Discovery of OSI-906, a Small-molecule Inhibitor of the Insulin-like Growth Factor-1 and Insulin Receptors

MARK J. MULVIHILL AND ELIZABETH BUCK

OSI Pharmaceuticals, Inc., Broadhollow Bioscience Park, 1 Bioscience Park Drive, Farmingdale, NY 11735, USA

4.1 Targeting the IGF-1R/IR Pathway for the Discovery of Anti-tumor Agents

It is well established that the insulin and insulin-like growth factor-1 receptors (IR and IGF-1R, respectively) can promote the growth and survival of human tumor cells, and therefore, these receptors have been attractive targets for anti-cancer drug discovery.[1,2] Both IGF-1R and IR are transmembrane receptor tyrosine kinases (RTKs), consisting of two extracellular α-subunits disulfide-bonded to two transmembrane-spanning β-subunits containing cytoplasmic tyrosine kinase activity.[3–7] IGF-1R and IR can exist as either homodimers or heterodimers and are activated upon binding of their cognate ligands through interaction with the extracellular α-subunits, which promotes a conformational shift to the intracellular catalytic domain, residing in the β-subunit.[7] IGF-1R is activated by either IGF-1 or IGF-2 ligands, while IR can be activated by either insulin or IGF-2. IGF-1R has been shown to transduce proliferation and

RSC Drug Discovery Series No. 4
Accounts in Drug Discovery: Case Studies in Medicinal Chemistry
Edited by Joel C. Barrish, Percy H. Carter, Peter T. W. Cheng and Robert Zahler
© Royal Society of Chemistry 2011
Published by the Royal Society of Chemistry, www.rsc.org

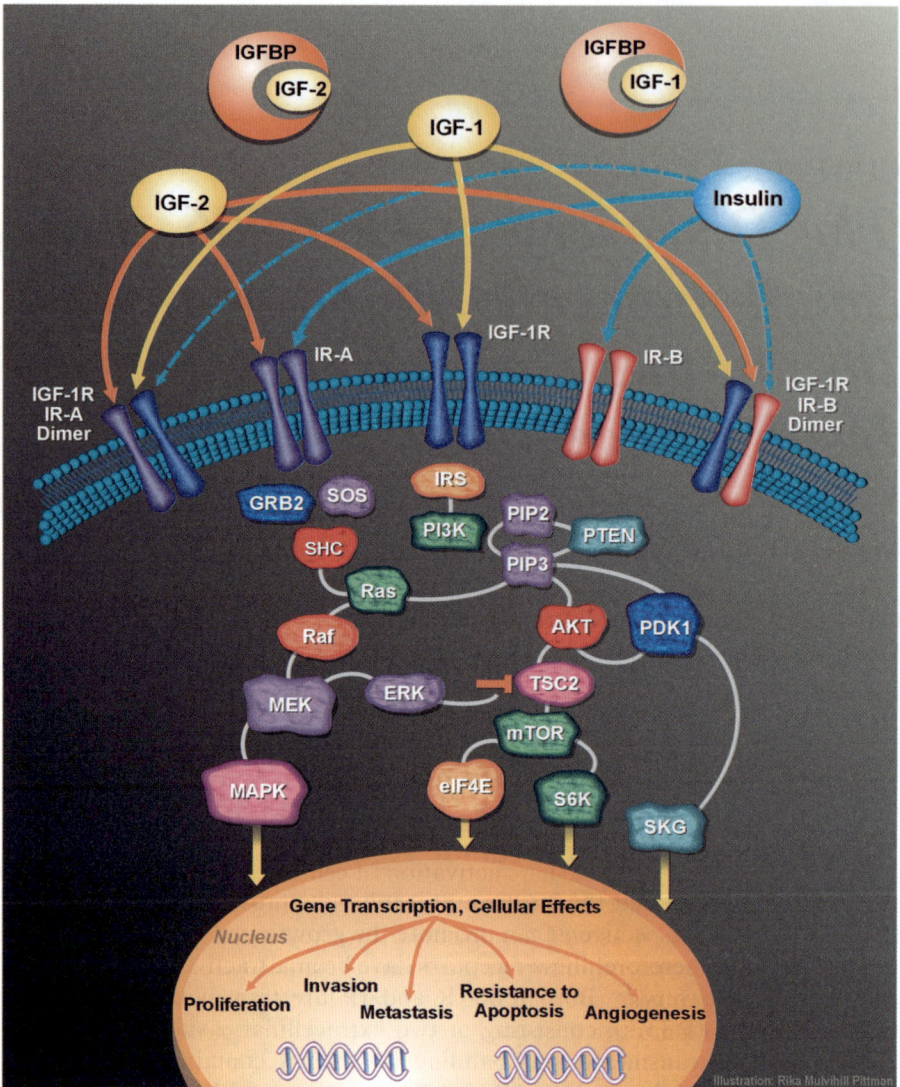

Figure 4.1 Insulin-like growth factor-1 receptor signaling pathway.

survival signaling through both the AKT and ERK pathways, and especially strong coupling between IGF-1R and the PI3K-AKT cascade has been observed in a multitude of tumor cell settings (Figure 4.1).[8–11]

The IGF-1R is required for normal embryonic development and for post-natal growth, and the ability of tumor cells to exploit this biology to maintain growth and survival signaling has been well established in preclinical models. IGF-1R is required for cellular transformation by a number of oncogenes, including Ras,[11,12] and expression of IGF-1R can overcome cellular dependence on other growth factors.[13] Ablation of IGF-1R signaling through targeted down-regulation of receptor expression using siRNA or inhibiting receptor activity using small-molecule tyrosine kinase inhibitors (TKIs), receptor neutralizing antibodies, or dominant negative constructs has been shown to inhibit tumor cell growth and survival both *in vitro* and *in vivo*.[14–19] The ability of tumor cells to exploit IGF-1R signaling is further underscored by the observed loss of imprinting (LOI) for the gene encoding IGF-2 ligand, leading to an increase in IGF-2 protein levels.[20–25] Such tumor cells that harbor an autocrine IGF loop have been found to be especially sensitive to agents targeting IGF-1R/IR.

In addition to its role in metabolic signaling, there is growing support for the involvement of IR in tumor mitogenic signaling. The IR can be transforming for both fibroblasts and mammary epithelial cells, and insulin can activate tumoral AKT signaling in both the *in vivo* and *in vitro* settings.[26,27] Epidemiological studies provide further support for the role of insulin signaling in tumor growth. For a number of tumor types, including breast and prostate, increased levels of insulin and C-peptide are associated with poor prognosis.[2,28,29] Furthermore, it has recently been reported that inhaled insulin is associated with increased risk of lung cancer.[30] These observations support the potential for enhanced and broader anti-tumor activity for IGF-1R/IR TKIs compared to selective IGF-1R inhibition using anti-IGF-1R antibodies.

4.2 IGF-1R/IR Inhibitors as Anchors for Combinatorial Anti-cancer Drug Strategies

IGF-1R/IR signaling can mediate activation of cellular survival in the presence of a multitude of other anti-tumor agents, including cytotoxic chemotherapeutics and radiation as well as molecular targeted therapies (MTTs).[31] The ability for IGF-1R/IR inhibitors to augment the efficacy for these agents has been extensively investigated in the preclinical setting and is the rational underpinning for a multitude of combinatorial drug strategies being explored in the clinical setting.

Herein, we describe the path leading to the discovery and development of OSI-906, a selective, orally bioavailable, low molecular weight, dual inhibitor of IGF-1R and IR. This discovery was enabled through the use of rational, structure-based drug design to optimize both the binding affinity for IGF-1R/IR and achieve a high degree of selectivity *versus* other kinases. Empirical medicinal chemistry in combination with high-speed analoging synthetic strategies were utilized to expedite the optimization process. Finally, a streamlined testing cascade of *in vitro* and *in vivo* IGF-1R and IR models were used to assess for pharmacokinetics (PK), pharmacodynamics (PD), efficacy, and safety, as well as to enable structure–activity development, compound stratification, and optimization. Finally, biomarker and combinatorial drug strategies were applied to facilitate a clinical development plan for this agent.

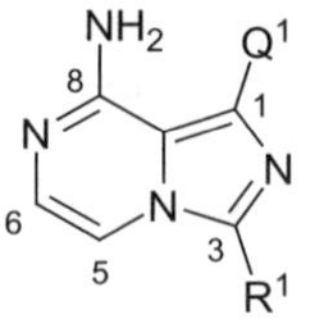

1

Figure 4.2   1,3-Disubstituted imidazopyrazine.

4.3   Rationale for Imidazopyrazine-derived IGF-1R Inhibitors

Our interest in identifying small-molecule IGF-IR kinase inhibitors for oncology-directed applications centered on the underexploited imidazo[1,5-a]pyrazine core (Figure 4.2). This novel 6/5 heteroaryl template offers several advantages as a priviledged kinase scaffold, as it retains key pharmacophoric donor/acceptor interactions with the kinase hinge region through the N7 and 8-NH$_2$ moieties, mimicking the adenine core of the natural ATP substrate. Additionally, the imidazo[1,5-a]pyrazine template facilitates efforts to build upon structure–activity relationships (SARs) derived from programs targeting various kinase inhibitors. For example, the 6/5 bicyclic heteroaryl class, such as the thieno-, pyrrolo-, and pyrazolopyrimidines, have served as adenine hinge-binding mimetics in a number of kinase inhibitor programs.[18,32] The imida-zopyrazine scaffold benefits from (1) synthetic flexibility at C3, where sub-stituents (R^1) are derived from common carboxylic acids, (2) stable attachment of both hetero and carbon atoms to C3 directly or through a methylene linker, and (3) the inclusion of C1 substituents (Q^1) either in the first step through a linear route or at the last step through a convergent route. As such, we report herein the progression of a novel series of 1,3-disubstituted 8-aminoimidazo [1,5-a]pyrazines (**1**) IGF-IR TKIs from hit to lead to investigational new drug (IND), with the ultimate discovery of OSI-906 as a potent, selective, orally bioavailable dual IGF-1R/IR inhibitor that is currently progressing through clinical trials.

4.4   Synthetic Routes to Imidazopyrazine-derived IGF-1R Inhibitors

Two routes, linear and convergent, were employed to allow for modifications at C1 and C3 of the imidazopyrazine core.[33,34] Scheme 4.1 illustrates the generic linear route involving the direct *ortho*-metallation of 2-chloropyrazine (**2**) in the presence of a preformed solution of lithium tetramethylpiperidide, followed by quenching with various aldehydes to afford alcohol **3**. Conversion of alcohol **3** to amine **4** via the Gabriel synthesis proceeded through a Mitsunobu reaction with phthalimide, followed by unmasking of the amine through reaction with

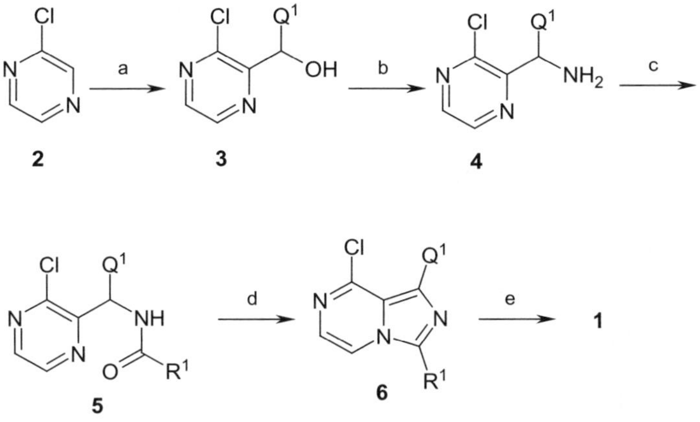

Scheme 4.1 Reagents and conditions: (a) 2 M *n*BuLi in hexanes, tetramethyl-piperidine, −78 °C; then Q^1-CHO; (b) phthalimide, DIAD, PPh$_3$, THF; then NH$_2$NH$_2$, EtOH/CH$_2$Cl$_2$ (3:1), 16 h; (c) R^{1}CO$_2$H, EDC, HOBT, CH$_2$Cl$_2$ or R^{1}COCl, DIEA, CH$_2$Cl$_2$; (d) POCl$_3$, MeCN, 80 °C, 24 h; (e) NH$_3$, *i*PrOH, 110 °C, 24 h.

hydrazine. Acylation of amine **4** with various acid chlorides (R^{1}COCl) or through coupling with a variety of carboxylic acids (R^{1}CO$_2$H) afforded amide **5**, which when treated with POCl$_3$, cyclized to afford 8-chloroimidazo[1,5-a]pyrazine (**6**). Treatment of compound **6** with ammonia in isopropanol in a Parr reactor at 110 °C for 24 h afforded the desired 8-aminoimidazopyrazine products **1**.

Scheme 4.2 illustrates the generic convergent route for the conversion of 2-chloropyrazine (**2**) to alcohol **7** via directed *ortho*-metallation as described above, followed by quenching with DMF and subsequent treatment with NaBH$_4$ in methanol. Conversion of alcohol **7** to its respective phthalimide via the Mitsunobu reaction, followed by deprotection with hydrazine to amine **8** (Gabriel synthesis), and subsequent acylation, afforded amide **9**. Treatment of amide **9** with POCl$_3$ afforded 8-chloroimidazopyrazine **10**, which, when subjected to NIS, afforded 1-iodoimidazopyrazine **11**. Ammonolysis of compound **11** afforded the desired late-stage intermediate 8-amino-1-iodoimidazopyrazine **12**, which underwent Suzuki coupling with various arylboronic acids/esters to afford the desired final 1-aryl-8-aminoimidazopyrazines **1**.

4.5 Series I: Benzyloxyphenyl-derived Imidazopyrazine IGF-1R Inhibitors

The initial *in vitro* testing cascade employed an IGF-1R biochemical and mechanistic assay. To determine biochemical inhibitory activity, compounds were tested using a GST-tagged recombinant kinase domain derived from human IGF-IR and assayed using poly(Glu:Tyr) (4:1) as the substrate at an

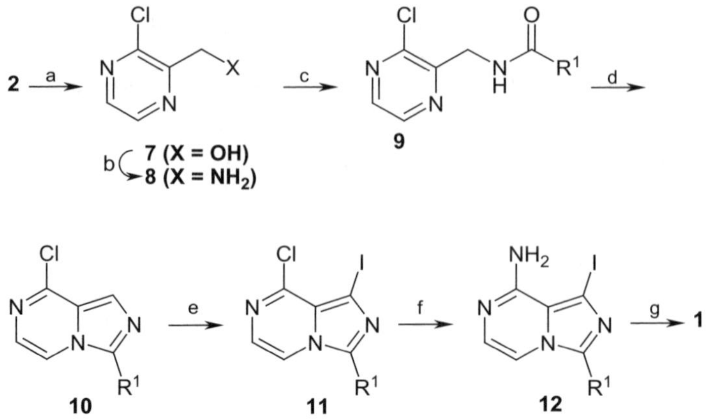

Scheme 4.2 Reagents and conditions: (a) 2 M *n*BuLi in hexanes, tetramethyl-piperidine, −78 °C, then DMF, MeOH, and NaBH$_4$; (b) phthalimide, DIAD, PPh$_3$, THF; then NH$_2$NH$_2$, EtOH/CH$_2$Cl$_2$ (3:1), 16 h; (c) R^{1}CO$_2$H, EDC, HOBT, CH$_2$Cl$_2$ or R^{1}COCl, DIEA, CH$_2$Cl$_2$; (d) POCl$_3$, MeCN, 80 °C, (e) NIS, DMF, rt, 16 h; (f) NH$_3$, Pr^{i}OH, 110 °C, 24 h; (g) Pd(PPh$_3$)$_4$, DME/H$_2$O (4:1), 100 °C, K$_2$CO$_3$, Q^1-B(OH)$_2$/Q^1-B(pin).

ATP concentration of 100 µmol L^{-1}. Compounds which displayed biochemical activities below 1 µM were further tested for inhibition of IGF-I-stimulated receptor autophosphorylation in intact cells, where an NIH-3T3 line stably overexpressing full-length human IGF-IR (LISN) was employed in a capture ELISA assay.

With two synthetic methods in place to afford 1,3-disubstituted 8-amino-imidazopyrazines, analoging efforts focused on the synthesis of a series of C1- and C3-substituted imidazopyrazines. From those efforts emerged 3-BnOC$_6$H$_4$-1-cyclobutylimidazopyrazine (**1a**) (IGF-1R biochemical IC$_{50}$ = 606 nM) (Table 4.1).[33] Further SAR development around this early hit revealed that the 3-BnOC$_6$H$_4$ moiety was preferred at the C1 position of the imidazo-pyrazine core when compared to phenyl (**1b**), 3-MeOC$_6$H$_4$ (**1c**), 4-MeOC$_6$H$_4$ (**1d**), 4-BnOC$_6$H$_4$ (**1e**), or 4-PhOC$_6$H$_4$ (**1f**). The preliminary SAR around the C3 position of the imidazopyrazine core when Q^1 = 3-BnOC$_6$H$_4$ suggested that critical mass was required (R^1 = H, compound **1k**, IC$_{50}$ > 10 µM), preferably a cycloalkyl group (**1a**, **1g**, **1i**, or **1j**) and specifically cyclobutyl (**1a**). Additionally, phenyl (**1h**) was tolerated, but the larger naphthyl group (**1l**) was not.

To further expand the SAR around initial hit **1a**, the role of the benzyloxy moiety was assessed. In short, the unsubstituted benzyloxy moiety was preferred, as substitution of the terminal phenyl ring with cycloalkyl or heteroaryl (Table 4.2) or substitution on the terminal phenyl ring (Table 4.3) resulted in slight or extreme losses in biochemical and/or cellular IGF-1R potency.[33]

With the establishment of this initial SAR, the next phase of SAR development explored substitution on the cycloalkyl moieties at position C3 of the

Table 4.1 IGF-IR biochemical potencies for compounds **1a–l**

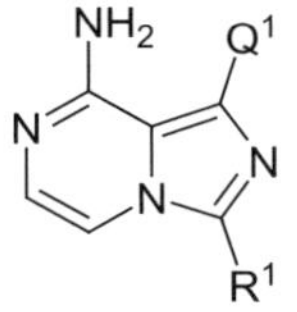

1

Compound	Q^1	R^1	IC_{50} (μM)
1a	3-BnOC$_6$H$_4$	cyclobutyl	0.606
1b	Ph	cyclobutyl	>10.0
1c	3-MeOC$_6$H$_4$	cyclobutyl	>10.0
1d	4-MeOC$_6$H$_4$	cyclobutyl	>10.0
1e	4-BnOC$_6$H$_4$	cyclobutyl	1.97
1f	4-PhOC$_6$H$_4$	cyclobutyl	>10.0
1g	3-BnOC$_6$H$_4$	cyclopentyl	1.05
1h	3-BnOC$_6$H$_4$	phenyl	1.68
1i	3-BnOC$_6$H$_4$	cyclohexyl	3.51
1j	3-BnOC$_6$H$_4$	cycloheptyl	3.79
1k	3-BnOC$_6$H$_4$	H	>10.0
1l	3-BnOC$_6$H$_4$	1-naphthyl	>10.0

Table 4.2 IGF-IR biochemical potencies for compounds **13a–d**

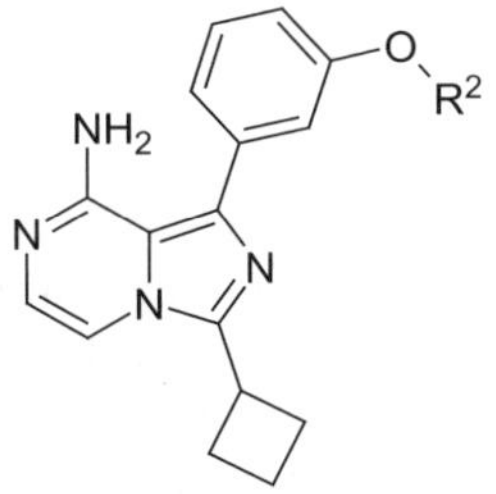

13

Compound	R^2	IC_{50} (μM)
13a	CH$_2$-cyclopropyl	2.27
13b	CH$_2$-cyclohexyl	1.11
13c	CH$_2$-2-pyridyl	1.09
13d	CH$_2$-3-pyridyl	>10.0

imidazopyrazine ring while maintaining the preferred 3-BnOC$_6$H$_4$ moiety at C1. A range of substituted C4-cyclohexyl and C3-cyclobutyl analogs were synthesized, specifically, a series of amides and aminomethyl derivatives (Table 4.4). The *trans*-(aminomethyl)cyclohexyl derivative **15c** displayed biochemical and

Table 4.3   IGF-IR biochemical and cell potencies for compounds **1a** and **14a–m**

14

Cpd	R^3	IC$_{50}$ (μM)	
		Biochemical	*Cell*
1a	H	0.606	1.16
14a	2-F	0.224	2.06
14b	3-F	0.510	3.29
14c	4-F	1.23	–
14d	2-Cl	0.343	–
14e	3-Cl	2.12	–
14f	4-Cl	0.980	> 10.0
14g	2-OCF$_2$H	3.28	–
14h	3-OCF$_2$H	5.78	–
14i	4-OCF$_2$H	2.82	–
14j	2-CN	> 10.0	–
14k	4-CN	> 10.0	–
14l	3-CHCONH$_2$	> 10.0	–
14m	3-NHCOCH$_3$	> 10.0	–

cellular IGF-1R kinase IC$_{50}$ values of 119 and 534 nM, respectively. In the cyclobutyl series, the dimethylamino derivative **16b** displayed the best balance of biochemical and cellular potencies, with respective IC$_{50}$ values of 166 and 191 nM. Select lead compounds were profiled in mice to gain an early understanding of the PK properties associated with this series (Table 4.5). Overall, the compounds had favorable oral bioavailabilities but displayed clearances exceeding mouse liver blood flow (>90 mL min^{-1} kg^{-1}). Metabolic profiling of compound **15a** revealed two metabolites: hydrolysis of the parent primary amide to the carboxylic acid and the acyl-glucuronidate metabolite of the acid. Interestingly, the slightly bulkier methylamide **15b** displayed a significantly lower rate of clearance than the primary amide **15a**, which correlated with a slower rate of metabolic hydrolysis to the acid. Moreover, removal of the amide altogether to afford aminomethyl derivative **15c** resulted in the lowest rate of clearance amongst these cyclohexyl analogs.

In order to gain insight to the selectivity profile associated with this class of compounds, a representative compound, **15c**, was screened for inhibitory activity against 15 additional purified protein kinases from the tyrosine and serine/threonine kinase families. Less than 50% inhibition at 10 μM of the

Table 4.4 IGF-IR biochemical and cell potencies for compounds **15** and **16**

15   **16**

Cpd	R^4/R^5	IC_{50} (μM)	
		Biochemical	*Cell*
15a	$CONH_2$	0.221	0.898
15b	CONHMe	0.105	1.72
15c	CH_2NH_2	0.119	0.534
15d	CH_2NEt_2	0.115	0.440
15e	CH_2-azetidinyl	0.081	0.398
15f	CH_2-pyrrolidinyl	0.103	0.547
15g	CH_2-morpholino	0.091	0.651
16a	CH_2NH_2	0.060	0.690
16b	CH_2NMe_2	0.166	0.191
16c	$CONH_2$	0.554	2.08

Table 4.5 Mouse iv and oral pharmacokinetics for compounds **1a**, **15a**, **15b**, and **15c**.

Compound	*1a*	*15a*	*15b*	*15c*
iv				
Dose $(mg\,kg^{-1})$	1.00	1.00	5.00	1.00
$t_{1/2}$ (h)	1.30	1.30	3.15	2.29
V_{ss} $(L\,kg^{-1})$	11.50	4.02	4.57	9.58
Cl $(mL\,min^{-1}\,kg^{-1})$	221	215	116	93
Oral				
Dose $(mg\,kg^{-1})$	50.0	50.0	50.0	50.0
C_{max} (μM)	0.86	2.22	4.15	2.85
$AUC_{0-\infty}$ $(ng\,h\,mL^{-1})$	1756	1323	3666	3926
Oral bioavailability $(\%F)$	47	34	51	47

compound was noted for the following enzymes: Abl, Cdk2/CyclinA, Cdk2/CyclinE, Chk2, CK2, c-Raf, Fes, IKK-β, MAPK2, p70S6K, PDGFR-β, PDK1, PKA, PKBα and PKCα. Inhibition of IR was similar to that observed for IGF-IR, reflecting the high homology of the IGF-1R and IR ATP binding domains.

4.6 Series II: Quinolinyl-derived Imidazopyrazine IGF-1R Inhibitors

The aforementioned medicinal chemistry analoging efforts provided insights into a novel series of 1,3-disubstituted 8-aminoimidazo[1,5-α]pyrazine IGF-IR inhibitors. While the $BnOC_6H_4$ derivatives such as (aminomethyl)cyclohexyl and -cyclobutyl analogs **15** and **16**, respectively, provided early leads, a significant increase in potency and an improvement in drug metabolism and pharmacokinetic (DMPK) properties were required to afford molecules that would justify clinical development. A key breakthrough in the program came with an X-ray co-crystal structure of IR and cyclohexyl analog **15a** (Figure 4.3, panel A). From the co-crystal structure, several key binding determinants were identified (Figure 4.3, panel B): (1) the 8-amino group and the N7 nitrogen are making critical hydrogen-bonding interactions with the hinge backbone residues E1080 and M1082, respectively; (2) the methylene and ether oxygen are almost completely coplanar with the proximal phenyl ring; (3) the oxygen from the benzyl ether moiety accepts an H-bond from K1033; (4) a modest hydrophobic cavity exists below the proximal phenyl ring and benzylic carbon; (5) the terminal phenyl ring is buried deep in a hydrophobic pocket, making critical van der Waals contacts with the DFG motif and F1037 and A1051 from the α-helix C; (6) the cyclohexyl moiety makes van der Waals interactions with the P-loop near V1013 and L1005 above the ribose binding pocket; and (7) the cyclohexyl amide extends into the ribose binding pocket. Based on the identification of these key binding determinants, structure-based design (SBD) efforts ensued. Efforts focused on locking the benzyloxyphenyl moiety into the bioactive conformation via exploiting the near co-planarity between the –O–CH$_2$– and proximal phenyl ring while simultaneously filling the unoccupied hydrophobic pocket below the proximal phenyl and ether moiety. This was envisioned to be accomplished through replacing the benzyloxyphenyl moiety with 2-phenylquinolin-7-yl (Figure 4.3, panels C and D), while maintaining all other key pharmacophoric elements associated with binding (hinge-binding motif, substituents occupying the ribose binding pocket, and H-bond acceptor to K1033 in the form of the quinoline nitrogen). To our delight, the introduction of a constrained quinolinyl moiety to provide compound **17** led to a 20-fold increase in IGF-1R potency when compared directly to its $BnOC_6H_4$ counterpart **15a**. Additionally, an overall improvement in oral PK properties (%F, clearance, C_{max}, and AUC) was observed (Table 4.6). We also learned that salt form and formulation could play a significant role in influencing oral bioavailability and exposure. For example, the bis-HCl salt of compound **17**,

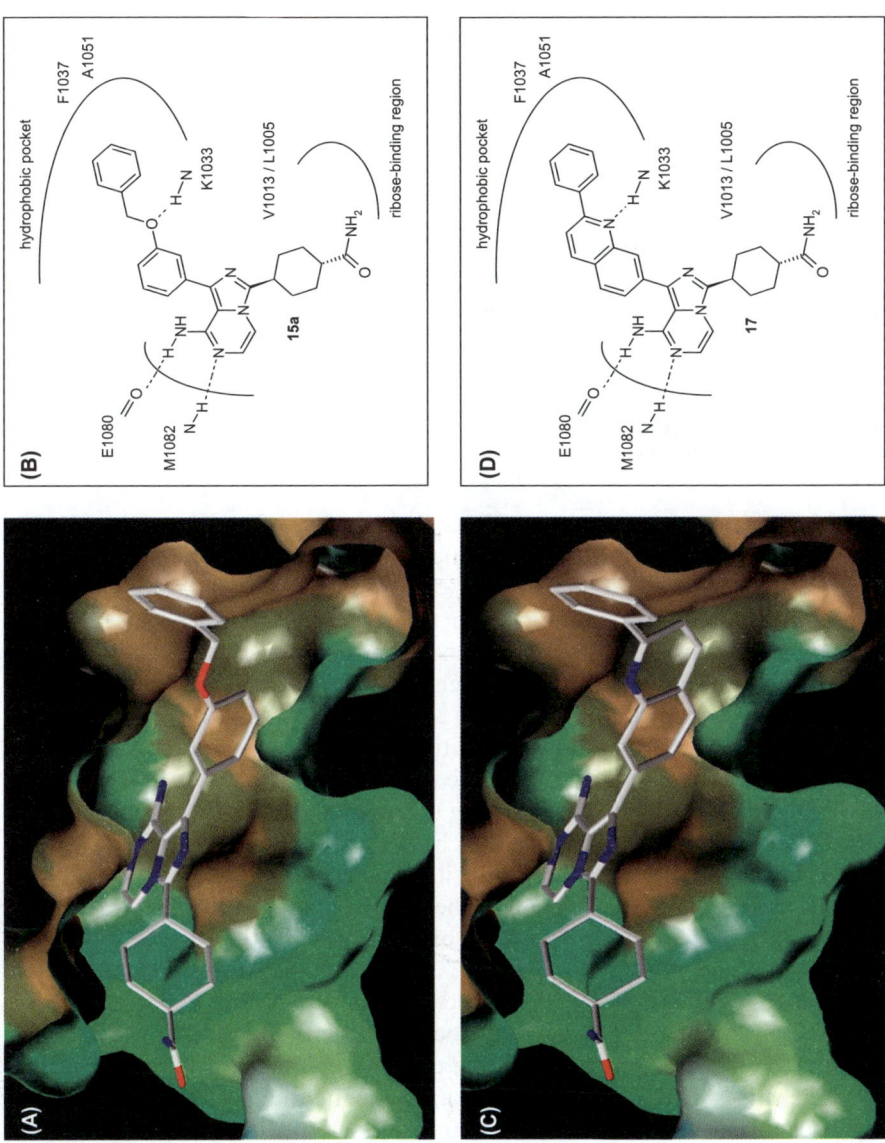

Figure 4.3 (A) IR/compound **15a** co-crystal structure. (B) Key binding determinants of compound **15a**. (C) Direct quinoline analog of compound **15a** docked into IR co-crystal structure. (D) Key binding determinants of quinoline analog **17**.

Table 4.6 Pharmacokinetic comparison of benzyoxyphenyl derivative **15a** to 2-phenylquinolinyl derivative **17**.[a]

Compound		**15a**			**17**		
iv	Dose (mg kg^{-1})	1			5		
	Vehicle[b]	A			C		
	C_{max} (µM)	0.52			8.67		
	Terminal $t_{1/2}$ (h)	0.63			0.80		
	V_{ss} (L kg^{-1})	4.02			1.38		
	Cl (mL min^{-1} kg^{-1})	215			41.0		
Oral	Dose (mg kg^{-1})	5	50	5	50	50	50[c]
	Vehicle[b]	B	B	B	B	C	D
	C_{max} (µM)	0.13	2.22	0.55	3.93	9.66	11.35
	Terminal $t_{1/2}$ (h)	1.56	2.56	1.79	N/C	2.02	2.46
	AUC$_{0-last}$ (ng h mL^{-1})	76	1211	454	4954	10318	12554
	%F	20	34	22	24	51	77

[a]Cl = clearance, V_{ss} = volume of distribution at steady state, $t_{\frac{1}{2}}$ = elimination half-life, C_{max} = maximum measured plasma concentration, C_{24h} = plasma concentration at 24 h, AUC$_{0-last}$ = area under the concentration–time curve extrapolated to the last measured sampling time, N/C = not calculated, %F = oral bioavailability.
[b]Vehicle: (A) PEG 400 50% v/v in water, (B) 50:50 v/v PEG 400:citric acid 5% w/v, (C) pH 2 saline at a dosing volume of 4 mL kg^{-1}, (D) water.
[c]Bis-HCl salt of parent compound.

formulated in water (or compound **17** as the free base formulated in pH 2 saline) provided better overall PK properties than formulating the free base in 50% v/v PEG 400:citric acid 5% w/v.

With significant improvements in both target potency and PK properties, efforts were then focused exclusively on the quinolinyl series.[35–39] The overall SAR was similar to that observed in the benzyloxy series and was supported by an IGF-1R co-crystal structure with the advanced lead, PQIP (Figure 4.5), which confirmed the key binding interactions as noted with the earlier IR/benzyloxyphenyl-derived imidazopyrazine co-crystal structure.[40] A few key SAR/SBD highlights include the following (Figure 4.4): (1) Substitution on the 8-amino moiety was disfavored since such substitution would interfere with critical hinge interactions. (2) The C2 position of the quinoline ring showed a strong preference for an unsubstituted phenyl ring over smaller alkyl groups (*i.e.* methyl or ethyl), heteroaryl groups, or hydrogen. In the case of replacement of the terminal phenyl ring with H, a complete loss in IGF-1R activity (IC$_{50}$ > 10 µM) was observed. This complete loss in activity can be rationalized based on the binding mode, as the space occupied by the terminal phenyl ring cannot be fully occupied by hydrophobic collapse of the nearby protein residues. (3) The substituent at C3 of the imidazopyrazine ring occupied the ribose binding pocket and influenced both potency and PK properties. (4) The quinolinyl moiety was critical, as replacement with naphthyl resulted in a complete loss in IGF-1R activity (IC$_{50}$ > 10 µM). The significant loss in potency was attributed to the loss of the H-bond acceptor to K1033.

With the SAR from the quinolinyl series corresponding to that observed in the earlier BnOC$_6$H$_4$ series, efforts shifted towards further expanding the SAR

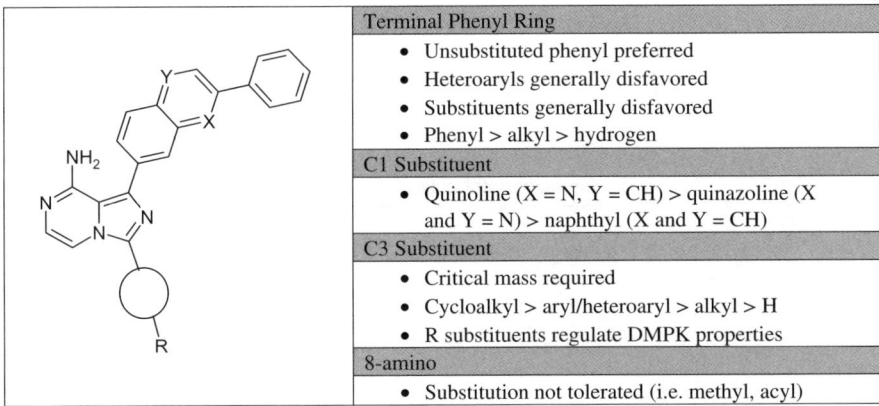

Terminal Phenyl Ring		
• Unsubstituted phenyl preferred		
• Heteroaryls generally disfavored		
• Substituents generally disfavored		
• Phenyl > alkyl > hydrogen		
C1 Substituent		
• Quinoline (X = N, Y = CH) > quinazoline (X and Y = N) > naphthyl (X and Y = CH)		
C3 Substituent		
• Critical mass required		
• Cycloalkyl > aryl/heteroaryl > alkyl > H		
• R substituents regulate DMPK properties		
8-amino		
• Substitution not tolerated (i.e. methyl, acyl)		

Figure 4.4   SAR summary of quinolinyl-derived imidazopyrazine IGF-1R inhibitors.

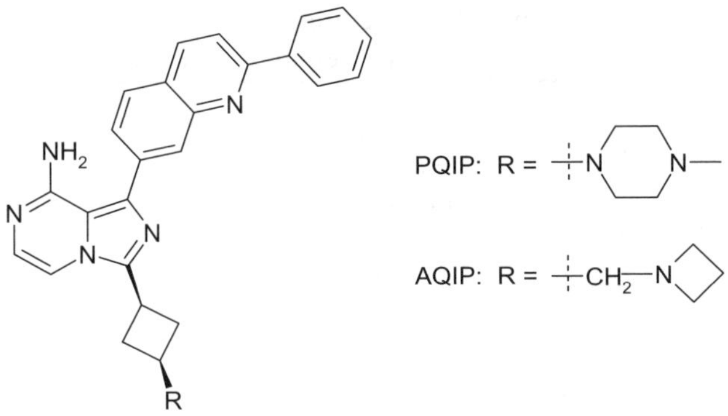

PQIP: R =

AQIP: R =

Figure 4.5   PQIP and AQIP.

around the imidazopyrazine C3 substituents with the aim of improving the DMPK shortcomings of early leads, *i.e.* the high clearance associated with cyclohexylamide **15a**. As noted earlier, the substituents at C3 of the imidazo-pyrazine ring occupy the ribose binding pocket of IGF-1R and also regulate the overall DMPK properties of the molecule. While exploring the SAR in this region, we discovered that unsubstituted cycloalkyl derivatives, such as compound **18** (Scheme 4.3), were potent IGF-1R inhibitors but displayed poor metabolic stability in the presence of liver microsomes, with an extraction ratio[41] of 0.9 for both mouse and human (Table 4.7). PK analysis in mice revealed that this *in vitro* instability in liver microsomes was reflected by a clearance rate equal to the liver blood flow in mice ($Cl = 90\,\mathrm{mL\,min^{-1}\,kg^{-1}}$) (Table 4.7). Following the oral administration of $100\,\mathrm{mg\,kg^{-1}}$ of compound **18** in mice, both the parent compound and a major $M + 16$ metabolite were detected in the plasma samples (Table 4.8). *In vivo* metabolite identification studies determined that hydroxylation at the C3 position of the cyclobutyl ring

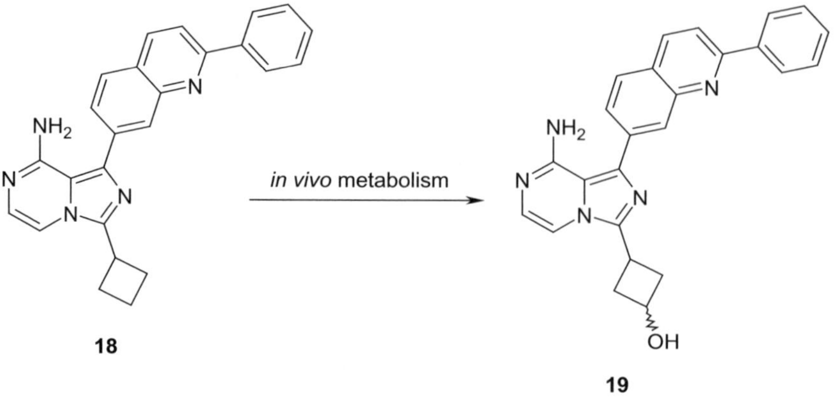

in vivo metabolism

18

19

Scheme 4.3 Metabolic transformation of compound **18** to alcohol **19**.

Table 4.7 IGF-1R biochemical and cellular potency, metabolic stability, and pharmacokinetic properties (female CD-1 mice) for compound **18**.[a]

IGF-1R IC$_{50}$ and microsomal stability		
IGF-1R IC$_{50}$ (μM)	Enzyme	0.064
	Cell (3T3/huIGF-1R)	0.059
Microsomal stability *in vitro*	Human	0.900
extraction ratio	Mouse	0.900

Pharmacokinetic parameters			
Route	*iv*	*po*	
Nominal dose (mg kg^{-1})	5.00	5.00	50.00
C_{max} (ng mL^{-1})	1660	122	1050
C_{max} (μM)	4.24	0.31	2.68
AUC$_{0-\infty}$ (ng h mL^{-1})	929	371	5170
Terminal $t_{1/2}$ (h)	0.98	2.05	1.65
Cl (mL min^{-1} kg^{-1})	90	N/A	N/A
V_{ss} (L kg^{-1})	6.25	N/A	N/A
Bioavailability (%F)	100	40	56

[a]Formulation: PEG 400:0.2% citric acid (50:50). Cl = clearance, V_{ss} = volume of distribution at steady state, $t_{1/2}$ = elimination half-life, C_{max} = maximum measured plasma concentration, AUC$_{0-\infty}$ = area under the concentration–time curve extrapolated to infinity, N/A = not available, %F = oral bioavailability.

afforded a mixture of the *cis*- and *trans*-alcohols **19**[38] (Scheme 4.3), as well as a minor transient keto oxidation $M + 14$ product, 3-[8-amino-1-(2-phenylquinolin-7-yl)imidazo[1,5-a]pyrazin-3-yl]cyclobutanone, an intermediate in the epimerization to alcohols **19**.

In order to block this metabolic hotspot in the unsubstituted cyclobutyl analog **18**, efforts focused on exploring additional imidazopyrazine C3 substituents while incorporating polar groups to help improve overall solubility.

Table 4.8 Plasma concentration–time course for parent compound **18** and $M + 16$ metabolite following a single oral dose of $100 \, mg \, kg^{-1}$ of compound **18** in the female *nu/nu*-CD-1 mouse.

Compound	Time (h)	Plasma concentration (µM)
18	1	8.19
	4	4.62
	8	7.95
	16	2.64
$M + 16$ metabolite	1	11.3
	4	8.99
	8	11.4
	16	5.80

Table 4.9 *In vitro* profile of PQIP and AQIP including inhibition of pIGF-1R (enzymatic and mechanistic), CYP3A4 IC_{50}, and microsomal stability.

IGF-1R inhibitor		PQIP	AQIP
IGF-1R IC_{50} (µM)	Enzyme	0.024	0.035
	Cell (3T3/huIGF-1R)	0.019	0.020
CYP3A4 IC_{50} (µM)		8.30	>20.0
Microsomal stability (ER)	Human	0.59	0.28
	Mouse	0.53	0.52
	Rat	0.21	0.70
	Dog	0.50	0.42

Along these lines, a series of C3-substituted cyclobutyl analogs were synthesized, from which advanced leads PQIP and AQIP emerged (Figure 4.5).[35–37]

Both PQIP and AQIP displayed improved potency and microsomal metabolic stability (as noted by lower ERs) (Table 4.9), and the latter translated into lower *in vivo* clearance and good oral bioavailability in both mice and rats (and dogs, data not shown) (Table 4.10). Despite these positive drug-like attributes associated with both PQIP and AQIP, in mouse and rat models, both compounds displayed a low C_{max} and high V_{ss}. The high V_{ss} associated with these basic analogs results in the need for relatively high doses to achieve efficacious plasma levels, and these high doses combined with relatively long half-lives increase the risk of accumulation and toxicity upon subacute or chronic dosing. Upon further structural examination of the PQIP and AQIP series, we noted a trend where the high V_{ss}, long $t_{\frac{1}{2}}$, and low clearance was associated with the basic moiety at C3 of the cyclobutyl ring. As a result, we focused efforts on further exploring the effect of non-basic substituents at C3 of the cyclobutyl ring, including a series of C3 cyclobutyl alcohols. Those studies led to the evaluation of the *cis*- and *trans*-secondary alcohols as well as a series of tertiary cyclobutyl alcohols,[38] from which the tertiary alcohol OSI-906 (Figure 4.6) emerged as the best in the series based upon multiple factors, including potency, efficacy, physical properties, and overall ADME/DMPK properties.

Table 4.10   Mouse and rat iv and oral pharmacokinetics of PQIP and AQIP.[a]

Compound Route of administration	Species	PQIP		AQIP	
		Mouse	Rat	Mouse	Rat
iv	Dose $(mg\,kg^{-1})$	10	5	5	1
	$T_{1/2}$ (h)	5	20	5.1	19.9
	V_{ss} $(L\,kg^{-1})$	8	21	10.0	14.0
	Cl $(mL\,min^{-1}\,kg^{-1})$	19	16	24	10
Oral	Dose $(mg\,kg^{-1})$	25	20	25	50
	C_{max} (μM)	2.2	1.35	3.4	1.6
	C_{24h} (μM)	1.04	0.74	0.4	0.9
	$T_{1/2}$ (h)	23.6	28.7	N/A	N/A
	$AUC_{0-\infty}$ $(ng\,h\,mL^{-1})$	34546	28705	25582	37429
	%F	100	100	100	38

[a]Cl = clearance, V_{ss} = volume of distribution at steady state, $t_{1/2}$ = elimination half-life, C_{max} = maximum measured plasma concentration, C_{24h} = plasma concentration at 24 h, $AUC_{0-\infty}$ = area under the concentration–tme curve extrapolated to infinity, N/A = not available, %F = oral bioavailability.

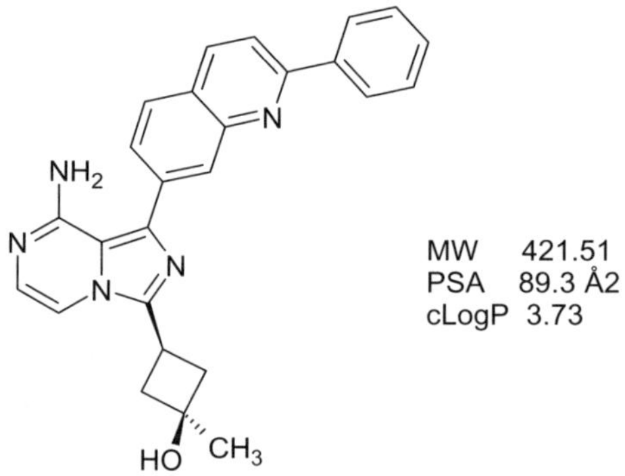

MW   421.51
PSA   89.3 Å2
cLogP   3.73

Figure 4.6   OSI-906.

4.7   Synthesis of OSI-906

The linear synthesis of OSI-906 (Scheme 4.4) began with the treatment of commercially available 7-methylquinoline (**20**) with phenyllithium, followed by refluxing with sulfur in methanol to afford 7-methyl-2-phenylquinoline (**21**). Oxidation of the 7-methyl group with selenium dioxide afforded 2-phenylquinoline-7-carbaldehyde (**22**) in 42% over three steps. Directed *ortho*-metallation of 2-chloropyrazine with a preformed solution of LiTMP followed by quenching with aldehyde **22** afforded alcohol **23** in 76% yield. Conversion of alcohol **23** to its respective chloride by treatment with thionyl chloride, followed by reaction with potassium phthalimide and then subsequent hydrazine-mediated phthalimide deprotection, afforded the desired amine **24** in 72% over three steps. Coupling of 3-methylenecyclobutanecarboxylic acid with amine **24**

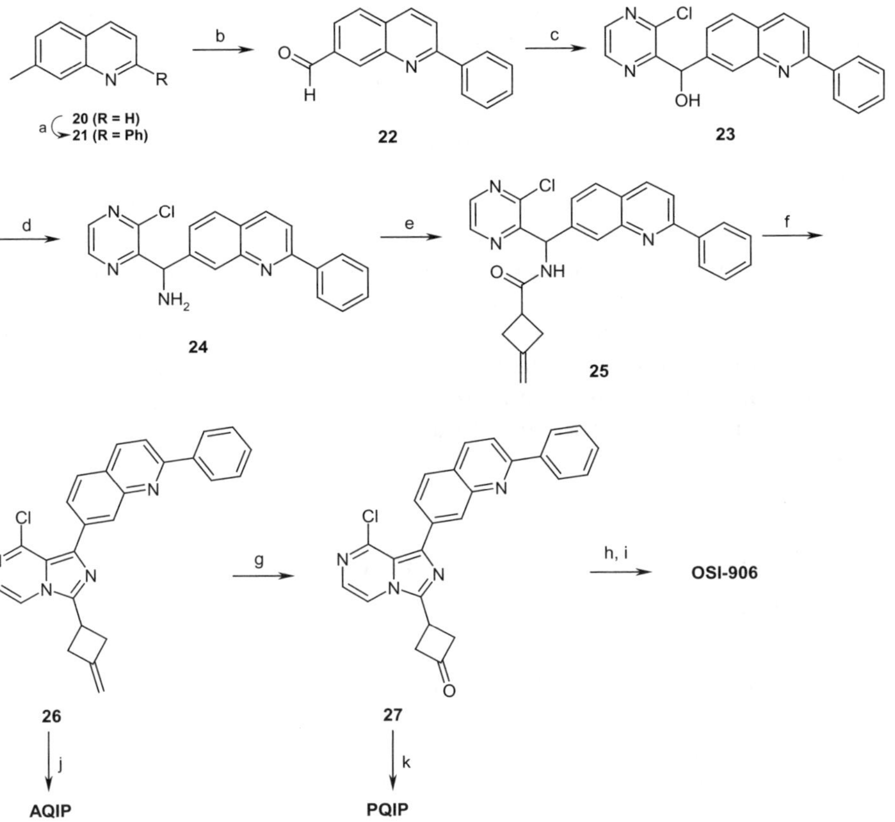

Scheme 4.4 Linear synthesis of imidazo[1,5-*a*]pyrazine-derived IGF-IR inhibitors: AQIP, PQIP, and OSI-906.(a) PhLi, THF, 0 °C; then sulfur, MeOH, 70 °C; (b) SeO$_2$, 160 °C (42% three-step yield); (c) *n*BuLi, tetramethylpiperidine, 2-chloropyrazine, THF, − 78 °C; (d) SOCl$_2$, CHCl$_3$, rt; then potassium phthalimide; then NH$_2$NH$_2$, EtOH, rt; (e) 3-methylenecyclobutanecarboxylic acid, EDC, HOBT, DIEA, CH$_2$Cl$_2$; (f) POCl$_3$, DMF, MeCN, 55 °C; (g) NMO, cat. K$_2$OsO$_4$, THF/H$_2$O, rt; then NaIO$_4$, THF/H$_2$O (3:1), 0 °C → rt (88% two-step yield); (h) 3 M MeMgCl in THF, N$_2$, − 78 °C → rt, 1 h; (i) NH$_3$, *i*PrOH, stainless steel Parr apparatus, 110 °C, 15 h (61% two-step yield); (j) 9-BBN, THF, 0 °C → rt (62%); then NH$_3$, *i*PrOH, 110 °C, 24 h (51%); then Ts$_2$O, pyridine, CH$_2$Cl$_2$, − 40 °C → rt (53%); then azetidine, THF, 50 °C; (k) Na(OAc)$_3$BH, THF, *N*-methylpiperizine, DCE, rt; then NH$_3$, *i*PrOH, 110 °C, 24 h.

in the presence of EDC/HOBT afforded amide **25** in 88% yield, which was subsequently cyclized to imidazopyrazine **26** in 93% yield with POCl$_3$ and DMF in acetonitrile. Imidazopyrazine **26** proved to be a versatile intermediate and was used to synthesize several analogs, including AQIP, PQIP, and OSI-906.[39,35–37] For the synthesis of AQIP, hydroboration of 8-chloroimidazopyrazine **26** with 9-BBN, followed by oxidative cleavage, afforded a 5:1 ratio of *cis/trans*-(hydroxymethyl)cyclobutyl isomers. This *cis/trans* mixture was treated with ammonia/isopropanol in a Parr reactor at 110 °C to displace the 8-chloro group, followed by tosylation of the primary alcohol to afford the *cis/trans* mixture of tosylates. At this stage, the isomers were readily separable by silica gel chromatography and treatment of the *cis*-tosylate with azetidine afforded AQIP. For the synthesis of PQIP, compound **26** was converted to cyclobuta-none derivative **27** *via* osmium tetraoxide-mediated dihydroxylation of the exocyclic alkene moiety followed by sodium periodate oxidative cleavage of the resulting diol. Cyclobutanone **27** was subjected to reductive amination with *N*-methylpiperazine followed by NH_3 displacement of the 8-chloro group to afford PQIP. For the synthesis of OSI-906, reaction of cyclobutanone inter-mediate **27** with 3 M MeMgCl in THF stereoselectively afforded the *cis*-tertiary alcohol, which was subsequently treated with ammonia/isopropanol in a stainless steel Parr apparatus to afford OSI-906 in a yield of 61% over two steps.

4.8 OSI-906: Biochemical and Cellular IGF-1R Activity

The mature *in vitro* through *in vivo* drug discovery cascade (Figure 4.7) focused primarily on two cell lines, a 3T3/huIGF-1R cell line (LISN) and a GEO col-orectal line.[35–37,39,42,43] The LISN line represents an engineered IGF-1R over-expressing driven mechanism in which signaling pathways and functional effects of IGF-1R inhibition could be cleanly assessed. The GEO line harbors an IGF-1R/IGF-2 autocrine loop and therefore represents a naturally occur-ring IGF-1R ligand-driven tumor line. Moreover, both lines form tumors

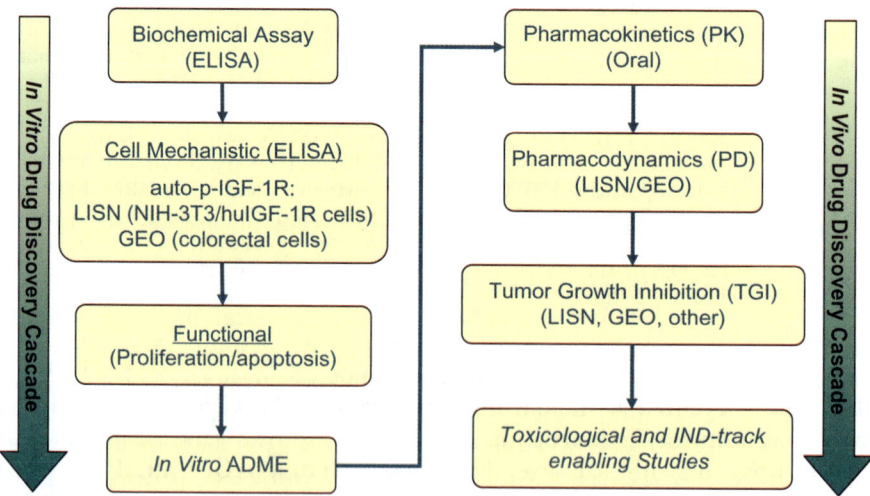

Figure 4.7   General *in vitro* and *in vivo drug* discovery cascade.

LISN Cells (0.5% FCS starved)

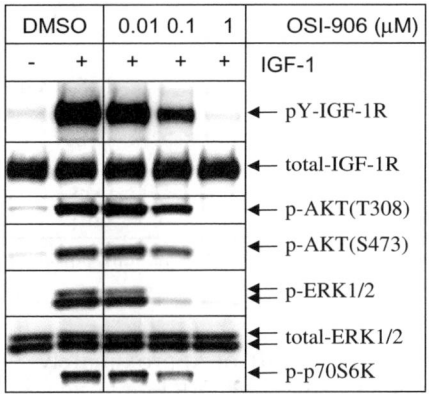

GEO Cells (10% FCS)

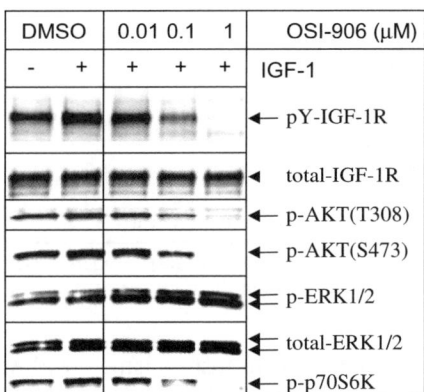

Figure 4.8   Inhibition of IGF-1R and cellular signaling pathways by OSI-906 in LISN and GEO cells.

in vivo and were therefore used as proof-of-concept models for tumor growth inhibition (TGI) studies where the degree and duration of *in vivo* inhibition of tumor IGF-1R phosphorylation could be correlated to *in vivo* efficacy. This allowed for a critical *in vivo* correlation to be established between drug exposure (PK), inhibition of pIGF-1R (PD), and overall tumor growth inhibition.

OSI-906 is a potent inhibitor of the tyrosine kinase activity of IGF-1R. In biochemical assays of enzyme activity using poly(Glu:4Tyr) as the substrate, recombinant human IGF-1R kinase domain was inhibited by OSI-906 with an IC_{50} value of 35 nM. In both LISN and GEO cell lines, OSI-906 fully blocked IGF-1-stimulated receptor phosphorylation (Figure 4.8). To note, the GEO line, because of its IGF2 autocrine loop, displays high levels of p-IGF-1R prior to stimulation with exogenous IGF1. In contrast, the LISN line displays high levels of total IGF-1R which requires ligand stimulation for IGF-1R activation. Immunoprecipitation and immunoblotting analysis of cell lysates also demonstrated that IGF-1-stimulated phospho-AKT, phospho-ERK, and phospho-p70S6K signaling pathways are inhibited as a result of IGF-1R inhibition by OSI-906 in the LISN cells. In the GEO cells, OSI-906 effectively inhibited AKT activation in the presence of 10% FCS, suggesting that survival signaling is mediated predominantly by IGF-1R in this background. IC_{50} values for the cellular inhibition in each respective cell line of IGF-1R autophosphorylation and activation of the downstream signaling proteins AKT, ERK1/2, and S6 kinase are shown in Table 4.11. We assessed the influence of plasma protein binding for varied species on the biological activity of OSI-906 in order to allow for the prediction of efficacious *in vivo* plasma levels across species. Whole plasma (90%) from mouse, rat, dog, monkey, or human was included in the IGF-1R cellular autophosphorylation assay. Table 4.11 summarizes the mouse and human plasma protein binding effects on OSI-906 in this mechanistic assay. The greatest protein binding potency shift was seen in

Table 4.11 *In vitro* profile of OSI-906 in LISN and GEO cells.

Assay	IC_{50} (μM)	Cell line
IGF-1R biochemical	0.035	–
IGF-1R	0.024	LISN/GEO
pERK1/2	0.028	LISN
p-70S6K	0.060	LISN
pAKT (Ser473)	0.130/0.110	LISN/GEO
IGF-1R + 90% mouse plasma	1.275^a	GEO
IGF-1R + 90% human plasma	0.213^a	GEO

^{a}Percent bound drug by ultracentrifugation was 98.4% (mouse) and 96.7% (human), correlating with the IC_{50} shifts in the presence of the respective plasmas.

the presence of mouse plasma, leading to an IC_{50} shift of approximately 79-fold, whereas the least protein-binding effect was observed with human plasma, approximately 4.7-fold. These differential plasma protein-binding shifts of OSI-906 were consistent with direct measurements of the binding of OSI-906 to plasma proteins from various species, as measured by % bound drug by ultracentrafugation, and suggest that the activity of the compound to inhibit IGF-1R *in vivo* will be greater in humans than in mice.

4.9   OSI-906: Selectivity Profile

As evidenced in Figure 4.8 and Table 4.11, OSI-906 is a potent inhibitor of IGF-1R in both biochemical and mechanistic assays. To determine the overall kinase selectivity profile of OSI-906, the compound was assayed *in vitro* for inhibitory activity against a panel of kinases.[39] OSI-906 at a concentration of 1 μM did not show greater than 50% inhibition against 88 additional purified protein kinases representing the tyrosine and serine/threonine kinase families (Figure 4.9, Table 4.12). A follow-up study was conducted to determine IC_{50} values against a key subset of kinases, where OSI-906 displayed IC_{50} values of > 10 μM (Table 4.13). The exquisite selectivity of OSI-906 might be rationalized from the isosteric PQIP structure.[40] OSI-906 targets an intermediate conformation of the protein that features an inactive orientation of the C-helix (0P form) but an orientation of the activation loop more associated with the phospho-protein (3P form) (Figure 4.10). OSI-906 enables this intermediate conformation primarily *via* its interactions with the C-helix. Since induction of this intermediate conformation appears quite rare in kinases and residues are generally less conserved in the C-helix region across different kinases, specificity ensues. Naturally, for the > 80% homologous kinase domains of insulin receptor (IR) and insulin receptor-related kinase (IRR), similar affinities as for the IGF-1R kinase domain are observed.

Recent reports suggest that signaling through the IGF-1R axis, specifically through IGF-1R/IR hybrid receptors and/or IR, may also produce tumor promoting effects.[44] IR-A has been implicated in malignant transformation[45–47] and shown in select instances to be driven by the IGF-2 ligand in an autocrine

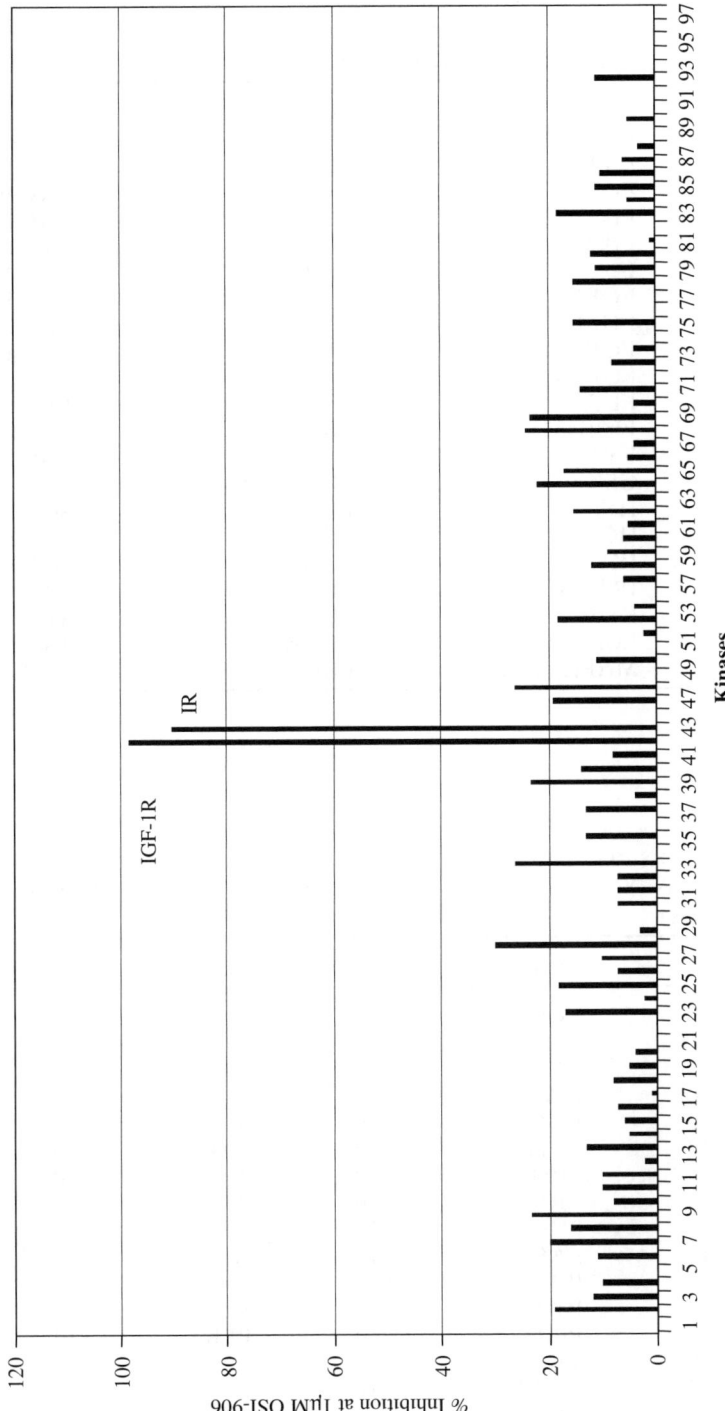

Figure 4.9 Selectivity profile of OSI-906 at a concentration of 1 μM when profiled using the ProfilerPro™ Kinase Selectivity Assay Kit (Caliper Life Sciences) against protein kinases representing the tyrosine and serine/threonine kinase families.

Table 4.12 Kinases which were not inhibited by greater than 50% by OSI-906 at a concentration of 1 μM when profiled using the ProfilerPro™ Kinase Selectivity Assay Kit (Caliper Life Sciences).

ABL	c-TAK1	MAPKAPK3	PKC-ε
AKT1	DCAMKL2	MARK1	PKC-η
AKT2	DYRK1a	MARK2	PKC-γ
AKT3	EGFR	MET	PKC-θ
AMPK	Erk1	MSK1	PKC-z
Arg	Erk2	MSK2	PKD1
AurA	Fer	MST2	PKD2
AurC	FGFR1	NEK2	PKD3
BMX	FGFR2	p38a	PKG1-β
BRSK1	FGFR3	p38-β2	PKGa
BRSK2	FGFR4	p38-δ	PRAK
BTK	FLT3	p38-γ	ROCK2
CAMK2	Flt3(D835Y)	p70S6K	RSK1
CAMK4	FYN	PAK2	RSK2
CaMKII-β	GSK3-β	PDGFR-α	RSK3
CaMKII-γ	Hck	PIM1	SGK1
CDK2	HGK	PIM2	SGK2
CHK1	IKK-β	PKA	SGK3
CHK2	IRAK4	PKC-α	SRC
CK1d	KDR	PKCb2	SYK
c-Raf	LCK	PKC-β1	TSSK1
CSNK1A1	MAPKAPK2	PKC-δ	TSSK2

Table 4.13 Selectivity profile of OSI-906 against a panel of kinases (IC_{50} values).

Kinase	IC_{50} (μM)	Kinase	IC_{50} (μM)	Kinase	IC_{50} (μM)
IGF-IR	0.035	IRAK4	>10	RON	>10
IR	0.075	JAK2	>10	ROS1	>10
IRR	0.075	KDR	>10	SRC	>10
Abl	>10	Lck	>10	TEK	>10
ALK	>10	LYN A	>10	MEK1	>10
BTK	>10	LYN B	>10	PAK3	>10
EGFR	>10	MET	>10	TAOK2	>10
EPHA1	>10	TRKA	>10	AKT1	>10
EPHB1	>10	PDGFR-α/-β	>10	AKT2	>10
FES	>10	FAK	>10	Aurora A	>10
FGFR-1/-2	>10	BRK	>10	Aurora B	>10
FYN	>10	RET	>10	PDK1	>10
PKA	>10	PIM1	>10	PLK1	>10
PKC-α	>10	CDK1	>10	P38α	>10
ROCK1	>10	DYRK3	>10	ERK1	>10
p70S6K	>10	GSK3α	>10	CAMK1D	>10

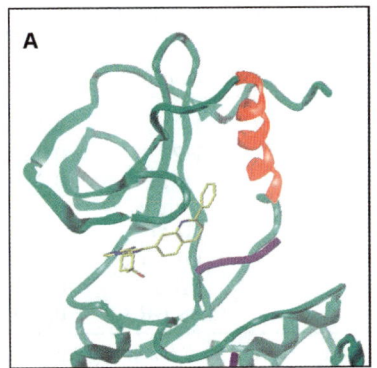 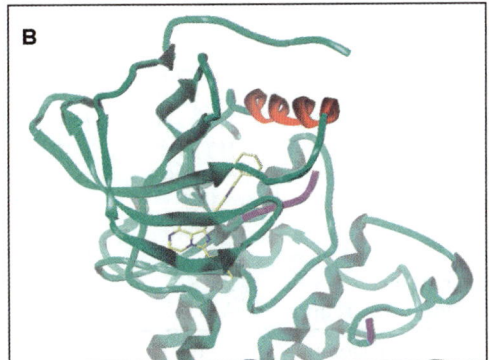

Figure 4.10 OSI-906 docked in IGF-1R. View into the ATP-binding pocket: (A) from the side opposite the hinge motif and (B) from above the N-terminal lobe. The C{α} trace of IGF-1R is given in green, with the C-helix in red and the termini of the activation loop in purple. OSI-906 features yellow carbons. The C-helix is positioned in nearly identical orientation relative to the remainder of the N-terminal lobe as in 0P structures of IGF-1R or IR. The activation loop has a trajectory that goes out and away from the ATP binding site before returning to the C-terminal lobe, a trajectory distinct from those observed in 0P structures of IGF-1R and IR.

Table 4.14 Microsomal stability and CYP450 profiles of OSI-906.

Inhibition of human pIR in cells			
Cell line		IC_{50} (µM)	
Hepa-1		0.039	
Microsomal stablilty of OSI-906			
Species	Mouse	Rat	Dog	Human
Extraction ratio	0.54	0.41	0.49	0.55
Cytochrome P450 profile of OSI-906			
CYP isoform		IC_{50}	
3A4		>10	
2D6		>10	
1A2		>10	

fashion.[48–51] With emerging data supporting a role for the insulin receptor, alone and in combination with IGF-1R in cancer, we further profiled OSI-906 for activity against the insulin receptor in the cellular setting. The effects of OSI-906 against human pIR inhibition in cells was measured in a murine-derived hepatoma cell line (Hepa-1) and afforded an IC_{50} value of 39 nM (Table 4.14). This result, in conjunction with a high degree of selectivity against a wide panel of kinases, establishes OSI-906 as a highly selective dual inhibitor of both IGF-1R and IR. In addition to its high degree of selectivity *versus* a

Table 4.15   Antiproliferative effects of OSI-906 *versus* sensitive tumor cell lines.[a]

Tumor type	Cell line	EC_{50} $(\mu M)^b$	Tumor type	Cell line	EC_{50} $(\mu M)^b$
LISN	NIH-3T3	0.078	NSCLC	H358	0.117
Colorectal	HT29	0.072		H292	0.235
	Colo205	0.098		H322	0.319
	GEO	0.021		Calu1	0.830
	SW620	0.021		H441	0.152
	SW480	0.191		DU4475	0.086
Pancreatic	BxPC3	0.150	Breast	MCF7	0.177
Rhabdomyosarcoma	A673	0.153		SKBR3	0.635

[a]Proliferation monitored using the CellTiterGlo™ luciferase-based cellular ATP assay kit (Promega). Conditions: 0.5% FCS + IGF-1.
[b]EC_{50} values ranged $\pm 0.025\,\mu M$.

wide array of kinases, OSI-906 was screened for microsomal stability against mouse, rat, dog, monkey, and human liver microsomes, as well as for P450 drug–drug interactions (Table 4.14). Overall, the microsomal stability of OSI-906 is consistent and favorable across all species tested. The CYP450 profile for OSI-906 is also favorable, with no anticipated CYP related drug–drug interactions.

4.10   OSI-906: Antiproliferative Effects

In cell-based phenotypic assays, OSI-906 inhibited IGF-1- or IGF-2-mediated proliferation (and induced apoptosis, data not shown) in cell lines representing a range of tumor types.[39] A variety of tumor cell lines exhibit differential sensitivity to OSI-906, with EC_{50} values ranging from 0.021 to 0.830 µM (Table 4.15). These data suggest that the activated IGF-1R within these cells is required to maintain proliferation and survival and can be inhibited by treatment with OSI-906. To further validate that both IR and IGF-1R are co-inhibited in tumor cells sensitive to OSI-906, the effects of OSI-906 on the phosphorylation of IR and IGF-1R was determined by RTK capture array. For both HT-29 and Colo205 CRC cells, OSI-906, at a concentration of 1 µM, fully inhibits both IR and IGF-1R phosphorylation (Figure 4.11).

4.11   OSI-906: Pharmacokinetic Properties

The pharmacokinetic parameters obtained from non-compartmental modeling following iv administration of OSI-906 in the female CD-1 mouse, female Sprague Dawley rat, and the male beagle dog are summarized in Table 4.16.[39] In the mouse and the rat, the clearance is relatively low, with elimination half-lives of 2–3 h, and in the dog the clearance is 39 mL min^{-1} kg^{-1}, which is approximately equal to the liver blood flow, with a shorter half-life around 1 h. The volume of distribution at steady state (V_{ss}) is less than 5 L kg^{-1} in each species, significantly lower than that observed with PQIP and AQIP,

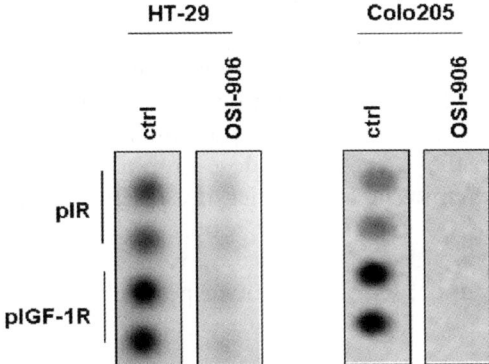

Figure 4.11 Inhibition of the phosphorylation of both IR and IGF-1R in colorectal tumor cells HT-29 and Colo205 by OSI-906 (signal realized by capture array).

Table 4.16 Median pharmacokinetic parameters for OSI-906 following iv and oral administration in different species.[a]

	iv		
Species	*Mouse (F)*	*Rat (F)*	*Dog (M)*
Dose (mg kg^{-1})	5	5	2.5
Elimination $t_{1/2}$ (h)	2.14	2.64	1.18
V_{ss} (L kg^{-1})	2.05	0.79	4.30
Cl (mL min^{-1} kg^{-1})	12	4	39
AUC$_{0-\infty}$ (ng h mL^{-1})	6954	23123	1066

	Oral						
Species	*Mouse (F)*				*Rat (F)*		*Dog (M)*
Dose (mg kg^{-1})	5	25	75	250	12.5	100	5
C_{max} (µM)	3.42	16.04[b]	16.80[c]	41.99	10.01	34.64	1.20
AUC$_{0-last}$ (ng h mL^{-1})	6770	26741	87455	317487	42832	424076	1328
%F	98	100	84	>100	74	92	64

[a]Cl = clearance, V_{ss} = volume of distribution at steady state, $t_{1/2}$ = elimination half-life, and AUC$_{0-\infty}$ = area under the concentration–time curve extrapolated to infinity. C_{max} = maximum measured plasma concentration, C_{24h} = plasma concentration at 24 h, and AUC$_{0-last}$ = area under the concentration–time curve extrapolated to the last measured sampling time. The dose vehicle was 25 mM tartaric acid, except for the 5 mg kg^{-1} dose in the mouse where saline pH 2 was used.
[b]C_{24h} = 0.02 µM.
[c]C_{24h} = 2.92 µM.

highlighting the difference in PK properties when comparing molecules with basic substituents *versus* a neutral alcohol in this series. The PK parameters following oral administration have been evaluated in the female CD-1 mouse, female Sprague Dawley rat, and the male beagle dog (Table 4.16) and were subsequently used for an allometric extrapolation to humans (data not shown); in each species, the oral bioavailability is high. In the mouse, the AUC$_{0-last}$ and

C_{max} are approximately linear between 5 and 250 mg kg^{-1}; however, the C_{max} is not dose proportional. This lack of C_{max} dose-proportionality is speculated to be due to dissolution rate-limited absorption at the higher doses. Consequently, at higher doses, the plasma concentration at 24 h is disproportionately higher. Similarly in the rat, the AUC_{0-last} is approximately linear between 12.5 and 100 mg kg^{-1}, while the C_{max} is not dose-proportional in this range. Upon subchronic dosing in mice in TGI studies, it is interesting to note that while the C_{max} values for the 25 and 75 mg kg^{-1} doses are very similar at 16.04 and 16.80 μM, respectively, the AUC and C_{24h} values at the 75 mg kg^{-1} dose are approximately 2.5- and 146-fold higher, respectively.

4.12   OSI-906: Antitumor Effects

OSI-906 was evaluated first for *in vivo* efficacy in TGI studies in the LISN xenograft model.[39] In this study, OSI-906 was dosed orally at 25 and 75 mg kg^{-1} once per day over 12 days. There was a dose-dependent effect on tumor growth inhibition, with the 75 mg kg^{-1} dose resulting in 100% TGI and 55% regression, while the 25 mg kg^{-1} dose afforded marginal TGI of 60% with no regression (Figure 4.12). Both doses showed statistically significant anti-tumor effects with $p < 0.001$ *vs.* control treated animals. There was $< 10\%$ body weight loss (BWL) with the animals treated with the 25 mg kg^{-1} dose and an average of 16% BWL with the 75 mg kg^{-1} group in this study. OSI-906 was profiled in other xenograft models, including the GEO colorectal and Colo-205 lines where it displayed significant TGI ($\geq 90\%$) when dosed orally at 60 mg kg^{-1} once-a-day for 14 days (data not shown).[42,43] In the GEO study, the mice were

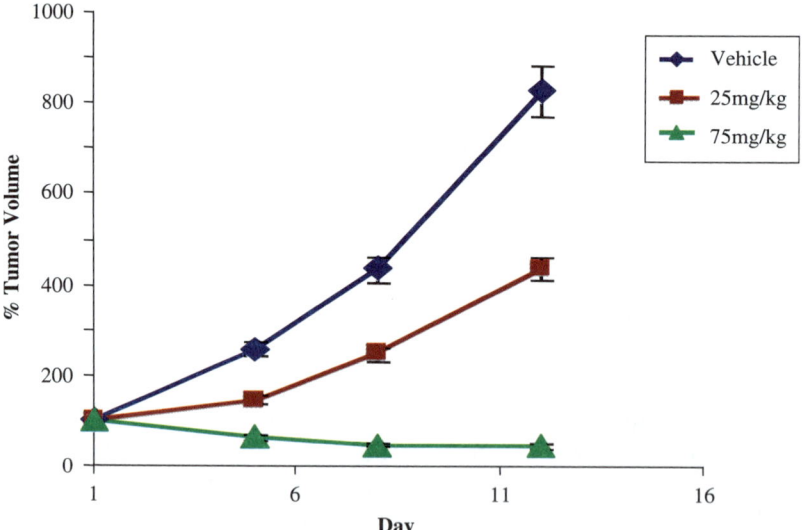

Figure 4.12   Antitumor activity of orally dosed OSI-906 in the LISN xenograft model.

evaluated for hyperglycemia (data not shown) and transient hyperglycemia was noted at the end of the 14-day dosing period for the first 8 h, with glucose levels returning to normal by 24 h.

4.13 OSI-906: Pharmacokinetic/Pharmacodynamic Correlation

To assess the drug exposure required for inhibition of tumor IGF-1R *in vivo*, a PD experiment was performed in LISN tumors in which the relationship between inhibition of IGF-1R phosphorylation (PD) and drug exposure (PK) was evaluated (Figure 4.13).[39] Dose escalating studies revealed that maximal inhibition of IGF-1R phosphorylation (80%) by OSI-906 at 75 mg kg^{-1} was achieved between 4–24 h, with plasma drug concentrations of 26.6 and 4.77 µM at the 4 and 24 h timepoints, respectively. The continued target inhibition at 24 h correlated with plasma drug exposures at 24 h being greater than the IGF-1R cell + mouse plasma (mP) IC$_{50}$ of 1.28 µM. The lower dose of 25 mg kg^{-1} provided initial target coverage from 4 to 16 h, but failed to provide adequate IGF-1R coverage out to 24 h (data not shown), paralleling the aforementioned PK profile where the C_{24h} for the 25 mg kg^{-1} dose is 64-fold below the cell + mP IC$_{50}$. Thus maximal inhibition of IGF-1R phosphorylation, maintained for at least a full 24 h with OSI-906 at 75 mg kg^{-1}, as shown in Figure 4.13, is consistent with the robust anti-tumor activity observed in the LISN xenograft model. These observations, together with the substantially greater TGI observed with 75 mg kg^{-1} relative to 25 mg kg^{-1}, suggest that maximal anti-tumor efficacy in this model requires maintenance of plasma levels sufficient to chronically suppress IGF-1R phosphorylation in tumors. A similar PK/PD correlation was observed in the GEO and Colo-205 xenograft studies. Such a correlation, in conjunction with the IGF-1R and IR cell + human plasma IC$_{50}$

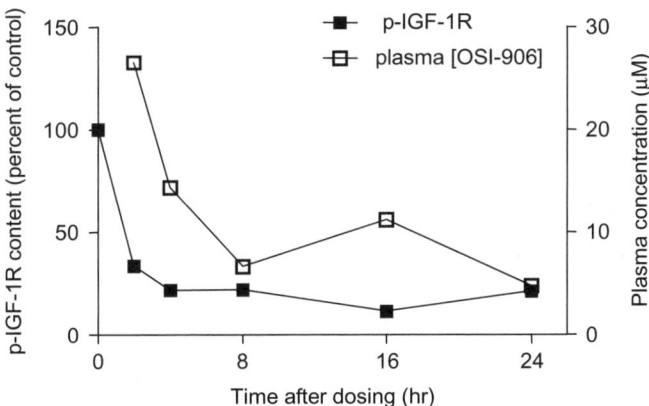

Figure 4.13 Pharmacokinetic/pharmacodynamic relationship of 75 mg kg^{-1} orally dosed OSI-906 in the LISN tumor xenograft model.

values, can be utilized in predicting OSI-906 plasma exposures required for sustained inhibition of one or both targets in humans.

4.14   OSI-906: Clinical Experience

Phase I studies evaluating OSI-906 as a single agent in advanced solid tumors commenced in June 2007. Trials were designed to evaluate both continuous once-daily dosing as well as intermittent dosing at QD 1–3/5/7 schedules every 14 days.[52,53] Based on preclinical data indicating that maintenance of plasma levels necessary to chronically suppress IGF-1R activity translated to improved anti-tumor efficacy, OSI-906 was additionally evaluated on a twice a day (BID) dosing schedule. Preliminary analysis has shown that OSI-906 PK is approximately dose-proportional.[52,53] The median time to C_{max} is 2–4 h after dosing, and the median terminal half life is 2–4 h. A dose of 300 mg OSI-906 administered once-daily generates plasma concentrations of OSI-906 in excess of 1 µM for 20–24 h. Preliminary PK data for BID dosing indicates that doses greater than 75 mg lead to plasma concentrations greater than 1 µM following continuous once-daily treatment.

Intermittent dosing schedules were evaluated, as this may facilitate optimal combination drug regimens with other agents. Preliminary PK analysis indicates approximately dose-proportional exposure of OSI-906, and doses greater than 450 mg led to plasma concentrations greater than 1 µM for 20–24 h after dosing. Collectively, these data indicate that, based upon preclinical models, the predicted efficacious concentration in plasma could be achieved in cancer patients.

Although the primary goal for the Phase I single-agent studies was to determine the maximum tolerated dose and recommended Phase II dose, preliminary evidence for anti-tumor activity was observed. Long-term stable disease greater than 24 weeks was observed for select patients receiving both the once-daily and intermittent dosing schedules.[52] Additionally, a partial response was observed for a patient with adrenocortical carcinoma receiving 450 mg OSI-906 administered on an intermittent schedule of QD 1–3 every 14 days. Preliminary anti-tumor activity in ACC is consistent with the scientific rationale for this being a tumor type driven by IGF-2 signaling.

4.15   Summary

In summary, OSI-906 is a potent, highly selective, and orally bioavailable dual inhibitor of both IGF-1R and IR. The discovery of OSI-906 was enabled through the use of rational SBD and empirical medicinal chemistry to optimize potency and overall drug-like properties, as well as achieving a high degree of selectivity. A streamlined testing cascade of IGF-1R and IR *in vitro* and *in vivo* models allowed for the assessment of PK, PD, efficacy, and safety. Upon once-per-day oral dosing, OSI-906 showed robust anti-tumor activity in multiple mouse xenograft models. The anti-tumor activity correlated with the degree and duration of inhibition of tumor IGF-1R phosphorylation achieved *in vivo* by OSI-906. The exceptional selectivity profile of OSI-906, in conjunction with its ability to inhibit both IGF-1R and IR, provides a unique opportunity to fully target the IGF-1R/IR axis. OSI-906 is currently being evaluated in the clinic.

Acknowledgements

We would like to thank Rika Mulvihill Pittman for kindly providing the wonderful illustration in Figure 4.1.

References

1. J. M. Knowlden, I. R. Hutcheson, D. Barrow, J. M. Gee and R. I. Nicholson, *Endocrinology*, 2005, **146**, 4609.
2. M. Pollak, *Nat. Rev. Cancer*, 2008, **8**, 915.
3. T. E. Adams, V. C. Epa, T. P. Garrett and C. W. Ward, *Cell Mol. Life Sci.*, 2000, **57**, 1050.
4. C. W. Ward, T. P. Garrett, N. M. McKern, M. Lou, L. J. Cosgrove, L. G. Sparrow, M. J. Frenkel, P. A. Hoyne, T. C. Elleman, T. E. Adams, G. O. Lovrecz, L. J. Lawrence and P. A. Tulloch, *Mol. Pathol.*, 2001, **54**, 125.
5. M. C. Lawrence, N. M. McKern and C. W. Ward, *Curr. Opin. Struct. Biol.*, 2007, **17**, 699.
6. P. De Meyts, *Trends Biochem. Sci.*, 2008, **33**, 376.
7. P. De Meyts and J. Whittaker, *Nat. Rev. Drug Discovery*, 2002, **1**, 769.
8. T. E. Adams, N. M. McKern and C. W. Ward, *Growth Factors*, 2004, **22**, 89.
9. L. Laviola, A. Natalicchio and F. Giorgino, *Curr. Pharm. Des.*, 2007, **13**, 663.
10. D. LeRoith and C. T. Roberts Jr, *Cancer Lett.*, 2003, **195**, 127.
11. F. Peruzzi, M. Prisco, M. Dews, P. Salomoni, E. Grassilli, G. Romano, B. Calabretta and R. Baserga, *Mol. Cell Biol.*, 1999, **19**, 7203.
12. C. Sell, M. Rubini, R. Rubin, J. P. Liu, A. Efstratiadis and R. Baserga, *Proc. Natl. Acad. Sci. U. S. A.*, 1993, **90**, 11217.
13. Z. Pietrzkowski, R. Lammers, G. Carpenter, A. M. Soderquist, M. Limardo, P. D. Phillips, A. Ullrich and R. Baserga, *Cell Growth Differ.*, 1992, **3**, 199.
14. E. Buck, A. Eyzaguirre, M. Rosenfeld-Franklin, S. Thomson, M. Mulvihill, S. Barr, E. Brown, M. O'Connor, Y. Yao, J. Pachter, M. Miglarese, D. Epstein, K. K. Iwata, J. D. Haley, N. W. Gibson and Q. S. Ji, *Cancer Res.*, 2008, **68**, 8322.
15. F. L. Chen, W. Xia and N. L. Spector, *Clin. Cancer Res.*, 2008, **14**, 6730.
16. C. L. Chernicky, L. Yi, H. Tan, S. U. Gan and J. Ilan, *Cancer Gene Therapy*, 2000, **7**, 384.

17. B. D. Cohen, D. A. Baker, C. Soderstrom, G. Tkalcevic, A. M. Rossi, P. E. Miller, M. W. Tengowski, F. Wang, A. Gualberto, J. S. Beebe and J. D. Moyer, *Clin. Cancer Res.*, 2005, **11**, 2063.

18. C. Garcia-Echeverria, M. A. Pearson, A. Marti, T. Meyer, J. Mestan, J. Zimmermann, J. Gao, J. Brueggen, H. G. Capraro, R. Cozens, D. B. Evans, D. Fabbro, P. Furet, D. G. Porta, J. Liebetanz, G. Martiny-Baron, S. Ruetz and F. Hofmann, *Cancer Cell*, 2004, **5**, 231.

19. E. K. Maloney, J. L. McLaughlin, N. E. Dagdigian, L. M. Garrett, K. M. Connors, X. M. Zhou, W. A. Blattler, T. Chittenden and R. Singh, *Cancer Res.*, 2003, **63**, 5073.

20. C. L. Chen, S. M. Ip, D. Cheng, L. C. Wong and H. Y. Ngan, *Clin. Cancer Res.*, 2000, **6**, 474.

21. H. Cui, M. Cruz-Correa, F. M. Giardiello, D. F. Hutcheon, D. R. Kafonek, S. Brandenburg, Y. Wu, X. He, N. R. Powe and A. P. Feinberg, *Science*, 2003, **299**, 1753.

22. C. Fottner, A. Hoeflich, E. Wolf and M. M. Weber, *Horm. Metab. Res.*, 2004, **36**, 397.

23. T. J. Giordano, R. Kuick, T. Else, P. G. Gauger, M. Vinco, J. Bauersfeld, D. Sanders, D. G. Thomas, G. Doherty and G. Hammer, *Clin. Cancer Res.*, 2009, **15**, 668.

24. N. Nonomura, K. Nishimura, T. Miki, N. Kanno, Y. Kojima, M. Yokoyama and A. Okuyama, *Cancer Res.*, 1997, **57**, 2575.

25. Z. Wang, Y. B. Ruan, Y. Guan and S. H. Liu, *World J. Gastroenterol.*, 2003, **9**, 267.

26. L. Frittitta, R. Vigneri, M. R. Stampfer and I. D. Goldfine, *J. Cell Biochem.*, 1995, **57**, 666.

27. F. Giorgino, A. Belfiore, G. Milazzo, A. Costantino, B. Maddux, J. Whittaker, I. D. Goldfine and R. Vigneri, *Mol. Endocrinol.*, 1991, **5**, 452.

28. M. Jenab, E. Riboli, R. J. Cleveland, T. Norat, S. Rinaldi, A. Nieters, C. Biessy, A. Tjonneland, A. Olsen, K. Overvad, H. Gronbaek, F. Clavel-Chapelon, M. C. Boutron-Ruault, J. Linseisen, H. Boeing, T. Pischon, D. Trichopoulos, E. Oikonomou, A. Trichopoulou, S. Panico, P. Vineis, F. Berrino, R. Tumino, G. Masala, P. H. Peters, C. H. van Gils, H. B. Bueno-de-Mesquita, M. C. Ocke, E. Lund, M. A. Mendez, M. J. Tormo, A. Barricarte, C. Martinez-Garcia, M. Dorronsoro, J. R. Quiros, G. Hallmans, R. Palmqvist, G. Berglund, J. Manjer, T. Key, N. E. Allen, S. Bingham, K. T. Khaw, A. Cust and R. Kaaks, *Int. J. Cancer*, 2007, **121**, 368.

29. J. Ma, E. Giovannucci, M. Pollak, A. Leavitt, Y. Tao, J. M. Gaziano and M. J. Stampfer, *J. Natl. Cancer Inst.*, 2004, **96**, 546.

30. J. Kling, *Nat. Biotechnol.*, 2008, **26**, 479.

31. M. Pollak, *Curr. Opin. Pharmacol.*, 2008, **8**, 384.

32. Y. Dai, Y. Guo, R. R. Frey, Z. Ji, M. L. Curtin, A. A. Ahmed, D. H. Albert, L. Arnold, S. S. Arries, T. Barlozzari, J. L. Bauch, J. J. Bouska, P. F. Bousquet, G. A. Cunha, K. B. Glaser, J. Guo, J. Li, P. A. Marcotte, K. C. Marsh, M. D. Moskey, L. J. Pease, K. D. Stewart, V. S. Stoll,

P. Tapang, N. Wishart, S. K. Davidsen and M. R. Michaelides, *J. Med. Chem.*, 2005, **48**, 6066.

33. M. J. Mulvihill, Q. S. Ji, D. Werner, P. Beck, C. Cesario, A. Cooke, M. Cox, A. Crew, H. Dong, L. Feng, K. W. Foreman, G. Mak, A. Nigro, M. O'Connor, L. Saroglou, K. M. Stolz, I. Sujka, B. Volk, Q. Weng and R. Wilkes, *Bioorg. Med. Chem. Lett.*, 2007, **17**, 1091.

34. P. A. Beck, C. Cesario, M. Cox, H. Dong, K. Foreman, J. M. Mulvihill, A. Nigro, L. Saroglou, A. G. Steinig, Y. Sun, Q. Weng, D. Werner, R. Wilkes and J. Williams, *US Pat. Appl.* 0084654 A1, 2006.

35. Q. S. Ji, M. J. Mulvihill, M. Rosenfeld-Franklin, A. Cooke, L. Feng, G. Mak, M. O'Connor, Y. Yao, C. Pirritt, E. Buck, A. Eyzaguirre, L. D. Arnold, N. W. Gibson and J. A. Pachter, *Mol. Cancer Ther.*, 2007, **6**, 2158.

36. M. J. Mulvihill, Q. S. Ji, H. R. Coate, A. Cooke, H. Dong, L. Feng, K. Foreman, M. Rosenfeld-Franklin, A. Honda, G. Mak, K. M. Mulvihill, A. I. Nigro, M. O'Connor, C. Pirrit, A. G. Steinig, K. Siu, K. M. Stolz, Y. Sun, P. A. Tavares, Y. Yao and N. W. Gibson, *Bioorg. Med. Chem.*, 2008, **16**, 1359.

37. Y. Wang, Q. S. Ji, M. J. Mulvihill and J. A. Pachter, *Recent Results Cancer Res.*, 2007, **172**, 59.

38. L. Arnold and M. J. Mulvihill, *US Pat.* 7 534 797 B2, 2009.

39. M. J. Mulvihill, A. Cooke, M. Rosenfeld-Franklin, E. Buck, K. Foreman, D. Landfair, M. O'Connor, C. Pirrit, Y. Sun, Y. Yao, L. Arnold, N. W. Gibson and Q. S. Ji, *Future Med. Chem.*, 2009, **1**, 1153.

40. J. Wu, W. Li, B. P. Craddock, K. W. Foreman, M. J. Mulvihill, Q. S. Ji, W. T. Miller and S. R. Hubbard, *EMBO J.*, 2008, **27**, 1985.

41. A. Rane, G. R. Wilkinson and D. G. Shand, *J. Pharmacol. Exp. Ther.*, 1977, **200**, 420.

42. M. J. Mulvihill, Q. S. Ji, M. Rosenfeld-Franklin, E. Buck, A. Cooke, A. Eyzaguirre, L. Feng, K. Foreman, D. Landfair, G. Mak, M. O'Connor, C. Pirritt, S. Silva, R. Turton, Y. Yao, L. Arnold, R. Wild, J. A. Pachter and N. W. Gibson, presented at the 99th Annual Meeting of the American Association for Cancer Research, 2008, abstr. 4893.

43. M. Rosenfeld-Franklin, A. Cooke, C. Pirritt, D. Landfair, S. Silva, R. Turton, L. Feng, M. J. Mulvihill, E. Buck, J. A. Pachter, Q. S. Ji and R. Wild, presented at the AACR-NCI-EORTC International Conference: Molecular Targets and Cancer Therapeutics, 2007, abstr. B244.

44. H. Hartog, J. Wesseling, H. M. Boezen and W. T. van der Graaf, *Eur. J. Cancer*, 2007, **43**, 1895.

45. P. Rose, J. Carroll, P. Carroll, V. DeFilippis, M. Lagunoff, A. Moses, C. Roberts, Jr. and K. Fruh, *Oncogene*, 2007, **26**, 1995.

46. G. Milazzo, F. Giorgino, G. Damante, C. Sung, M. R. Stampfer, R. Vigneri, I. D. Goldfine and A. Belfiore, *Cancer Res.*, 1992, **52**, 3924.

47. S. Avnet, L. Sciacca, M. Salerno, G. Gancitano, M. F. Cassarino, A. Longhi, M. Zakikhani, J. M. Carboni, M. Gottardis, A. Giunti, M. Pollak, R. Vigneri and N. Baldini, *Cancer Res.*, 2009, **69**, 2443.

48. L. Sciacca, A. Costantino, G. Pandini, R. Mineo, F. Frasca, P. Scalia, P. Sbraccia, I. D. Goldfine, R. Vigneri and A. Belfiore, *Oncogene*, 1999, **18**, 2471.
49. V. Vella, G. Pandini, L. Sciacca, R. Mineo, R. Vigneri, V. Pezzino and A. Belfiore, *J. Clin. Endocrinol. Metab.*, 2002, **87**, 245.
50. L. Sciacca, R. Mineo, G. Pandini, A. Murabito, R. Vigneri and A. Belfiore, *Oncogene*, 2002, **21**, 8240.
51. R. Schillaci, A. Galeano, D. Becu-Villalobos, O. Spinelli, S. Sapia and R. F. Bezares, *Br. J. Haematol.*, 2005, **130**, 58.
52. C. P. Carden, S. Frentzas, M. Langham, I. Casamayor, A. S. Stephens, S. Poondru, J. Wheaton, S. M. Lippman, S. B. Kaye and E. S. Kim, presented at the ASCO 2009 Annual Meeting, 2009, abstr. 3544.
53. C. R. Lindsay, E. Chan, T. R. J. Evans, S. Campbell, P. Bell, A. S. Stephens, A. Franke, S. Poondru, M. L. Rothenberg and I. Puzanov, presented at the ASCO 2009 Annual Meeting, 2009, abstr. 2559.

CHAPTER 5

The Genesis of the Antibody Conjugate Gemtuzumab Ozogamicin (Mylotarg®) for Acute Myeloid Leukemia

PHILIP R. HAMANN[†]

Adjunct Professor, John Jay College of Criminal Justice, New York, USA

5.1 Acute Myeloid Leukemia

Although leukemia is commonly thought of as a disease of childhood, there are leukemias that affect primarily adults, the most common being acute myeloid leukemia (AML, average patient age 65), which has a US incidence of 2.4 per 100 000.[1] If left untreated, it is routinely fatal in a matter of weeks to months.[2] The initial symptom is usually fatigue, which arises from the replacement of normal marrow and blood cells by leukemic blasts. This leads to progressive anemia, neutropenia, and thrombocytopenia, which are associated with infection and hemorrhage during the course of the disease.

When a patient is strong enough to withstand the side effects, aggressive chemotherapy is the initial treatment, consisting of multiple rounds of daunorubicin and cytosine arabinoside (Ara-C) or similar agents. The common side effects include nausea, vomiting, hair loss, heart damage, and CNS toxicity. Remission rates vary from 50 to 80%; unfortunately, 60–80% of patients have a

---

[†]Formerly at Pfizer, 401 N. Middletown Rd., Pearl River, NY 10965, USA.

RSC Drug Discovery Series No. 4
Accounts in Drug Discovery: Case Studies in Medicinal Chemistry
Edited by Joel C. Barrish, Percy H. Carter, Peter T. W. Cheng and Robert Zahler
© Royal Society of Chemistry 2011
Published by the Royal Society of Chemistry, www.rsc.org

relapse after a complete response (CR) lasting an average of 15 months. Patients with refractory or relapsed disease have a poor prognosis, with only 30–35% achieving a second CR with follow-up chemotherapy with a median duration of relapse-free survival of only 3–12 months. For the 90–95% who do not have sustained disease-free survival from chemotherapy, the only choice is myeloablative therapy followed by hematopoietic stem cell transplantation (HSCT), although most patients over 55 do not qualify. Long-term survival for patients undergoing HSCT is 10–40%, with younger patients performing better.[3]

5.2 The CD33 Antigen

The earliest hematopoietic cells are CD34 positive and CD33 negative. When the lymphoid and the myeloid lineages diverge, the differentiating myeloid cells are CD33 positive. Although mature myeloid cells have only low levels, this antigen is used as one of the markers of AML, with over 90% of patients being positive. CD33, also known as Siglec-3, is a 67 kDa, type-1 transmembrane member of the immunoglobulin family that may be involved in cell adhesion and/or immune signaling.[4]

CD33 is an internalizing antigen,[5] which is essential for an antibody drug conjugate (ADC), as conjugates that are not internalized rarely show significant potency. It is also readily accessible from circulation, making this indication an ideal proving ground for ADCs. It was hypothesized that a properly constructed conjugate would rapidly target and eliminate the AML cells and largely spare most other normal cells. Differentiating CD33-positive myeloid cells would also be targeted and eliminated, but the CD34-positive pluripotent stem cells would likely be spared to a significant extent and repopulate the bone marrow faster than after traditional chemotherapy, which does not inherently spare CD34-positive cells. If accomplished, this would result in a therapy for AML that would be applicable to the patients too sick to take traditional chemotherapy and may give a treatment that is superior for all patients.

To make such a conjugate, an antibody is required that is selective for CD33. It should be of the IgG class and it should be chemically well-behaved, meaning that it is not prone to aggregation or denaturization and that it is readily amenable to chemical modification. It should ideally also be humanized, but could be initially examined as a murine antibody, the only kind readily available at the time this research was initiated. An ideal candidate was found in the murine anti-CD33 antibody of Bernstein *et al.*[6]

5.3 The Calicheamicins

The calicheamicins are a family of natural products isolated from *Micromonospora echinospora*, ssp. *calichensis*.[7] They consist of an enediyne headpiece and a modified carbohydrate tail consisting of four sugar units and an unusually substituted aromatic ring (Figure 5.1). The carbohydrate tail exists naturally in a somewhat helical conformation in solution and fits neatly into the minor groove of DNA.[8]

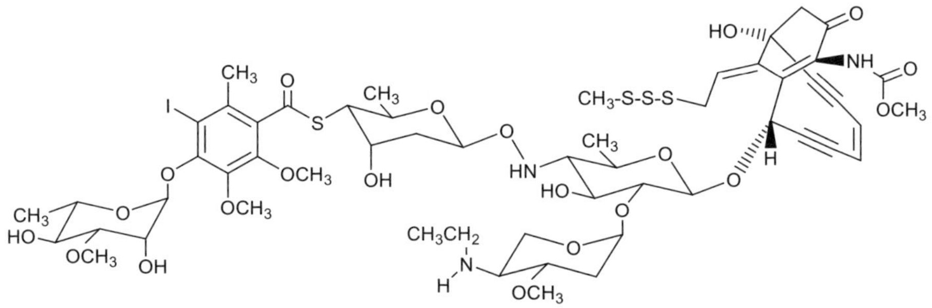

Figure 5.1 Calicheamicin γ_1^I. γ-Calicheamicin is the parent of this class of unusually complex natural products. Unique are the methyl trisulfide, the three sugar units closest to the bicyclic core, the per-substituted aromatic ring containing iodine, and, especially, the embedded enediyne that is responsible for its ability to cleave DNA.

Studies have shown that it prefers to bind to DNA sequences that have sufficient flexibility to bend slightly and widen the minor grove to accommodate binding.[8]

The calicheamicin headpiece has an unusual enediyne unit that is kept stable by a bicyclic framework containing a bridgehead alkene. This bridgehead alkene limits the flexibility of the enediyne and does not allow it to undergo its normal mode of reaction, Bergman cyclization to a diradical. However, reduction of the appended methyl trisulfide, whether chemically or biologically, leads to an intramolecular Michael reaction that removes the constraining bridgehead alkene (Figure 5.2). This then allows the Bergman cyclization to occur with an estimated half-life of 4.5 s.[9] When this diradical is situated in the minor grove of DNA, where calicheamicin prefers to bind, it rapidly abstracts hydrogens from opposite strands of the deoxyriboses, resulting in damage to both strands of DNA simultaneously.[10] The double-stranded nature of the DNA damage is at least in part responsible for the extreme potency of the calicheamicins. Such potency is important for conjugates, as the limited amount that can be internalized by the available antigen needs to constitute a lethal dose.

Previous work with the anti-PEM antibody CTM01 showed the feasibility of making conjugates of the calicheamicins.[11] The maximum therapeutic window for these initial conjugates was obtained with derivatives of calicheamicin that contained an amino sugar modified by acetylation (N-acetyl-γ-calicheamicin) and that contained a stabilized dimethyl disulfide instead of the naturally occurring trisulfide. It is unknown why the acylation (or removal) of the amino sugar is beneficial. A poorly understood chemical or biological role of the basic nitrogen, reduction of the potency of the conjugate to a level that spares non-target cells that take up small amounts of conjugate non-specifically, or the ability to increase the dose of the conjugate to an amount that allows for better pharmacokinetics (PK) are three possibilities.

Aside from the antibody and the cytotoxic agent, ADCs contain a linker that attaches these components. The linker needs to be stable in circulation for extended periods of time, but the ADC needs to be cleaved intracellularly,

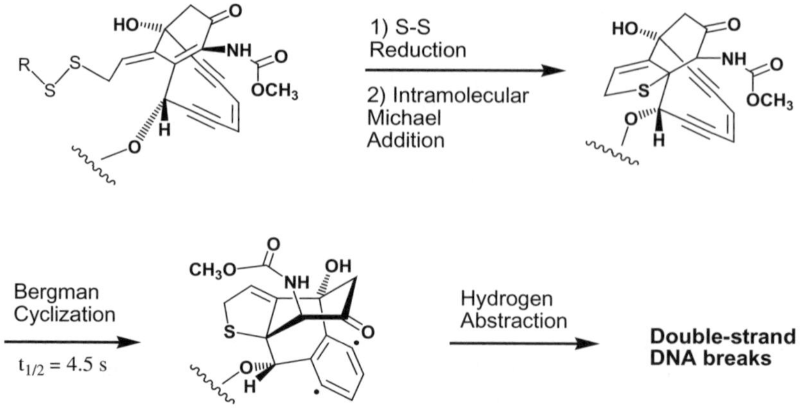

Figure 5.2 Mechanism of DNA damage. The bridgehead alkene prevents Bergman cyclization until reduction of the S–S bond at an undetermined intracellular location allows the resultant thiol to eliminate it by Michael addition. The half-life of 4.5 s allows the Michael adduct to sample the DNA and, perhaps, transit from the cytoplasm to the nucleus.

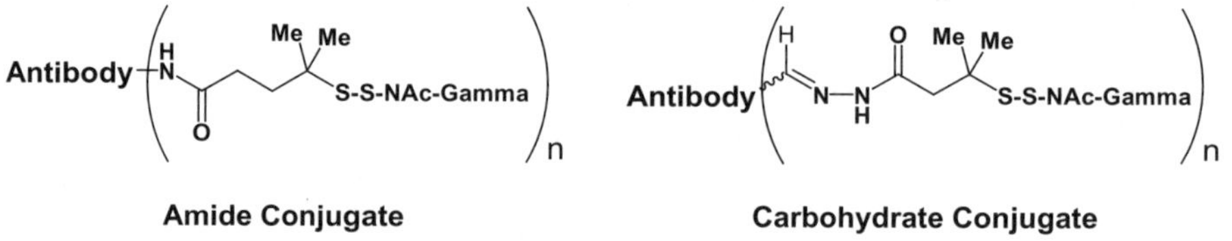

Amide Conjugate   **Carbohydrate Conjugate**

Figure 5.3 Initial conjugate types. These conjugates differ primarily in the hydrolytic release afforded by the hydrazone in the carbohydrate conjugate. Disulfide reduction of the highly stabilized disulfide could play a role in the release of calicheamicin, but the primary route of release of a calicheamicin derivative from the amide conjugate is likely to be lysosomal proteolysis of the antibody to release a zwitterionic derivative of lysine.[13]

usually in the lysosomes, in a way that releases an active cytotoxic agent. Two different types of linkers were effective with the CTM01 antibody (Figure 5.3), one that attached to oxidized carbohydrates on the antibody through hydrazone formation, and a second attached to lysines on the antibody.[12] These are loosely referred to as the carbohydrate conjugates and the amide conjugates. Initial work with the anti-CD33 antibody P67.6 showed that the *N*-acetylcalicheamicin was a preferred derivative and that stabilizing the disulfide by the introduction of two flanking methyl groups was again beneficial, but interesting differences in activity that depended on the nature of the linker were seen.

5.4 The Original Anti-CD33 Conjugate

Oxidation of the P67.6 antibody with periodate follow by reaction with *N*-acetyl-γ-calicheamicin dimethylhydrazide (DMH) gave a carbohydrate conjugate that can release the calicheamicin in the acidic environment of the lysosomes, where this antibody is trafficked.[14] This conjugate was potently cytotoxic toward the CD33-positive HL-60 cell line ($<0.006\,\mathrm{ng\,mL^{-1}}$ of calicheamicin), and was much less cytotoxic against an antigen-negative cell line ($0.79\,\mathrm{ng\,mL^{-1}}$). An alternate conjugate was made where a comparable calicheamicin derivative was attached to the lysines on the antibody, giving an amide conjugate that can be proteolytically degraded in the lysosomes to release the calicheamicin (Figure 5.3). This conjugate had greatly reduced potency against the HL-60 cell line ($44\,\mathrm{ng\,mL^{-1}}$). This is in marked contrast to the previous work with the CTM01 antibody and indicates an apparent need to include a site of hydrolysis in the final conjugate.

The conjugates under study at this point were made with murine antibodies that would very likely be immunogenic in humans. A humanization was, therefore, undertaken to humanize the antibody. Both IgG$_2$ and IgG$_4$ versions of the antibody were produced by Celltech Ltd. These isotypes were chosen at the time to minimize Fc interactions with effector cells that might interfere with the desired antigen targeting. It was felt that any of the natural mechanisms of antibody cytotoxicity would be inoperative at the anticipated doses of calicheamicin conjugates in any case. However, all attempts to make the desired carbohydrate conjugate from the selected humanized IgG$_4$ antibody resulted in conjugates that were almost completely devoid of antigen binding.[15]

5.5 Conjugate Redesign

Although work was begun on a new humanized antibody that might be amenable to oxidation, an attempt was also made to make use of the original humanized antibody. The conjugate was redesigned to incorporate a bifunctional linker that took advantage of lysine attachment and also contained a designed site of hydrolysis (Figure 5.4).[15] The antibody retained antigen binding when acylated, and including a site of hydrolytic release in the linker had the advantage over the carbohydrate attachment in that the rate of hydrolysis could be adjusted to optimize activity. The same calicheamicin derivative used previously was retained to make the hydrazone.

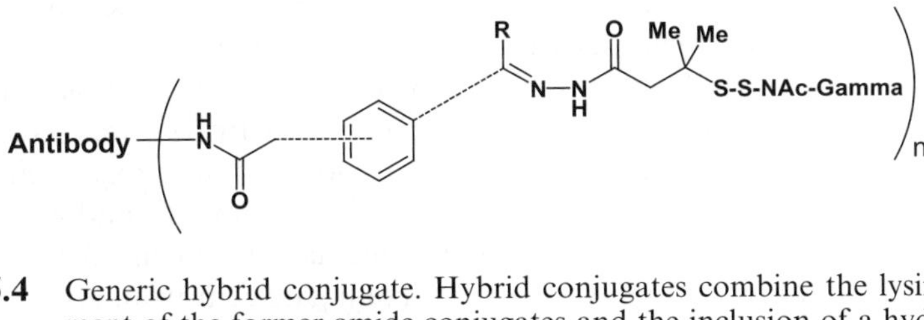

Figure 5.4 Generic hybrid conjugate. Hybrid conjugates combine the lysine attachment of the former amide conjugates and the inclusion of a hydrazone as in the carbohydrate conjugates. The necessary bifunctional linker can be modified to give the optimum balance of hydrolysis rate for serum stability and targeted cytotoxicity.

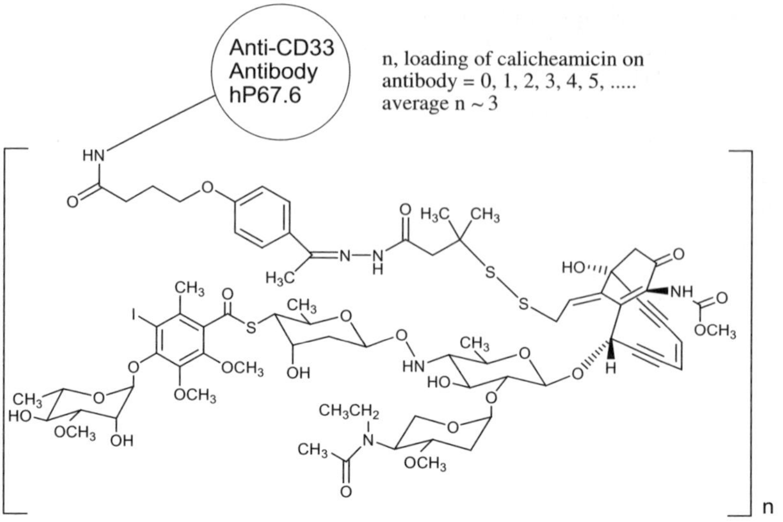

Figure 5.5 Structure of gemtuzumab ozogamicin. The anti-CD33 antibody is acylated with the bifunctional AcBut linker, which allows for hydrazone formation with the calicheamicin derivative. This derivative, *N*-acetyl-γ-calicheamicin dimethylhydrazide (DMH), has a disulfide stabilized by both the two methyl groups and the calicheamicin bicyclic core, and an acetylated amino sugar.

Numerous bifunctional linkers were explored, with aliphatic aldehydes and ketones giving hydrazones that were too unstable (*cf.* the well-known BR96-DOX conjugate, which has marginal stability in circulation[16]). Aromatic aldehydes gave conjugates with reasonable stability and activity, but acetophenones, which release the calicheamicin derivative faster at acidic pH's, such as found in lysosomes, gave the best activity with an acceptable degree of stability. The final choice of the bifunctional linker was 4-(4-acetylphenoxy) butanoic acid ("AcBut"), which resulted in the conjugate shown in Figure 5.5 as the structure of gemtuzumab ozogamicin (hP67.6-NAc-gamma calicheamicin DMH AcBut conjugate).

The hydrolytic behavior of the new hydrazone bond in this conjugate was examined in two ways. First, the carboxylic acid without the antibody, but containing the calicheamicin derivative attached to the bifunctional AcBut linker, was subjected to hydrolysis in buffer. At pH 7.4, 6% hydrolysis was seen in 24 h at 37 °C, while at pH 4.5 (lysosomal pH) 97% hydrolysis was seen. Studies were also conducted on a related conjugate, inotuzumab ozogamicin (CMC-544), which is in Phase 3 trials for non-Hodgkin's lymphoma[17] and which has an anti-CD22 antibody conjugated to the same calicheamicin derivative and bifunctional AcBut linker that is used in the conjugate gemtuzumab ozogamicin. This antibody conjugate was studied in human plasma and serum and showed a slower than expected rate of release of 2–3% per day at pH 7.4 (Figure 5.6),[18] which corresponds to an estimated half-life of hydrolysis of 23–34 days.

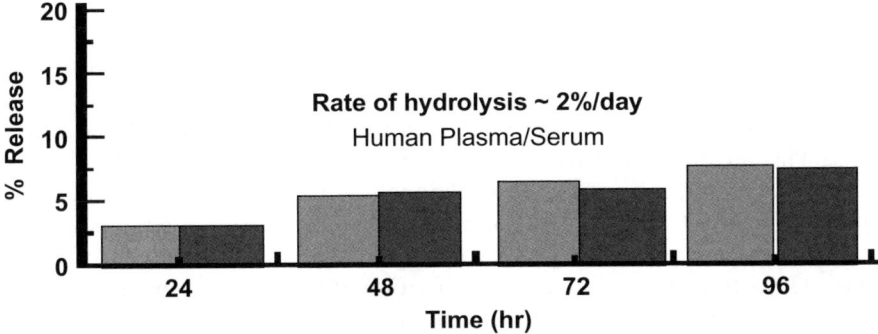

Figure 5.6 Linker hydrolysis. Inotuzumab ozogamicin was incubated in human plasma and serum at 37 °C and samples were analyzed at the indicated time for free *N*-acetyl-γ-calicheamicin DMH.

The difference in hydrolysis between the unconjugated and conjugated forms (6% *vs.* 2%) may be due to a lipophilic interaction between the calicheamicin and the antibody in the conjugate, resulting in an increase in stability; it could be due to the microenvironment of the hydrazone in the conjugated form; or it could be the result of different media used (serum/plasma *vs.* buffer containing an organic co-solvent). It is not known if the hydrolysis rate differs *in vivo* from that seen in plasma and serum. It has been difficult to study this issue because gemtuzumab ozogamicin, like most conjugates, is a mixture of species. If the antibody and calicheamicin conjugate clear at different rates *in vivo*, it is hard to discern if the linker is showing increased hydrolysis or if the higher loaded species are clearing faster than the lower loaded species, which would alter the ratio of calicheamicin to antibody without linker hydrolysis.

Gemtuzumab ozogamicin is a mixture of species due to variation in antibody glycosylation as well as to the sites and degree of modification, in this case the identity of the lysines acylated. However, gemtuzumab ozogamicin has two unusual properties as a conjugate in its constitution.[18] First, the calicheamicin is not randomly distributed on a large number of lysines scattered around the antibody, as might be expected and as has been shown for another lysine-based conjugate,[19] but resides primarily on eight lysines, four on each of the two heavy chains. Second, the distribution is bimodal; about half of the antibody carries little if any calicheamicin while the other half has twice the average loading with very little at lower loadings of 1–3 calicheamicins/antibody. This all-or-nothing effect might be due to the aggregated state of the highly lipophilic calicheamicin, leading to a locally high concentration near the site of any initial conjugate; or it may be due to a conformational change in the hinge region where these sites reside, resulting in increased reactivity; or it may be due to other factors that have yet to be explored.

These two unusual phenomena appear to be unrelated, as a previous calicheamicin conjugate, CMB-401, which lacks the AcBut linker (the amide conjugate), showed selectivity for the same lysines (as do all the calicheamicin conjugates that have been examined), but not the all-or-nothing effect, having a relative normal distribution of species.[20] With regard to the bimodal distribution, if one ignores any effects on pharmacokinetics and biodistribution properties, the average loading alone should determine the potency and selectivity of the conjugate due to the statistical mixture each cell will internalize. This means the *in vitro* studies are likely to be reliable, in spite of a wide loading distribution, but leads to questions for *in vivo* studies, as loading can significantly affect the pharmacokinetics of conjugates. Understanding and overcoming these effects is still an active area of research in the conjugate field.[21] Unfortunately, the loading distribution of gemtuzumab ozogamicin was not appreciated until it was in clinical trials. However, any PK effects are probably less significant for this anti-leukemia conjugate, as binding to the antigen occurs rapidly.[i]

During the discovery stage of gemtuzumab ozogamicin, its cytotoxic effects were first examined *in vitro* on a pair of cell lines: HL-60 acute promyelocytic leukemia (APL, a sub-type of AML) cells that express CD33 and Raji non-Hodgkin's lymphoma cells that do not (Table 5.1). Gemtuzumab ozogamicin showed an IC_{50} of 0.31 pM versus the target cell line with 150 000-fold selectivity *versus* the non-target cell line.[15] It is 20 000-fold more potent than the calicheamicin derivative that it releases by hydrolysis.

Gemtuzumab ozogamicin was very efficacious against the HL-60 cells when grown as xenografts in nude mice. Treatment of established 150–200 mg subcutaneous tumors with three doses given on days 1, 5, and 9 led to disappearance of the tumor mass in either 4/5 or 5/5 animals at doses of 50–300 µg kg^{-1} of calicheamicin, with no regrowth up to 100 days. At a dose of 25 µg kg^{-1} there was 80% inhibition of tumor growth *versus* buffer-treated control animals. A non-binding control conjugate showed a non-significant 20% inhibition at the top dose. See Figure 5.7 for a representative experiment.

Gemtuzumab ozogamicin was also used to treat bone marrow samples from AML patients in a colony-forming assay and compared to the number of colonies seen with a non-binding conjugate (Figure 5.8). As indicated by the bar graph, approximately one-third to one-half of the samples showed positive responses, depending of the cutoff used.

Table 5.1 Gemtuzumab ozogamicin *in vitro* activity.

Treatment	IC_{50} (nM) HL-60	IC_{50} (nM) Raji	Selectivity index[a]
Gemtuzumab ozogamicin	0.000 31	24	150 000
N-Ac-γ-calicheamicin DMH	6.1	3.1	[1]
Ratio	20 000	0.13	–

[a]Ratio of activity corrected for the difference in sensitivity of the two cell lines to calicheamicin.

[i]The related conjugate inotuzumab ozogamicin (CMC-544) has little if any free antibody due to its higher loading.

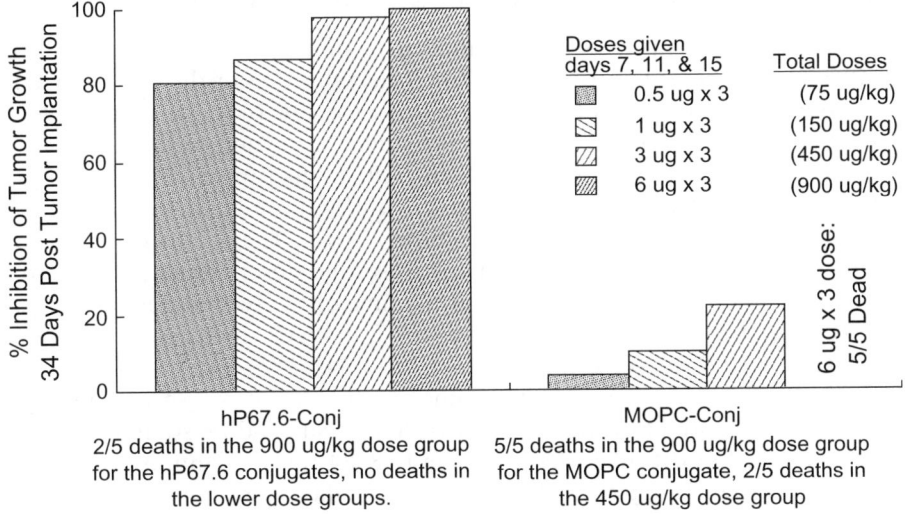

Figure 5.7 Gemtuzumab ozogamicin *in vivo* activity. Subcutaneous tumor implants were allowed to grow to an average size of about 150 mg prior to treatment as indicated. Results shown are at day 34. Tumors were absent in 4/5 animals at the lowest three doses and 5/5 at the highest dose of hP67.6-Conj (gemtuzumab ozogamicin) and did not regrow for over 100 days in the surviving animals. (The top dose of 900 µg kg^{-1} was well tolerated in most other experiments.) No animals were tumor-free in the MOPC-Conj group, a non-binding control conjugate.

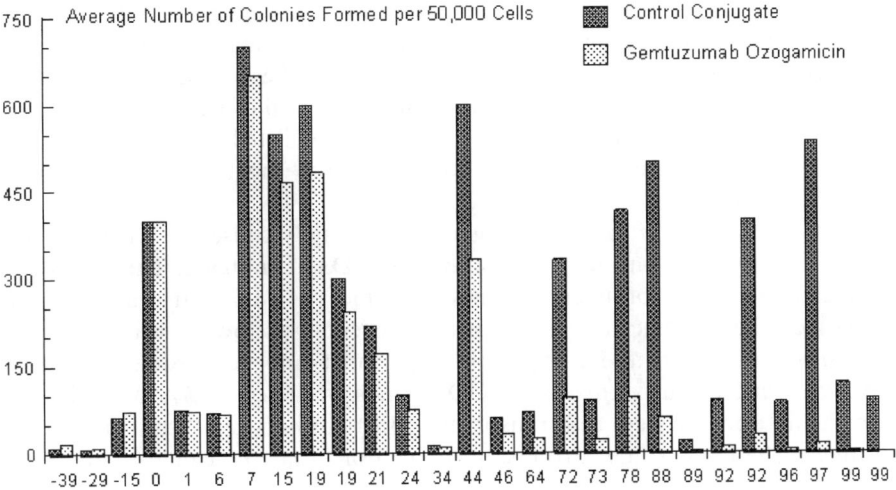

Figure 5.8 Gemtuzumab ozogamicin *ex vivo* activity. Bone marrow samples from AML patients were plated in the presence of either gemtuzumab ozogamicin or an isotype-matched, non-binding control conjugate. The numbers of colonies formed were counted for each sample and are plotted in order of increasing ratio, expressed as a percent.

5.6 Pharmacokinetic and Toxicology Studies

Pharmacokinetic properties were examined in non-tumor bearing rats using gemtuzumab ozogamicin tritiated on the N-acetyl group of the calicheamicin. At a dose of 11.6 mg m^{-2} of protein there was a C_{max} of 160 mg L^{-1}, an AUC of 920 mg h L^{-1}, and a $t_{1/2}$ of 95 h. Similar studies in cynomolgus monkeys at 16.3 mg m^{-2} gave a C_{max} of 64 mg L^{-1}, an AUC of 2000 mg h L^{-1}, and a $t_{1/2}$ of 119 h. Free calicheamicin was less than 2% during both studies. Pharmacokinetic studies in humans were done in Phase 2 trials. The results depended on whether the patient was receiving the first or second dose, presumably due to a lower leukemia burden in circulation at the time of the second dose. The C_{max} of 2.9 mg L^{-1} for the first dose increased to 3.7, the AUC increased from 123 to 239 mg h L^{-1}, and the $t_{1/2}$ increased from 72 to 94 h.[22]

Toxicology was examined in the same two species in single and multiple (weekly×6) dose studies with iv administration.[23] Single-dose studies in the rat were conducted at 0.4–12 mg kg^{-1} protein (2.8–84 mg m^{-2}) and in monkeys at 3–6 mg kg^{-1} (5.6–21 mg m^{-2}). Multiple-dose studies were conducted at 0.1–1.2 mg kg^{-1} (0.7–8.4 mg m^{-2}) in rats and 0.2–1.8 mg kg^{-1} (2.5–22 mg m^{-2}) in monkeys. The no-observed-adverse-effect dose with multiple doses was 0.7 mg kg^{-1} in the rat and 2.5 mg kg^{-1} in the monkey. There were numerous signs of toxicity, as is expected for administration of a cytotoxic conjugate.[ii] Organs that were affected in both studies included the liver, spleen, kidney, and bone marrow. These effects were reflected in general myelosuppression and elevation of liver enzymes, both of which tended to be reversible. The hP67.6 antibody in gemtuzumab ozogamicin is not cross-reactive in rats and monkeys; therefore, a tolerance study was performed in three chimpanzees at a single dose of 0.5 mg m^{-2}, which caused no significant drug-related effects.

The administration of gemtuzumab ozogamicin in humans resulted in a well-precedented infusion-related set of symptoms.[24] These symptoms included fever, chills, hypertension, nausea, and hypotension, all of which resolved after a couple of hours with supportive care that consisted of acetaminophen, diphenhydramine, and iv fluids, and can be mostly prevented with corticosteroids.[25]

Treatment with gemtuzumab ozogamicin routinely caused neutropenia and thrombocytopenia, as anticipated for an anti-CD33 conjugate. There was also significant elevation of liver enzymes in some patients that was generally reversible. The incidence of serious bleeding events was low, as was the rate of mucositis, infections, and pneumonia. The most unusual toxicity, although rare, was veno-occlusive disease (VOD), which occurred in 2% of patients.[26] Prior hematopoietic stem cell transplant appears to be a significant risk factor, although there is some promising indication that it can be prevented

---

[ii] It is not know whether toxicity is due to the internalization of conjugate or from the slow release of free calicheamicin. However, there is no *a priori* reason to assume that toxicity is from free calicheamicin, as proteins (as well as ADCs) are certainly taken up by bone marrow, liver, and other affected cells. Conjugates that do not release free calicheamicin to any detectable extent are somewhat better tolerated, but they have the same overall toxicity profile.

prophylactically.[27] Most toxicities appear to be diminished with fractionated dosing ($3 \times 3 \, \mathrm{mg \, m^{-2}}$ over one week) with no apparent loss in activity.[28]

5.7 Clinical Trials

A Phase 1 clinical trial was conducted in patients with relapsed AML, for which there was no approved treatment. Although there was very significant myelo-suppression consistent with the depletion of CD33-positive cells, formal dose-limiting toxicity was not reached.[29] A dose of $9 \, \mathrm{mg \, m^{-2}}$ was chosen as the final dose, based on the observed near saturation of circulating blasts.[30] There were a significant number of responders starting at a dose of only $1 \, \mathrm{mg \, m^{-2}}$ of conjugate, indicating that further clinical trials were warranted.

A total of 277 patients were treated in three similar Phase 2 clinical trials, all of which were open-label, single-arm studies.[31] The responders fell into two groups of about equal size, corresponding to patients with or without full recovery of platelets ($> 100 \, 000 \, \mu\mathrm{L^{-1}}$), which were designated CR and CRp. To qualify for the CRp group a patient had to be transfusion independent, and these two groups were almost identical in their responses other than having lower platelet levels. The combined remission rate was 26% and the medium recurrence-free survival in this heavily pre-treated patient population was 5.5 months. Based on these data, gemtuzumab ozogamicin was approved by the FDA in May 2000 for use in patients with CD33-positive AML in first relapse who are 60 years of age or older and are not candidates for cytotoxic chemotherapy. The most significant advantage of gemtuzumab ozogamicin is the ability to use it for patients who are not likely to survive the toxicities of traditional chemotherapy. This is reflected in the fact that many gemtuzumab ozogamicin patients are treated on an out-patient basis, which reduces hospitalization costs[32] and which is virtually unheard of for traditional chemotherapy for AML due to the need to monitor for toxicities.

Gemtuzumab ozogamicin has been used in many different experimental settings in oncology research centers. In the studies in Japan that were similar to the studies used to approve gemtuzumab ozogamicin in the US, particularly striking results were obtained, with 4/40 patients surviving for over 44 months.[33] Gemtuzumab ozogamicin appears to be particularly active in APL (M4 sub-group of AML),[34] and has shown activity in AML patients with myeloid sarcoma,[35] as well as in CD33-positive ALL.[36,37] Fractionated dosing appears to be promising.[28] Gemtuzumab ozogamicin has also been examined in pediatric AML.[38]

Although gemtuzumab ozogamicin is active as a single agent as initial therapy in AML,[39] it appears to be more promising as part of combination chemotherapy, with many of the possible combinations having been explored at the MD Anderson Cancer Center. These include topotecan/Ara-C,[40] all-*trans*-retinoic acid,[23] interleukin 11,[41] liposomal daunorubicin/Ara-C/cyclosporine,[42] daunorubicin/Ara-C/thioguanine,[43] daunorubicin/Ara-C,[43] fludarabine/Ara-C/G-CSF/idarubicin,[43] fludarabine/Ara-C/cyclosporine,[44] oblimersen,[45] fludarabine/Ara-C/idarubicin,[45] Ara-C/mitoxantrone,[46] Ara-C/G-CSF,[47] and hydroxyurea/azacitidine.[48]

The most significant study is a comparison of daunorubicin ($45 \, \text{mg m}^{-2}$ on days 1, 2, and 3) and cytarabine ($100 \, \text{mg m}^{-2}$ on days 1 through 7), with or without gemtuzumab ozogamicin (at a reduced dose of $6 \, \text{mg m}^{-2}$ on day 4). A CR rate of 84% with gemtuzumab ozogamicin *vs.* 55% without has been seen so far. This combination with gemtuzumab ozogamicin has been well tolerated with no incidence of VOD. At 6 months the median duration of CR had not been reached.[18] Combinations have also been used in pediatric AML.[49]

5.8 Mechanistic Studies

Several laboratory support studies have been performed to help define the parameters for response and resistance for gemtuzumab ozogamicin. The role of CD33 is the most important of these parameters. Although CD33 expression does correlate with response,[50] the correlation is not as strong as might be expected. Indeed, there is evidence that response can be seen with no CD33 in cells that have high endocytic capacity.[51] It has also been shown that excessive levels of CD33 in circulation diminish response, presumably due to the difficulty of overcoming this antigen sink and accessing the sites of leukemia in the bone marrow.[52] Fortunately, polymorphisms in CD33 appear to have no effect on the response to gemtuzumab ozogamicin.[53]

Inherent sensitivity to calicheamicin is another obvious possible parameter that has been explored.[54] A 100 000-fold difference in sensitivity between the most- and least-sensitive bone marrow samples was seen. Some of this difference has been shown to be due to P-glycoprotein-mediated drug resistance.[55] Several attempts have been made both in laboratory studies and in clinical trials to overcome resistance,[56,57] but practical success will require further clinical work.[44]

Other resistance mechanisms have been explored. BCRP[58] and MRP2 do not confer resistance but MRP1 does.[55] As expected, apoptosis related pathways have been shown to be important, specifically Bak/Bax,[59] mitochondrial apoptotic pathways activated by PK11195,[57] and low levels of Chk1 and Chk2 phosphorylation.[60] Syk down-regulation has also been shown to correlate with resistance.[61]

5.9 Summary and Future Prospects

Gemtuzumab ozogamicin is still the first and only antibody-directed chemotherapeutic agent approved for use. The research that led to this agent was critical in the conjugate field in making some key observations. It was the first conjugate to show that modifying the structure of the cytotoxic agent, in this case by going from γ-calicheamicin conjugates to *N*-acetyl-γ- conjugates, can lead to a significant improvement in the therapeutic window in preclinical *in vivo* models. It and the calicheamicin conjugate CMB-401, which was being explored at the same time, showed that a single linker–cytotoxic agent combination was not necessarily the best for all antibodies, and that considerations need to be made as to the antigen being targeted and the properties of cells expressing that antigen.

Gemtuzumab ozogamicin was probably the easiest of all possible calicheamicin conjugates to start clinical investigations with, even though other conjugates have failed in this setting.[62] The target cells are readily accessible from circulation and they are known to be prone to apoptosis induced by DNA damage. However, clinical experience with this conjugate has amply demonstrated some of the pitfalls that can be encountered in conjugate development. Although a relatively high proportion of the conjugate binds to target cells, the toxicities of traditional chemotherapy are not altogether avoided, even though there is significant improvement. The response rate is quite meaningful, but less than 100% due to various resistance mechanisms, such as Pgp and resistance to apoptosis, likely making combination therapy the ultimate best use of gemtuzumab ozogamicin. Much more remains to be learned in on-going and future clinical trials, with the obvious goal of discerning the optimal use of gemtuzumab ozogamicin in all stages of AML, especially first indication.

As to the other applications of calicheamicin conjugates, inotuzumab ozogamicin (CMC-544), which targets the CD22 antigen, has advanced to Phase 3 clinical trials for non-Hodgkin's lymphoma; considering some of the similarities of these hematopoietic malignancies, it is reasonable to expect that inotuzumab ozogamicin would be successful, although the target cells are not as accessible as leukemic cells. The next obvious goal for calicheamicin conjugates is activity against solid tumors. CMB-401 failed in two small Phase 2 trials, possibly due to shed antigen.[63] The anti-Lewis^Y conjugate CMD-193 failed in Phase 1, with unexpected and extended liver uptake.[64] Although the cause of this unfortunate phenomenon is unknown, it should be noted that CMD-193 is the only calicheamicin conjugate to be advanced with an IgG$_1$ antibody, the rest having IgG$_4$ antibodies like gemtuzumab ozogamicin.

These initial failures reflect the difficulty of targeting solid tumors, especially in the Phase 1 setting of advanced disease. However, there has been renewed interest in conjugates due to the on-going success of the tubulin-active maytansine conjugate of trastuzumab (T-DM1).[65] The antibody in this conjugate is almost ideal, with restricted antigen expression, high antigen expression, and rapid internalization. It is hoped that conjugates of different classes of cytotoxic agents with antibodies to other such antigens will also find clinical success against solid tumors.

References

1. B. Lowenberg and B. A. Downing, Jr., *New Engl. J. Med.*, 1999, **341**, 1051.
2. J. L. Shipley and J. N. Butera, *Exp. Hematol.*, 2009, **37**, 649.
3. G. Schiller, S. A. Feig, M. Territo, M. Wolin, M. Lill, T. Belin, L. Hunt, S. Nime, R. Champlin and J. Gajewski, *Br. J. Haematol.*, 1994, **88**, 72.

4. P. R. Crocker and P. Redelingshuys, *Biochem. Soc. Trans.*, 2008, **36**, 1467.
5. R. B. Walter, B. W. Raden, D. M. Kamikura, J. A. Cooper and I. D. Bernstein, *Blood*, 2005, **105**, 1295.
6. R. G. Andrews, J. W. Singer and I. D. Bernstein, *J. Exp. Med.*, 1989, **169**, 1721.
7. M. D. Lee, G. A. Ellestad and D. B. Borders, *Acc. Chem. Res.*, 1991, **24**, 235.
8. S. Walker, J. Murnick and D. Kahne, *J. Am. Chem. Soc.*, 1993, **115**, 7954.
9. J. J. De Voss, J. J. Hangeland and C. A. Townsend, *J. Am. Chem. Soc.*, 1990, **112**, 4554.
10. J. J. Hangeland, J. J. De Voss, J. A. Heath, C. A. Townsend, W. D. Ding, J. S. Ashcroft and G. A. Ellestad, *J. Am. Chem. Soc.*, 1992, **114**, 9200.
11. L. M. Hinman, P. R. Hamann, A. T. Menendez, R. Wallace, F. E. Durr and J. Upeslacis, *Cancer Res.*, 1993, **53**, 3336.
12. P. R. Hamann, L. M. Hinman, C. F. Beyer, L. M. Greenberger, C. Lin, D. Lindh, A. T. Menendez, R. Wallace, F. E. Durr and J. Upeslacis, *Bioconjug. Chem.*, 2005, **16**, 346.
13. H. K. Erickson, W. C. Widdison, M. F. Mayo, K. Whiteman, C. Audette, S. D. Wilhelm and R. Singh, *Bioconjug. Chem.*, 2010, **21**, 84.
14. P. R. Hamann, L. M. Hinman, C. F. Beyer, D. Lindh, J. Upeslacis, D. A. Flowers and I. Bernstein, *Bioconjug. Chem.*, 2002, **13**, 40.
15. P. R. Hamann, L. M. Hinman, I. Hollander, C. F. Beyer, D. Lindh, R. Holcomb, W. Hallett, H.-R. Tsou, J. Upeslacis, D. Shochat, A. Mountain, D. A. Flowers and I. Bernstein, *Bioconjug. Chem.*, 2002, **13**, 47.
16. N. P. Barbour, M. Paborji, T. C. Alexander, W. P. Coppola and J. B. Bogardus, *Pharm. Res.*, 1995, **12**, 215.
17. J. F. DiJoseph, M. M. Dougher, L. B. Kalyandrug, D. C. Armellino, E. R. Boghaert, P. R. Hamann, J. K. Moran and N. K. Damle, *Clin. Cancer Res.*, 2006, **12**, 242.
18. P. R. Hamann, presented at the 8th International Congress on Recombinant Antibodies, Cologne, Germany, June 2009.
19. L. Wang, G. Amphlett, W. A. Blattler, J. M. Lambert and W. Zhang, *Protein Sci.*, 2005, **14**, 2436.
20. M. M. Siegel, K. Tabei, A. Kunz, I. J. Hollander, P. R. Hamann, D. H. Bell, S. Berkenkamp and F. Hillenkamp, *Anal. Chem. Symp. Ser.*, 1997, **69**, 2716.
21. K. J. Hamblett, P. D. Senter, D. F. Chace, M. M. C. Sun, J. Lenox, C. G. Cerveny, K. M. Kissler, S. X. Bernhardt, A. K. Kopcha, R. F. Zabinski, D. L. Meyer and J. A. Francisco, *Clin. Cancer Res.*, 2004, **10**, 7063.
22. J. A. Dowell, J. Korth-Bradley, H. Liu, S. P. King and M. S. Berger, *J. Clin. Pharmacol.*, 2001, **41**, 1206.
23. E. H. Estey, F. J. Giles, M. Beran, S. O'Brien, S. A. Pierce, S. H. Faderl, J. E. Cortes and H. M. Kantarjian, *Blood*, 2002, **99**, 4222.
24. U. Winkler, M. Jensen, O. Manzke, H. Schulz, V. Diehl and A. Engert, *Blood*, 1999, **94**, 2217.

25. F. J. Giles, J. E. Cortes, T. A. Halliburton, S. J. Mallard, E. H. Estey, T. A. Waddelow and J. T. Lim, *Ann. Pharmacother.*, 2003, **37**, 1182.
26. L. H. Leopold, M. S. Berger and J. Feingold, *Clin. Lymphoma*, 2002, **2**(S1), S29.
27. B. Versluys, R. Bhattacharaya, C. Steward, J. Comish, A. Oakhill and N. Goulden, *Blood*, 2004, **103**, 1968.
28. A. L. Taksin, O. Legrand, E. Raffoux, T. de Revel, X. Thomas, N. Contentin, R. Bouabdallah, C. Pautas, P. Turlure, O. Reman, C. Gardin, B. Varet, S. de Botton, F. Pousset, H. Farhat, S. Chevret, H. Dombret and S. Castaigne, *Leukemia*, 2007, **21**, 66.
29. E. L. Sievers, F. R. Appelbaum, R. T. Spielberger, S. J. Forman, D. Flowers, F. O. Smith, K. Shannon-Dorcy, M. S. Berger and I. D. Bernstein, *Blood*, 1999, **93**, 3678.
30. V. H. J. van der Velden, J. G. te Marvelde, P. G. Hoogeveen, I. D. Bernstein, A. B. Houtsmuller, M. S. Berger and J. J. M. van Dongen, *Blood*, 2001, **97**, 3197.
31. R. A. Larson, E. L. Sievers, E. A. Stadtmauer, B. Lowenberg, E. H. Estey, H. Dombret, M. Theobald, D. Voliotis, J. M. Bennett, M. Richie, L. H. Leopold, M. S. Berger, M. L. Sherman, M. R. Loken, J. J. M. van Dongen, I. D. Bernstein, F. R. Appelbaum, M. Boogaerts, S. Castaigne, P. Huijgens, R. Spielberger, M. Tallman, C. Bernasconi, J. L. Harousseau, C. Karanes, A. List, D. C. Roy, A. Goldstone, F. Mandelli, M. Schuster, M. Gobbi, P. Mineur, S. Tarantolo, M. Andre, A. Burnett, N. Cambier, P. Cassileth, J. Esteve, M. Gramatzki, G. Heil, G. Juliusson, S. Tura, G. Ehninger, G. A. Granena, J. Karp, J. P. Marie, A. Parreira, C. Paul, K. Rai, G. Schiller, J. Sierra, B. Simonsson, L. Stenke, M. Wernli, R. Willemze, M. Aglietta, D. Clausen, J. Conde, S. Coutre, M. N. Fernandez, D. Fiere, U. Hess, H. A. Horst, W. Linkesch, D. Mediavilla, M. Minden, F. Nobile, D. Schenkein and A. Wahlin, *Cancer*, 2005, **104**, 1442.
32. K. Lang, J. Menzin, C. C. Earle and R. Mallick, *Am. J. Health. Syst. Pharm.*, 2002, **59**, 941.
33. Y. Kobayashi, K. Tobinai, A. Takeshita, K. Naito, O. Asai, N. Dobashi, S. Furusawa, K. Saito, K. Mitani, Y. Morishima, M. Ogura, F. Yoshiba, T. Hotta, M. Bessho, S. Matsuda, J. Takeuchi, S. Miyawaki, T. Naoe, N. Usui and R. Ohno, *Int. J. Hematol.*, 2009, **89**, 460.
34. F. Lo-Coco, G. Cimino, M. Breccia, N. I. Noguera, D. Diverio, E. Finolezzi, E. M. Pogliani, E. di Bona, C. Micalizzi, M. Kropp, A. Venditti, A. Tafuri and F. Mandelli, *Blood*, 2004, **104**, 1995.
35. P. P. Piccaluga, G. Martinelli, M. Rondoni, M. Malagola, S. Gaitani, A. Isidori, A. Bonini, L. Gugliotta, M. Luppi, M. Morselli, G. Sparaventi, G. Visani and M. Baccarani, *Leuk. Lymphoma*, 2004, **45**, 1791.
36. A. Balduzzi, V. Rossi, L. Corral, S. Bonanomi, D. Longoni, A. Rovelli, V. Conter, A. Biondi and C. Uderzo, *Leukemia*, 2003, **17**, 2247.
37. J. Golay, N. Di Gaetano, D. Amico, E. Cittera, A. M. Barbui, R. Giavazzi, A. Biondi, A. Rambaldi and M. Introna, *Br. J. Haematol.*, 2005, **128**, 310.

38. R. J. Arceci, J. Sande, B. Lange, K. Shannon, J. Franklin, R. Hutchinson, T. A. Vik, D. Flowers, R. Aplenc, M. S. Berger, M. L. Sherman, F. O. Smith, I. Bernstein and E. L. Sievers, *Blood*, 2005, **106**, 1183.
39. C. Nabhan, L. M. Rundhaugen, M. B. Riley, A. Rademaker, L. Boehlke, M. Jatoi and M. S. Tallman, *Leuk. Res.*, 2005, **29**, 53.
40. J. Cortes, A. M. Tsimberidou, R. Alvarez, D. Thomas, M. Beran, H. Kantarjian, E. Estey and F. J. Giles, *Cancer Chemother. Pharmacol.*, 2002, **50**, 497.
41. E. H. Estey, P. F. Thall, F. J. Giles, X.-M. Wang, J. E. Cortes, M. Beran, S. A. Pierce, D. A. Thomas and H. M. Kantarjian, *Blood*, 2002, **99**, 4343.
42. E. Apostolidou, J. Cortes, A. Tsimberidou, E. Estey, H. Kantarjian and F. J. Giles, *Leuk. Res.*, 2003, **27**, 887.
43. W. J. Kell, A. K. Burnett, R. Chopra, J. A. L. Yin, R. E. Clark, A. Rohatiner, D. Culligan, A. Hunter, A. G. Prentice and D. W. Milligan, *Blood*, 2003, **102**, 4277.
44. A. Tsimberidou, J. Cortes, D. Thomas, G. Garcia-Manero, S. Verstovsek, S. Faderl, M. Albitar, H. Kantarjian, E. Estey and F. J. Giles, *Leuk. Res.*, 2003, **27**, 893.
45. J. Moore, K. Seiter, J. Kolitz, W. Stock, F. Giles, M. Kalaycio, D. Zenk and G. Marcucci, *Leuk. Res.*, 2006, **30**, 777.
46. P. Chevallier, J. Delaunay, P. Turlure, A. Pigneux, M. Hunault, R. Garand, T. Guillaume, H. Avet-Loiseau, N. Dmytruk, S. Girault, N. Milpied, N. Ifrah, M. Mohty and J.-L. Harousseau, *J. Clin. Oncol.*, 2008, **26**, 5192.
47. L. Fianchi, L. Pagano, F. Leoni, S. Storti, M. T. Voso, C. G. Valentini, S. Rutella, A. Scardocci, M. Caira, G. Gianfaldoni and G. Leone, *Ann. Oncol.*, 2008, **19**, 128.
48. S. Nand, J. Godwin, S. Smith, K. Barton, L. Michaelis, S. Alkan, R. Veerappan, K. Rychlik, E. Germano and P. Stiff, *Leuk. Lymphoma*, 2008, **49**, 2141.
49. R. Aplenc, T. A. Alonzo, R. B. Gerbing, B. J. Lange, C. A. Hurwitz, R. J. Wells, I. Bernstein, P. Buckley, K. Krimmel, F. O. Smith, E. L. Sievers and R. J. Arceci, *J. Clin. Oncol.*, 2008, **26**, 2390.
50. R. B. Walter, T. A. Gooley, V. H. J. van der Velden, M. R. Loken, J. J. M. van Dongen, D. A. Flowers, I. D. Bernstein and F. R. Applebaum, *Blood*, 2007, **109**, 4168.
51. I. Jedema, R. M. Y. Barge, V. H. J. van der Velden, B. A. Nijmeijer, J. J. M. van Dongen, R. Willemze and J. H. F. Falkenburg, *Leukemia*, 2004, **18**, 316.
52. V. H. J. van der Velden, N. Boeckx, I. Jedema, J. G. te Marvelde, P. G. Hoogeveen, M. Boogaerts and J. J. M. van Dongen, *Leukemia*, 2004, **18**, 983.
53. J. K. Lamba, S. Pounds, X. Cao, J. R. Downing, D. Campana, R. C. Ribeiro, C. H. Pui and J. E. Rubnitz, *Leukemia*, 2009, **23**, 402.
54. B. F. Goemans, C. M. Zwaan, S. J. H. Vijverberg, A. H. Loonen, U. Creutzig, K. Haehlen, D. Reinhardt, B. E. S. Gibson, J. Cloos and G. J. L. Kaspers, *Leukemia*, 2008, **22**, 2284.

55. R. B. Walter, B. W. Raden, T. C. Hong, D. A. Flowers, I. D. Bernstein and M. L. Linenberger, *Blood*, 2003, **102**, 1466.
56. R. Tang, A.-M. Faussat, J.-Y. Perrot, Z. Marjanovic, S. Cohen, T. Storme, H. Morjani, O. Legrand and J.-P. Marie, *BMC Cancer*, 2008, **8**, 51.
57. R. B. Walter, B. W. Raden, M. R. Cronk, I. D. Bernstein, F. R. Appelbaum and D. E. Banker, *Blood*, 2004, **103**, 4276.
58. R. B. Walter, B. W. Raden, J. Thompson, D. A. Flowers, H. P. Kiem, I. D. Bernstein and M. L. Linenberger, *Leukemia*, 2004, **18**, 1914.
59. P. Haag, K. Viktorsson, M. L. Lindberg, L. Kanter, R. Lewensohn and L. Stenke, *Exp. Hematol.*, 2009, **37**, 755.
60. D. Amico, A. M. Barbui, E. Erba, A. Rambaldi, M. Introna and J. Golay, *Blood*, 2003, **101**, 4589.
61. L. Balaian and E. D. Ball, *Leukemia*, 2006, **20**, 2093.
62. O. Legrand, M. B. Vidriales, X. Thomas, C. Dumontet, A. Vekhoff, R. Morariu-Zamfir, J. Lambert, J. F. S. Miguel and J.-P. Marie, *Blood (ASH Annual Meeting Abstracts)*, 2007, **110**, 1850.
63. S. Y. Chan, A. N. Gordon, R. E. Coleman, J. B. Hall, M. S. Berger, M. L. Sherman, C. B. Eten and N. J. Finkler, *Cancer Immunol. Immunother.*, 2003, **52**, 243.
64. R. A. Herbertson, N. C. Tebbutt, F.-T. Lee, D. J. MacFarlane, B. Chappell, N. Micallef, S.-T. Lee, T. Saunder, W. Hopkins, F. E. Smyth, D. K. Wyld, J. Bellen, D. S. Sonnichsen, M. W. Brechbiel, C. Murone and A. M. Scott, *Clin. Cancer Res.*, 2009, **15**, 6709.
65. G. D. Lewis Phillips, G. Li, D. L. Dugger, L. M. Crocker, K. L. Parsons, E. Mai, W. A. Blaettler, J. M. Lambert, R. V. J. Chari, R. J. Lutz, W. L. T. Wong, F. S. Jacobson, H. Koeppen, R. H. Schwall, S. R. Kenkare-Mitra, S. D. Spencer and M. X. Sliwkowski, *Cancer Res.*, 2008, **68**, 9280.

CHAPTER 6

Novel Androgen Receptor Antagonists for the Treatment of Prostate Cancer

ASHVINIKUMAR V. GAVAI*, WILLIAM R. FOSTER, AARON BALOG AND GREGORY D. VITE

Bristol-Myers Squibb Research and Development, P.O. Box 4000, Princeton, NJ 08543, USA

6.1 Introduction

The androgen receptor (AR), a transcription factor in the steroid and nuclear hormone receptor superfamily, is a key molecular target in the etiology and progression of prostate cancer.[1] AR is expressed in many target tissues, with the highest expression being observed in testis, adrenal gland, and the prostate.[2] Binding of steroid hormones, such as testosterone (T) or dihydrotestosterone (DHT), to the ligand binding domain (LBD) of the AR promotes its dissociation from heat shock proteins, allowing its dimerization, phosphorylation, and subsequent translocation into the nucleus, where it binds to androgen-response elements located within the regulatory regions of a variety of genes. In conjunction with AR-associated proteins (AR co-activators or co-repressors), the AR complex can regulate the transcription of genes, many of which are involved in the growth, survival, and differentiation of prostate cells.[3] Options for targeting prostate cancer include surgical castration or luteinizing hormone-releasing hormone agonists that deplete serum testosterone concentrations, often in combination with an AR antagonist such as bicalutamide.[2] Despite an

RSC Drug Discovery Series No. 4
Accounts in Drug Discovery: Case Studies in Medicinal Chemistry
Edited by Joel C. Barrish, Percy H. Carter, Peter T. W. Cheng and Robert Zahler
© Royal Society of Chemistry 2011
Published by the Royal Society of Chemistry, www.rsc.org

initial favorable response, most tumors advance to a hormone-refractory prostate cancer (HRPC) stage in an average of 18 months. The designation hormone-refractory is misleading, since there is clear dependence on AR signaling even in the most advanced forms of HRPC. Microarray-based profiling of isogenic hormone-sensitive and hormone-refractory human prostate cancer xenograft mouse pairs indicated that AR mRNA was upregulated in all HRPC models.[4] Serum prostate specific antigen (PSA), the secreted protein product of an AR-regulated gene, is the most prognostic biomarker for prostate cancer.[1] PSA declines after initiation of hormone depletion therapy, and a subsequent rise is commonly the first sign of disease progression. This indicates that reactivation of AR signaling accompanies the development of HRPC. Analyses of tumor samples from patients with HRPC have suggested several mechanisms used by tumor cells to reactivate AR signaling, including gene amplification and increased expression of the AR protein, and selection of point mutations in the AR LBD that can result in activation by non-androgenic ligands. Taken together, these findings suggest that more potent AR pan-antagonists than bicalutamide may provide additional clinical benefits in the treatment of prostate cancer.

6.2   Bristol-Myers Squibb Approach

6.2.1   Strategy

The goal of the Bristol-Myers Squibb (BMS) AR antagonist program was to identify a new generation of anti-androgens with improved binding and functional AR antagonist potency over nonsteroidal clinical anti-androgens, such as bicalutamide, nilutamide, and flutamide.[2] 2-Hydroxyflutamide is the major metabolite of flutamide, with higher binding affinity for AR than the parent compound. Clinical utility of flutamide and nilutamide has been restricted due to the observed hepatotoxicity after prolonged administration. Bicalutamide is the preferred clinical anti-androgen because of its long half-life and significantly reduced liver toxicity. However, patients on bicalutamide are also susceptible to the anti-androgen withdrawal syndrome (AWS), a clinical phenomenon described in terms of decreasing PSA values and clinical improvement.[5] AR gene mutation to a form that can accommodate the anti-androgen in an agonist conformation has been offered as a plausible explanation for the observed AWS. Indeed, bicalutamide has been shown to work as an agonist for the W741C and W741L mutant variations of AR.[5b] Therefore, the program objective was to develop full antagonists against both wild-type as well as known mutant AR isoforms. Since bicalutamide is frequently employed as part of the initial treatment regimen for prostate cancer, it was important to demonstrate efficacy for refractory prostate tumors previously treated with bicalutamide. Existing experience suggested that complete and sustained AR antagonism for the entire duration of treatment was essential in order to observe a robust anti-tumor response. Indeed, mean steady-state concentration

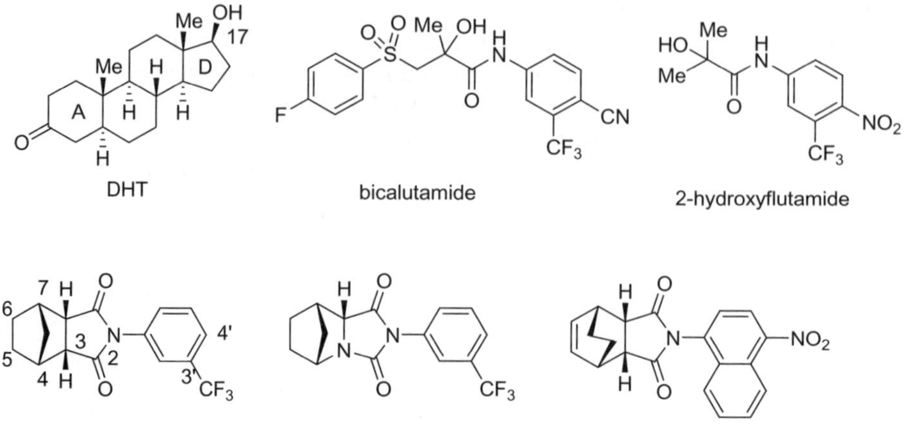

Figure 6.1 DHT and initial AR antagonist leads.

of (R)-bicalutamide at the clinically efficacious dose of bicalutamide was 21.6 mg L^{-1} (50 μM).[6] Therefore, the desired development candidate needed to exhibit a robust pharmacokinetic (PK) profile upon oral administration, and be well tolerated at adequate multiples of the efficacious concentration.

Structure-based tools for the rational design of novel AR antagonists were not available at the start of the program because co-crystal structures of ligands with the AR LBD had not been solved. In 2001, BMS reported the first crystal structures of the LBD of the AR, and its T877A mutant, complexed with the natural agonist DHT.[7a] In these structures, ring A of DHT (Figure 6.1) is engaged in hydrogen bond interactions with the side chains of R752 and Q711. In addition, the hydroxyl group of DHT is within hydrogen bonding distance of N705 and T877. In the mutant LBD, replacement of T877 with alanine creates additional space near the D ring of DHT to accommodate a larger substituent near position 17. This provided a rationale for the promiscuous ability of the T877A mutant AR present in the LNCaP prostate cells to bind additional hormones, such as progesterone, in the agonist conformation. The X-ray crystal structure of (R)-bicalutamide bound in an agonist conformation to the W741L mutant AR LBD has been reported.[7b] These crystal structures provided molecular rationale for the design of novel AR pan-antagonists.

6.2.2 Biological Assays

The general paradigm employed in this program and the primary assay information have been recently published.[8] A radioligand AR-competitive binding assay was performed using [^{3}H]-DHT and human breast adenocarcinoma MDA-MB-453 cells expressing endogenous wt AR. The effect of compounds on the transcriptional activity of AR was determined in cell lines expressing either endogenous wt (MDA-MB-453) or mutant (CWR-22rv1 or LNCaP) forms of AR that were transiently transfected with constructs consisting of secreted alkaline phosphatase or luciferase reporter genes driven by both the immediate promoter and enhancer of the human PSA gene. Effects of human prostate cell proliferation were assessed in LNCaP cells, or LNCaP cells engineered to overexpress wt AR. Pharmacodynamic (PD) effects on the growth of ventral prostate and seminal vesicles induced by exogenous testosterone propionate supplementation were evaluated in sexually immature male rats in a 3-day assay (immature rat prostate weight or IRPW assay). Weights of the prostates and seminal vesicles, expressed as percent control of the full body weight, were obtained to gauge the extent of PD effects. Anti-tumor activity was measured in prostate tumor xenograft mouse models derived from patients with advanced prostate carcinoma. The CWR-22-BMSLD1 (H874Y mutant AR) human tumors are derived from prostate and retain sensitivity to endogenous circulating levels of androgens, but are only marginally responsive to bicalutamide. Thus, this model was appropriate to investigate anti-tumor activity in a bicalutamide-refractory setting. MDA PCa 2b is a tumor line derived from bone metastasis of a prostate cancer patient and harbors two AR mutations (L701H and T877A) which confer decreased AR sensitivity and increased susceptibility to AR agonist activity of hormones, such as cortisol and cortisone. LuCaP 23.1 and LuCaP 35 are human tumor xenografts derived from lymph node metastases which express wt AR with gene amplification and are expected to provide a good measure of activity against HRPC.

6.2.3 Screening Hits and Early Structure–Activity Relationships

An *in silico* pharmacophore-driven screening approach based upon clinical anti-androgens, such as bicalutamide[2] and 2-hydroxyflutamide,[2] led to the identification of novel bicyclic imides[9] and hydantoins[10] (*e.g.*, **1** and **2**, Figure 6.1) that demonstrated potent binding affinity to and functional antagonism of the endogenous wt AR in MDA-MB-453 cells (Table 6.1). Attempts to co-crystallize ligands in an antagonist conformation with the AR LBD were unsuccessful due to aggregation of the protein. Compound **3** provided a unique opportunity since it was an antagonist of wt AR, but exhibited an agonist profile toward the T877A mutant AR in LNCaP cells.[11] Co-crystal structures of the T877A AR LBD with **3** provided significant insight into binding of the lead series to AR at the molecular level. Several key structural features were elucidated upon comparison of this structure to the available crystal structure of DHT with T877A AR LBD.[11] The 4-nitro group on the naphthalene ring occupies the same space as the C3 carbonyl of DHT, making contacts with R752 and Q711. F764 forms an edge–face interaction with the first aromatic ring of the naphthalene. The lipophilic bicyclic portion of **3** occupies the same position in the LBD as the C and D rings of DHT, but does not carry substituents to form hydrogen bonds similar to those forged by the C17 hydroxyl of DHT. Neither the carbonyls nor the nitrogen atom of the

Table 6.1 Profile of first-generation AR antagonists.

Compound	AR Binding Ki (nM)	AR Transactivation IC_{50} (nM)
1	360	152
2	21	130
3	19	224
4	3	> 5000
5	1170	2860
6	16	20
BMS-949	40	41
BMS-305	8	10
BMS-511	14	44
7	3	23
8	5	18
BMS-641988	2	16
2-hydroxyflutamide	43	26
bicalutamide	37	173

imide portion of **3** generate significant interactions with any residues in the AR LBD. This result was surprising since the known non-steroidal anti-androgens contain an aryl amide functionality isosteric with the imide system of these BMS AR antagonists. Perhaps the primary role of the imide system is to constrain the bicyclic and aromatic sections of the molecule into a geometry that optimizes binding to AR. Modeling studies suggest a steric clash between the bicyclic portion of **3** and the T877 residue in wt AR, thus preventing it from being accommodated in an agonist conformation.

Structure–activity relationship (SAR) trends around the aniline moiety of **1** established that electron-withdrawing groups at the 4'-position had the greatest impact on binding to AR, with an additive effect observed for an additional electron-withdrawing substituent at the 3'-position. Both the bicyclic portion as well as the aniline appeared to contribute to the observed agonist/antagonist profile in MDA-MB-453 (wt AR) or LNCaP (T877A AR) cells. Despite significant advances in *in vitro* potency, this series was eventually de-emphasized since lead compounds exhibited unsatisfactory PK properties related to rapid clearance and poor oral bioavailability.[12]

6.3 Advanced Preclinical Leads

6.3.1 Identification of BMS-949

Introduction of an oxygen atom in the bicyclic portion of the first lead series **1** generated novel oxabicyclo[2.2.1] imides (Figure 6.2) that showed significant improvement in oral absorption,[12] in addition to potent binding to the AR (Table 6.1). Attempts to replace the 4-nitronaphthyl group in **4** led to **5** with a significant drop in AR binding, but improved overall functional antagonist activity. Modeling studies noted the presence of two methionine residues (M780 and M895) in the AR LBD adjacent to the C4 and C7 positions of the

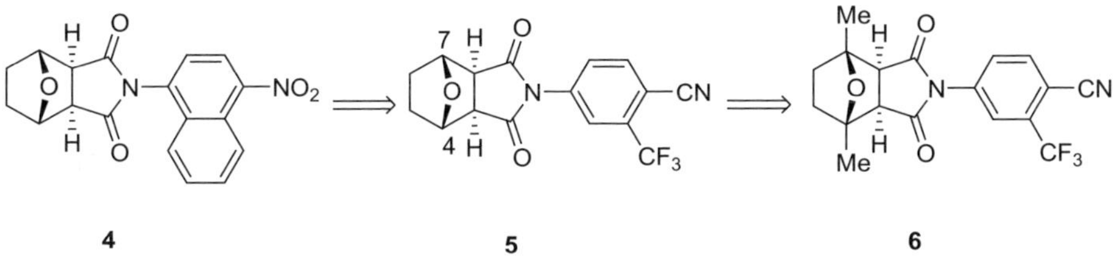

Figure 6.2 Hit-to-lead progress in the oxabicyclo[2.2.1] imide series.

oxabicyclic ring system. Appending methyl groups from the two bridgehead positions would be expected to result in positive lipophilic interactions with each of the methionine residues, thus generating improved binding affinity to the AR. Indeed, the dimethyl analog **6** showed almost a 100-fold improvement in binding affinity and a commensurate increase in antagonist activity as compared to **5**.

Upon daily oral administration of just a 1 mg kg^{-1} dose, **6** exhibited excellent PD effects in the immature rat prostate weight (IRPW) assay. However, the extent of the observed *in vivo* potency of this compound did not correlate with the very low plasma exposures of this compound (below the IC$_{50}$ in the transactivation assay) that were achieved in this study. Since this result was suggestive of the presence of an active metabolite, a rapid bioassay-guided method was designed to generate and detect potential active metabolites.[13] This method confirmed the presence of a single oxidative metabolite with potent AR antagonist activity. Based on the reported metabolite profile for a similar oxabicycle,[14] hydroxylation at either the C5 or C6 position of **6** was hypothesized. Concurrent efforts in chemical synthesis and LC-NMR analysis confirmed that BMS-949 (Figure 6.3) was the active metabolite of **6**. Chiral HPLC analysis showed that the oxidative metabolism was highly enantioselective (>99% enantiomeric excess), resulting in the exclusive formation of the *exo*-alcohol. Subsequently, BMS-949 was shown to be the major metabolite formed by CYP 3A4 mediated oxidation of **6** in both rodent and human liver microsomes (Figure 6.3).

BMS-949 demonstrated potent binding to, and antagonism of, the AR (Table 6.1). Importantly, it exhibited excellent oral bioavailability (>85%) and long half-life (10 h in mice to 28 h in dogs) in animals. At a low oral dose of 1 mg kg^{-1}, BMS-949 provided PD effects comparable to castration control in the IRPW assay. In the CWR-22-BMSLD1 human tumor xenograft model, BMS-949 (30 and 90 mg kg^{-1}, po, qd) showed superior efficacy compared to the highest achievable dose of bicalutamide (150 mg kg^{-1} day^{-1}).

6.3.2 Preclinical Safety Evaluation of BMS-949

In the *in vitro* competitive binding safety pharmacology assays of 25 off-target transporters, ion channels, and receptors, BMS-949 did not exhibit significant

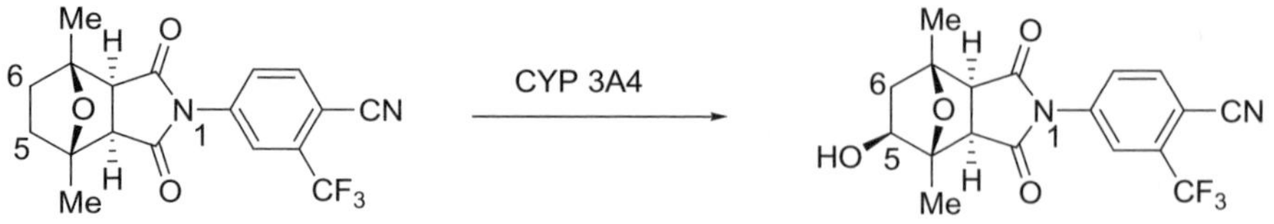

Figure 6.3 Oxidative metabolism of **6**.

activity at 10 µM. In a two-week exploratory toxicity study in male rats, BMS-949 was well tolerated at daily oral doses of 2.5 and 25 mg kg^{-1} with dose-proportional exposure and no adverse non-mechanism based pharmacological changes. However, convulsions were observed at the highest dose tested of 250 mg kg^{-1} at approximately T_{max} following three days of dosing. In a 10-day exploratory toxicity study in male dogs, convulsions occurred following five to six days of daily oral dosing at 25 and 50 mg kg^{-1}. In one-month rat and dog studies with maximum doses selected to be below those producing convulsions in the prior two-week rat and 10-day dog studies, pharmacology (prostate atrophy) without toxicity was observed. These studies established a no adverse effect level (NOAEL) at systemic exposures that were 2- to 4-fold (rat and dog, respectively) of the efficacious exposure required in CWR-22-BMSLD1 tumor xenograft studies. Convulsions can be life threatening both directly (*e.g.*, by interrupting respiration) and indirectly (*e.g.*, by accidents following loss of consciousness while driving a car or while walking on stairs). Because of the serious consequences of convulsions, the lack of predictive symptoms preceding their onset, and the often steep exposure–convulsion relationship, regulatory guidance on the starting human dose is to increase the safety margins for human testing when convulsion is an identified hazard.[15] In light of the low safety multiples for the observed CNS effects and the expected chronic treatment regimen for an AR antagonist, human testing of BMS-949 was not pursued.

Following discontinuation of the development of BMS-949, a series of studies were designed to investigate the underlying cause of the convulsions and to improve the screening strategy for the identification of new AR antagonists. These studies attempted to answer the following. (a) Were convulsions specific to the structural class under investigation? (b) Were the convulsions an on-target effect of AR modulation? (c) Was there potential explanatory off-target pharmacology for the convulsions? (d) Were the convulsions due to seizure in the central nervous system (CNS) or were they due to a peripheral effect, such as in the muscle or at the neuromuscular junction? The outcome of these studies has been described in detail elsewhere.[16] In brief, deaths and/or convulsions were observed in male mice following oral administration of a structurally diverse set of AR antagonists, including the clinical anti-androgens nilutamide and flutamide. Convulsions were also observed in studies with AR knockout

(KO) mice, thus demonstrating that convulsion was not a necessary consequence of the intended pharmacology. Electroencephalography in the rat established that the convulsions were caused by classical seizure, and drug concentration measurements in brain and plasma across a structurally diverse set of AR antagonists established the generality of steep dose–responses for convulsion and the importance of brain penetration. *In vitro* proconvulsive target screening studies revealed that the common off-target pharmacology of $GABA_A$ inhibition was shared by all AR antagonists tested ($IC_{50} = 7$–$47\,\mu M$) and spanning a broad range of chemical structures (*e.g.*, including the AR antagonists bicalutamide, 2-hydroxyflutamide, nilutamide, and the imide-based BMS anti-androgens). These results led to the hypothesis that antagonism of the $GABA_A$ receptor was causal for convulsions when combined with high brain drug concentrations and successful drug competition for the steroid sites occupied by endogenous $GABA_A$ modulators.

The outcome of these studies posed significant challenges for modifications in the screening strategy. Efforts to identify potent AR antagonists that were devoid of measurable $GABA_A$ antagonist activity were not successful. A potential approach to reduce seizure risk would be to identify compounds with limited distribution to the CNS. However, no high throughput assay was available to reliably predict distribution of compounds to the CNS at high doses. Therefore, compound selection had to rely on assessment of low throughput *in vivo* properties determinative of safety margin, such as plasma and brain drug concentrations corresponding to the observed CNS events, and projected human C_{max} for the efficacious dose. Therefore, the previously described screening paradigm was altered to include assessment of CNS effects in mouse single dose toxicokinetic and tolerability studies prior to evaluation of compound activity in repeat-dose testing in the CWR-22-BMSLD1 tumor xenograft mouse model. In these studies, clinical signs were continuously monitored for four hours following administration of high doses (500–$1000\,mg\,kg^{-1}$) of the AR antagonists and drug concentrations in plasma and brain were determined at four hours post-dose. Plasma and brain exposure data from the IRPW study were also utilized to get an early read on the brain penetration potential of the lead compounds.

6.3.3 Improved CNS Profile with BMS-305

Further optimization using the modified screening strategy identified BMS-305, a close analogue of BMS-949 in which the trifluoromethyl group at the 3′-position of the aniline ring was replaced by an iodide (Figure 6.4). As shown in Table 6.1, this simple change provided a 4-fold increased potency in the *in vitro* assays. BMS-305 was efficacious in the CWR-22-BMSLD1 tumor model when administered orally at a low dose of $10\,mg\,kg^{-1}\,day^{-1}$ ($C_{max} = 26\,\mu M$).[12] Despite achievement of very high brain exposure ($293\,\mu M$), BMS-305 did not produce significant CNS effects in mice upon daily oral administration of a $1000\,mg\,kg^{-1}$ dose. No convulsions were observed in exploratory rat and dog

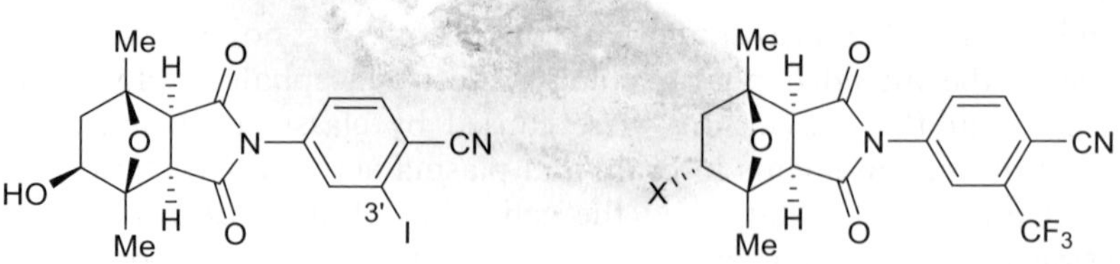

Figure 6.4   Lead compounds in the oxabicyclo[2.2.1] imide series.

toxicity studies. A less than dose-proportional increase in exposure limited maximum achievable systemic exposures in both species. Unfortunately, alterations in the levels of thyroid hormones T_3 and T_4, as well as significant quantities of the de-iodinated metabolite of BMS-305 in plasma, were detected following two weeks of dosing in rats and following 10 days of dosing in dogs. Subsequent studies demonstrated that BMS-305 was a substrate of 5′-deiodinase, leading to interference with thyroid hormone homeostasis. Although further development of BMS-305 was terminated, these studies demonstrated that it was possible to identify oxabicyclo[2.2.1] imide-based potent AR antagonists that lacked significant CNS liabilities.

6.4   Clinical Leads

6.4.1   Discovery of BMS-641988

At this stage, attention was focused on optimizing functional groups on the C5–C6 bridge of the bicyclic scaffold. Modeling studies suggested that the bridging oxygen atom can forge a hydrogen bonding interaction with N705 in the AR LBD. *Endo* substitution at C5 or C6 of the core would result in a steric clash and a possible displacement of helix-12, similar to the proposed antagonist conformation of bicalutamide.[2] A variety of functional groups, including amides, carbamates, sulfamides, ureas, and sulfonamides, were tolerated at the C5 or C6 positions.[17] Three compounds with highly desirable *in vitro* potency profiles (Table 6.1) are shown in Figure 6.4. With daily oral administration of a 10 mg kg^{-1} dose, these compounds produced PD effects similar to castration in the three-day IRPW assay.

Exploratory studies in mice with BMS-949 suggested that convulsions can occur at lower doses with repeat *versus* single-dose regimen. This finding may be due to a combination of drug accumulation, poor tolerability, or time-dependent increases in seizure sensitivity. Therefore, the selection process was modified to include a five-day mouse behavioral study for lead compounds that passed seizure screening with a single high dose. In the five-day mouse toxicity study at 450 mg kg^{-1} day^{-1}, amide **7** and sulfonamide **8** were lethal to one or more mice and their development was discontinued. In the same study, BMS-641988 showed clinical signs of decreased activity but no convulsions or death over the entire dosing period. Findings of decreased activity were absent by the end of the study and correlated with a 5-fold drop in maximal exposure with repeat versus single dosing, likely due to CYP 3A enzyme induction since hepatic transcript abundance for CYP 3A1 increased 10-fold and *in vitro* reaction phenotyping studies with human P450s identified oxidation by CYP 3A4 as the principal metabolic route.

BMS-641988 exhibits higher AR binding affinity and significantly increased potency as an antagonist of both wt and mutant AR transcriptional activity compared with bicalutamide.[8] Although BMS-641988 showed potent antagonist activity in transactivation assays done with LNCaP cells transiently transfected with the artificial reporter constructs, it also promoted the proliferation of these cells. This T877A mutant AR agonist activity has been observed with other potent anti-androgens, such as nilutamide and flutamide.[2] In a 14-day PD study in mature male rats, BMS-641988 provided a significant reduction in the weights of ventral prostate and seminal vesicles as well as in the percentage of nuclei stained positively for the proliferation marker Ki-67 in prostate sections compared with the vehicle control. At daily oral doses of $10\,\text{mg}\,\text{kg}^{-1}$ and beyond, BMS-641988 demonstrated significant anti-tumor activity in the CWR-22-BMSLD1 human xenograft mouse model. The tumors were only marginally sensitive to $150\,\text{mg}\,\text{kg}^{-1}\,\text{day}^{-1}$ bicalutamide, which provided 3-fold higher steady-state exposures in mice than those obtained with its clinical dose in humans. Animals receiving $90\,\text{mg}\,\text{kg}^{-1}\,\text{day}^{-1}$ of BMS-641988 exhibited tumor stasis during treatment (97% tumor growth inhibition) and a 3-fold delay in target tumor progression on cessation of treatment compared with those receiving the highest dose of bicalutamide. Furthermore, BMS-641988 decreased serum PSA levels, a surrogate marker for tumor burden in humans,[18] up to 80% compared to the vehicle controls, whereas bicalutamide caused less than 50% reduction in PSA. In addition, the efficacy of BMS-641988 was demonstrated in the androgen-dependent LuCaP 23.1 xenografts driven by wt AR.

The effectiveness of BMS-641988 as a second-line hormonal therapy was assessed in CWR-22-BMSLD1 xenografts failing bicalutamide treatment. Mice receiving daily bicalutamide at $150\,\text{mg}\,\text{kg}^{-1}$ showed a 3-fold increase in median tumor size after 14 days of treatment. At this point, one group was switched to treatment with BMS-641988 $(90\,\text{mg}\,\text{kg}^{-1})$, and the remaining animals continued the treatment with $150\,\text{mg}\,\text{kg}^{-1}$ of bicalutamide. As shown in Figure 6.5, animals switched to BMS-641988 showed minimal tumor growth, whereas tumors from bicalutamide treated animals continued to grow progressively. These data indicate that tumors that fail bicalutamide treatment retain their sensitivity to BMS-641988.

BMS-641988 exhibited good oral bioavailability (64–95%) in mouse, rat, dog, and cyno PK studies. Plasma half-life ranged from 2.4 h in mice to 21.8 h in dogs. Based on low metabolic rates in human liver microsomes and hepatocytes, as well as allometric scaling in comparison to other species, the human

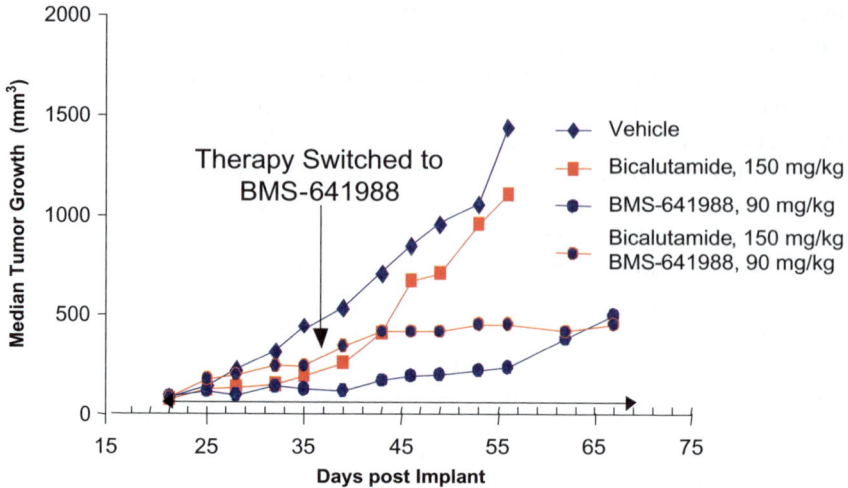

Figure 6.5 Anti-tumor activity of BMS-641988 in CWR-22-BMSLD1 xenografts after failure to bicalutamide therapy. Reprinted with permission from *Cancer Res.*, 2009, **69**, 6522.

clearance and plasma elimination half-life were predicted to be 0.75 mL min^{-1} kg^{-1} and 31 h, respectively. Given the significant discrepancy in the observed half-life in mice and the predicted human half-life, BMS-641988 was administered to CWR-22-BMSLD1 tumor-bearing mice *via* subcutaneous mini-osmotic pumps to determine the sustained minimum exposure required for efficacy. Greater than 50% tumor growth inhibition (TGI) was achieved with a peak concentration of 0.53 μM, and an average steady-state concentration of 0.32 μM of BMS-641988. Analysis of the plasma samples from this efficacy study revealed metabolism of BMS-641988 to BMS-949, with an average steady-state concentration of about 2 μM BMS-949. Reaction phenotyping, using recombinant human isoforms and human liver microsomes, determined that metabolism occurs *via* CYP 3A4 mediated oxidation of the C5 position of BMS-641988, followed by elimination of the sulfonamide to generate BMS-511, which is rapidly reduced by cytosolic reductase to BMS-949 (Figure 6.6). Both BMS-511 and BMS-949 are active metabolites of the parent drug since they exhibit potent binding and functional antagonism to AR (Table 6.1). *In vivo* characterization of BMS-511 could not be accomplished since it was readily converted to BMS-949 in all preclinical species. The formation of BMS-949 was of particular concern due to its ability to induce seizures in rodents and dogs at high concentrations (>150 μM). However, a comfortable safety margin was anticipated since only 2 μM steady-state concentrations of BMS-949 were observed in the minimum efficacious exposure study with BMS-641988.

Due to the presence of a base-labile imide functionality, the parent form of BMS-641988 is in a pH-dependent equilibrium with two regioisomers of the

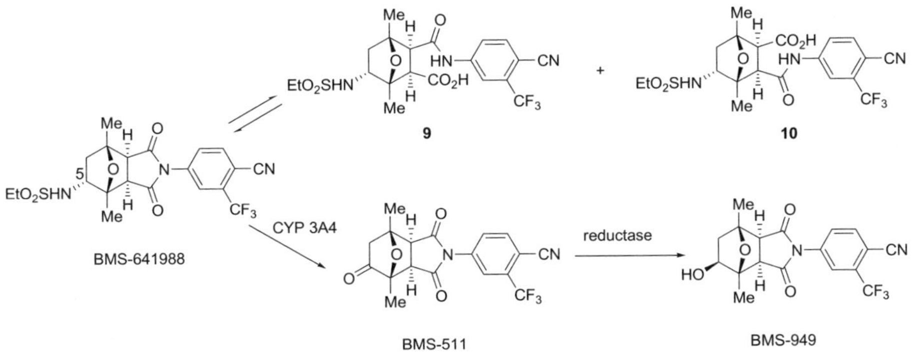

Figure 6.6 BMS-641988: stability and metabolism.

open imide form (Figure 6.6). In all animals investigated, BMS-641988 was the favored species, with the open imide isomers being generated in concentrations of 8–15% of the parent. While this equilibrium did not raise any additional safety or efficacy concerns, it necessitated additional analytical steps to monitor levels of the parent drug and its metabolites in subsequent studies.

6.4.2 Preclinical Safety Profile of BMS-641988

In a one-month toxicity study in rats, pharmacologic effects were observed in the prostate at all doses and there were no behavioral changes at any dose. Exposures decreased by 50% at the high dose of $250 \, \text{mg} \, \text{kg}^{-1} \, \text{day}^{-1}$ from day 1 to day 29, while liver weights increased by 50%. Exposures of BMS-641988 at this dose were approximately 50-fold higher than the minimum exposure required for efficacy in xenograft models. BMS-949 was the most abundant metabolite in circulation, but only at low exposure levels, thus providing a comfortable margin with respect to the CNS threshold.

In a one-month toxicity study in dogs, pharmacologic effects were observed in the prostate at all doses. Clinical signs of toxicity observed at and above the $25 \, \text{mg} \, \text{kg}^{-1} \, \text{day}^{-1}$ dose included ataxia, decreased activity, and loss of righting reflex, but not convulsions. At the lowest dose, where no clinical signs of toxicity were observed, plasma concentrations of BMS-641988 were approximately 20-fold of the efficacious exposure in tumor models. In contrast to BMS-949, which is highly brain penetrant, whole brain concentrations of BMS-641988 did not exceed 10% of plasma values in dogs at 24 h post-dose. Very low levels of BMS-949 were detected in the dog toxicity studies, resulting in a high safety multiple (>100) for seizure effects.

Binding data for cardiac ion channels and *in vitro* electrophysiology studies suggested low potential for cardiovascular liabilities with BMS-641988. However, BMS-641988 produced dose- and exposure-related reversible QT prolongation in telemetrized dogs with a no-effect concentration of 5-fold the projected maximal exposure required for efficacy. BMS-641988-dependent QT prolongation was not predicted by *in vitro* assessments of hERG inhibition since QT prolongation was observed with plasma drug concentrations of $2\,\mu M$ free drug while the whole-cell patch-clamp hERG IC_{50} of BMS-641988 was $46\,\mu M$ (free drug). Lack of activity in sodium or calcium channel whole-cell patch assays, absence of any significant effects on action potential duration in Purkinje fiber assays, or QT effects in suspended rabbit hearts were also in contrast to the observed *in vivo* cardiovascular effects. The observed QT prolongation was unlikely to be due to the active metabolites BMS-949 and BMS-511, since their combined concentrations did not exceed 2% of parent drug C_{max} in dogs. Furthermore, BMS-949 did not cause QT prolongation in dogs at concentrations significantly higher than those achieved in the BMS-641988 dog cardiovascular safety study.

In summary, BMS-641988 did not cause convulsions at any dose. It is an antagonist of the $GABA_A$ receptor, but it did not induce seizures in dogs up to the maximum dose tested ($75\,\mathrm{mg\,kg^{-1}\,day^{-1}}$) with a plasma drug concentration of $80\,\mu M$ ($13\,\mu M$ free). It showed limited brain exposure across all species (brain to plasma ratio of 0.02 to 0.15). Concentrations of the seizurogenic active metabolite BMS-949 were higher in rodents than in dogs. A significant margin (>20-fold) was established, based on extensive preclinical seizure studies and the predicted human exposure of BMS-949 at the efficacious dose. However, the therapeutic index for the observed QT prolongation was relatively low (5-fold). In light of its impressive efficacy and tolerable safety profile, BMS-641988 was selected for clinical evaluation of safety and efficacy in humans. Simultaneously, a backup program was initiated to address the potential liabilities of this clinical candidate.

6.4.3   Second-generation AR Antagonist: BMS-779333

The goal of the backup program was to identify an AR pan-antagonist with an improved safety profile compared to BMS-641988 and with an equivalent or superior *in vivo* efficacy in tumor models. The desired compound would exhibit a significantly improved safety margin for cardiovascular effects, would be devoid of chemical stability issues under physiologic conditions, and would have no major active circulating metabolites in preclinical studies. The sulfonamide group in BMS-641988 was implicated in the formation of both the major metabolites, whereas the imide functionality of BMS-641988 was responsible for its chemical equilibrium with the ring-open forms. Based on extensive SAR, neither of these two functionalities appeared to be essential for potent AR antagonist activity in the lead series. Analysis of the available crystal structures of the T877A AR LBD with imides in agonist mode further confirmed that the imide moiety generated no significant interactions within the LBD. Therefore, initial medicinal chemistry efforts were focused on the tricyclic lactams and sultams represented by the lead compounds **11** and **12**, respectively (Figure 6.7). However, both the lactam and sultam series were eventually dropped, primarily due to suboptimal PK properties leading to a lack of robust *in vivo* efficacy in tumor xenograft models.

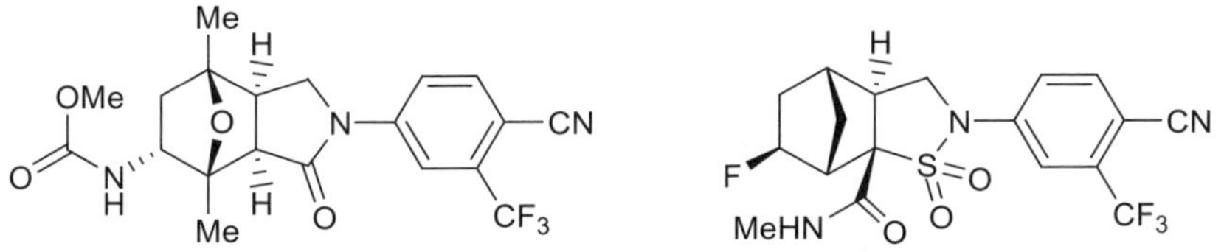

Figure 6.7 Lead compounds in the lactam and sultam series.

Significant differences in SAR between the imide and lactam series suggested that the imide carbonyls may be important to enhance the AR antagonist profile of these compounds. Extensive medicinal chemistry efforts to improve the overall profile of these related series included detailed functional group manipulation at each position of the tricyclic core. One such avenue of exploration to modify one of the quaternary methyl groups of the tricyclic lactam yielded an unanticipated result. Attempted reduction of a hydroxyethyl-substituted tricyclic imide **13** led to the novel tetracyclic compound **15**, instead of a corresponding tricyclic lactam **14** (Figure 6.8). Compound **15** showed high level of binding and potent antagonism to AR (Table 6.2). In spite of the presence of an "aminal-like" functionality in these molecules, pH-dependent solution-state stability studies established that the tetracyclic scaffold was stable over a broad pH range (2.0–7.4) for two weeks at 50 °C. The observed stability may be related to the significantly reduced basicity of the aniline nitrogen atom. In addition, the constrained pyran ring may provide added chemical stability to these compounds.

SAR studies suggested that a variety of functional groups (*e.g.*, alcohol, carbamate, sulfonamide, *etc.*) were tolerated on the C2–C3 two-carbon bridge of the tetracyclic ring system. As shown in Table 6.2, compound **16** exhibited an excellent *in vitro* profile and provided castration level PD effects at a low daily dose of $3 \, \text{mg} \, \text{kg}^{-1}$ in the IRPW assay (Figure 6.9). However, it failed to elicit a significant level of efficacy in the CWR-22-BMSLD1 tumor xenograft model, even at a daily dose of $150 \, \text{mg} \, \text{kg}^{-1}$. Analysis of the plasma samples from the tumor study suggested that the observed lack of activity may be related to the inability of the compound to provide sustained AR antagonism for the duration of time between doses. Plasma exposures of the compound were excellent ($>25 \, \mu\text{M}$) for the first 8 hours after dosing, but the trough exposures ($0.06 \, \mu\text{M}$, 87% protein binding) were not sufficient to completely antagonize the androgen receptor, given the compound's antagonist IC_{50} value of 226 nM. This result highlighted the need to improve PK properties in this lead series. Biotransformation studies confirmed that the *endo*-carbamate moiety was the primary site of metabolism for **16**. *N*-Dealkylation of the carbamate moiety *via* oxidation at the carbon bearing the carbamate nitrogen led to the formation of inactive ketone and alcohol metabolites. Importantly, these studies indicated that the tetracyclic framework offered a satisfactory level of metabolic stability under physiologic conditions.

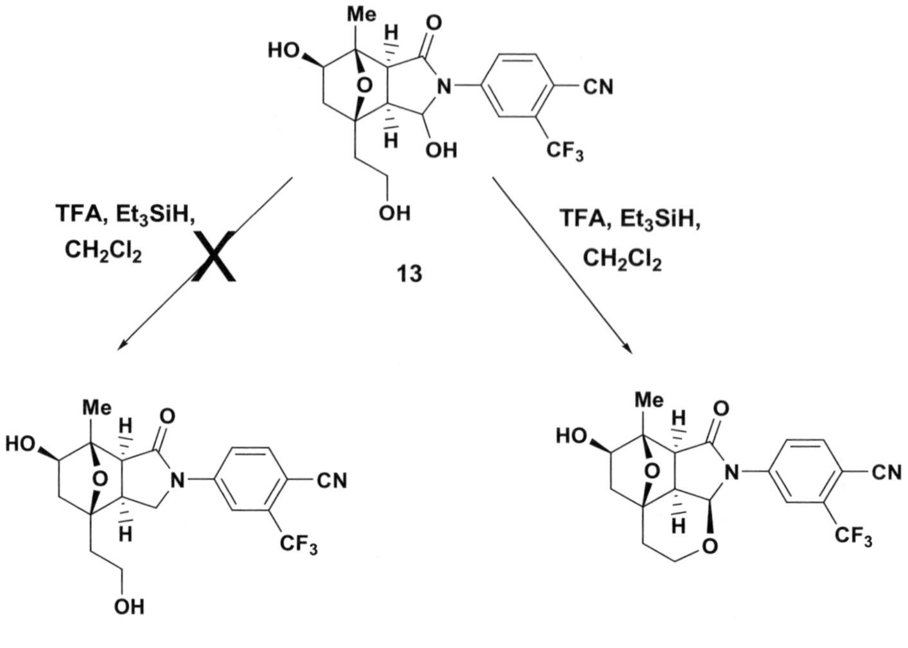

Figure 6.8 Serendipitous discovery of novel tetracyclic AR antagonists.

Table 6.2 Profile of tetracyclic AR antagonists.

Compound	AR Binding Ki (nM)	AR Transactivation IC$_{50}$ (nM)
15	10	225
16	23	226
BMS-779333	18	148
BMS-641988	10	56
bicalutamide	38	534

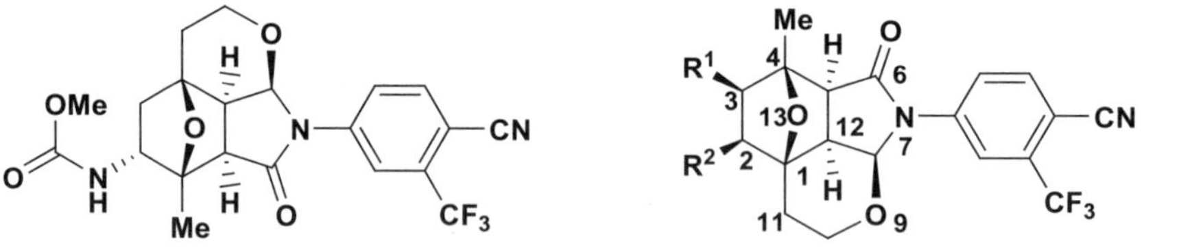

16

15, R^1 = OH, R^2 = H
BMS-779333, R^1 = H, R^2 = OH

Figure 6.9 Lead compounds for the AR backup program.

Further optimization of SAR and PK properties led to the identification of BMS-779333, a potent AR antagonist with potential to meet the criteria set for the AR backup program. As shown in Table 6.2, BMS-779333 exhibited AR binding potency and functional antagonism that are superior to bicalutamide and comparable to the clinical candidate, BMS-641988.[19] The activity of BMS-779333 on different mutant AR forms was compared against bicalutamide, BMS-641988, and BMS-949 in PC3 cells (AR negative) co-transfected with either the wt or variant AR expression constructs and an AR-dependent (PSA promoter driven) luciferase reporter construct (Figure 6.10). As expected, the natural ligand DHT stimulated luciferase expression of all the AR (wt and mutant) forms tested. BMS-779333 did not show significant agonist activity for the wt AR, the mutant AR expressed in the LNCaP cells (T877A), or the CWR-22-BMSLD1 human prostate cancer cells (H874Y), or the two mutants previously described to be activated by bicalutamide (W741C and W741L).[1c] In contrast to BMS-641988, no agonistic activity for Flag-AR-LNCaP cell proliferation was observed for BMS-779333. Docking studies suggested that regardless of the size of the amino acid residue at the 877 position (*i.e.*, T877 *vs.* A877), the hydroxyl group of BMS-779333 may clash with the backbone of amino acid 877 in the LBD of AR (Figure 6.11). Thus, BMS-779333 is unlikely to be accommodated in the agonist conformation and may promote partial unfolding of AR. This offers a plausible explanation for the observed pan-antagonist profile of BMS-779333.

In the CWR-22-BMSLD1 tumors, BMS-779333 exhibited efficacy superior to bicalutamide and comparable to that of BMS-641988. Tumors that failed bicalutamide treatment were shown to retain their sensitivity to respond to BMS-779333. In the LuCaP 23.1 tumor xenografts, BMS-779333 produced significant delay in tumor growth and time to progression to target size along with a major reduction of PSA levels compared to the control group. In this model, the minimum efficacious dose was determined to be $4.2\,\mathrm{mg\,kg}^{-1}$

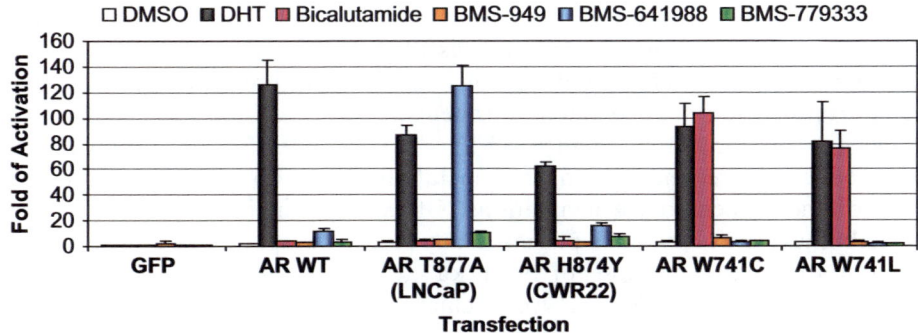

Figure 6.10 Comparison of BMS-779333, BMS-641988, BMS-949, bicalutamide, and DHT in transactivation assays in PC3 cells expressing exogenous wt or various mutant forms of AR (all compounds at 1 μM concentration).

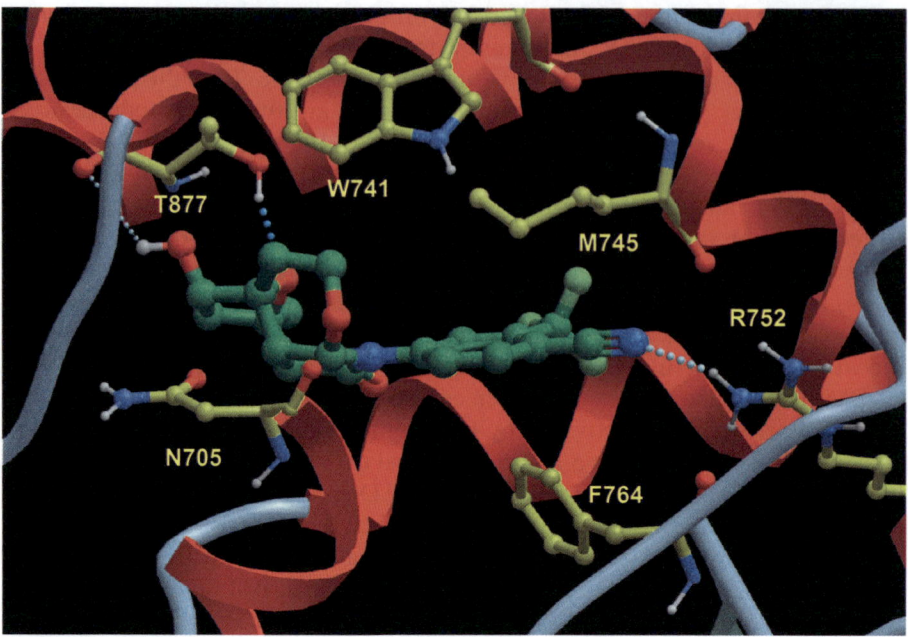

Figure 6.11   BMS-779333 docked in the AR LBD.

(po, bid), and it corresponded with 71% tumor growth inhibition ($AUC_{0-24h} =$ 74 µM.h). In LuCaP 35 tumor xenografts that overexpress wt AR with gene amplification, BMS-779333 consistently elicited an anti-tumor response that was marginally superior to BMS-641988 and bicalutamide. Importantly, by day 7 of treatment, transcriptomic changes induced by BMS-779333 were closer to castration than those produced by other drug treatments. MDA PCa 2b is a tumor line derived from bone metastasis of a prostate cancer patient and harbors two AR mutations (L701H and T877A) which confer decreased AR sensitivity and increased susceptibility to AR agonism by other hormones, such as cortisol and cortisone. Remarkably, BMS-779333 elicited a robust anti-tumor response in these tumors, whereas BMS-641988 was inactive.

BMS-779333 exhibited good oral bioavailability in preclinical species, ranging from 65% in dogs to 100% in rodents. Plasma half-life was significantly longer in dogs (35 h) than in mice (5.8 h) or rats (8 h). BMS-779333 showed low metabolic turnover and no significant metabolites were observed in human liver microsome or hepatocyte incubations. Of note, no major circulating active metabolites were observed in plasma in any preclinical species. In a telemetrized dog study at an oral dose of 50 mg kg^{-1}, high exposures of BMS-779333 ($C_{max} = 109$ µM or 52 µM free) were achieved with no documented changes in the QT interval. These results are consistent with the lack of significant hERG inhibition (6% at 30 µM free) observed in the patch-clamp assay. Minimal increases in blood pressure were observed at the 50 mg kg^{-1}, but not at the 25 mg kg^{-1} dose ($C_{max} = 55$ µM).

Compared to BMS-641988, BMS-779333 showed higher brain penetration, with a brain to plasma ratio of 0.7 in mice receiving high oral doses ($>150\,\text{mg}\,\text{kg}^{-1}$). In mice, a no-effect plasma drug concentration for convulsion of 220 μM was established for BMS-779333. In a two-week rat study, no convulsions were observed with maximum drug concentrations of 200 μM. In a four-day dog study, convulsions were noted with repeat dosing (no-effect concentration = 170 μM). The margins between the projected maximal exposure at efficacious dose in humans and the no-effect concentrations for the identified liabilities in animal testing were 20-fold for seizure and 10-fold for minimal blood pressure elevation. These values were comparable to BMS-641988, where margins based on early clinical PK were projected to be 20-fold for seizure due to mean BMS-949 concentrations and 5-fold for QT prolongation. The advantages of BMS-779333 were that: (1) minimal blood pressure elevation was viewed as being less serious than QT prolongation; (2) the margin for seizure was more certain as it arose from the parent drug instead of a mixture of parent and active metabolites, including BMS-511, which could not be evaluated for seizurogenic potential; and (3) BMS-779333 was an AR pan-antagonist against all available AR mutations, while BMS-641988 was an agonist against certain AR mutations.

In first-in-human enabling dog toxicology studies of longer duration (30 days) and with more animals per group, BMS-779333 unfortunately produced convulsions at plasma drug concentrations that were less than 10-fold the projected C_{max} for the efficacious dose in humans. Given the critical importance of adequate margins to support human testing,[18] the development of BMS-779333 was discontinued.

6.4.4 Clinical Experience with BMS-641988

The human PK, safety, and efficacy of BMS-641988 was evaluated in Phase I clinical studies.[20] *In vitro* incubation experiments had suggested that BMS-641988 was metabolized by human liver microsomes to BMS-511 via oxidative *N*-dealkylation. No metabolites were detected from incubation of BMS-641988 with human hepatocytes. Preclinical studies had established BMS-949 as the major circulating active metabolite of BMS-641988 in mice, rat, and monkeys (Table 6.3). Small amounts of BMS-511 were also observed in rats and

Table 6.3 Ratios at steady state of plasma concentrations of the parent drug and active metabolites upon daily oral administration of BMS-641988.

Species	BMS-641988	BMS-949	BMS-511
mouse	1	9	0
rat	1	0.8	0.2
monkey	1	0.2	0.2
dog	1	0.025	0.02
human	1	2.5	3.3

monkeys. In contrast to the findings in dogs and monkeys, as well as the predictions of *in vitro* metabolism studies, steady-state concentrations of the active metabolites BMS-511 and BMS-949 exceeded those of the parent drug in humans (Table 6.3).[20] As discussed before, the high observed concentrations of BMS-949 were cause for concern because this compound was known to induce seizures in rodents and dogs. Significant concentrations of BMS-511 raised new safety issues since *in vivo* toxicity information in animals (*e.g.*, seizurogenicity, QT prolongation) could not be generated for this compound due to its rapid conversion to BMS-949 in all preclinical species. *In vitro* proconvulsive target screening (*e.g.*, benzodiazepine, $GABA_A$, and chloride channel) indicated that BMS-511 shares a profile similar to BMS-949. Clinical investigators and research subjects were informed of the clinical PK results and precautions for seizures were instituted, including restrictions on driving automobiles and operating heavy machinery. Dose escalation intervals were recalculated to accommodate variability in inter-patient exposure and to maintain human exposure significantly below safe levels in animals (*i.e.*, levels in animals where no seizures occurred).

Two Phase 1 studies enrolling patients with castration-resistant prostate cancer were implemented, with dose cohorts escalating from a starting dose of 5 mg. At doses of 40 mg and beyond, higher exposures of BMS-641988 and BMS-949 were achieved than those required for significant tumor growth inhibition in the tumor models. A complete description of the criteria and clinical trial results will be reported separately.[20] Significant adverse events included one incidence of QTc prolongation in one subject and incidence of seizure in another subject. Both adverse events resolved without significant sequelae. Development of BMS-641988 was discontinued based on a review of the clinical findings on safety and efficacy.

6.5   Conclusion

Recent research by Sawyers and colleagues,[4] that an increase in AR mRNA and protein is both necessary and sufficient for the development of resistance to anti-androgen therapy, has reinvigorated interest in the discovery of more potent AR antagonists for the treatment of advanced prostate cancer. A structure-based design approach led to the identification of BMS-641988 with significantly increased potency and efficacy compared with bicalutamide. However, BMS-641988 exhibited a cardiovascular liability related to QT prolongation. In animal testing, an off-target effect of seizure was identified that significantly slowed drug development. Other AR antagonists have also been reported to induce seizures in preclinical studies,[16] as well as in patients.[21] Even though BMS-641988 did not induce seizures in animals, one of its major active metabolites (BMS-949) was associated with seizure risk in preclinical studies. A different active metabolite (BMS-511) was identified as the primary metabolite species in humans. Preclinical toxicological studies of BMS-511 could not be interpreted due to its immediate metabolism to BMS-949 in all animal species.

In the clinic, BMS-641988 failed to provide evidence of efficacy despite achievement of plasma exposures significantly higher than the minimum efficacious concentration in xenograft models. Two significant adverse events were documented: prolongation of QTc interval in a patient receiving 40 mg BMS-641988, and seizure in a patient on a 60 mg dose. This combination of safety and efficacy findings led to the discontinuation of clinical development of BMS-641988.

Further structural modifications to address metabolism issues, pan-AR antagonism, and cardiac safety challenges provided BMS-779333, which exhibited broad spectrum efficacy in four human prostate tumor models dependent on wt or mutant AR. Despite an encouraging preliminary safety profile, advancement of BMS-779333 was halted when it induced seizures in a 30-day dog toxicology study with a margin that was less than 10-fold relative to the exposure needed for efficacy in the tumor xenograft models.

Several important lessons can be gleaned from this research experience. Despite a rational approach, the empirical nature of drug discovery can require navigating an unpredictable and diverse set of liabilities that may not be related to the biological target, such as convulsions, QT prolongation, thyroid hormone disruption, and blood pressure elevation. Increased investigation of active metabolites is warranted, especially when the lead compounds suffer from severe toxicity without premonitory signs, such as seizure. High-dose testing of animals has a potential to correctly identify a hazard, but the identified safety margin may be significantly lower in longer duration studies in animals or in the clinic. Even adjusted for the distinct pattern of human metabolism relative to that observed in animals, safety margin predictions from preclinical studies were poorly predictive of clinical margins, perhaps due to important differences in seizure sensitivity between young, healthy animals and elderly patients with advanced disease.

The unsatisfactory clinical efficacy of BMS-641988 warrants further investigation. In the clinic, serum concentrations of PSA decreased for some subjects after discontinuation of BMS-641988, suggesting that BMS-641988 has partial AR agonist properties. Anti-androgen withdrawal syndrome has been well documented with clinical AR antagonists.[5] The BMS AR program relied heavily on the bicalutamide-resistant CWR-22-BMSLD1 tumor xenograft model expressing mutant AR. The relevance of this AR mutation in HRPC has not been established. Recently, development of a novel anti-androgen has been reported, based on a screen for retention of AR antagonist activity in the setting of increased AR expression.[22] Such a screen, employing LNCaP human prostate cancer cells engineered to express higher amounts of wt AR, may be more representative of the clinical scenario of castration resistant disease. Translocation of AR to the nucleus is an essential step for downstream transcription of genes involved in proliferation of prostate cells. Therefore, alternative novel strategies that involve degradation of AR, or prevent its translocation to the nucleus, offer significant opportunities for treatment of HRPC.

References

1. (a) S. M. Dehm and D. J. Tindall, *Mol. Endocrinol.*, 2007, **21**, 2855; (b) Y. Chen, C. L. Sawyers and H. I. Scher, *Curr. Opin. Pharmacol.*, 2008, **8**, 440; (c) Y. Chen, N. J. Clegg and H. I. Scher, *Lancet*, 2009, **10**, 981.
2. W. Gao, C. E. Bohl and J. T. Dalton, *Chem. Rev.*, 2005, **105**, 3352.
3. Y. Shang, M. Myers and M. Brown, *Mol. Cell*, 2002, **9**, 601.
4. C. D. Chen, D. S. Welsbie, C. Tran, S. H. Baek, R. Chen, R. Vessella, M. G. Rosenfeld and C. L. Sawyers, *Nat. Med.*, 2004, **10**, 33.
5. (a) R. Paul and J. Breul, *Drug Safety*, 2000, **23**, 381; (b) T. Hara, J. Miyazaki, H. Araki, M. Yamaoka, N. Kanzaki, M. Kusaka and M. Miyamoto, *Cancer Res.*, 2003, **63**, 149.
6. K. Wellington and S. J. Keam, *Drugs*, 2006, **66**, 837.
7. (a) J. S. Sack, K. F. Kish, C. Wang, R. M. Attar, S. E. Kiefer, Y. An, G. Y. Wu, J. E. Scheffler, M. E. Salvati, S. R. Krystek Jr, R. Wienmann and H. M. Einspahr, *Proc. Natl. Acad. Sci. U. S. A*, 2001, **98**, 4904; (b) C. E. Bohl, W. Gao, D. D. Miller, C. E. Bell and J. T. Dalton, *Proc. Natl. Acad. Sci. U. S. A.*, 2005, **102**, 6201.
8. R. M. Attar, M. Jure-Kunkel, A. Balog, M. E. Cvijic, J. Dell-John, C. A. Rizzo, L. Schweizer, T. E. Spires, J. S. Platero, M. Obermeier, W. Shan, M. E. Salvati, W. R. Foster, J. Dinchuk, S.-J. Chen, G. Vite, R. Kramer and M. M. Gottardis, *Cancer Res.*, 2009, **69**, 6522.
9. M. E. Salvati, A. Balog, D. D. Wei, D. Pickering, R. M. Attar, J. Geng, C. A. Rizzo, J. T. Hunt, M. M. Gottardis, R. Weinmann and R. Martinez, *Bioorg. Med. Chem. Lett.*, 2005, **15**, 389.
10. A. Balog, M. E. Salvati, W. Shan, A. Mathur, L. W. Leith, D. D. Wei, R. M. Attar, J. Geng, C. A. Rizzo, C. Wang, S. R. Krystek, J. S. Tokarski, J. T. Hunt, M. M. Gottardis and R. Weinmann, *Bioorg. Med. Chem. Lett.*, 2004, **14**, 6107.
11. M. E. Salvati, A. Balog, W. Shan, D. D. Wei, D. Pickering, R. M. Attar, J. Geng, C. A. Rizzo, M. M. Gottardis, R. Weinmann, S. R. Krystek, J. Sack, Y. An and K. Kish, *Bioorg. Med. Chem. Lett.*, 2005, **15**, 271.
12. M. E. Salvati, A. Balog, W. Shan, R. Rampulla, S. Giese, T. Mitt, J. A. Furch, G. D. Vite, R. M. Attar, M. Jure-Kunkel, J. Geng, C. A. Rizzo, M. M. Gottardis, S. R. Krystek, J. Gougoutas, M. A. Galella, M. Obermeier, A. Fura and G. Chandrasena, *Bioorg. Med. Chem. Lett.*, 2008, **18**, 1910.
13. A. Fura, A. Balog, R. Hanson, V. Vyas, T. Harper, G. Vite, R. Attar, A. Apedo, J. Geng and M. Salvati, *Drug Metab. Dispos.*, 2010, manuscript under preparation.
14. R. C. Dockens, *Drug Metab. Dispos.*, 2000, **28**, 973.
15. United States Food and Drug Administration Center for Drug Evaluation and Research, *Guidance for Industry: Estimating the Maximum Safe Starting Dose in Clinical Trials for Therapeutics in Adult Healthy Volunteers*, 2005; accessed at http://www.fda.gov/downloads/Drugs/GuidanceComplianceRegulatoryInformation/Guidances/UCM078932.pdf.

16. W. R. Foster, H. Shi, S.-J. Chen, M. T. Obermeier, G. Cornelius, M. Huang, J. Gan, W. Chen, J. Price, J. Dell-John, D. Fernando, Y. Callejas, B. Lehman, V. Livanov, E. Fitzpatrick, J. C. Arezzo, H. E. Janke, S. S. Powlin, F. Myers, J. E. Dinchuk, D. Norris, A. Balog, M. Salvati, R. Attar, M. Gottardis and B. D. Car, *Prostate*, 2010, submitted.
17. A. Balog, R. Rampulla, G. S. Martis, S. Krystek, R. Attar, J. Dell-John, J. DiMarco, D. Fairfax, L. B. Fleming, J. Gougoutas, C. L. Holst, J. T. Hunt, A. Nation, C. Rizzo, L. M. Rossiter, L. Schweizer, W. Shan, S. Spergel, T. Spires, M. Gottardis, G. Trainor, G. Vite and M. E. Salvati, *ACS Med. Chem. Lett.*, 2010, submitted.
18. F. M. Chybowski, J. J. Keller, E. J. Bergstralh and J. E. Oesterling, *J. Urol.*, 1991, **145**, 313.
19. The assay conditions were modified by increasing the amount of DHT from 1 nM to 3 nM. This resulted in changes in the potency of BMS-641988 and bicalutamide compared to the previous values reported in Table 6.1.
20. D. Rathkopf, G. Wilding, M. Carducci, H. Scher, M. Gottardis and S. Cheng, *J. Clin. Oncol.*, 2010, manuscript under preparation.
21. H. I. Sher, T. M. Beer, C. S. Higano, M. Taplin, E. Efstathiou, A. Anand, D. Hung, M. Hirmand and M. Fleisher, *J. Clin. Oncol.*, 2009, **27**, 5011.
22. C. Tran, S. Ouk, N. J. Clegg, Y. Chen, P. A. Watson, V. Arora, J. Wongvipat, P. M. Smith-Jones, D. Yoo, A. Kwon, D. Wasielewska, D. Welbe, C. D. Chen, C. S. Higano, T. M. Beer, D. T. Hung, H. I. Scher, M. E. Jung and C. L. Sawyers, *Science*, 2009, **324**, 787.

CHAPTER 7

The Discovery of UK-390957: the Challenge of Targeting a Short Half-life, Rapid T_{max} SSRI

MARK D. ANDREWS AND DONALD S. MIDDLETON

Worldwide Medicinal Chemistry, Pfizer Global Research and Development, Ramsgate Road, Sandwich, Kent CT13 9NJ, UK

7.1 Introduction: SSRIs and Premature Ejaculation

Since the launch of the first selective serotonin re-uptake inhibitors (SSRIs) in the 1980s, this class of drugs has become extremely important for the management of depression and other psychiatric disorders, and indeed has become one of the most widely prescribed class of drugs in the developed world.[1] Although SSRIs are generally extremely well tolerated they do cause some sexual side effects, with delayed ejaculation and absent or delayed orgasm being the most commonly reported.[2–4] While this is clearly a problem for people with normal sexual function, this can be an advantage for men suffering from premature ejaculation (PE).

PE is defined in the *Statistical Manual of Psychiatry* as "persistent or recurrent onset of orgasm and ejaculation with minimal sexual stimulation before, upon or shortly after penetration and before the person wishes it".[5] With prevalence estimates varying widely from 5 to 40%,[6,7] it is the most common male sexual disorder across all age groups and is associated with "marked distress or

RSC Drug Discovery Series No. 4
Accounts in Drug Discovery: Case Studies in Medicinal Chemistry
Edited by Joel C. Barrish, Percy H. Carter, Peter T. W. Cheng and Robert Zahler
© Royal Society of Chemistry 2011
Published by the Royal Society of Chemistry, www.rsc.org

interpersonal difficulty . . . ''. Although there are several plausible physiological factors (*e.g.* anxiety), which may cause both lifelong and acquired PE, no psychological profile or psychosocial feature is either consistent or pervasive.

Several studies have investigated the use of the marketed SSRI antidepressants for the off-label treatment of PE. In the clinical data that has been published, fluoxetine **1**, paroxetine **2**, sertraline **3** and citalopram **4** all appear to be well tolerated and effective, with paroxetine often being reported as the most effective.[2–4] Fluvoxamine **5** has also been investigated but did not show effects that were significantly different to placebo.

7.2 Targeting a Rapid T_{max}, Short Half-life Agent

While the marketed SSRIs are clearly effective treatments for PE, it was recognised that there was an opportunity to develop a rapid-onset, short half-life SSRI for the treatment of premature ejaculation and other conditions where on-demand (*prn*) dosing might be a preferred option. The antidepressant activity of SSRIs is only evident after several weeks of chronic dosing, involving continual exposure to free drug, as it is believed that desensitisation of 5-HT$_{1A}$, and possibly also 5-HT$_{1B}$, autoreceptors is required for sustained increase of central serotonin (5-HT) levels.[8,9] On-demand dosing with a short half-life agent would not be expected to cause this desensitisation and would not therefore be expected to cause an antidepressant effect. On-demand dosing with a short half-life agent would also minimise exposure to the drug and so could result in an improved side-effect profile.

Encouragingly, just after the project to discover a rapid-onset, short half-life SSRI had started, support for the concept of on-demand dosing of SSRIs was provided by the report, from McMahon and Touma, that paroxetine was an effective treatment for PE when taken as needed, 3–4 hours prior to anticipated intercourse, in a single-blind, placebo-controlled crossover study.[10]

The marketed SSRI antidepressants lack suitability for on-demand dosing as they are lipophilic amines (p$K_a \approx 9$) with high volumes of distribution (V_D)

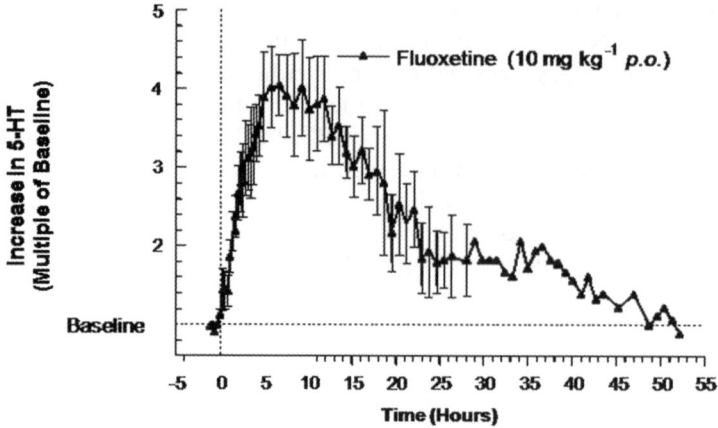

Figure 7.1   Time course of CNS effect following oral dosing.

(8–$43\,L\,kg^{-1}$), resulting in long elimination half-lives (16 to $>30\,h$) and slow time to maximal compound concentration (T_{max}) (typically 4–8 h after oral dosing).[11] Oral CNS microdialysis studies in rodents conducted at Pfizer demonstrated that systemic T_{max} broadly correlated with time to peak central 5-HT levels, as shown for fluoxetine in Figure 7.1,[12] suggesting that reducing T_{max} could lead to more rapid elevation in central 5-HT, consistent with our goal of achieving rapid onset of effect.

The delayed T_{max} of lipophilic amines such as the SSRIs has been explained as being due to the extensive distribution of high V_D compounds into the liver after they are absorbed,[13] which results in a slow hepatic transit time. Targeting compounds with lower V_D should therefore lead to compounds with more rapid T_{max}. As the half-life of a compound *in vivo* is described by eqn (7.1), where Cl_P is plasma clearance, reducing the V_D would also help achieve a shorter $t_{1/2}$:[14]

$$t_{1/2} = 0.693\frac{V_D}{Cl_P} \qquad (7.1)$$

Another way of decreasing the half-life is to increase clearance, but there is a limit to how high the clearance can be without adversely affecting bioavailability (F), as shown by eqn (7.2) (Cl_H is hepatic clearance and Q_h is hepatic blood flow):

$$F = 1 - \frac{Cl_H}{Q_h} \qquad (7.2)$$

Using these equations helped us set a realistic target for both volume of distribution and clearance. Aiming for a minimum oral bioavailability of 50% and assuming a compound is fully absorbed and mainly cleared by the liver ($Cl_P = Cl_H$) provides a maximum acceptable clearance number of $10\,mL\,min^{-1}\,kg^{-1}$ (taking a liver blood flow value of $20\,mL\,min^{-1}\,kg^{-1}$ for man). Using eqn (7.1) and aiming for a half-life of <10 hours, this results in a target V_D of less than $9\,L\,kg^{-1}$.

In addition to the pharmacokinetic (PK) targets above, our other project goals were to find a potent, CNS penetrant SSRI with an SRI (serotonin re-uptake inhibition) $IC_{50} < 10\,nM$ and >100-fold selectivity over dopamine (DRI) and noradrenaline re-uptake inhibition (NRI). In addition, we required the absence of significant off-target activity, as measured by screening against a panel of other enzymes and receptors, no significant CYP inhibition,[1] and, in order to avoid large exposure difference between individuals, the compound should not be predominantly metabolised by CYP 2D6.[15]

7.3 Reducing Volume Starting from a Precedented SSRI Template

Our initial strategy to achieve the desired PK profile was to take one of the marketed SSRIs as a starting point and seek to reduce its lipophilicity (clog P and log D) by incorporating polar groups (Figure 7.2). While all of the factors that underlie volume of distribution are not understood, it is clear that lipo-philicity plays a significant role, as higher lipophilicity results in higher tissue and membrane affinity and therefore higher V_D.[16] Non-basic electron-with-drawing groups were particularly targeted as it was anticipated that these would reduce the pK_a of the basic nitrogen and potentially reduce membrane affinity by reducing the energetically favourable ionic interaction between the proto-nated amine and the negatively charged phospholipid phosphate head-group.

By starting from a marketed drug, we knew that many of the properties we would need (*e.g.* SRI potency, selectivity and CNS penetration) should be achievable from that template. In addition, it was felt that starting with a chemotype with clinical precedent would help mitigate development risks such as toxicity. However, the known SSRIs are all highly lipophilic, and it was not obvious that the polarity would be tolerated, either from a potency or CNS penetration point of view. In terms of CNS penetration, we recognised the critical need to balance our strategy of introducing polar groups to reduce V_D

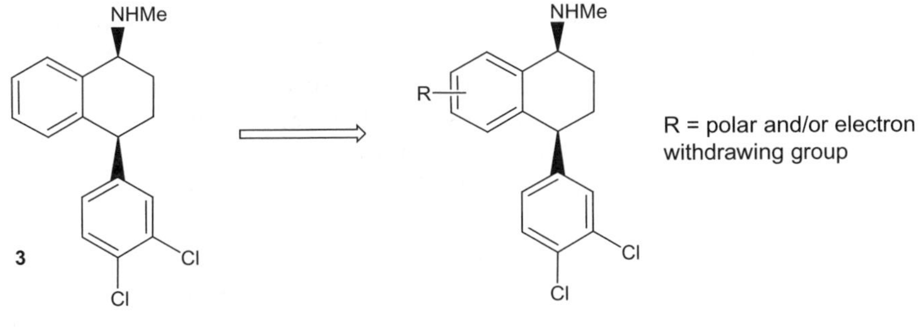

Figure 7.2

with maintaining properties that did not compromise the excellent CNS penetration of the known agents. The majority of CNS drugs have a low number of hydrogen bond donors (HBD) (all of the top 25 selling CNS drugs from 2004 had <3 HBD), lower polar surface area (PSA) (96% of the top 25 CNS drugs had TPSA <90 Å^2) and relatively high clog P (68% of the top 25 CNS drugs had clog P values between 2 and 5).[17] As a result, we focused on compounds that had the following properties: MW <400, log D = 2–3, HBD count <3 and TPSA <80 Å^2.

While, in principle, this strategy is equally applicable to all lipophilic SSRIs, several factors led us to select compound **3** as our starting point: (1) **3** has proven efficacy in PE clinical trials; (2) published[18] and in-house structure–activity relationships (SAR) in this series gave us a clear understanding of the structural features essential for potency; (3) the template is amenable to synthetic modifications, allowing rapid SAR development; (4) **3** has low CYP inhibition and low CYP2D6 affinity,[19] minimising the risk of drug–drug interactions and variable inter-subject metabolism.

Based on the known SAR around **3**, we did not try to vary the dichlorophenyl or *cis*-methylamino substituents, but concentrated on substituting on the aromatic part of the tetrahydronaphthalene ring.[20,21] We quickly found that polar substitution is well tolerated, with substitution at C8 (**6–8**), C7 (**9–18**) and C6 (**19–23**) leading to potent SSRIs (selected examples shown in Table 7.1). Interestingly, the acid intermediates (**6, 9, 19**) used to make amides at the three positions all retain significant SRI activity. As these compounds are zwitterionic, they would be expected to have an inherently low V_D,[22] but unfortunately they did not meet our potency criteria and, ultimately, were not investigated further. The acidic substituents also raised concerns as to how CNS penetrant they would be. Substitution at C8 resulted in a loss in selectivity over DRI (*e.g.* 36-fold for **7**), so focus was given to investigating substitution at C7 and C6. In general, substitution at C7 gave compounds with the best SRI activity, while substitution at C6 gave compounds with the best selectivity over DRI and NRI (compare **14** with **23**, **11** with **21**, **12** with **22**). As the C7 position was synthetically the most accessible and many analogues were found with acceptable selectivity *versus* DRI and NRI, this became the focus of the project's efforts and ultimately led to the compounds that were progressed for further studies.

Both the primary and secondary amides (**10, 11**) are potent SSRIs, but the tertiary amide lost some SRI potency and as a result lost selectivity over DRI and NRI. Nitrile substitution (**13**) resulted in an SSRI which is more lipophilic than our starting point **3** (log D 3.4 *vs.* 3.1) and so did not meet our goal of reducing log D. This is presumably a result of the electron-withdrawing nitrile group reducing the basicity of the amine and therefore increasing the log D. Primary and secondary sulfonamides (**14, 15**) are both potent SSRIs, with the primary sulfonamide being one of the most potent compounds we identified (1 nM). As with the amides, the tertiary sulfonamide **16** is a weaker SSRI than the primary and secondary sulfonamides, but in this case the drop-off was greater than 10-fold. The N-linked sulfonamide **17** is also potent and polar, but has unacceptably high levels of DRI and NRI activity (less than 20-fold selective). Polar heterocycles such as triazole **18** also give potent SSRIs but result in unacceptable levels of CYP2D6 inhibition, a problem that was not seen with the other analogues prepared.

More than one polar substituent could be introduced onto the tetra-hydronaphthalene ring to further reduce clog *P*, and this did give some SSRIs with reasonable potency, such as amide/sulfonamide **24**. This strategy was not pursued further, partly because of the difficult syntheses involved and, more importantly, because the high TPSA and HBD count were outside the targets we had set to maximise the chance of achieving good CNS penetration.

7.3.1 Key Polar Analogues in the Tetrahydronaphthalene Template

Compounds **10**, **11**, **14** and **15** were then selected for further progression into permeability and CNS penetration studies based on their potency, selectivity over DRI and NRI (at least 50-fold) and synthetic accessibility (Table 7.2). This was despite **10** and **14** not meeting our goal of having <3 HBD, putting them at risk of impaired brain penetration. As anticipated, it was found that incorporating a polar, strongly electron-withdrawing group onto the tetra-hydronaphthalene core both attenuates the pK_a of the amine and reduces the log *D*, meeting the goals we had set. Despite the increased polarity, all of the analogues tested show good flux and no efflux in a Caco-2 assay. In fact, like compound **3** they actually show a degree of influx, possibly indicating apical-to-basal active transport across the membrane. In rat CNS penetration studies, most of the analogues showed excellent brain exposure, with amide **10** and sulfonamide **14** displaying CSF (cerebrospinal fluid) to free blood levels of 1:1 and 2.6:1 respectively, despite both having 3 HBD and **14** having a TPSA of 81 Å^2, outside our original "ideal" target. While the *N*-methyl amide **11** also showed good CSF levels, the *N*-methyl sulfonamide **15** showed somewhat reduced CSF levels, despite being more lipophilic and having a HBD less than **14**. This highlights that there are factors other than physicochemical properties, perhaps such as transporter affinity, which determine the level of CNS penetration and following property guidelines too strictly (*e.g.* HBD<3) could lead to interesting compounds being overlooked.

3, 10, 11, 14, 15

Table 7.1 Activities and physicochemical properties of substituted tetrahydronaphthalenes

3, 6-24

Cpd	R^1	R^2	R^3	SRI $(nM)^a$	DRI $(nM)^a$	NRI $(nM)^a$	TPSA $(\mathring{A}^2)$	clog P	log D
3	H	H	H	3	310	825	12	5.4	3.1
6	CO_2H	H	H	31	540	880	49	2.8	–
7	$CONH_2$	H	H	8	290	1150	55	3.9	–
8	CONHMe	H	H	31	420	1500	41	4.1	–
9	H	CO_2H	H	90	–	–	49	2.8	–

No.									
10	H	CONH$_2$	H	2	170	410	55	3.9	2.3
11	H	CONHMe	H	3	450	440	41	4.1	2.8
12	H	CONMe$_2$	H	9	270	270	32	3.8	—
13	H	CN	H	3	100	3200	36	4.8	3.4
14	H	SO$_2$NH$_2$	H	1	90	770	81	3.5	2.3
15	H	SO$_2$NHMe	H	5	270	1250	67	4.1	2.8
16	H	SO$_2$NMe$_2$	H	70	1500	2700	58	4.5	3.1
17	H	NHSO$_2$Me	H	3	55	60	67	4.2	2.2
18	H	1,2,4-triazol-1-yl	H	4	250	760	43	4.4	2.9
19	H	H	CO$_2$H	60	16700	20300	49	2.8	—
20	H	H	CONH$_2$	13	1300	1100	55	3.9	—
21	H	H	CONHMe	25	3300	5000	41	4.1	—
22	H	H	CONMe$_2$	7	920	11500	32	3.8	—
23	H	H	SO$_2$NH$_2$	6	2500	2100	81	3.5	—
24	H	NHSO$_2$Me	CONH$_2$	24	3600	2100	110	2.6	—

[a]The inhibitory potency at human serotonin, dopamine and noradrenaline transporters was quantified through measurement of a compound's ability to inhibit transport and subsequent intracellular accumulation of radiolabelled ligand in an *in vitro* radiometric assay using HEK-293 cells stably expressing the appropriate transporters. All assay determinations are $\geq n = 2$.

Table 7.2   Permeability and brain penetration data for key polar analogues of **3**.

Compound	R^2	pK_a	Caco-2 (A-B/B-A)	Rat CNS studies brain:blood	Rat CNS studies CSF:free blood
3	H	9.3	21/8	14:1	1:1
10	$CONH_2$	8.5	–	–	1:1
11	CONHMe	8.8	26/11	–	1:1
14	SO_2NH_2	8.4	18/7	16:1	2.6:1
15	SO_2NHMe	–	–	–	0.3:1

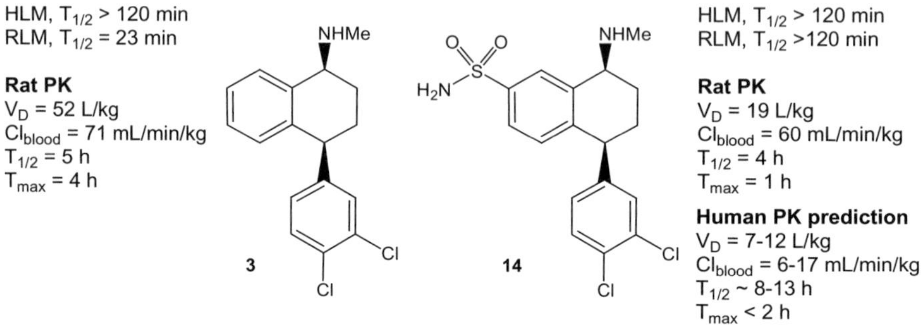

HLM, $T_{1/2}$ > 120 min
RLM, $T_{1/2}$ = 23 min

Rat PK
V_D = 52 L/kg
Cl_{blood} = 71 mL/min/kg
$T_{1/2}$ = 5 h
T_{max} = 4 h

HLM, $T_{1/2}$ > 120 min
RLM, $T_{1/2}$ >120 min

Rat PK
V_D = 19 L/kg
Cl_{blood} = 60 mL/min/kg
$T_{1/2}$ = 4 h
T_{max} = 1 h

Human PK prediction
V_D = 7-12 L/kg
Cl_{blood} = 6-17 mL/min/kg
$T_{1/2}$ ~ 8-13 h
T_{max} < 2 h

Figure 7.3

Compound **14** was progressed to rat PK studies and, as hoped, it showed a significantly reduced volume of distribution relative to **3** ($19 \, L \, kg^{-1}$ *vs.* $52 \, L \, kg^{-1}$, unbound volume $660 \, L \, kg^{-1}$ *vs.* $1900 \, L \, kg^{-1}$) (Figure 7.3). Very importantly, this also translates to a significantly reduced T_{max} relative to **3** (1 h *vs.* 4 h). Despite being stable in rat liver microsomes (RLM) ($t_{1/2} > 120 \, min$), **14** shows moderate clearance *in vivo*, resulting in a half-life of 4 h. A dog PK study was also carried out on sulfonamide **14** and gave results consistent with those seen in the rat ($V_D = 12 \, L \, kg^{-1}$, $Cl_{blood} = 11 \, mL \, min^{-1} \, kg^{-1}$, $t_{1/2} = 7 \, h$). Scaling to man gave a prediction of a volume of $7–12 \, L \, kg^{-1}$ and a T_{max} of less than 2 h. Because turnover is not seen in rat, dog or human microsomes, clearance in rat and dog appears to be the result of an active clearance process rather than metabolic turnover. As a result, clearance in man was predicted based on scaling from rat and dog, leading to a half-life prediction of 8–13 h.

7.3.2   Clinical Data for UK-373911

Although predictions for volume were at the upper end of what we had targeted and the higher end of the predicted clearance range presented a risk of low bioavailability in man, compound **14** (UK-373911) was considered to be suitable for progression and was selected for clinical development. Compound **14** showed no significant off-target pharmacology, successfully completed

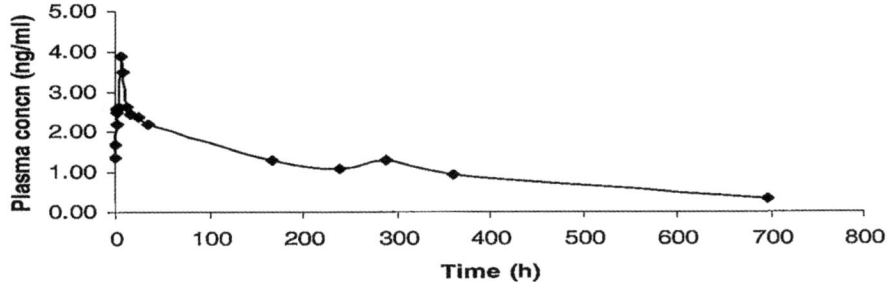

Figure 7.4 Free drug levels of UK-373911 following oral dosing in man.

pre-clinical toxicology studies, and was progressed to first-in-human (FIH) studies. These studies showed that while we had achieved our target of reducing T_{max} (<2 h) relative to compound **3**, we had failed to deliver a compound with a short half-life. Compound **14** has exceptionally low clearance in man, which gives it a half-life of >240 h (Figure 7.4).

In contrast to rat and dog, clearance in man appears to be predominantly due to CYP-mediated metabolism to the circulating metabolite, primary amine **25** (SRI $IC_{50} = 50$ nM). A retrospective analysis in human liver microsomes (HLM) showed that monitoring for appearance of the primary amine **25** gives a good estimate of the low clearance of **14** observed in man.

25

7.4 Optimising an Inherently Lower Volume Template: Diphenyl Ethers

Although we managed to reduce the V_D in the tetrahydronaphthalene template by introducing polarity and reducing the pK_a, we began to realise that our initial medicinal chemistry strategy of starting with a marketed SSRI was flawed. While polarity and basicity do affect V_D, chemical structure plays a very important, if poorly understood, role in determining the volume. As the marketed SSRIs represent chemical structures with high V_D, it will be inherently

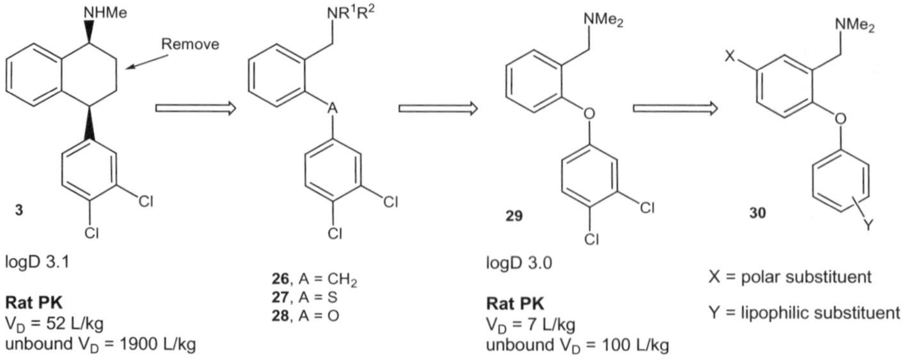

Figure 7.5

difficult to reduce the V_D by the amount needed to achieve a short T_{max} and $t_{1/2}$. In order to achieve a significantly lower V_D, we concluded that we would need to start with a series that had a lower V_D. Therefore, prior to the FIH read-out of compound **14**, the decision was made to switch to a structurally distinct series to provide a back-up to replace **14** should it not have the profile we were seeking, as turned out to be the case.

In order to increase our confidence in achieving a moderate half-life in the clinic, we set a lower V_D target of $<5 \, L \, kg^{-1}$. In addition, to increase our confidence in human clearance predictions, we targeted compounds with predominantly metabolic clearance mechanisms in preclinical species and humans, rather than the presumed active transport observed with **14**.

Work within Pfizer had shown that the diphenyl ether template **28**, which can be viewed as a simplified version of the tetrahydronaphthalene template **3** with an ethylene bridge removed (Figure 7.5), could give compounds with potent SRI activity, although not necessarily good selectivity over DRI and NRI.[23] For example, the dichloro compound **29** is a potent SRI (11 nM) with 12-fold and 26-fold selectivity over DRI and NRI, respectively; however, **29** was shown to have a significantly lower V_D in the rat than **3** ($7 \, L \, kg^{-1}$ *vs.* $52 \, L \, kg^{-1}$, unbound volume $100 \, L \, kg^{-1}$ *vs.* $1900 \, L \, kg^{-1}$), despite being equally lipophilic (log D 3.0 *vs.* 3.1).

The corresponding diphenylmethane **26** and diphenyl sulfide **27** systems were also investigated and found to be potent SSRIs,[24,25] but the diphenyl ethers were preferred due to their synthetic accessibility and slightly lower lipophilicity.

7.4.1 SAR Around the Phenoxy Ring

Given the similarity between the diphenyl ether and tetrahydronaphthalene templates, we reasoned that introducing polarity onto the benzylamine ring was likely to be well tolerated and, as seen previously, also likely to reduce the V_D.

Table 7.3 Potency and selectivity SAR for the phenoxy ring of diphenyl ethers.

Compound	Y	X	SRI IC$_{50}$ (nM)^a	Selectivity over DRI	Selectivity over NRI
29	3,4-Cl$_2$	H	11	12	26
31	2-Cl	H	293	14	0.2
32	3-Cl	H	192	4.7	0.3
33	4-Cl	H	22	4.7	8.5
34	4-F	H	136	NT	NT
35	4-Et	H	6	1067	361
36	4-SMe	H	9	324	71
37	4-CF$_3$	H	32	600	93
38	4-OMe	H	25	192	28
39	4-OCF$_3$	H	7	827	956
40	4-SO$_2$NH$_2$	H	29	1722	788
41	4-CONH$_2$	H	180	28	3.5
42	4-SOMe^b	F	599	46	26
43	4-SO$_2$Me	F	744	NT	NT

^{a}The inhibitory potency at human serotonin, dopamine and noradrenaline transporters was quantified through measurement of a compound's ability to inhibit transport and subsequent intracellular accumulation of radiolabelled ligand in an *in vitro* radiometric assay using HEK-293 cells stably expressing the appropriate transporters. All assay determinations are $\geq n = 2$.
^{b}Racemic mixture of sulfoxide enantiomers.

Before embarking on optimisation of polar substitution on the benzylamine ring, however, we quickly explored the SAR around the benzylamine and the phenoxy ring (Table 7.3). It soon became apparent that little variation of the amine group was tolerated and that the dimethylamine is generally optimal for potency. Monomethylamines are only slightly less active, but activity tends to drop off rapidly if larger alkyl groups are present. For example, mono-ethylamine loses more than 10-fold in potency. Cyclic amines were also investigated, and while azetidine is well tolerated, pyrrolidine and morpholine derivatives are essentially inactive.

Simple chloro substitution on the phenoxy ring highlighted some interesting trends, with both the 2-chloro **31** and 3-chloro **32** derivatives showing poor SRI IC$_{50}$ values, 293 nM and 192 nM, respectively. They are in fact selective inhibitors of noradrenaline re-uptake. Reasonable activity (22 nM) is achieved with the 4-chloro isomer **33**, but this compound has significant activity against the other two monoamine transporters. Better selectivity over DRI and NRI can be achieved, to varying degrees, by placing slightly larger hydrophobic substituents at the 4-position of the phenoxy ring. The 4-ethyl derivative **35** is very potent (6 nM) and has greater than 300-fold selectivity over DRI and NRI. The 4-methylsulfanyl system **36** is less selective over NRI (71-fold), but again has excellent SRI activity (9 nM). The 4-trifluoromethyl compound **37** has good levels of selectivity over DRI and NRI, but somewhat disappointing SRI activity (32 nM). The smaller 4-fluoro substituted analogue **34** has much less SRI activity, with an IC$_{50}$ value of only 136 nM. The 4-methoxy analogue

38 has moderate SRI activity (25 nM), but insufficient selectivity over NRI. The 4-trifluoromethoxy derivative **39** is particularly interesting, as it combines excellent potency (7 nM) with greater than 800-fold selectivity over DRI and NRI.

29, 31-43

Although a number of hydrophobic groups are well tolerated at the 4-position of the phenoxy ring, we wished to explore the possibility of introducing polarity at this position with the aim of potentially introducing polar substituents on both rings. We were encouraged to find that the sulfonamide **40** has good potency (29 nM), excellent selectivity and is relatively polar (log $D = 0.5$). It therefore seemed like an ideal starting point for generating a low-volume and potent SSRI. Unfortunately, we were unable to increase potency in this series, as adding either polar or lipophilic substituents to the benzylamine ring resulted in a loss of activity. The activity of **40** suggested that alternative polar substituents might be well tolerated on the phenoxy ring, but we were unable to find any with sufficient potency. For example, the amide **41** has activity of only 180 nM while the sulfoxide **42** and sulfone **43** are even less active.

7.4.2   Introducing Polarity in the Diphenyl Ether Template

Having explored the SAR around the phenoxy ring and the amine group, we then investigated the effect of introducing polar groups onto the benzylamine ring.[24,26] By analogy with the tetrahydronaphthalene template, we anticipated that polar substitution would be well tolerated at the 5-position of the benzylamine ring of the (2-phenoxybenzyl)amine template. We initially looked at compounds with a trifluoromethyl or trifluoromethoxy substituent on the phenoxy ring, as these substituents had shown good selectivity and were anticipated to be metabolically stable, which at this stage we believed was important (Table 7.4).

We were quickly able to show that adding small polar substituents, such as primary *S*-linked sulfonamide (**47**), *N*-linked sulfonamide (**44, 50**) and primary amide (**45, 51**), led to highly potent and selective compounds. It is interesting to note, however, that for the trifluoromethoxy-substituted series, the primary amide **45** is nearly 5-fold more active than the methanesulfonamide

Table 7.4 Activity and lipophilicity of diphenyl ethers **44–58**

44-58

Compound	Y	X	SRI IC$_{50}$ (nM)[a]	Selectivity over DRI	Selectivity over NRI	clog P	log D
44	OCF$_3$	NHSO$_2$Me	14	2028	565	3.9	–
45	OCF$_3$	CONH$_2$	3	12 387	1394	3.6	2.3
46	OCF$_3$	CONHMe	10	1441	275	3.8	–
47	OCF$_3$	SO$_2$NH$_2$	9	920	6283	3.3	2.2
48	OCF$_3$	SO$_2$NHMe	11	2056	444	3.9	–
49	OCF$_3$	CH$_2$NHSO$_2$Me	11	530	520	3.8	2.1
50	CF$_3$	NHSO$_2$Me	5	2375	166	3.8	2.3
51	CF$_3$	CONH$_2$	9	2217	109	3.5	–
52	CF$_3$	CONHMe	11	1523	114	3.7	–
53	CF$_3$	CONMe$_2$	39	698	11	3.4	–
54	CF$_3$	SO$_2$NHMe	9	265	265	3.7	–
55	CF$_3$	SO$_2$NMe$_2$	79	164	38	4.2	–
56	CF$_3$	CH$_2$OH	13	40	95	3.9	–
57	CF$_3$	CH$_2$NH$_2$	59	43	236	3.9	–
58	CF$_3$	CH$_2$NHSO$_2$Me	55	49	62	3.6	–

[a]The inhibitory potency at human serotonin, dopamine and noradrenaline transporters was quantified through measurement of a compound's ability to inhibit transport and subsequent intracellular accumulation of radiolabelled ligand in an *in vitro* radiometric assay using HEK-293 cells stably expressing the appropriate transporters. All assay determinations are $\geq n = 2$.

44 (3 nM *vs.* 14 nM), while for the trifluoromethyl-substituted series, the methanesulfonamide **50** has similar potency to the amide **51** (5 nM *vs.* 9 nM). This shows that there is cross-talk between the benzylamine ring and phenoxy ring SARs.

While mono *N*-alkylation of either the amides or sulfonamides was tolerated, dialkylation was detrimental to both activity and selectivity. For example, the tertiary amide **53** is 4-fold less active than the secondary amide **52** ($IC_{50} = 39$ nM *vs.* 11 nM) and only has 11-fold selectivity over NRI. Similarly, the dimethylsulfonamide **55** has very poor activity and selectivity in comparison with the methylsulfonamide **54** (79 nM *vs.* 9 nM). The effect of incorporating a methylene spacer between the aromatic ring and the polar group was also investigated. With a 4-trifluoromethyl group on the phenoxy ring, the hydroxymethyl derivative **56** is reasonably potent (13 nM) but lacks selectivity, particularly over DRI. Changing the benzyl alcohol to benzylamine **57** results in a 4-fold drop in potency (59 nM), while converting the benzylamine to a methylene spaced sulfonamide **58** has no effect on potency (55 nM) but results in a loss of selectivity over NRI. With a 4-trifluoromethoxy group on the phenoxy ring, however, the methylene-spaced sulfonamide **49** is potent (11 nM) and retains excellent selectivity over DRI and NRI (> 500-fold).

Based on the results above, the potent methylene-spaced sulfonamide **49** and *N*-linked sulfonamide **50** were investigated further to determine their potential to be used as on-demand agents for PE. These compound both have relatively low log *D* values of 2.1 and 2.3, respectively, but despite this still show good flux in a Caco-2 assay with no efflux (Figure 7.6). *In vivo* studies showed that, in the dog, both compounds have a rapid T_{max} (0.5 h) and low volume of distribution ($V_D = 6$ and $4.5 \, L \, kg^{-1}$, respectively, unbound $V_D = 88$ and $60 \, L \, kg^{-1}$). In combination with moderate clearance (16 and 14 mL min^{-1} kg^{-1}, respectively), this resulted in half-lives of 4 h for both compounds. The human PK for compound **50** was projected as possessing a T_{max} of 0.5 h and a half-life of about 5 h (based on predicted clearance of 5 mL min^{-1} kg^{-1} and V_D of 2 L kg^{-1}). The human oral bioavailability was also predicted to be ~75%.

Unfortunately, drug metabolism studies showed that these compounds are largely metabolised to the monomethylamines **59** and **60**. Monomethylamine

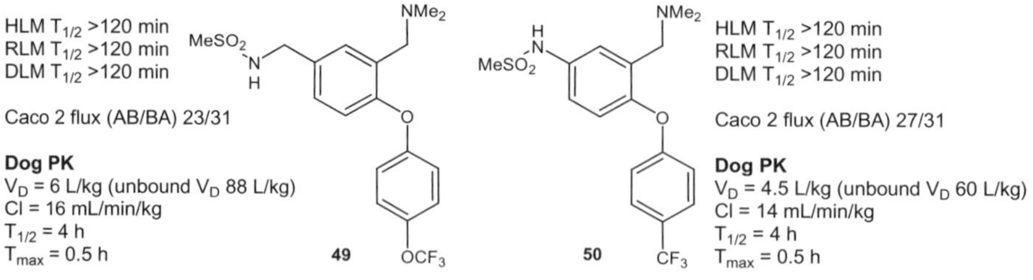

Figure 7.6

60 was profiled further and found to be 10-fold weaker than **50** (SRI IC$_{50}$ = 49 nM) and to have a much longer half-life (~10 h in rat and dog) than **50**. Therefore, dosing with **50** would result in continued exposure to a serotonin re-uptake inhibitor and, therefore, would not be significantly different to the use of existing SSRIs for the treatment of PE.

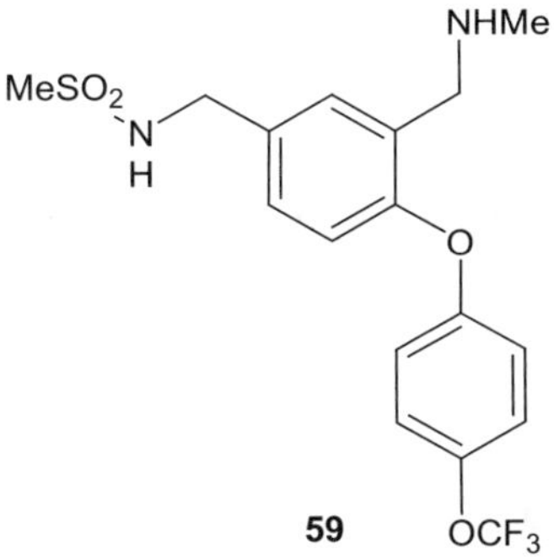

7.4.3 Strategies to Avoid Active Circulating Metabolites

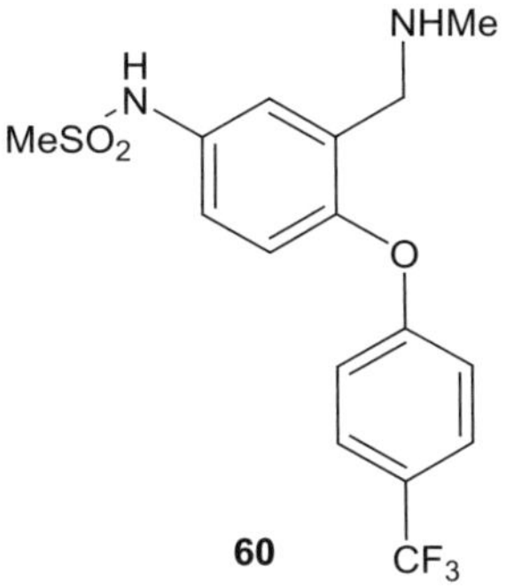

We initially considered two strategies to overcome this problem of a long-lived, active metabolite (Figure 7.7). One strategy was to modify the amine group such that metabolism around the amine would not generate an active meta-bolite. From our initial work varying the amine group, we knew that azetidines **61** retained SRI potency, but on investigating this further with polar sub-stituents on the benzylamine ring we found that these analogues tend to be weaker than desired and, in addition, have chemical stability issues.[27] We also investigated cyclising the amine into a pyrrolidine ring to give the compounds **62**. While this gives potent SSRIs,[28] the compounds made are also potent CYP 2D6 inhibitors. The introduction of a stereogenic centre made the synthesis of analogues more complicated, and we chose not to pursue this strategy.

The second strategy was to introduce a metabolically vulnerable group onto the template, so that metabolism would generate an inactive metabolite.[29]

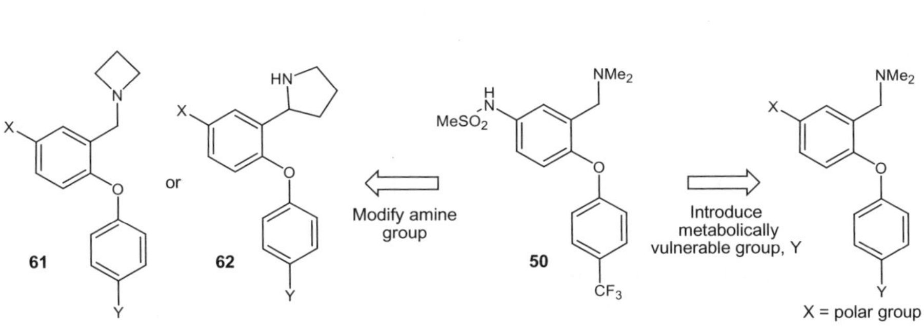

Figure 7.7

While there are clearly several routes of metabolism that could be targeted,[30] from our work on optimising the phenoxy ring substituent we knew that a methyl sulfide group in the 4-position gives good potency (**36**, 9 nM), while the more polar sulfoxides have very weak activity (**42**, 599 nM). Therefore, as long as metabolic oxidation of the sulfur occurs more rapidly than *N*-demethylation, which we considered likely, then a significantly less active metabolite would be formed. For these reasons, we made and profiled a range of 4-methylsulfanyl-substituted compounds (Table 7.5).

7.4.4   SAR for Methylsulfanyl-substituted Diphenyl Ethers

Although the parent methyl sulfide **36** only has moderate selectivity over NRI (71-fold), we were encouraged by results from the trifluoromethyl and tri-fluoromethoxy series, where incorporation of a small polar group on the benzylamine ring often led to significant increases in selectivity. This was found to be true for both the sulfonamide **63** and amide **64**, which are both very potent SSRIs with greater than 200-fold selectivity over DRI and NRI. The same is not true of the *N*-linked sulfonamide **65**, which only has modest selectivity *vs.* NRI (23-fold). Alkylation was detrimental to potency and selectivity, with both the secondary amide **66** and secondary sulfonamide **67** having less potency and poor selectivity over NRI. Introduction of a small, polar heterocycle into the 5-position of the benzylamine ring gives potent and selective compounds, *e.g.* 1-linked 1,2,3-triazole **68** (2 nM), but as in the tet-rahydronaphthalene template, these compounds have unacceptable levels of CYP inhibition. We looked again at the possibility of incorporating an alkyl spacer between the benzylamine ring and the polar functionality, but, as with the trifluoromethyl-substituted compounds described above, the resulting compounds have poor selectivity over DRI/NRI.

We also looked at pyridyl compounds such as **69** (and analogues with additional substituents on the phenoxy ring) and these were found to have excellent potency and selectivity in combination with low lipophilicity. Slightly surprisingly, given the CYP inhibition found with compounds such as **68**, compound **69** shows no significant CYP inhibition. Unfortunately, when their routes of metabolism were investigated *in vitro*, they showed a significant degree of *N*-demethylation in addition to the desired *S*-oxidation and so did not meet our objective of avoiding active circulating metabolites.[29]

7.4.5   Detailed Pre-clinical Profile of Sulfonamide 63

Sulfonamide **63** is very polar for an SSRI (log D 1.5) and contains three HBD, but still has high flux and no efflux in a Caco-2 assay (A-B/B-A = 40:41), indicating that permeability was not compromised. Compound **63** was there-fore progressed to rat and dog PK studies, where it was found to have an excellent *in vivo* profile (Figure 7.8). In both rat and dog, sulfonamide **63** has a low V_D (4 L kg^{-1} in both species; unbound $V_D = 74$ L kg^{-1} and 144 L kg^{-1}, respectively), a rapid T_{max} (0.5 h and 0.25 h) and moderate clearance

Table 7.5 Activity and lipophilicity of diphenyl ethers **63–70**

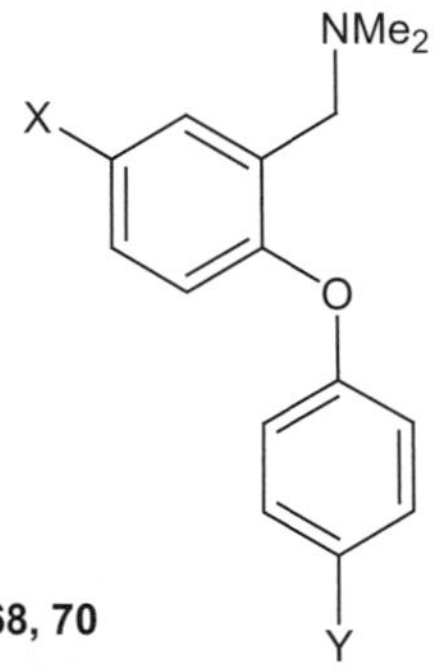

63-68, 70

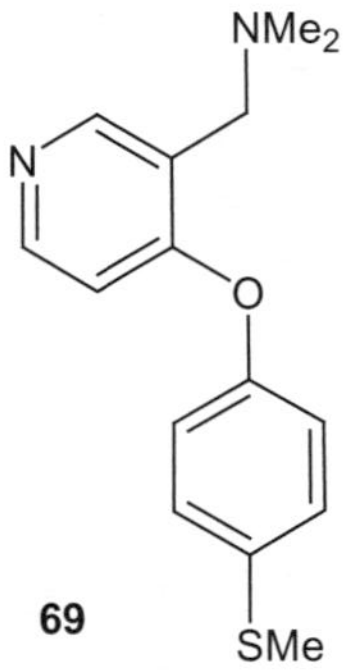

69

Compound	Y	X	SRI IC$_{50}$ (nM)^a	Selectivity over DRI	Selectivity over NRI	c log P	log D
63	SMe	SO$_2$NH$_2$	4	633	218	2.8	1.5
64	SMe	CONH$_2$	2	1351	320	3.2	2.2
65	SMe	NHSO$_2$Me	5	711	23	3.4	1.7
66	SMe	CONHMe	12	197	27	3.4	–
67	SMe	SO$_2$NHMe	9	60	25	3.4	–
68	SMe	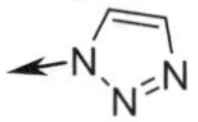 (tetrazole)	2	899	128	4.2	2.5
69	–	–	2	3600	500	3.1	1.5
70	SOMe	SO$_2$NH$_2$	>1000	–	–	1.0	<–2

^{a}The inhibitory potency at human serotonin, dopamine and noradrenaline transporters was quantified through measurement of a compound's ability to inhibit transport and subsequent intracellular accumulation of radiolabelled ligand in an *in vitro* radiometric assay using HEK-293 cells stably expressing the appropriate transporters. All assay determinations are ≥ $n = 2$.

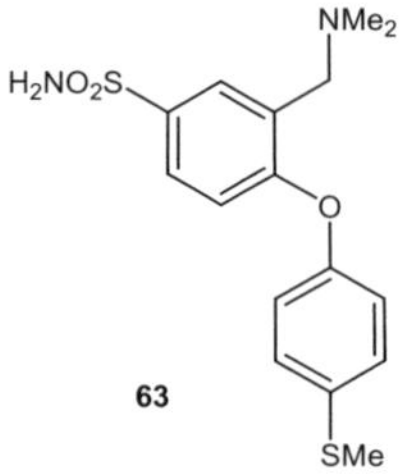

Figure 7.8

HLM $T_{1/2}$ = 40 min
RLM $T_{1/2}$ = 27 min
DLM $T_{1/2}$ = 13 min

Caco 2 flux (AB/BA) 40/41
pKa 8.4
logD 1.5

Rat PK
V_D = 4 L/kg (V_{DU} 74 L/kg)
Cl_{blood} = 19 mL/min/kg
$T_{1/2}$ = 3 h
T_{max} = 0.5 h
CSF:free blood 0.9:1

Dog PK
V_D = 4 L/kg (V_{DU} 144 L/kg)
Cl_{blood} = 20 mL/min/kg
$T_{1/2}$ = 2.4 h
T_{max} = 0.25 h

63

($Cl_{blood} = 19 \, \text{mL} \, \text{min}^{-1} \, \text{kg}^{-1}$ and $20 \, \text{mL} \, \text{min}^{-1} \, \text{kg}^{-1}$), resulting in a relatively short half-life of 3 hours and 2.4 hours in rat and dog, respectively. Total clearance is moderate and bioavailability for **63** is high in both rat and dog, despite its having relatively high turnover in microsomes and correspondingly high unbound clearance ($Cl_u = 377$–$730 \, \text{mL} \, \text{min}^{-1} \, \text{kg}^{-1}$). This is thought to be due to the protective effect of partitioning into red blood cells,[31] which reduces the blood clearance and also protects against high first-pass clearance. Scaling from rat and dog PK gave a human PK prediction of low V_D (2–$4 \, \text{L} \, \text{kg}^{-1}$), rapid T_{max} ($\sim 0.5 \, \text{h}$) and moderate blood clearance (4–$8 \, \text{mL} \, \text{min}^{-1} \, \text{kg}^{-1}$), resulting in high predicted bioavailability ($> 70\%$), a short half-life of 4–$10 \, \text{h}$, and low predicted dose ($< 20 \, \text{mg}$), exactly in keeping with our goals. Sulfonamide **63** was profiled against a range of receptors, enzymes and ion-channels (more than 30 Cerep assays) and was found to have no significant activity ($IC_{50} > 10 \, \mu\text{M}$ except at dopamine D1, $IC_{50} = 2.4 \, \mu\text{M}$, and opiate receptors, $IC_{50} = 2.9 \, \mu\text{M}$).

CNS penetration studies of **63** in the rat showed high drug levels in the brain (CSF : free blood ratio of 0.9). CNS microdialysis studies in the rat demonstrated that, after oral dosing, central serotonin levels rapidly increase (Figure 7.9), reaching a peak within 1 h at $3 \, \text{mg} \, \text{kg}^{-1}$, a slightly later time than the T_{max} of **63** in the plasma. This is in marked contrast to the data obtained for fluoxetine (Figure 7.1), where peak serotonin levels are not obtained until after more than 5 h.

Drug metabolism studies showed that **63** is almost completely metabolised to the sulfoxide **70** ($> 90\%$) in human liver microsomes and also *in vivo* in rat and dog. Not only is this compound completely inactive against all three monoamine transporters, but it is also essentially inactive in the CEREP screening panel (all $IC_{50} > 10 \, \mu\text{M}$). Compound **70** is extremely polar ($\log D < -2$) and is rapidly cleared *in vivo* ($t_{1/2} < 4 \, \text{h}$ in rat and dog).

While, in theory, oxidation of the sulfur in **63** could be mediated by several P450 isoforms, or by other oxidative enzymes such as flavin monooxygenases (FMO),[32] drug metabolism studies showed that **63** is predominantly metabolised by CYP2D6 and CYP2C9, with CYP2D6 appearing to be the major metabolising enzyme (50–60%). Based on this it was anticipated that roughly a 2-fold increase in human half-life would be seen in patients who do not express

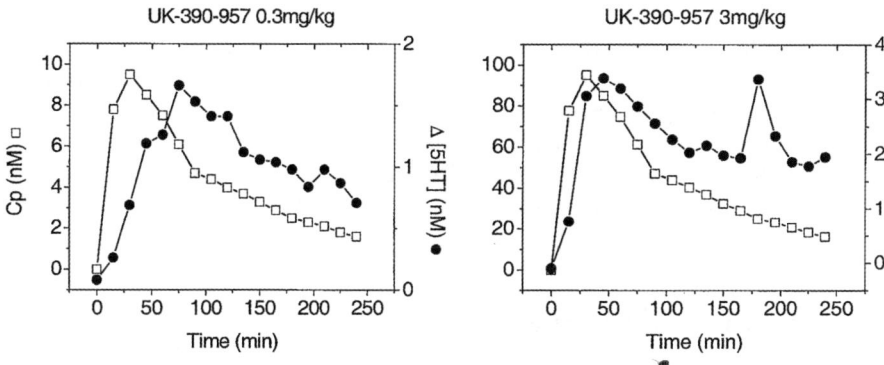

Figure 7.9

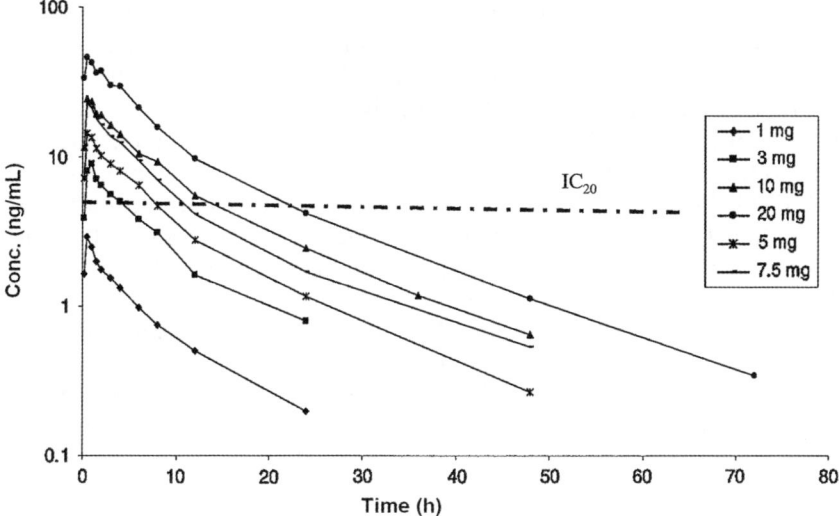

Figure 7.10   Free drug levels of UK-390957 following oral dosing in man.

this polymorphic enzyme. While this prediction did not quite meet the target that we had set, the overall profile of sulfonamide **63** (UK-390957) was compelling enough for it to be selected for clinical development.

7.4.6   Clinical Data for UK-390957

UK-390957 successfully progressed to FIH studies and gave exposure in man (Figure 7.10) that was linear with increasing dose from 1 to 20 mg. A dose of 5 mg UK-390957 exceeded the SRI IC_{20} for approximately 10 h. The PK parameters in man are very close to those predicted, with UK-390957 having excellent bioavailability and a rapid T_{max} (<1 h), although the estimated clearance is slightly lower than predicted ($Cl_{blood} \approx 2–3$ mL min^{-1} kg^{-1}), which results in a somewhat longer half-life of 8–12 h (initial elimination phase).

UK-390957 was progressed to Phase II studies for the treatment of PE and showed some efficacy in PE sufferers. It was not progressed further, however, and so its effectiveness as a treatment for PE in comparison to daily treatment with marketed SSRI antidepressants was not confirmed.

7.5 Profile of Dapoxetine

While UK-390957 was not taken beyond Phase II by Pfizer, Johnson and Johnson have progressed dapoxetine **71** (Figure 7.11) to Phase III studies for the treatment of PE. Dapoxetine is an SSRI that was originally discovered by workers at Eli Lilly in the late 1980s and progressed as a potential anti-depressant.[33] Subsequently it was recognised that, due to its short half-life (1.5 h) and relatively rapid T_{max} (~ 1 h), it had potential to be taken forward as an on-demand agent for the treatment of PE.[34,35] As with UK-390957, the more rapid T_{max} of dapoxetine is almost certainly driven by its low V_D (2.1 L kg^{-1}) relative to the marketed SSRIs. In terms of lipophilicity, dapoxetine is very similar to the marketed SSRI antidepressants and significantly more lipophilic than UK-390957 (clog P 5.3 *vs.* 2.8), and yet, despite this, dapoxetine has a very low V_D, like UK-390957 and unlike the other marketed SSRIs. This clearly shows that while lipophilicity does play a role in determining V_D, other poorly understood structural features sometimes play an even greater role.

As a consequence of its high lipophilicity, dapoxetine has very high plasma protein binding (99% bound), which is probably important in enabling it to maintain reasonable bioavailability (42%) by protecting it against extensive first-pass metabolism in the liver.[34] The major metabolite of dapoxetine *in vivo* is the inactive *N*-oxide, with the active *N*-demethyl analogue only being present at low levels,[35] so as with UK-390957, there is no major circulating active metabolite.

The development of dapoxetine has stimulated some debate as to whether a rapid onset, short half-life agent is the correct approach to treating PE. While it was proposed that on-demand dosing may be a preferred dosing regime for patients, recent studies indicate that patients actually prefer a daily treatment

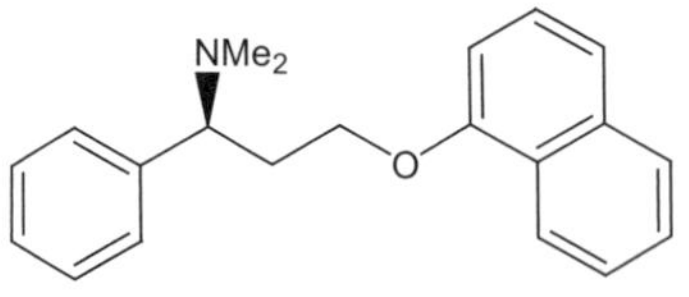

SRI IC$_{50}$ = 8 nM

Human PK
V_D = 2.1 L/kg
Plasma protein binding = 99%
F = 42%
$T_{1/2}$ = 1.5 h
T_{max} ~ 1 h

71

Figure 7.11

regime when treating sexual dysfunctions, as continued efficacy allows for a more spontaneous sexual life.[36,37] Secondly, while the short half-life could result in an improved side-effect profile due to reduced drug exposure, the efficacy of on-demand treatment has been questioned.[9,38] The argument is that on-demand treatment does not lead to the adaptive changes in brain chemistry that are seen with chronic (daily) treatment, which may be required for maximum efficacy. There is also some data to suggest that on-demand dosing with classical SSRIs such as paroxetine may show reduced efficacy relative to chronic dosing.[2]

Dapoxetine has shown efficacy in clinical trials for the on-demand treatment of PE,[4,39] and was approved for this indication under the trade name Priligy™ by several European countries in 2009. An earlier filing in the USA, however, was given a "not approvable" letter by the FDA in 2005.[40]

7.6 Summary

Initial work to identify a short $t_{1/2}$, rapid T_{max} SSRI focused on reducing V_D in the tetrahydronaphthalene template **3**. While a moderate reduction in V_D was achieved, this strategy was ultimately unsuccessful as sulfonamide UK-373911 has an extremely long $t_{1/2}$ in man. The key to finding compounds with the properties we desired was to switch to the diphenyl ether template, which has an inherently lower V_D. Early analogues had the desired low V_D, and resulting short $t_{1/2}$ and rapid T_{max}, but gave circulating active metabolites. Incorporation of a metabolically vulnerable sulfide resulted in metabolism to essentially inactive sulfoxide metabolites and led to the identification of the sulfonamide UK-390957 as a clinical candidate. Profiling in man showed that we had achieved our desired target of a rapid T_{max} and short $t_{1/2}$ and UK-390957 was progressed to Phase II studies for the treatment of PE. That structure can be more important than physicochemical properties in determining volume is supported by the fact that the SSRI dapoxetine has a low volume of distribution, comparable to that of UK-390957, despite being significantly more lipophilic.

References

1. A. Hemeryck and F. M. Belpaire, *Curr. Drug Metab.*, 2002, **3**, 13, and references cited therein.
2. A. J. Moreland and E. H. Makela, *Ann. Pharmacother.*, 2005, **39**, 1296.
3. M. D. Waldinger, *Expert Opin. Emerging Drugs*, 2006, **11**, 99.
4. K. Patel and W. J. G. Hellstrom, *Curr. Opin. Invest. Drugs*, 2009, **10**, 681.
5. American Psychiatry Association, *Diagnostic and Statistical Manual of Mental Disorders*, 4th edn, American Psychiatric Association, Washington, 1994, p. 509.
6. M. E. Metz, *J. Sex Marital Ther.*, 1997, **23**, 3.
7. E. O. Laumann, A. Paik and R. C. Rosen, *J. Am. Med. Assoc.*, 1999, **281**, 537.
8. Y. Huang and W. A. Williams, *Expert Opin. Ther. Pat.*, 2007, **17**, 889.

9. M. D. Waldinger, *Drug Discovery Today: Ther. Strat.*, 2005, **2**, 37.
10. C. G. McMahon and K. Touma, *J. Urol.*, 1999, **161**, 1826.
11. M. Hurst and H. M. Lamb, *CNS Drugs*, 2000, **14**, 51.
12. Microdialysis studies carried out by Hans Rollema's team, Pfizer Groton.
13. D. K. Walker, M. J. Humphrey and D. A. Smith, *Xenobiotica*, 1994, **24**, 243, and references cited therein.
14. B. Clark and D. A. Smith, *An Introduction to Pharmacokinetics,* Blackwell Science, Oxford, 2001.
15. B. C. Jones, D. S. Middleton and K. Youdim, *Prog. Med. Chem.*, 2009, **47**, 239.
16. R. S. Obach, *Annu. Rep. Med. Chem.*, 2007, **42**, 469.
17. S. A. Hitchcock and L. D. Pennington, *J. Med. Chem.*, 2006, **49**, 7559, and references therein.
18. W. M. Welch, *Adv. Med. Chem.*, 1995, **3**, 113.
19. S. H. Preskorn, *Clin. Pharmacokinet.*, 1997, **32**, 1.
20. D. S. Middleton and A. Stobie, *Pat. Appl.* WO 2000/051972-A1, 2000.
21. D. S. Middleton, M. Andrews, P. Glossop, G. Gymer, A. Jessiman, P. J. Johnson, M. MacKenny, M. J. Pitcher, T. Rooker, A. Stobie, K. Tang and P. Morgan, *Bioorg. Med. Chem. Lett.*, 2006, **16**, 1434.
22. For example, the zwitterionic H1 antagonist cetirizine has a significantly lower volume than the corresponding basic H1 antagonists: C. Chen, *Curr. Med. Chem.*, 2008, **15**, 2173.
23. M. L. Elliott, H. R. Howard, Jr., C. J. Schmidt and T. F. Seeger, *Pat. Appl.* WO 2000/050380-A1, 2000.
24. D. S. Middleton, M. Andrews, P. Glossop, G. Gymer, D. Hepworth, A. Jessiman, P. S. Johnson, M. MacKenny, M. J. Pitcher, T. Rooker, A. Stobie, K. Tang and P. Morgan, *Bioorg. Med. Chem. Lett.*, 2008, **18**, 4018.
25. M. D. Andrews, D. Hepworth and D. S. Middleton *Pat. Appl.* WO 2004/016593-A1, 2004.
26. M. D. Andrews, D. Hepworth, D. S. Middleton and A. Stobie, *Pat. Appl.* WO 2001/072687-A1, 2001.
27. Ring opening of the azetidine, particularly under acidic conditions.
28. M. D. Andrews, D. Hepworth, D. S. Middleton and A. Stobie, *Pat. Appl.* EP 1184372-A1, 2002.
29. D. S. Middleton, M. Andrews, P. Glossop, G. Gymer, D. Hepworth, A. Jessiman, P. S. Johnson, M. MacKenny, A. Stobie, K. Tang, P. Morgan and B. Jones, *Bioorg. Med. Chem. Lett.*, 2008, **18**, 5303.
30. T. F. Woolf, *Handbook of Drug Metabolism,* Dekker, New York, 1999.
31. D. A. Smith and R. M. Jones, *Curr. Opin. Drug Discovery Dev.*, 2008, **11**, 72.
32. S. K. Krueger and D. E. Williams, *Pharmacol. Ther.*, 2005, **106**, 357.
33. D. W. Robertson, D. C. Thompson and D. T. Wong, *Pat. Appl.* EP 0288188-A1, 1988.
34. K.-E. Andersson, J. P. Mulhall and M. G. Wyllie, *Br. J. Urol. Int.*, 2006, **97**, 311.

35. N. B. Modi, M. J. Dresser, M. Simon, D. Lin, D. Desai and S. Gupta, *J. Clin. Pharmacol.*, 2006, **46**, 301.
36. M. D. Waldinger, A. H. Zwinderman, B. Olivier and D. H. Schweitzer, *J. Sex. Med.*, 2007, **4**, 1028.
37. H. Porst, K. Hell-Momeni and H. Büttner, *Urologe A*, 2009, **48**, 1318.
38. M. D. Waldinger and D. H. Schweitzer, *J. Sex. Med.*, 2008, **5**, 966.
39. J. L. Pryor, S. E. Althof, C. Steidle, R. C. Rosen, W. J. G. Hellstrom, R. Shabsigh, M. Miloslavsky and S. Kell, *Lancet*, 2006, **368**, 929.
40. M. G. Wyllie, *Br. J. Urol. Int.*, 2006, **98**, 227.

CHAPTER 8

The Discovery of the Long-acting PDE5 Inhibitor PF-489791 for the Treatment of Pulmonary Hypertension

ANDREW S. BELL AND MICHAEL J. PALMER

Sandwich Laboratories, Pfizer Global Research and Development, Sandwich, Kent, CT13 9NJ, UK

8.1 Introduction

The cyclic nucleotides cyclic adenosine monophosphate (cAMP) and cyclic guanosine monophosphate (cGMP) play a vital role as the intracellular second messengers for several signal transduction pathways. Levels of the cyclic nucleotides are elevated in response to vascular stimuli, for example atrial natriuretic factor, stress or nitric oxide (NO) through activation of adenylate or guanylate kinases, then degraded through the action of cyclic nucleotide phosphodiesterase enzymes (PDEs). Since amplification of the effects of the cyclic nucleotides can produce a marked physiological response, the PDEs have been longstanding therapeutic targets for the pharmaceutical industry; notable examples are PDE3 inhibitors as inotropes and PDE4 inhibitors as anti-asthma treatments.[1]

The human PDE family has been shown to consist of 11 members coded for by 21 genes and, in addition, there are a number of different splice variants (Table 8.1).[2] The family members can be distinguished by their tissue

RSC Drug Discovery Series No. 4
Accounts in Drug Discovery: Case Studies in Medicinal Chemistry
Edited by Joel C. Barrish, Percy H. Carter, Peter T. W. Cheng and Robert Zahler
© Royal Society of Chemistry 2011
Published by the Royal Society of Chemistry, www.rsc.org

Table 8.1 Cyclic nucleotide specificity and discovery date for PDE isoforms.

PDE subtype	Specificity	Year of discovery
1A, 1B	cGMP > cAMP	1970
1C	cGMP = cAMP	1970
2A	cAMP ≫ cGMP	1970
3A, 3B	cAMP ≫ cGMP	1970
4A, 4B, 4C, 4D	cAMP ≫ cGMP	1970
5A	cGMP ≫ cAMP	1978
6A, 6B, 6C, 6D, 6G, 6H	cGMP ≫ cAMP	1985–2000
7A	cAMP ≫ cGMP	1993, 2000
8A, 8B	cAMP ≫ cGMP	1998
9A	cGMP ≫ cAMP	1998
10A	cGMP = cAMP	1999
11A	cGMP = cAMP	2000

distribution and their specificity for the cyclic nucleotides. Much of the early work in the PDE field was carried out prior to the characterisation of the entire PDE family; hence many early drug candidates were subsequently found to have suboptimal selectivity.

The discovery of the potent, selective PDE5 inhibitor sildenafil enabled the clinical investigation of the role of PDE5 in humans.[3] Since PDE5 is widely expressed in smooth muscle throughout the body, several potential indications have been identified for PDE inhibitors. After initial investigation as a potential anti-anginal, the observation of spontaneous maintained erections in early clinical trials lead to the development of sildenafil as a treatment for male erectile dysfunction (Viagra™). However, further studies identified an additional utility for PDE5 inhibitors as a new treatment option for pulmonary arterial hypertension (PAH), leading to the approval of an alternative dosage form of sildenafil (Revatio™).[4]

Following the disclosure of the first clinical results for sildenafil, interest in PDE5 inhibition was stimulated industry-wide. A number of different chemotypes were identified, but only three additional NCEs representing two chemotypes have been approved [vardenafil,[5] tadalafil[6] and udenafil[7] (Figure 8.1)]. Each of the chemotypes can be characterised by their selectivity profile over the other members of the PDE family. Thus the sildenafil/vardenafil/udenafil chemotype results in high selectivity over all PDEs other than PDE6, whereas tadalafil was found to have low selectivity over PDE 11. Inhibition of PDE6 is thought to contribute to undesired visual side effects,[8] while PDE11 inhibition has been implicated in spermatogenesis.[9]

Ongoing clinical and pre-clinical studies with sildenafil,[10] and subsequent clinical candidates from the same series, identified several additional indications for PDE5 inhibitors. Since these include disease indications (*e.g.* hypertension, diabetes) requiring chronic treatment and/or single-dose daily administration in order to be competitive, we sought novel agents with an excellent safety profile and, in particular, no potential for off-target activity against other PDE family members. Based on our clinical experience with

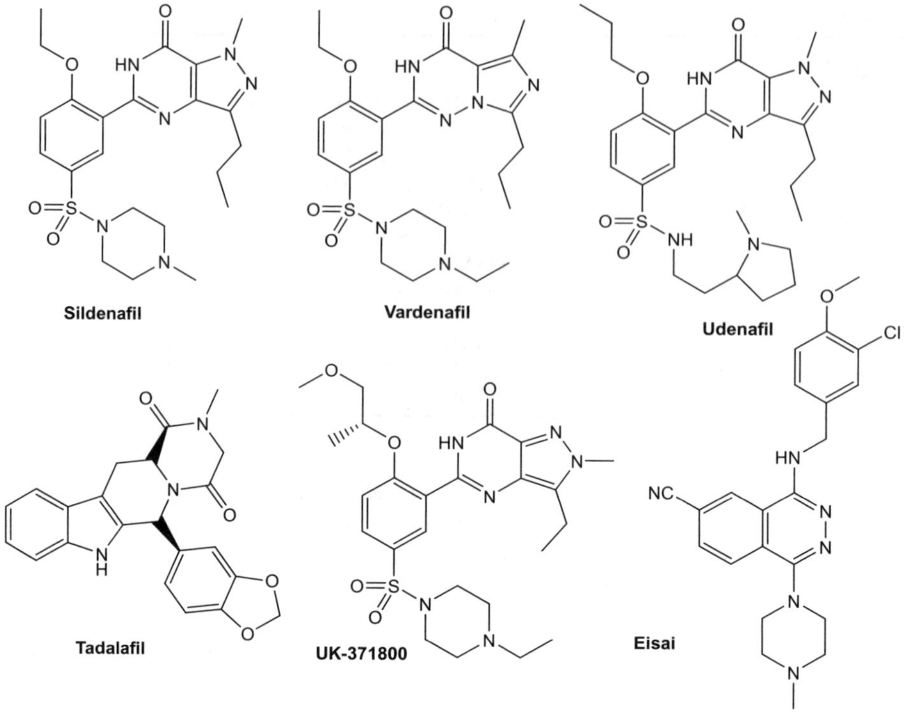

Figure 8.1 Structures of marketed and advanced PDE5 inhibitors.

sildenafil, this would require free drug levels below the corresponding PDEX IC_{50} at all times. We believed that this profile could be best achieved by a long-duration compound with low peak-to-trough concentrations.

8.2 X-Ray Crystallography and Structure-based Drug Design

Although not available at the time of the design of the first-generation PDE5 inhibitors, an in-house effort yielded a crystal structure of a truncated form of the enzyme complexed with sildenafil.[11] An independent study[12] also solved the structures of the catalytic domain of PDE5 complexed to sildenafil, vardenafil and tadalafil. We were also able to obtain bound structures of compounds from our sildenafil follow-on effort, including UK-371800,[13,14] as well as those for literature compounds (*e.g.* a close analogue of E-4010),[15] which we felt would provide additional insights into the promiscuity of the PDE5 active site (Figure 8.2). The availability of structural information enabled us to rationalise the structure–activity relationships (SAR) we had observed in our earlier synthetic efforts and thus direct further effort into regions of the active site we felt would yield more optimal interactions.

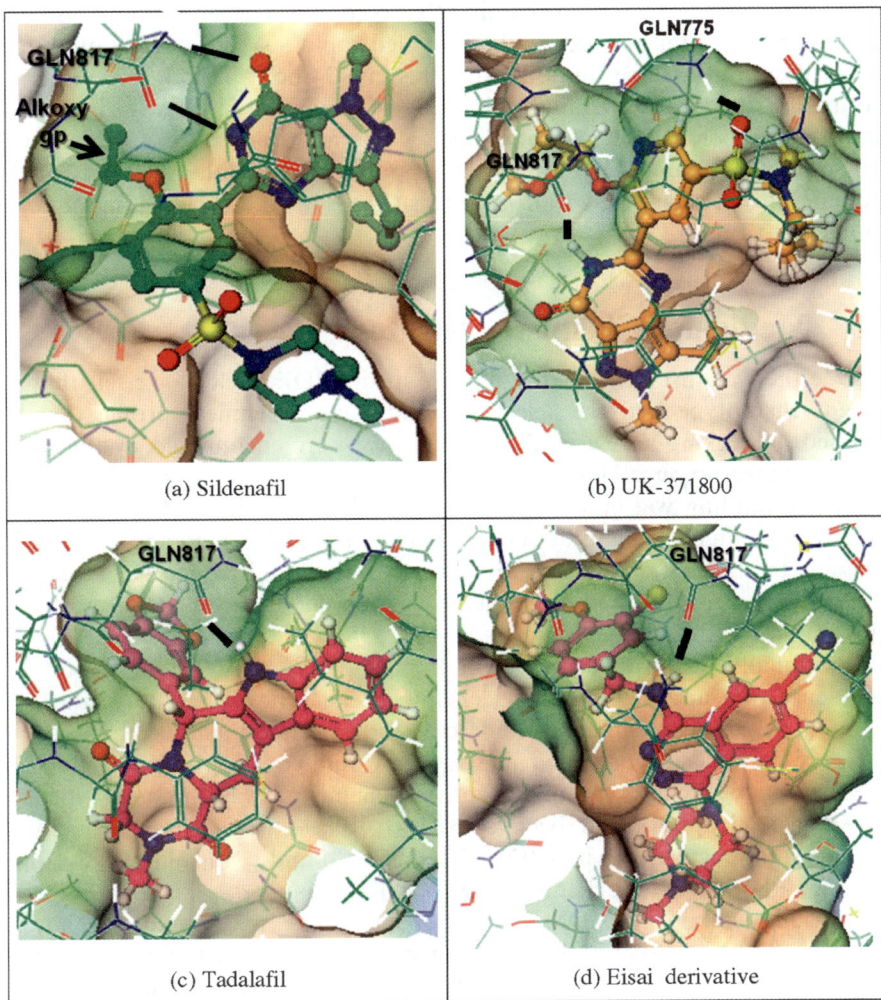

Figure 8.2   Co-crystal structures of varied PDE5 inhibitors with the PDE5 catalytic domain, illustrating key interactions and the diversity possible in binding mode. Solid renderings to PDE5 binding site surface are coloured by hydrophobicity. (a) Sildenafil with bidentate H-bond to GLN817 and the ethoxy group located in an orthogonal hydrophobic binding pocket, the "alkoxy" pocket. (b) UK-371800, which is "flipped" relative to sildenafil and making H-bonds with both GLN817 and GLN775. The extended alkoxy group occupies the orthogonal pocket. (c) Tadalafil, H-bonding to GLN817 and with the methylenedioxyaryl group occupying the "alkoxy" pocket. (d) An Eisai phthalazine derivative H-bonding to GLN817 and with a benzyl group occupying the orthogonal "alkoxy" pocket.

To facilitate the process of obtaining PDE5 co-crystal structures, we also had a standard operating procedure that enabled rapid soaking of new inhibitors and provision of high-resolution structures in a 24-hour period. Our intention was to use this readily available binding information for newly discovered inhibitors to assess series potential and, where appropriate, to guide potency and selectivity tuning. The latter was supported by homology models of the catalytic sites of each of the PDE isoforms (PDEs 6, 10, 11) closest in sequence similarity to PDE5.

8.3 Medicinal Chemistry Strategy

Despite the wealth of chemical literature in the PDE5 field,[16] we felt that no current chemotype had the selectivity and physicochemical profile that we required. In seeking a new inhibitor series, we were encouraged by the structural diversity amongst literature PDE5 inhibitors and the knowledge that active-site binding was not restricted to one binding mode. As a result, we undertook a high-throughput screen (HTS) of the Pfizer compound file. In assessing the hit matter that emerged, we set out to utilise our PDE structural knowledge, together with parallel chemistry, in order to accelerate progress and allow rapid optimisation.

Additionally, we proposed a set of key property criteria, designed to give confidence that any new emerging chemical lead series would have the capability of providing a clinical candidate with the desired highly PDE5-selective and once-daily dosing profile. This strategy would contrast with the sildenafil programme, wherein the required physicochemistry was pursued after the active pharmacophore had been discovered. These lead criteria principally comprised potency/selectivity (PDE5 $IC_{50} < 50$ nM and > 10-fold selective over all other PDEs), physicochemistry (MW < 400, clog $P < 4$, log $D = 1$–2) and absorption/metabolism (ADME) components (well fluxed, predicted human half-life > 12 h), together with wide chemical scope. These criteria were used to track progress during the hit-to-lead phase and throughout the subsequent lead optimisation programme. Ligand efficiency (LE)[17,18] and ligand lipophilicity efficiency values (LLE)[19,20] were used to aid the hit and lead assessment process. Binding free energy for LE was calculated from $-RT \ln K_{eq}$ using PDE5 IC_{50}, as the surrogate for dissociation constant and RT equal to 1.4. LLE reflects the minimally acceptable lipophilicity per unit of *in vitro* potency, giving an indication of specific hydrophobic binding requirement and was calculated from $-\ln K_{eq}$ minus clog P with dissociation constant defined as for LE.

8.4 Hit and Lead Optimisation

8.4.1 Diaminoquinazoline Hit Follow-up Strategy

A series of diaminoquinazoline-based hits were identified from high-throughput screening and displayed a number of positive attributes that encouraged

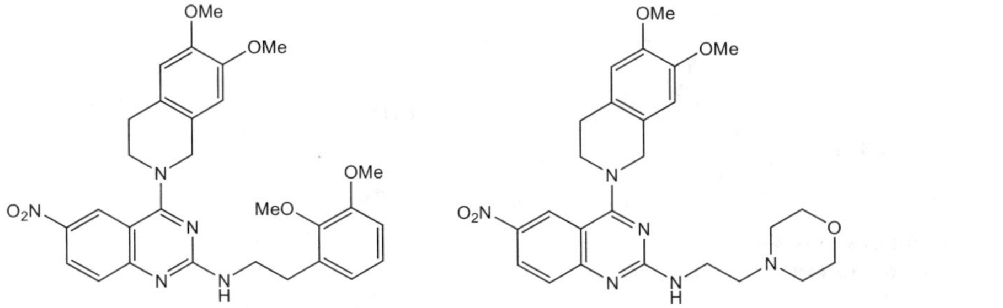

1
PDE5 IC50 20nM
cLogP 5.4, MWt 545
LE 0.27, LLE 2.3

6-nitro-2,4-diaminoquinazoline related to 1 (See Figure X4)
PDE5 IC$_{50}$ 137nM
cLogP 3.8, MWt 494
LE 0.27, LLE 3.05

Figure 8.3 Representative 6-nitro-2,4-diaminoquinazoline HTS hit **1**.

follow-up. The series displayed sub-100 nM level potency and was readily amenable to parallel chemistry. The nitro derivative **1** was typical, displaying 20 nM level potency (Figure 8.3), but unattractive physicochemistry (clog $P = 5.4$, MW $= 545$), resulting in LE and LLE values that fell short of the corresponding figures for established PDE5 agents. The nitro group also represented a potential toxicophore and gave us cause for concern with respect to safety.

Additionally, we had binding site information from an X-ray co-crystal structure that suggested significant scope for improvement (Figure 8.4a). In particular, whilst the nitro group was maintaining a bidentate hydrogen bond spanning two glutamine (GLN) residues (GLN817, GLN775) important to activity, hydrophobic interactions appeared far from efficient and the sildenafil "alkoxy" pocket was unoccupied.

Using our structural knowledge of the requirements for PDE5 inhibition,[11,21,22] we set out to replace the nitro group using the diaminoquinazoline core to target alternative interactions with GLN817. Additionally, we sought to improve the efficiency of the hydrophobic interactions. To this end, a library based on sequential displacement of dichloroquinazoline templates was designed and synthesised (Figure 8.5).[23] One of the products, compound **2**, featured an aminopyridine motif that represented a novel PDE5 pharmacophore and retained potency despite the significantly modified structure. The physicochemistry of **2** was vastly improved and reflected in a LE value that was much more in accord with established PDE5 inhibitors. LLE also showed a slight improvement.

With a more structurally desirable pharmacophore in hand, we designed a second library (Figure 8.6). Objectives were to improve potency, identify alternative diamino heterocycle templates that offered greater chemical scope for lead optimisation and to incorporate ionisable functionality to aid solubility, which in the case of a basic centre could also increase volume of distribution to aid pharmacokinetics. Whilst several interesting structures

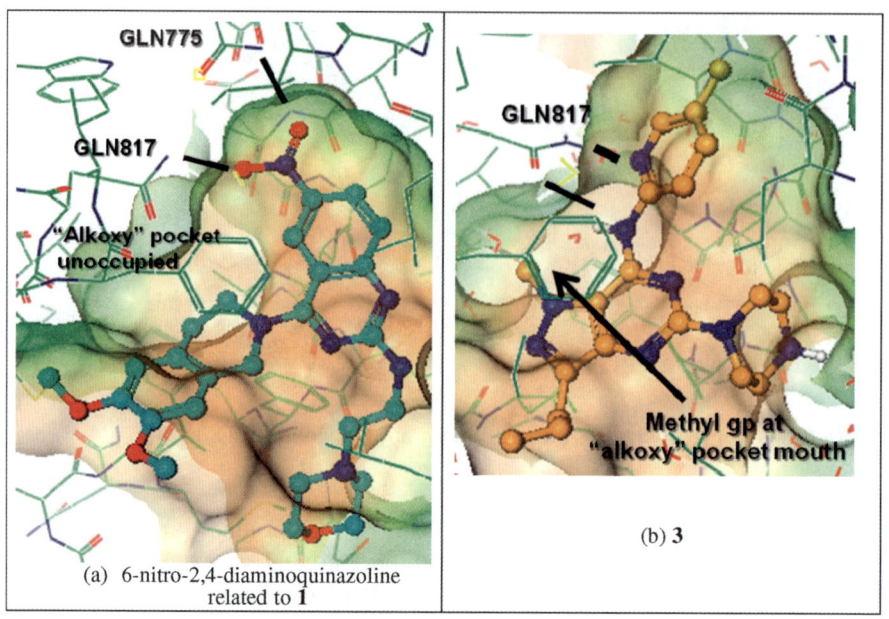

(a) 6-nitro-2,4-diaminoquinazoline
related to **1**

(b) **3**

Figure 8.4 (a) Co-crystal structure of a 2,4-diamino-6-nitroquinazoline related to **1** with the catalytic domain of PDE5. H-bonds to GLN817 and GLN775 illustrated, but no "alkoxy" pocket interaction. (b) Co-crystal structure of **3** illustrating the GLN817 bidentate H-bond from the aminopyridine motif and the N1 methyl group sitting at the mouth of the "alkoxy" pocket. Solid renderings to the PDE5 binding site surface are coloured by hydrophobicity.

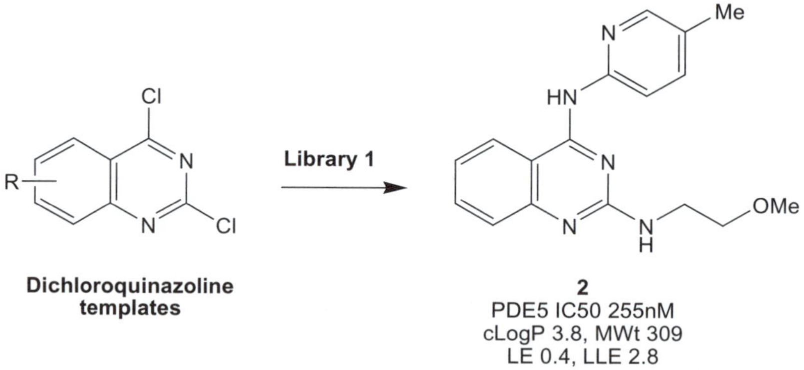

R—

CI

CI

N

N

**Dichloroquinazoline
templates**

Library 1 →

HN

N

N

N
H

Me

OMe

2
PDE5 IC50 255nM
cLogP 3.8, MWt 309
LE 0.4, LLE 2.8

Figure 8.5 Follow-up library based on HTS hit **1** leading to aminopyridine derivative **2**.

were identified, the pyrazolopyrimidine **3** stood out. The pharmacology in terms of PDE5 potency and selectivity against the other PDEs was close to our lead criteria, whilst the physicochemical and ADME properties were well inside our target range (Table 8.2). In particular, rat pharmacokinetics

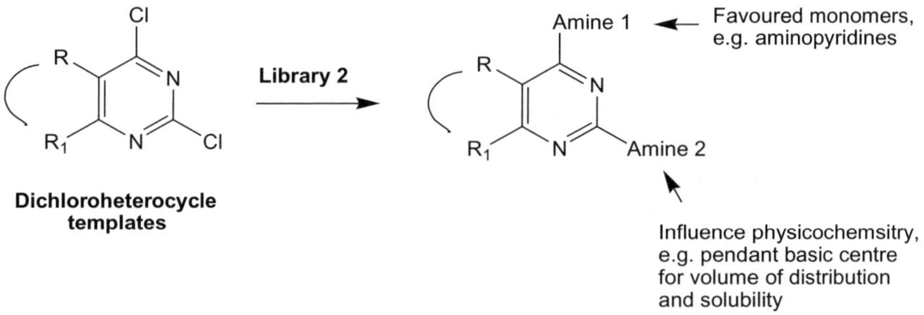

Figure 8.6 Design basis for second library.

Table 8.2 In-depth profile of lead compound **3**

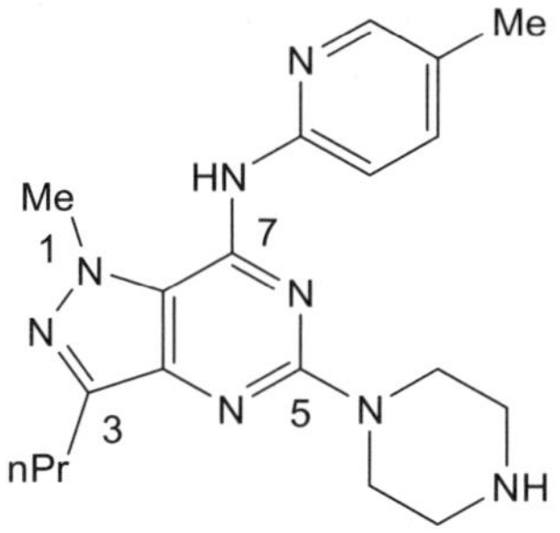

3

Criteria	Value
PDE5 IC$_{50}$ (nM)	71
PDE6 IC$_{50}$ (nM)	1330
PDE10 IC$_{50}$ (nM)	79
PDE11 IC$_{50}$ (nM)	210
All other PDE IC$_{50}$'s (nM)	> 1000
clog *P*	3.1
MWt	366
LE (PDE5)	0.37
LLE (PDE5)	4.0
Caco-2 flux	A-B $= 18 \times 10^{-6}$ cm s^{-1}, ER < 1
Rat pharmacokinetics	$V_D = 23$ L kg^{-1}, $t_{1/2} = 1.4$ h
Projected human half-life	> 12 h

indicated a high volume of distribution that enabled a projected human half-life of > 12 h. A further important advantage that **3** offered was wide chemical scope, with four substituent positions readily amenable to chemical modification.

A co-crystal structure was obtained for **3** and encouragingly indicated inhibitor binding in the PDE5 active-site domain to have significant scope for improvement (Figure 8.4b). The pyrazolopyrimidine template was flipped relative to the related pyrazolopyrimidinone core featured in sildenafil and made favourable hydrophobic interactions, whilst the aminopyridine restored the bidentate hydrogen bond with GLN817. However, the "alkoxy" hydrophobic binding pocket that located the sildenafil ethoxy substituent was unoccupied. Encouragingly, the N1 methyl substituent of **3** was located at the mouth of the "alkoxy" pocket, suggesting that extension would enable additional hydrophobic binding and improve potency. Additionally, the equivalent hydrophobic binding pocket in PDE10 is much shallower, based on homology and later upon the X-ray structure,[24] a difference that could promote improvements in selectivity over PDE10. In terms of PDE6 and PDE11 selectivity, binding site homology comparison pointed to residue differences in the region of the mouth of the "alkoxy" pocket. The C3 propyl substituent of **3** is located in this region of the active site and could potentially be exploited to further enhance PDE5 selectivity. Additionally, the second glutamine residue, GLN775, is a threonine and serine in PDE10 and PDE11, respectively. The C7 pyridine substituent of **3** sits in this region of the active site, offering a further selectivity opportunity.

Based on these considerations, compound **3** was selected as an advanced lead and follow-up initiated.

8.4.2   Lead Optimisation

The initial follow-up of **3** was duly guided by the co-crystal structure and homology studies, together with the capability to vary four positions, two of which were parallel chemistry enabled.[25] A summary of the initial structure–activity relationship obtained is detailed in Figure 8.7. Key findings were that

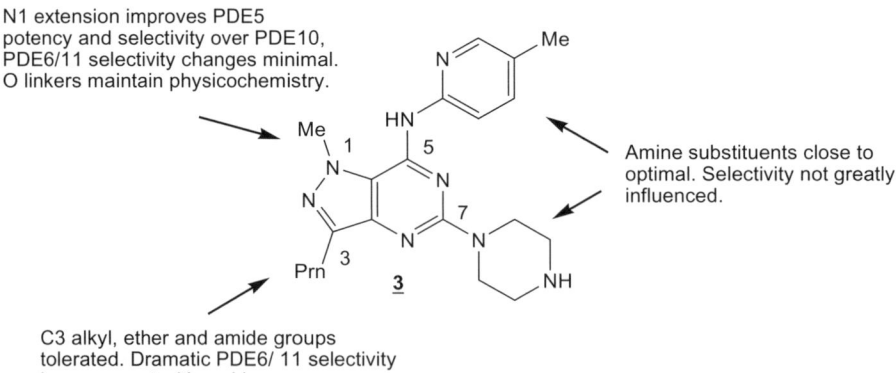

N1 extension improves PDE5 potency and selectivity over PDE10, PDE6/11 selectivity changes minimal. O linkers maintain physicochemistry.

Amine substituents close to optimal. Selectivity not greatly influenced.

C3 alkyl, ether and amide groups tolerated. Dramatic PDE6/ 11 selectivity improvement with amides.

Figure 8.7   Developing the series. Guided by X-ray, the four positions of variation were explored.

extension of the N1 methyl substituent did improve PDE5 potency and bulky substituents improved selectivity against PDE10 as anticipated. Overall, N1 alkoxyalkyl substituents provided the optimal balance between pharmacology improvement and retention of good physicochemistry. Secondly, parallel chemistry-based study of the C5 and C7 N-substituents indicated that the amine substituents discovered from the two initial libraries were close to optimal. No selectivity gains were achieved from variation of the pyridine group that sits close to GLN775. However, thirdly, and more crucially, variation of the C3 substituent indicated tolerance of a range of substituents and amide-based groups gave dramatic PDE5 selectivity improvements.

Two advanced leads emerged upon combining the SAR knowledge gained, compounds **4** and **5** (Table 8.3), both displaying excellent LE and LLE values. Both featured an N1 ethoxyethyl substituent and an isomeric 4-methylpyridine group relative to **3**, but differed in their respective C3 substituents. The pyrazolopyrimidine **4** featured a C3 methyl whilst **5** carried a C3 methylamide.[26] Whilst **4** had low nanomolar level PDE5 potency and retained physicochemistry compatible with good pharmacokinetics, the selectivity against the key PDE targets, PDEs 6, 10 and 11, remained moderate. The amide **5**, on the other

Table 8.3 Profiles of developed leads **4** and **5**

Criteria	4	5
PDE5 IC$_{50}$ (nM)	3.4	0.5
PDE6 selectivity fold	20	104
PDE10 selectivity fold	11	186
PDE11 selectivity fold	5	156
clog P	2.5	1.65
MW	396	439
LE (PDE5)	0.40	0.41
LLE (PDE5)	6.02	7.65
Caco-2 Papp A-B ($\times 10^{-6}$ cm s^{-1}), efflux ratio	18, 1	2, >10
TPSA	93	122
HBD, HBA	2, 9	3, 11

hand, displayed outstanding PDE5 potency and selectivity. However, the altered physicochemistry was sufficient to compromise permeability, as evidenced by a Caco-2 permeability assay. These findings were consistent with Lipinski rules.[27] In particular, the two additional heteroatoms from the amide motif raised both the H-bond donor/acceptor count (HBD/HBA) and the topological polar surface area (TPSA), potentially key influencing properties relative to permeability.

The PDE5 selectivity of the amide **5** can be rationalised from the co-crystal structure (Figure 8.8). Selectivity over PDE6 through the C3 amide substituent is thought to be driven primarily by the interactions with the residues Met816Leu(PDE6) and especially Leu804Met(PDE6) at the "alkoxy" pocket entrance. Selectivity over PDE10 can be rationalised *via* the extended N1 substituent interacting with the Ala783Tyr(PDE10) residue change, where the tyrosine in PDE10 occludes the "alkoxy" pocket. The C3 amide substituent has also influenced selectivity over PDE11, where residues equivalent to positions 804 and 816 are Ile and Leu, respectively. These key features of the catalytic domain binding mode of **5** were factored in as we sought to address the poor absorption.

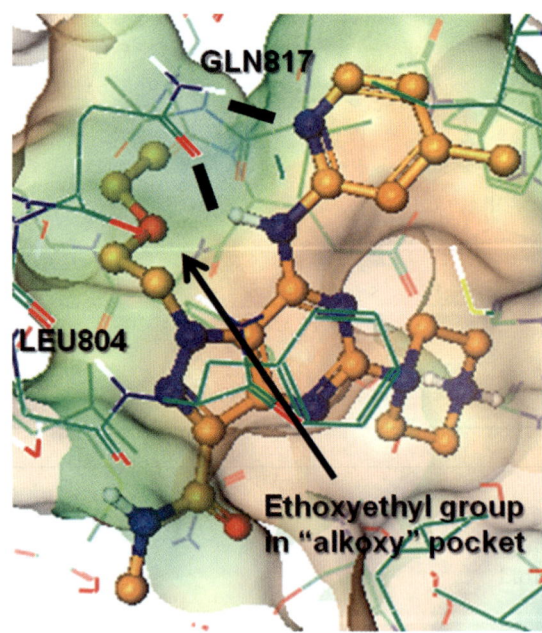

Figure 8.8   Co-crystal structure of amide derivative **5** with the catalytic domain of PDE5 and the solid rendering to PDE5 binding site surface coloured by hydrophobicity. The GLN817 bidentate H-bond is illustrated together with the LEU804 "isopropyl" motif believed to play a role in PDE selectivity, visible behind the C3 amide group.

8.4.3 Candidate Discovery

A range of basic amide derivatives of **5** were synthesised with an emphasis on reduction of H-bond donor count and TPSA and included appending the basic centre to the C3 amide group. Whilst good pharmacology could be maintained, permeability remained poor. A more productive approach was based upon the non-essential potency contribution made by the basic piperazine of **5**. Given that **5** was sub-nanomolar against PDE5, and a slight potency loss could be tolerated, lower mass non-basic derivatives were explored as a means of reducing H-bond donor count and TPSA. Potency was retained with ether, heterocycle and neutral amide derivatives, as exemplified by compounds **6**, **7** and **8** (Table 8.4),[28,29] which retained good LE and LLE. Whilst all three options restored permeability in addition to retaining excellent potency and selectivity, the amide derivative **8** appeared promising, lacking the human liver microsome (HLM) metabolic liabilities of the other two classes and the I_{Kr} potassium ion channel potency of the ether **6**, indicative of the potential for unfavourable cardiac activity.

Table 8.4 Profiles of representative neutral derivatives **6**, **7** and **8**

Criteria	6	7	8
PDE5 IC_{50} (nM)	0.83	0.80	0.84
PDE6 selectivity fold	82	169	116
PDE11 selectivity fold	104	155	54
clog P	2.8	2.1	2.8
MW	429	422	398
LE (PDE5)	0.40	0.40	0.43
LLE (PDE5)	6.28	7.00	6.28
Caco-2 Papp A-B ($\times 10^{-6}$ cm s^{-1}), efflux ratio	NA	NA	25, <2
TPSA	99	112	119
HBD, HBA	1, 10	1, 11	3, 10
I_{Kr} IC_{50} (nM)	526	>10 000	2610
hERG IC_{50} (nM)	–	–	80

Table 8.5 Representative basic derivatives **9** and **10**

Criteria	9	10
PDE5 IC$_{50}$ (nM)	1.76	1.61
PDE6 selectivity fold	57	53
PDE11 selectivity fold	88	43
clog P	3.06	3.16
MW	411	423
LE (PDE5)	0.41	0.39
LLE (PDE5)	5.69	5.63
Caco-2 Papp A-B ($\times 10^{-6}$ cm s^{-1}), efflux ratio	19, <1	15, 1
TPSA	93	93
HBD, HBA	2, 9	2, 9
I$_{Kr}$ IC$_{50}$ (nM)	3562	10 600
hERG	25% @ 0.1 µM	39% @ 1 µM

The neutral amide **8** was profiled in more detail. Pharmacokinetic study was undertaken (Table 8.6) and the projected human half-life of **8** was 12–36 h, underpinned by the lower predicted clearance for the compound relative to sildenafil. Unfortunately, detailed safety evaluation for **8** indicated significant hERG I$_{Kr}$ ion channel activity (IC$_{50}$ = 80 nM) and evidence of cardiac effects *in vivo*, precluding any further development due to the low therapeutic index.

Leveraging the synthetic scope of the series, alternative series types were explored. Firstly, the basic series typified by **4** was revisited and further rounds of automated chemistry undertaken. Efforts focused on improving PDE5 selectivity and the cardiovascular safety window, whilst retaining permeability and a long projected human half-life underpinned by a high volume of distribution. Compounds **9** and **10** were identified and featured alternative C5 and C7 amine moieties (Table 8.5).[23] Both derivatives displayed high V_D values in rat but the bicyclopiperazine **10** projected to the largest cardiovascular safety window and was selected for further progression. The projected human half-life

Table 8.6   Actual and projected human pharmacokinetic data for sildenafil, **8**, **10** and **13**.

Criteria	Sildenafil	8	10	13
Human clearance (mL min^{-1} kg^{-1})	6	0.1–0.3^a	2.3^a	0.2–0.3^b
Human V_D (L kg^{-1})	1.2	0.3^a	3–7^a	~0.2^b
Human $t_{1/2}$ (h)	3–5	11–44^a	15–35^a	12–14^b

^{a}Projected human pharmacokinetics.
^{b}Derived from human oral data.

for **10** indicated the potential for once-daily dosing (Table 8.6). However, whilst the hERG IC$_{20}$ of 600 nM indicated a substantial safety window against primary polypharmacology, further testing *in vivo* in a conscious dog indicated threshold cardiovascular effects at 190 nM and progression was halted.

Alongside revisiting the basic series, the synthetic scope of the series was further leveraged to explore acidic systems.[30] The co-crystal structure of the amide **5** supported the belief that acidic moieties would be tolerated at the C3 position. Given that cardiovascular safety issues with acidic molecules are extremely rare, presumably due to the negatively charged nature of the hERG ion channel,[31] this new sub-class represented an attractive option. In terms of retaining a long projected human half-life, ensuring low clearance would be essential as an acid would likely have a low V_D confined to blood plasma volume.

Various C3 carboxylic acids were explored and the isopropyl(methyl)amino derivative **11** emerged as a promising early lead (Table 8.7). Sub-nanomolar PDE5 potency, good selectivity and good physical/ADME properties encouraged further profiling. The rat pharmacokinetics of **11** indicated low clearance, but poor bioavailability was seen upon oral dosing. A portal-systemic rat study was undertaken and indicated high levels of the glucuronide metabolite, consistent with first-pass glucuronidation in the intestine.

Applying the rationale that acid isosteres would have a lower propensity for glucuronidation, whilst still retaining appropriate features for activity, a range of keto-oxadiazole and acylsulfonamide derivatives were made. Compounds **12** and **13** represented the best examples in each class (Table 8.7), displaying sub-nanomolar level PDE5 potency, good PDE5 selectivity and physicochemical/ADME properties that merited more detailed profiling. Whilst both compounds displayed improved oral bioavailability in rat pharmacokinetic studies relative to the carboxylic acid **11**, the acylsulfonamide derivative **13** stood out by virtue of very low clearance that resulted in a long 9.3 h half-life. Additionally, **13** displayed almost complete oral bioavailability. The resulting projected human pharmacokinetics for **13** pointed to a >10 h human half-life. Blood pressure lowering was observed in a spontaneously hypertensive rat with maximal effects at 5× the PDE5 *in vitro* IC$_{50}$ and the compound displayed an excellent safety profile. No significant cross activity was seen upon wide ligand profiling and the hERG signal was very weak (15% at 10 μM). Additionally, no cardiovascular effects were observed up to 221× the efficacious concentration

Table 8.7 Profile of acidic leads **11**, **12** and **13**

11 **12** **13**

Criteria	11	12	13
PDE5 IC_{50} (nM)	0.43	0.47	0.55
PDE6 selectivity fold	32	38	41
PDE11 selectivity fold	48	30	37
clog P	3.9	3.34	2.45
MW	413	439	476
LE (PDE5)	0.43	0.4	0.39
LLE (PDE5)	5.47	6.00	6.81
Caco-2 Papp A-B ($\times 10^{-6}\,cm\,s^{-1}$), efflux ratio	12, <2	34, <2	25, <2
TPSA	118	140	144
HBD, HBA	2, 10	2, 12	2, 13
Rat clearance (mL $min^{-1}\,kg^{-1}$)	1.8	10.9	0.54
Rat V_D (L kg^{-1})	0.3	1.3	0.48
Rat $t_{1/2}$ (h)	3.1	1.4	10.9
Rat $F\%$	18 (variable)	30	82

in a dog haemodynamic study. The acylsulfonamide **13**, PF-489791, was nominated for development and is currently in Phase II clinical studies for pulmonary arterial hypertension. Phase I first-in-man pharmacokinetic data indicated a genuine once-daily oral dosing profile with a human half-life in the 12–14 h range (Table 8.6). The compound was well tolerated and dose-proportional increases of cGMP were observed,[28] indicative of PDE5 inhibition.

8.5 Conclusions

In conclusion, a second-generation PDE5 agent has been discovered from an HTS initiative. Key elements' in discovery were the focus on physicochemistry and pharmacokinetics throughout the programme, together with the use of co-crystal structure data to guide design and identification of a parallel chemistry amenable, wide synthetic scope template. LE and LLE measurements served as important indicators of efficient binding. These elements enabled rapid discovery of a proprietary lead **3** with inherently good physicochemistry, a novel aminopyridine pharmacophore and PDE6 selectivity. Further optimisation improved potency and selectivity leading to **5**, whereupon leverage of the synthetic scope gave rise to advanced neutral **8** and basic **10** leads with improved absorption. Both **8** and **10** projected to long human half-lives on the basis of low clearance and moderate clearance/higher volume respectively (Table 8.6). Finally, a potent and selective acidic clinical candidate **13** with low clearance and an outstanding safety profile was identified. The excellent pre-clinical profile derived from the property-based strategy detailed above has successfully translated to once-daily oral pharmacokinetics in man.

References

1. V. Boswell-Smith, D. Spina and C. P. Page, *Br. J. Pharmacol.*, 2006, **147**(suppl. 1), S252.
2. S. P. H. Alexander, *Br. J. Pharmacol.*, 2009, **158**, S203.
3. N. K. Terrett, A. S. Bell, D. Brown and P. Ellis, *Bioorg. Med. Chem. Lett.*, 1996, **6**, 1819.
4. N. Galie, H. A. Ghofrani, A. Torbicki, R. J. Barst, L. J. Rubin, D. Badesch, T. Fleming, T. Parpia, G. Burgess, A. Branzi, F. Grimminger, M. Kurzyna and G. Simonneau, *New Engl. J. Med.*, 2005, **353**, 2148.
5. H. Haning, U. Niewohner, T. Schenke, S. M. Es, G. Schmidt, T. Lampe and E. Bischoff, *Biorg. Med. Chem. Lett.*, 2002, **12**, 865.
6. A. Daugan, P. Grondin, C. Ruault, A. -C. Le Monnier de Gouville, H. Coste, J. Kirilovsky, F. Hyafil and R. Labaudiniere, *J. Med. Chem.*, 2003, **46**, 4525.
7. H. J. Shim, Y. C. Kim, J. H. Lee, B. O. Ahn, J. W. Kwon, W. B. Kim, I. Lee and M. G. Lee, *J. Pharm. Pharmacol.*, 2004, **56**, 1543.
8. R. H. Cote, *Int. J. Impot. Res.*, 2004, **16**, S28.
9. A. Makhlouf, A. Kshirsagar and C. Niederberger, *Int. J. Impot. Res.*, 2006, **18**, 501.
10. H. A. Ghofrani, I. H. Osterloh and F. Grimminger, *Nat. Rev. Drug Discovery*, 2006, **5**, 689.
11. D. G. Brown, C. R. Groom, A. L. Hopkins, T. M. Jenkins, S. H. Kamp, M. M. O'Gara, H. J. Ringrose, C. M. Robinson and W. E. Taylor, *Pat. Appl.*, WO 2003/038080, 2003.
12. B. J. Sung, K. Y. Hwang, Y. H. Jeon, J. I. Lee, Y. S. Heo, J. H. Kim, J. Moon, J. M. Yoon, Y. L. Hyun, E. Kim, S. J. Eum, S. Y. Park, J. O. Lee, T. G. Lee, S. Ro and J. M. Cho, *Nature*, 2003, **425**, 98.
13. C. M. N. Allerton, C. G. Barber, K. C. Beaumont, D. G. Brown, S. M. Cole, D. Ellis, C. A. L. Lane, G. N. Maw, N. M. Mount, D. J. Rawson, C. M. Robinson, S. D. A. Street and N. W. Summerhill, *J. Med. Chem.*, 2006, **49**, 3581.
14. M. E. Bunnage, J. P. Mathias, A. Wood, D. Miller and S. D. A. Street, *Bioorg. Med. Chem. Lett.*, 2008, **18**, 6033.
15. N. Watanabe, H. Adachi, Y. Takase, H. Ozaki, M. Matsukura, K. Miyazaki, K. Ishibashi, H. Ishihara, K. Kodama, M. Nishino, M. Kakiki and Y. Kabasawa, *J. Med. Chem.*, 2000, **43**, 2523.

16. D. P. Rotella, *Nat. Rev. Drug Discovery*, 2002, **1**, 674.
17. I. D. Kuntz, K. Chen, K. A. Sharp and P. A. Kollman, *Proc. Natl. Acad. Sci. U.S.A.*, 1999, **96**, 9997.
18. A. L. Hopkins, C. R. Groom and A. Alex, *Drug Discovery Today*, 2004, **9**, 430.
19. A. R. Leach, M. M. Hann, J. N. Burrows and E. J. Griffen, *Mol. Biosyst.*, 2006, **2**, 430.
20. P. D. Leeson and B. Springthorpe, *Nat. Rev. Drug Discovery*, 2007, **6**, 881.
21. D. G. Brown, C. R. Groom, A. L. Hopkins, T. M. Jenkins, S. H. Kamp, M. M. O'Gara, H. J. Ringrose, C. M. Robinson and W. E. Taylor, *Pat. Appl.*, WO 2004/097010, 2004.
22. K. Y. Zhang, G. L. Card, Y. Suzuki, D. R. Artis, D. Fong, S. Gillette, D. Hsieh, J. Neiman, B. L. West, C. Zhang, M. V. Milburn, S. H. Kim, J. Schlessinger and G. Bollag, *Mol. Cell*, 2004, **15**, 279.
23. A. S. Bell, D. G. Brown, D. N. A. Fox, I. R. Marsh, A. I. Morrell, M. J. Palmer and C. A. Winslow, *Pat. Appl.*, WO 2004/096810, 2004.
24. J. Pandit, *Pat. Appl.*, US 2005/0202550, 2005.
25. M. J. Palmer, A. S. Bell, D. N. A. Fox and D. G. Brown, *Curr. Top. Med. Chem.*, 2007, **7**, 405.
26. A. S. Bell, D. G. Brown, D. Bull, D. N. A. Fox, I. R. Marsh, A. I. Morrell and M. J. Palmer, *Pat. Appl.*, WO 2005/049617, 2005.
27. C. A. Lipinski, *Drug Discovery Today Technol.*, 2004, **1**, 337.
28. M. J. Palmer, A. S. Bell, D. N. A. Fox and D. G. Brown, presented at ACS ProSpectives, Cambridge, MA, April 2009.
29. M. J. Palmer, A. S. Bell, D. N. A. Fox and D. G. Brown, presented at the 238th ACS National Meeting, Washington, 2009, ORGN 582.
30. A. S. Bell, D. G. Brown, K. N. Dack, D. N. A. Fox, I. R. Marsh, A. I. Morrell, M. J. Palmer and C. A. Winslow, *Pat. Appl.*, WO 2005/049616, 2005.
31. D. A. Price, D. Armour, M. de Groot, D. Leishman, C. Napier, M. Perros, B. L. Stammen and A. Wood, *Curr. Top. Med. Chem.*, 2008, **8**, 1140.

CHAPTER 9

From HTS to Market: the Discovery and Development of Maraviroc, a CCR5 Antagonist for the Treatment of HIV

CHRIS BARBER* AND DAVID PRYDE

Worldwide Medicinal Chemistry, Pfizer Ltd, Sandwich, Kent, CT13 9NJ, UK

9.1 AIDS and HIV

AIDS was recognised as a disease in 1981;[1] two years later, the causative pathogen human immunodeficiency virus 1 (HIV-1) was reported.[2,3] Since then, this disease has become a global pandemic which is estimated to have caused the death of 25 million people. A further 33 million individuals are currently infected. Significant research into this disease over the past 25 years has resulted in 25 approved drugs;[4] however, HIV can rapidly mutate to become resistant to a single drug treatment. As a consequence, treatment is typically a combination of at least three different drugs across a number of classes, a regime known as HAART (highly active antiretroviral therapy). While this has been successful in suppressing mutant viruses that may have resistance to any one drug, such combinations can result in complex dosing schedules and in severe side effects, which can reduce patient compliance and allow resistant strains to emerge. One study has shown that 5–10% of patients undergoing HAART therapy are infected with multidrug-resistant strains.[5] As

RSC Drug Discovery Series No. 4
Accounts in Drug Discovery: Case Studies in Medicinal Chemistry
Edited by Joel C. Barrish, Percy H. Carter, Peter T. W. Cheng and Robert Zahler
© Royal Society of Chemistry 2011
Published by the Royal Society of Chemistry, www.rsc.org

a consequence, the desire to identify novel classes of drugs with antiviral activity continues to be strong.

9.2 HIV Infection and the Choice to Target CCR5 Antagonism

The HIV virus enters the host cell through a cascade of events that result in fusion of viral and host cellular membranes. The process is believed to be initiated by binding of viral surface protein (gp120) to host cell CD4, resulting in conformational changes that enable interaction with a co-receptor (CCR5 or CXCR4); this exposes a second viral protein (gp41), which mediates host/virus membrane fusion.[6] It is generally only following interaction with the co-receptor that infection by primary HIV-1 strains can take place. Viral strains that are transmitted and establish new infections tend to require the presence of CCR5 (these are termed R5-tropic strains). Viral entry *via* CXCR4 (X4-tropic strains) is typically seen in later stages of disease progression and has been associated with rapid CD4 T-lymphocyte decline.

There are many potential points to intervene in the viral life-cycle (Figure 9.1)[7–9] and the majority of these steps have been targeted as potential intervention points against the HIV virus. The choice to block a host receptor to prevent viral infection is at first sight an unusual one, since the pharmacological consequences of intervention of host cellular function would normally preclude this option. The risks become even higher when one considers that an effective antiviral agent would require chronic dosing at levels to ensure permanent and complete blockade of CCR5. In this case, however, such safety concerns were neatly addressed by key data that also demonstrated the importance of CCR5 for viral infection. Approximately 1% of north European Caucasians[i] carry a homozygous 32 base-pair deletion in the gene encoding CCR5 (CCR5-Δ32), resulting in the lack of functional CCR5. These individuals are highly resistant to HIV-1 infection. Heterozygotes for this allele show low cell-surface expressions of CCR5 receptors and exhibit delayed disease progression. The vast majority of individuals[12] either homo- or heterozygous for CCR5-Δ32 are healthy.[ii] This then suggests that CCR5 is a redundant receptor and that blockade by an antagonist to prevent cellular infection by HIV-1 will be safe during chronic treatment. Interestingly, CCR5 agonists have also been proposed as potential antiviral agents, as they could both directly block the receptor and accelerate its clearance from the cell surface by triggering receptor internalization mechanisms.[14] However, to be pharmacologically benign,

---

[i] The emergence of a significant population of north European Caucasians carrying CCR5-Δ32 is believed to be a result of selective pressure from a long-term and widespread epidemic. Population modelling suggests that this could be the result of endemic smallpox which infected the vast majority of children for around 700 years in Europe. This had a fatality rate of ∼30% and could have resulted in the levels of CCR5 -/- observed today.[10,11]

[ii] It has been reported that individuals expressing homozygous CCR5-Δ32 have reduced resistance to West Nile virus.[13]

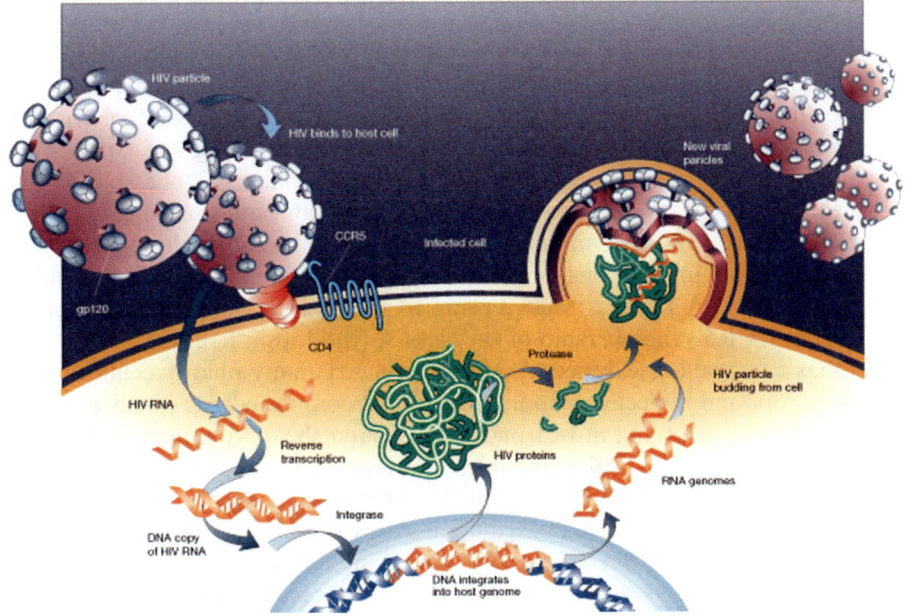

Figure 9.1 HIV viral life-cycle. Reproduced with permission from Weiss.[7]

internalisation would need to occur before significant receptor signalling, and this has yet to be demonstrated *in vivo*.

9.3 CC Chemokine Receptor 5

CC Chemokine receptor 5 (CCR5) is a member of the GPCRs (G protein-coupled receptors), which are targets for 30% of marketed drugs. This observation suggests that it should be possible to develop small-molecule antagonists to CCR5.[15] The three extracellular loops (ECLs 1–3) together with the N-terminus are the targets for chemokine binding and for interactions with the viral gp120 protein. The three intracellular loops and C-terminus participate in G protein-mediated signal transduction. CCR5 is expressed on the cell surface of a wide range of cell types, predominantly on T cells, macrophages, microglia and dendritic cells.[16] To date, it has been found to bind to a number of different chemokines (MIP1α, MIP1β, MCP-2, RANTES, MCP3 and MCP4), but only MIP1β is selective for CCR5.[17] Activation of CCR5 results in a number of typical GPCR responses, including Ca^{2+} release and cAMP production, which result in the activation and trafficking of leukocytes to sites of inflammation.[18] There is no X-ray structural information for CCR5, although many groups have proposed homology models based upon the crystal structure of rhodopsin.[19]

9.4 Identifying Leads with Affinity for CCR5

Since the natural ligands for CCR5 are known, it may be possible to develop drugs from these. However, such an approach offers two significant challenges. Firstly, the "conversion" of peptidic leads into drugs with good pharmaco-kinetic properties continues to be difficult, as it requires designing compounds that mimic the key interactions of an unknown conformation of a flexible peptide while simultaneously improving metabolic stability and intestinal permeability. Secondly, these chemokines interact with CCR5 *via* extracellular loops through a series of complex interactions which it may not be possible to replicate with a small molecule.[20] A more attractive route to small-molecule CCR5 antagonists stems from CCR5's membership of the class A (rhodopsin-like) GPCRs, a class which has been well-researched. As a consequence, many pharmaceutical companies' compound collections contain drug-like compounds designed to target related GPCRs, including muscarinics, $5HT_{1A}$ and the α1-adrenoceptor.[21] Like many other pharmaceutical companies, we embarked on a strategy of high-throughput screening (HTS) to identify suitable leads for a drug discovery programme. We chose a screen based upon stably expressed CCR5 on HEK-293 cells.[22] As noted above, the one known chemokine selective for CCR5 was MIP-1β; consequently, the assay was configured to measure inhibition of radiolabelled MIP-1β0 binding. While we were confident that this assay would select ligands that block the binding of MIP-1β, we were less certain if these hits would also block binding by viral gp120 protein. As a consequence, additional assays were developed during the course of the programme to more accurately assess inhibition of viral protein binding to the receptor.

At that time, the Pfizer compound collection contained ~0.5 million compounds. Following screening, the "hits" were triaged based upon inhibitor potency. The most potent hits from HTS are often the largest and most lipophilic compounds. However, these frequently tend to be the least tractable hits for conversion into drugs. Consequently, ligand efficiencies were calculated for hits from the CCR5 screen to re-bias hits towards those showing the most efficient binding for their size. Ligand efficiency (LE) is quickly obtained by scaling the binding energy by the number of heavy atoms in the molecule (eqn 9.1).[23] More recently, other scaling parameters have been developed, including lipophilic ligand efficiency (LipE)[24,25] (which tries to compensate for lipophilicity-driven binding) and "fit quality score"[26] (which further scales ligand efficiency to prioritise smaller fragments which may otherwise be overlooked).

$$
\begin{aligned}
\text{Ligand efficiency} &= \frac{\Delta G(\text{binding})}{\sum \text{heavy atoms}} = \frac{-RT \log(Ki)}{\sum \text{heavy atoms}} \\
&\approx \frac{-1.4 \log(Ki)}{\sum \text{heavy atoms}} \ (\text{kcalmol})
\end{aligned}
\tag{9.1}
$$

Two of the leading compounds to emerge from the HTS were **1** (UK-107543, binding $IC_{50} = 0.34\,\mu M$, LE $= 0.29$) and **2** (UK-179645, binding $IC_{50} = 1.1\,\mu M$,

LE $= 0.21$).[27] These are shown in Figure 9.2,[27–32] alongside a number of other CCR5 antagonist leads found by others through screening campaigns. Comparison of these structures show that they possess similar molecular weights, lipophilicities and, in most cases, a central basic centre.

In addition to their lipophilicity and relatively high molecular weight, these hits showed only weak affinity for the CCR5 receptor and no antiviral activity. While clearly desirable, the initial lack of antiviral activity was not critical to progression, since we had little understanding of the dynamic range of the antiviral assay. However, if antiviral activity had continued to be lacking with more potent CCR5 binders, then a re-evaluation of this approach would have become essential.

9.5 Defining Objectives and an Appropriate Screening Sequence

The programme was initiated with the objectives of identifying a safe, orally effective agent that would be administered twice (but preferably once) daily. To be safe, a compound would need to be a selective CCR5 antagonist that avoided drug–drug interactions when dosed in conjunction with other agents used to treat HIV (see HAART above). This strategic objective was translated into a lab profile as follows:

- antiviral (AV) $IC_{90} < 10$ nM
- > 100-fold selective over other targets (specifically other GPCRs)
- low turnover in human liver microsomes (HLM), $t_{1/2} > 30$ min
- apical to basal flux in caco-2 cells > 5 (believed predictive of oral absorption)
- no cytochrome P450 inhibition (to minimise potential drug–drug interactions)

While antiviral assays were available which measured the ability of compounds to block HIV_{Bal} from infecting PM-1 or PBL cells, these were low throughput and somewhat capricious, certainly not able to deliver sufficient data to drive structure–activity relationships (SAR). However, without recourse to regular antiviral screening, we ran the risk of optimising binding to a region of the CCR5 receptor which had no influence on the binding of viral gp120. This drove the parallel development of additional higher throughput assays that would give better translation to antiviral activity than our initial MIP-1β binding assay. The solution was a cell fusion assay run in a 384-well plate format. It consisted of two cell lines: (1) HeLa P4 expressing CCR5, CD4 and HIV-1-LTR-β-Gal and (2) a CHO cell line expressing viral gp160 and TAT. If cell fusion takes place, then the TAT in the CHO cell would transactivate the HIV-1LTR present in HeLa cell, leading to the measurable expression of the β-galactosidase enzyme.[33] A retrospective comparison of the MIP-1β binding, cell fusion and antiviral assays (Figure 9.3) shows an

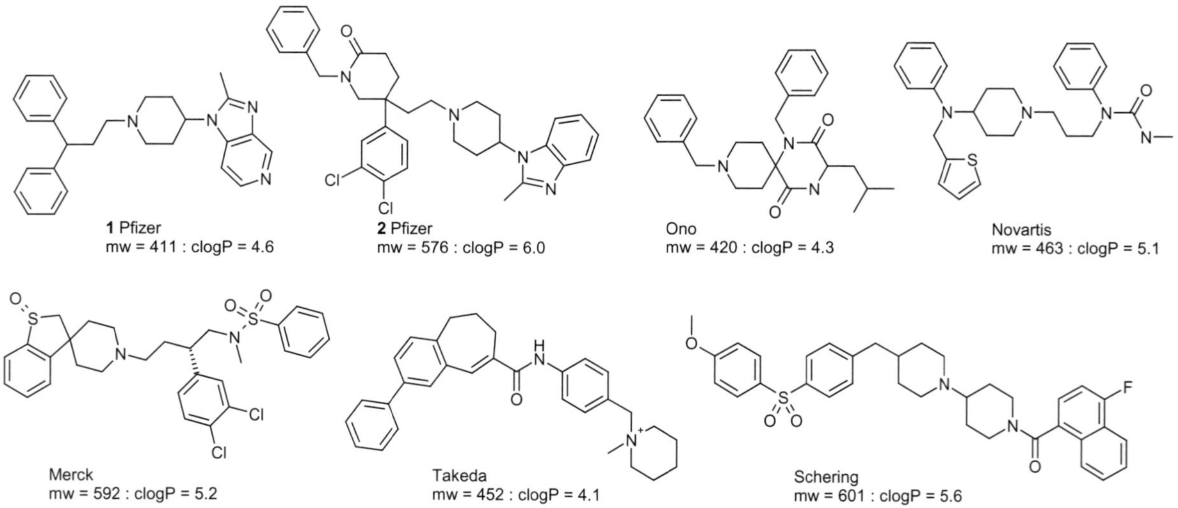

Figure 9.2 Leads reportedly used to initiate CCR5 antagonist programs.

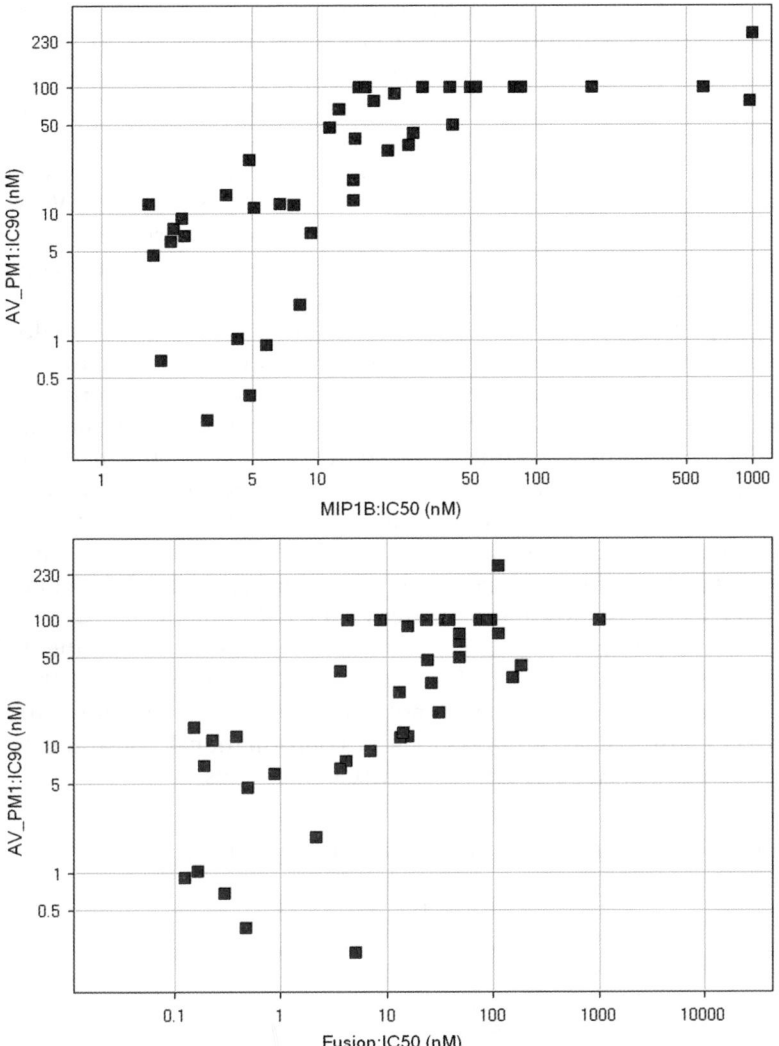

Figure 9.3 Correlations between antiviral activity [IC$_{90}$ (nM)] and (*top*) MIP-1β binding and (*bottom*) a cell fusion-based assay.

improved correlation with the fusion assay, and a greater dynamic range with which antiviral activity can be predicted. Throughout this chapter we will refer to these assays as binding, fusion or antiviral (AV).

9.6 Developing SAR

Before embarking on the production of a wide range of analogues to follow **1**, a broader activity profile was determined to identify key off-target activities that would need addressing while establishing SAR. This immediately flagged two key issues. Strong activity was observed against the cardiac hERG (human ether-a-go-go) potassium ion channel ($IC_{50} < 316$ nM). Activity against the hERG channel was particularly concerning, given the risk of developing a potentially fatal arrhythmia through prolongation of the QT interval, an effect that had lead to the withdrawal of terfenadine and cisapride.[34] In addition, **1** also exhibited strong inhibition of the P450 cytochrome 2D6 isoform ($IC_{50} = 40$ nM). Inhibition of this enzyme would affect drug levels for compounds metabolised by CYP2D6, which could not be tolerated in a compound likely to be administered in combination with other agents. In short, a reduction in lipophilicity, together with a concomitant increase in potency, decreased hERG affinity and reduced CYP2D6 inhibition, would all have to be rapidly demonstrated to give confidence for further investment in this series.

Modelling of the CYP2D6 binding pocket suggested a binding mode in which the pyridyl nitrogen could coordinate to the heme; this has often been seen with nitrogen-containing heterocycles.[35] Replacing the pyridyl nitrogen with a carbon would thus be expected to reduce CYP2D6 inhibition, and from **2** was a change expected to maintain CCR5 binding affinity. This substitution gave **3** (Table 9.1), which did indeed show a significant reduction in affinity for CYP2D6 ($IC_{50} = 710$ nM) and an increase in CCR5 binding.

Reduction in lipophilicity remained critical for this series and the diphenyl methylene moiety was targeted. It was proposed to remove one of the phenyl rings and replace it with either a polar heterocycle (should aromaticity be required) or just a small group (should a conformational constraint be sufficient). Table 9.1 shows a selection of these compounds.

Table 9.1 Replacing one phenyl with amino derivatives

Compound	R	binding IC_{50} (LE^a)	clog P^b
3	Ph	1.8 nM (0.38)	5.7
4	NHBoc	40 nM (0.30)	4.8
5	NH_2	7.9 µM (0.27)	3.0
6	1,3,4-triazolyl	70 nM (0.32)	2.1
7	NHCOPh	13 nM (0.31)	4.8

^{a}LE = ligand efficiency (see eqn 1).
^{b}clog P = calculated log P.

Table 9.2 Adding a bridge across the piperidine; minimised energy conformations

8	**9**	**10**
Binding $IC_{50} = 4$ nM AV $IC_{90} = 440$ nM	Binding $IC_{50} = 6$ nM AV $IC_{90} = 4$ nM	Binding $IC_{50} = 2$ nM AV $IC_{90} = 5$ nM

While heterocycles both reduced lipophilicity and appeared to be tolerated (*e.g.* **6**), they were synthetically less accessible than simple amides and carbamates (*e.g.* **4** and **7**), which also possessed good ligand efficiency (>0.3). Even more encouragingly, **7** demonstrated antiviral activity (AV $IC_{90} = 0.4 \mu M$) and consequently a wider range of less lipophilic amides were prepared in a parallel manner.[36] This work culminated in the identification of cyclobutyl derivative **8** (*S* enantiomer), which demonstrated excellent ligand efficiency (0.36) and antiviral activity (Table 9.2).

Compound **8** was screened across a wide panel of targets to identify potential selectivity issues. Some CYP2D6 activity was observed ($IC_{50} = 5 \mu M$), prompting a fresh analysis of the CYP binding model. Molecular modelling suggested the central amine could coordinate Asp301, which literature had concluded was critical for activity (Figure 9.4 shows a model of the binding of **8**).[37] Removal of this basic centre was not investigated, as SAR suggested it was essential for CCR5 binding potency (a conclusion later supported by the commonality of a central basic amine in other CCR5 antagonists; Figure 9.2). Instead, attempts to modulate the pK_a, sterically encumber the basic centre or present it differently were undertaken.[27,36] The most effective of these strategies was to add a bridge across the piperidine to give a tropane ring (Table 9.2).

9.7 Exploration of the Tropane Core

The addition of this bridge (from **8** to **9** and **10**) was pivotal in the development of this series. In a single step, it completely ablated activity against CYP2D6 with negligible impact on CCR5 binding affinity (no activity against CYP2D6 at $2.3 \mu M$). Most dramatically, it increased antiviral activity by two orders of

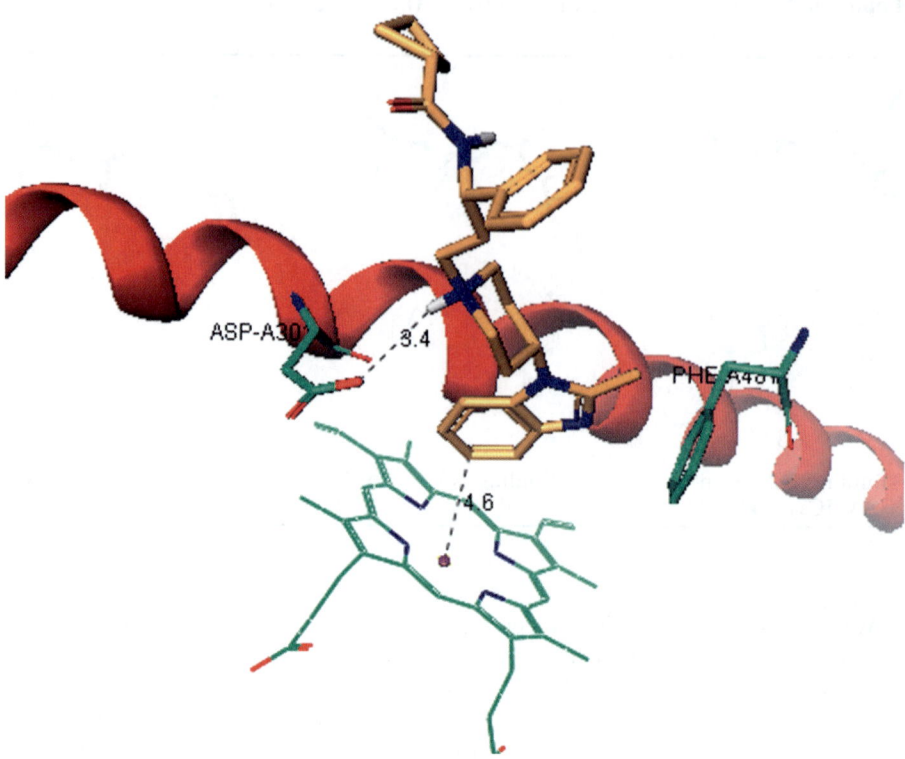

Figure 9.4   Model of the interaction of **8** with CYP2D6.

magnitude, highlighting a disconnect between binding and antiviral activity. This discrepancy has also been observed by a number of other groups.[38] There are a number of possible reasons for this disconnect, including differences in the off-rates of inhibitors from the CCR5 receptor, or the engagement of different binding sites on CCR5 for chemokines, small molecules and the viral protein. Initially, it had not been expected that both *endo* and *exo* isomers would show such similar levels of activity, but energy-minimised structures show that the tropane ring can switch from chair to boat conformations to minimise steric clashes, thereby giving a comparable disposition of substituents (Table 9.2).

Since incorporation of the tropane core represented a key change in chemotype for the project, **9** and **10** were profiled against a wide range of targets and submitted to *in vitro* metabolic stability assays. Unsurprisingly for a compound with a log $D > 3.6$, compound **9** showed poor stability in human liver microsomes ($t_{1/2} = 8$ min). A more pressing concern, however, was the activity that both compounds showed against the hERG ion channel (**9**, 99% inhibition at $1\,\mu$M; **10**, 80% inhibition at 300 nM). Consequently, a prime objective for programme became the reduction of hERG activity and lipophilicity.

Activity against the hERG channel was not unexpected, as the original lead (1) had significant activity and also because the chemotype conforms to generic descriptors predicting hERG activity.[39] Other companies have subsequently reported hERG activity for their CCR5 analogues, including Schering[40] and Merck.[41] Docking 9 into a model of the hERG ion channel suggested that adding polarity around the amide or the benzimidazole ring could reduce hERG affinity (Figure 9.5).[42,43] Having said that, the SAR for hERG activity is often highly subtle, and the ability to rapidly generate SAR within the series of interest is of greater value than a generic pharmacophore-based model. A high-throughput hERG binding assay was used to generate this SAR.[44]

Introduction of polarity around the amide bond was the synthetically most accessible option, and a wide number of polar-containing analogues were prepared in parallel. Within this dataset of analogues, it became clear that the *endo* isomer was preferred. A number of key (*endo*) analogues are shown in Table 9.3.

Significant changes in overall polarity (as measured by log *D*) were not necessary to modulate hERG activity; careful placement of polar groups in the amide substituent could dramatically reduce ion channel activity. This empirical SAR identified compounds that retained good antiviral activity whilst exhibiting essentially no hERG activity, including 14 and 15, although

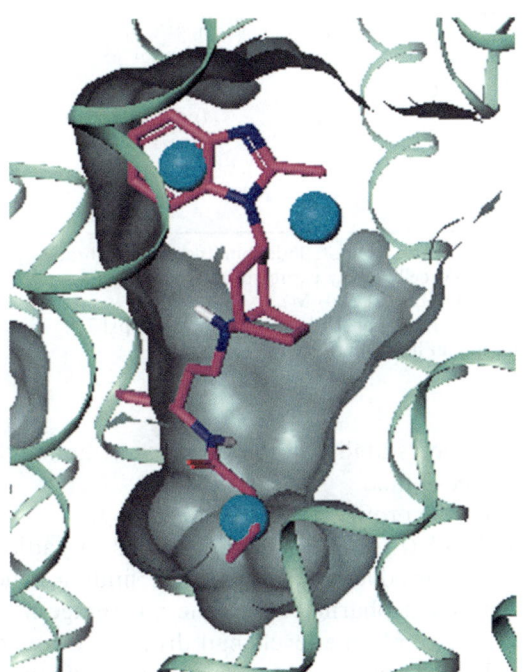

Figure 9.5 General pharmacophore for hERG (*blue balls* highlight regions requiring lipophilicity) overlaid onto a model of 9 (*purple*) docked into the K$^+$ channel binding pocket (*grey surface*) and protein structure (*green helices*).

Table 9.3 Incorporating polarity into the amide (*endo*-tropanes)

Compound	R	Binding IC_{50} (nM)	AV IC_{90} (nM)	hERG binding[a]	log D^b	HLM $t_{1/2}$ (min)[c]
11		7	4.5	755 nM	3.0	100
12		5	27	(8%)	2.0	1
13		6	16	(87%)		12
14		0.8	1.5	>10 µM	2.5	–
15		8	0.6	(0%)	2.7	77

[a]hERG = competition binding of [³H]dofetilide binding in membranes prepared from human embryonic kidney (HEK293) cells stably expressing human ether-a-go-go related gene (HERG) K$^+$ channels.[38] IC_{50} (% inhibition at 300 nM).
[b]Partition coefficient measured in octan-1-ol/aqueous buffer at pH 7.4.
[c]HLM $t_{1/2}$ = half-life in human liver microsomes.

only the latter maintained stability in the presence of human liver microsomes and showed good Caco-2 flux.

Given the initial *in vitro* profile, **15** was further progressed. It demonstrated a slow off-rate from CCR5 ($t_{1/2}$ = 3.5 h, as determined by radiolabelled washout studies).[45] This was believed to be a significant finding, since slow receptor disassociation could give a pharmacodynamic advantage *in vivo* by retaining antiviral activity after free plasma levels had dropped below efficacious levels, and could help explain the previously discussed discrepancy between binding and antiviral assays *in vitro*.

When **15** was studied *in vivo*, the oral bioavailability in the dog was low (<10%), which was ascribed to high first-pass clearance based upon clearance rates approximately equal to liver blood flow ($Cl = 49$ mL min^{-1} kg^{-1}). Robust clearance combined with a modest volume of 4.9 L kg^{-1} resulted in an elimination $t_{1/2}$ of 1.2 h. Reducing the lipophilicity seemed to offer the best chance of improving metabolic stability. Given the molecular weight, a strategy of removing lipophilicity was undertaken rather than adding polarity.[36] Such an approach was supported by Lipiniski's analysis of the probability of achieving good oral bioavailability with compounds with a molecular weight under 500.[46] The benzimidazole fragment became a focus for this phase of the optimisation.

9.8 Optimising the Heterocyclic Group

Previous work within the piperidine series around **8** had highlighted a number of alternative heterocycles that retained efficacy (Table 9.4).[43]

Initially the phenyl was considered important and amide and heterocyclic expressions that retained a phenyl ring were prepared (**16–19**); these tended to show similar levels of binding. However, when the phenyl was replaced with smaller substituents, potency (and ligand efficiency) was largely retained (*cf.* **17** and **20**). This observation was then applied to the tropane core, and a range of five-membered ring aromatic heterocycles prepared and resulted in the identification of the 1,3,4-triazoles. Both *exo-* and *endo-*1,3,4-triazoles were prepared, with potency showing little sensitivity to the substituents but favouring the *exo* analogues (**21–24,** Table 9.5).

Compound **23** showed a good balance of antiviral activity and HLM stability and was thus profiled *in vivo*. Bioavailability was low in the rat as a result of poor absorption (only 20%), but improved absorption in the dog (>80%) resulted in a bioavailability of 43%. Allometric scaling from rat gave a human dose prediction of 350 mg *bid*, while the superior dog PK profile led to a human dose prediction of 100 mg *bid*. However, the binding affinity for the hERG ion channel appeared significant, and so **23** was further profiled in a cellular patch-clamp assay, and then in dog Purkinjie fibre. These three assays were in agreement: the risk of QT prolongation from plasma levels likely to be necessary for antiviral effects precluded the further development of this compound.[43]

9.9 Closing in on Maraviroc

Despite the hERG activity shown by **23**, the project was encouraged by the human dose-predictions and the knowledge that modifications to the amide could attenuate hERG ion channel activity. Parallel chemistry allowed rapid access to an array of amides, a selection of which is shown in Table 9.6.

Carbon homologation of **23** gave cyclopentyl analogue **28**, with increased antiviral efficacy, but at the cost of metabolic stability. The tetrahydropyranyl substituent (see **25**), which had shown good antiviral activity in the benzimidazole series (**15**, Table 9.3), lost antiviral activity when applied to the triazoles.

Table 9.4 Potency is maintained with a range of 4-substituted piperidines, even when the phenyl ring is removed

Compound	R	Binding IC$_{50}$	LE	log D
8		4 nM	0.36	2.5
16		15 nM	0.33	1.8
17		2 nM	0.36	nd
18		55 nM	0.29	2.9
19		2 nM	0.36	3.1
20		10 nM	0.36	> 3.6

Table 9.5 Potency remained high through a series of 2-alkyl-1,3,4-triazoles

Compound		Binding IC_{50}	AV IC_{90}	hERG binding at 300nM	log D	HLM $T_{1/2}$ mins
21		12 nM	13	8%	1.2	nd
22		9 nM	9	34%	1.1	nd
23		7 nM	8	30%	1.6	55
24		75 nM	> 100	45%	2.3	22

Fluorination improved metabolic stability, despite a modest rise in lipophilicity (compare **26** and **27**), leading to the design of the difluorocyclohexyl fragment. This was incorporated into **29**, which exhibited good antiviral activity, metabolic stability and hERG selectivity (no hERG binding at 300 nM). Fluorination has often been used as a blocking group at sites of potential metabolism;[47,48] however, the use of 4,4-difluorocyclohexanecarboxylic acid had been rarely described prior to disclosure of these compounds.[49] Since then, it has become a more widely used fragment, with 74 reports in 2008 alone.

Table 9.6 Balancing potency, microsomal stability and hERG affinity across a range of amides containing a 1,3,4-triazole

Compound	R	Binding IC_{50} nM	AV IC_{90} nM	hERG binding at 300 nM	HLM $T_{1/2}$ min	logD
25		27	120	0%	>120	0.9
26	Pr^n	5	1.3	0%	95	1.5
27		5	14	14%	>120	1.8
28	cyclopentyl	5	0.2	0%	21	2.1
29		0.2	0.7	0%	22	2.0

9.10 Profiling Maraviroc

Despite the low levels of activity of **29** in the hERG binding assay, it was further profiled in both the patch-clamp assay (no activity at 200 nM) and in dog-isolated Purkinje fibres. Finally, it was studied in the conscious dog, and no effects were seen at 179 nM free plasma concentrations (>100-fold predicted levels required for efficacy).[33]

Compound **29** was also profiled against a wide range of targets *in vitro* and showed no significant activity at concentrations up to 10 μM. Importantly, it also retained a lack of activity against CYP P450s (IC_{50} values > 50 μM), affording it a significant advantage over existing HIV therapies.[50]

When tested *in vivo*, it demonstrated a modest PK profile in rat (absorption 23%, bioavailability 6%, $t_{1/2} = 0.9$ h) and dog (absorption 70%, bioavailability 42%, $t_{1/2} = 2.3$ h). It displayed limited tissue distribution ($V_D = 6.5$ and 4.3 L kg^{-1} in rat and dog, respectively), commensurate with a moderately lipophilic weak base ($pK_a = 7.7$). In human liver microsomes, compound **29** was a substrate for both CYP3A4 and CYP2D6, giving primary metabolites from hydroxylation of the difluorocyclohexyl and phenyl rings. These metabolites were only seen at low levels ($<5\%$ of parent) in rat or dog, suggesting that they would not reach significant levels in humans.

The physical properties of **29** were suitable for an oral agent. The free base was crystalline and non-hygroscopic, and exhibited a melting point (197 °C) which was sufficiently high to allow easy processing. It showed good chemical stability both as a solid and in solution, along with aqueous solubility $>1.1\,\text{mg}\,\text{mL}^{-1}$ across a wide pH range (2–13), which should avoid solubility-limited absorption (either by rate or extent). Caco-2 permeability was lower than initially targeted (absorptive P_{app} value $= 2.6 \times 10^{-6}\,\text{cm}\,\text{s}^{-1}$), but efflux was not significant (as determined in PGP knock-out mice). While a low Caco-2 flux does suggest an increased risk of partial absorption (as seen in the rat), it does not guarantee it, especially in the case of highly soluble compounds which are not strong substrates for PGP efflux transporters when administered at moderate doses.

Scaling the rat data to man predicted a $t_{1/2}$ of 3 h. An oral dose of 100 mg *bid* was projected to provide free drug levels in excess of the antiviral IC$_{90}$ at C_{min}. Compound **29** was efficacious against a wide range of CCR5-tropic HIV-1 clinical isolates and was not cytotoxic. Receptor offset studies using tritiated **29** determined that it exhibited non-competitive kinetics as a result of slow off-rates from the receptor ($t_{1/2} > 8\,\text{h}$).[51]

Compound **29** passed into development at the end of 2000. This was the culmination of about $2\frac{1}{2}$ years work and almost 1000 analogues.

9.11 Synthesis

The original (Discovery) synthesis is shown in Scheme 9.1.[52,53]

Reduction of the oxime prepared from tropinone (**30**)[54] with sodium metal gave the thermodynamically favoured *exo*-tropane **31**, which was acylated to give **32**. Treatment with PCl$_5$ yielded an iminochloride, which upon reaction with *N*-acetylhydrazine cyclised to form the 1,3,4-triazole ring, delivering **33** after *N*-debenzylation. β-Amino ester **34** was *N*-protected and the ester was converted over three steps to aldehyde **35**, which was reductively aminated with **33** to afford **36**. Deprotection and amide formation furnished **29**.

9.12 Results from the Clinic

Maraviroc was well tolerated in three clinical Phase 1 studies at doses up to 1200 mg, with postural hypotension being the dose-limiting event.[55] In Phase IIa studies, treatment-naïve patients infected with CCR5-tropic virus received drug monotherapy at doses ranging from 25 mg *qd* up to 300 mg *bid* over 10 days and experienced viral load reductions of up to 1.6 log$_{10}$ copies mL^{-1},[56] providing positive proof-of-concept for the compound.

The safety and efficacy of maraviroc was confirmed in three Phase IIb/III randomised and double-blind clinical studies. MOTIVATE-1 and -2 (Maraviroc

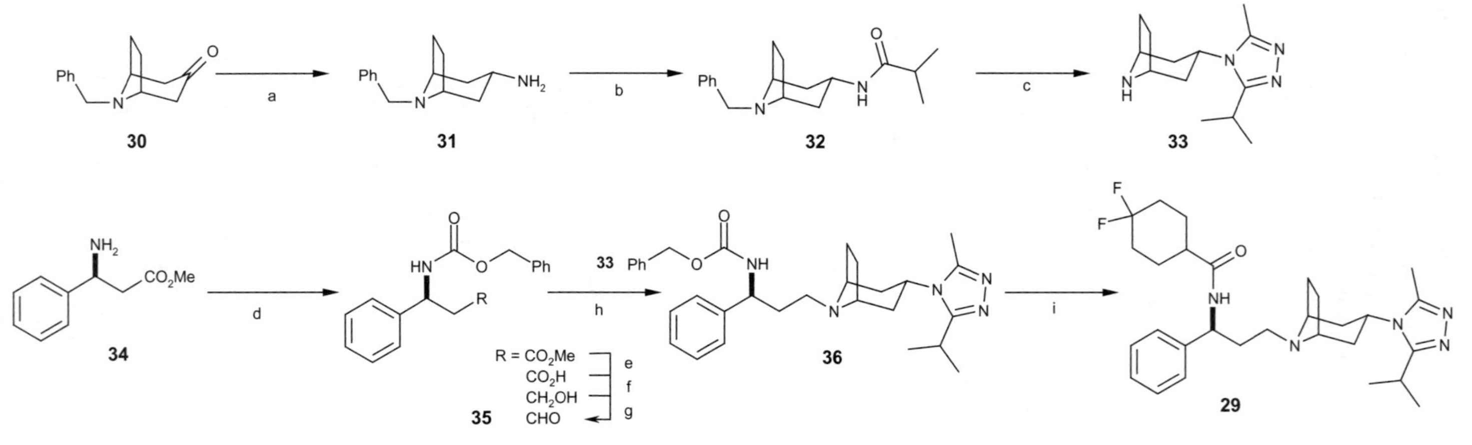

Scheme 9.1 The initial synthetic route to **29** (maraviroc). (a) H₂NOH; then Na, pentanol, 130 °C; (b) ᶦPrCOCl, NaHCO₃; (c) PCl₅; then H₂NNCOMe; then H₂, Pd, tosic acid; (d) CbzCl; (e) NaOH, MeOH; (f) BH₃; (g) SO₃.py; (h) NaBH(OAc)₃; (i) H₂, Pd/C; then *N*-(3-dimethylaminopropyl)-*N*-ethylcarbodiimide, CH₂Cl₂, hydroxybenztriazole, 4,4-difluorocyclohexanecarboxylic acid.

plus Optimised Therapy In Viremic Antiretroviral Treatment Experienced patients) were superiority studies and enrolled antiretroviral-treatment experienced patients with CCR5-tropic virus and ongoing viremia.[57,58] A total of 1049 patients were randomised and received either a 300 mg equivalent maraviroc dose once daily (*qd*), twice daily (*bid*) or placebo, each in combination with an "optimised background regimen" comprising a cocktail of existing antiretroviral drugs. The third study, MERIT (Maraviroc versus Efavirenz Regimens as Initial Therapy), enrolled treatment-naïve patients with CCR5-tropic virus and was a non-inferiority study comparing maraviroc to efavirenz, each given in combination with two nucleoside reverse transcriptase inhibitors.[59] The primary endpoints of all three studies were based on maraviroc's safety and efficacy at 48 weeks, with long-term follow-up of 2–5 years.

In the MOTIVATE trials, patients who received maraviroc as 300 mg *qd* or *bid* within an optimised background regimen achieved on average an approximately 1 $\log_{10}$ drop in viral load over background therapy alone.[60] Maraviroc was well tolerated, with no compound-related effects on immune function or hepatic transaminase levels.

Initial analysis of the MERIT trial showed that maraviroc was non-inferior to efavirenz using the plasma viral load cut-off of <400 copies mL^{-1}, but failed to meet the pre-specified non-inferiority criteria using the more stringent test of <50 copies mL^{-1}.[61] Patients with exclusively CCR5 virus had been selected for the study[62] using a viral tropism test (Trofile™) which had a reported 100% sensitivity for detecting CXCR4-using virus when it accounted for at least 10% of the viral population, but only $\sim 83\%$ when the viral population fell below 5%.[iii] Monogram Biosciences, the company providing the tropism test, have since replaced their original version with an enhanced Trofile™ assay, which achieves 100% sensitivity in the detection of CXCR4-using strains at just 0.3% incidence.[63] Following discussion with regulatory authorities, the screening samples from MERIT were re-analysed with the enhanced version, and as a result virus from a significant number (9.6%) of patients in the maraviroc treatment group were reclassified as CXCR4-using at baseline. Re-analysis using this new data demonstrated that maraviroc was shown to be non-inferior to efavirenz.[64]

Maraviroc received accelerated regulatory approval in the US and Europe in 2007 as the first prescribed CCR5 antagonist for the treatment of CCR5-tropic confirmed HIV-1 infection in treatment-experienced patients. Maraviroc is sold under the tradename Selzentry™ (Celsentri™ in Europe and Canada).[65]

9.13 Resistance to Maraviroc

A fully-mapped binding pocket model of CCR5 has been recently described for five different chemotypes.[66] This depicts the binding site for these

---

[iii] This highlights that the ability to detect X4-using virus depends on the proportion of minority strains in the overall viral population.

small-molecule antagonists as a lipophilic pocket buried within the trans-membrane region of the receptor, while interaction of gp120 with the receptor occurs on the extracellular surface at the N-terminus of the receptor.[67,68]

A strain of maraviroc-resistant HIV-1 was selected by the serial passage of a primary viral isolate grown in peripheral blood lymphocytes in the presence of sub-EC_{50} levels of maraviroc.[69] Sequencing studies showed that resistance was associated with two amino acid mutations, A316T and I323V (confirmed by reverse mutation), both in the V3 loop of gp120. Mutations in the V3 loop have also been associated with virologic failure in the MOTIVATE clinical trials with maraviroc. Plateaux in the maximal levels of inhibition were identified as a marker of resistance, demonstrating the ability of resistant virus to use compound-bound receptors to gain cell entry.[70] A second theoretical mechanism by which mar-aviroc resistance can emerge is through tropism shifting from CCR5-tropic to CXCX4-tropic or dual-tropic virus. Again, in the MOTIVATE trials, dual- or mixed-tropic virus was observed among patients who experienced treatment failure. Phylogenetic analysis of viral envelopes indicated that the emergent strains represented an outgrowth (under the selective pressure of a CCR5 antagonist) of virus which was initially present, but not detected, at initial screening.[71]

From the *in vitro* data obtained to date, maraviroc-resistant strains retain sensitivity to other CCR5 antagonists, indicating that not only do they still rely on CCR5 for entry, but that their ability to gain access to antagonist-associated receptor is limited to the compound used to select for them in the first instance.[70,72]

9.14   Identifying Back-up Agents

The identification of a compound that matches laboratory objectives, allows for the prediction of an appropriate PK profile in man and has a demonstrable safety window over toxic side-effects is a significant achievement for a drug discovery team. However, attrition remains a significant risk before the drug can reach the market; typically only 1 in 10 of compounds entering development will reach the market.[73] Ideally, feedback from the clinical programme would drive the design of backup agents, but such information is rarely timely. Consequently, assembled project teams will often attempt to identify further compounds to address perceived risks with the prototype before expertise, screening cascades and resources are disbanded. The greatest attrition risk is from toxicity, and often a backup programme will attempt to maximise struc-tural differentiation from the prototype to mitigate structure- or chemotype-based toxicity. The CCR5 team devoted effort to identify additional agents within the same chemotype and more widely to develop new structural classes.

Work within the same structural class led to the identification of **37** (PF-232798; Table 9.7), which progressed into Phase 2 clinical trials in 2007.[74] This agent retains the tropane core, shows a similar primary and selectivity/safety profile to maraviroc, and with an improved PK profile allowing for once-daily dosing.[75] Interestingly, despite its clear structural relationship to maraviroc, it

Table 9.7 Profile of **37** (PF-232798), a second tropane-based clinical agent

37 PF-232798
AV $IC_{90} = 2.0$ nM
hERG $IC_{50} > 10 \mu M$
Ph 1 PK (man)
250 mg *qd* gave free plasma levels at 24 h in excess of those seen for maraviroc (300 mg *bid*).

also exhibits efficacy against lab-generated mutant strains of HIV that are resistant to maraviroc. This activity has been proposed to be as a result of additional interactions with the extracellular loop ECL2.[76,77]

In parallel with identification of further agents from within the same chemotype, it was proposed to also replace the tropane core. This appeared to be a significant challenge, given the previously described advantages it offered (reduction of CYP2D6 activity and enhanced antiviral efficacy). Analysis suggested that the bridging unit's influence could result from a number of factors. It could change the energetically-accessible global conformational minima, increase steric hindrance around the central N and increase the pK_a of the central N by approximately half a log unit. Designing other ways to achieve the same stereoelectronic consequences resulted in a series of α-methylpiperidines.[78] Development of this series led to the identification of further compounds (**38** and **39**) with similar *in vitro* profiles to maraviroc (Table 9.8).[79] It is worth noting the significant change in log *D* between the isomeric triazoles is primarily a result of a change in pK_a of the central nitrogen of ~0.8 units. This was ascribed to the interaction between the nitrogen lone pair and the triazole's dipole.[80] It is also likely that this increased lipophilicity was the direct cause of the modest activity seen against the CYP3A4 isozyme.[81] No further development of this series has yet been disclosed.

A third backup strategy was triggered by reports of a wide range of chemotypes also showing activity against CCR5, including E-913 (related to aplaviroc),[83] ancriviroc[40] and TAK-779[84] (see Section 9.15).

Overlays of these structures into an active-site model suggested a common central basic nitrogen atom was required to anchor the ligand to Glu283 within the transmembrane region of the CCR5 receptor, and a proximal lipophilic group was required to interact with Tyr108. It was proposed that these apparently diverse compounds could occupy distinct but partially overlapping regions within the CCR5 receptor, a theory later supported by site-directed mutagenesis.[66] The team sought to develop a new chemotype through the screening of libraries of compounds based upon the framework shown in Figure 9.6.[85,86] The design focused upon fixing the substituents and varying the central core, which would contain one basic nitrogen (calculated pK_a 6–9) and one or more ring(s) to

Table 9.8 Replacing the tropane bridge in maraviroc with a side-chain α-methyl group resulted in similar profiles both *in vitro* and *in vivo*.[82]

	29	**38**	**39**
Fusion IC$_{50}$ (nM)	0.2	0.5	3.1
AV IC$_{90}$ (nM)	2.8	1.8	2.6
log D	1.9	2.2	3.0
HLM $t_{1/2}$ (min)	22	12	15
MDCK^a	Papps = 0.5; er = 2	Papps = 0.9; er = 21	Papps = 23; er = 2
hERG IC$_{50}$ (μM)	>10	>10	>10
CYP2D6 inhib^b	23% @ 3 μM	>30 μM	>30 μM
CYP3A4 inhib^b	23% @ 3 μM	>30 μM	3.4 μM
Rat PK			
Bioavailability	6%	6%	57%
Cl (mL min^{-1} kg^{-1})	74	92	30
V_D (L kg^{-1})	6.5	10	2.5
$t_{1/2}$ (h)	0.9	1.3	1.0

^{a}MDCK = flux through Madin-Darby canine kidney cells;[74] Papps = AB (10^{-6} cm s^{-1}), er = efflux ratio A-B/B-A.
^{b}Inhibition of turnover of cytochrome isozyme.
^{c}Most active stereoisomer; absolute configuration not determined.

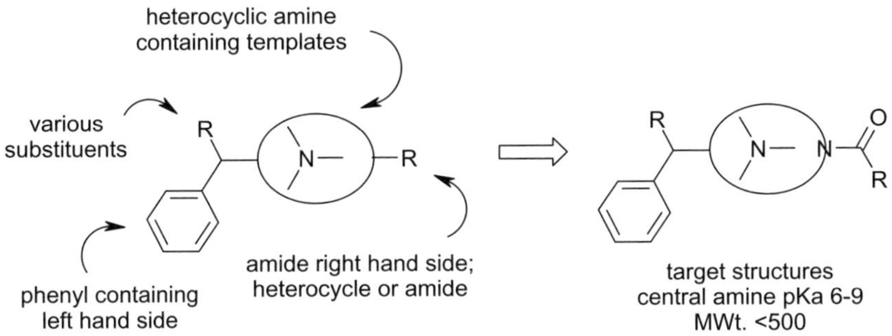

Figure 9.6 Design strategy for a novel series of CCR5 antagonists.

Table 9.9 *In vitro* profiles of two piperidine leads showing antiviral efficacies but high lipophilicities, resulting in little metabolic stability

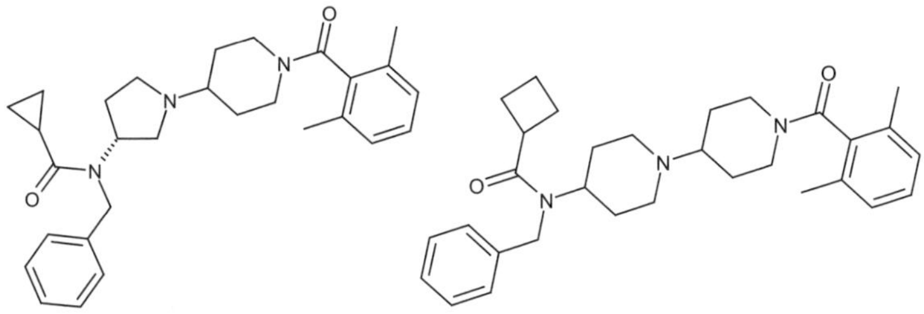

Compound	**40**	**41**
Fusion IC$_{50}$ (nM)	1	3
log D	2.6	3.5
HLM $t_{1/2}$ (min)	4	4
AV IC$_{90}$ (nM)	2	40
Binding IC$_{50}$ (nM)	8	nd
hERG IC$_{50}$ (nM)	9800	3410
Caco AB/BA	23/26	nd

give some rigidity. After two iterations, two lead series were identified, represented by compounds **40** and **41** (Table 9.9). Both compounds showed poor metabolic stability in hepatic microsomes, presumably *via* a facile benzylic oxidation. Further evolution of this series has not been reported.

9.15 Other CCR5 Antagonists

A number of other companies have invested significant work to identify CCR5 antagonists for HIV and for other indications, such as transplant rejection.[87]

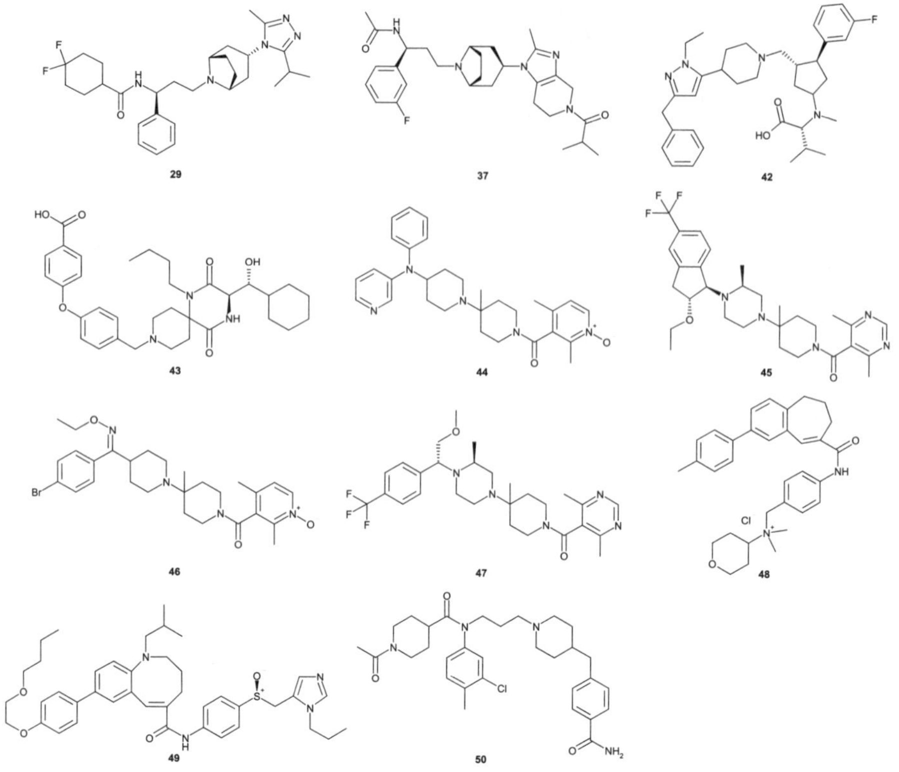

Figure 9.7 Structures of leading CCR5 antagonists.

The story described here for maraviroc was typical for many other programmes. Internal high-throughput file screening identified leads that were developed using conventional SAR with a concomitant realisation of the potential for activity at the hERG ion channel, which was later reduced during series development. Figure 9.7 shows a selection of disclosed small-molecule antagonists that have reached clinical trials or represent the most advanced agents disclosed by companies.

Merck's leading agent **42** (MrkA) showed high levels of antiviral potency (EC$_{95}$<8 nM) and a good cross-species pharmacological profile (F>40% rat, dog, monkey), but was hepatotoxic in rat.[88] Compound **43** (aplaviroc, GSK) showed high levels of antiviral activity (EC$_{50}$ = 0.1–0.6 nM) across a range of clinical isolates, oral bioavailability and a $t_{1/2}$ in man of ∼3 h.[89] However, it was withdrawn in 2006 during Ph 2b/c trials as a result of hepatotoxicity, raising concerns for mechanism-based toxicity, which the later success of maraviroc now tends to suggest is unlikely.[90] Compound **44** (NIBR-1282, Novartis) was not designed as an anti-HIV agent, but to suppress kidney rejection; it showed a large window over the hERG ion channel and a good pharmacokinetic profile (monkey $F = 43\%$, MRT 8 h).[87] Incyte's **45** (INCB-9471) has the potential to become best-in-class, given a high level of antiviral potency (EC_{90} 8–9 nM) and an extremely long $t_{1/2}$ in man of 60 h. Treatment (200 mg *qd*) led to a 2.1-log drop in viral load by day 20 in a Phase 2a clinical trial, but this compound has not yet progressed further.[91] Compound **46** (ancriviroc, Sch-C) progressed into Phase 1 on the basis of good antiviral activity and a strong oral PK profile (92% bioavailable, $t_{1/2} = 10$ h in dog).[32] The sterically encumbered amide bond resulted in restricted rotation around both the N–CO bond ($t_{1/2} = 5.6$ h) and the CO–aryl bond ($t_{1/2} = 23$ h), giving rise to an equilibrium mixture of four isomers.[92] Compound **47** (vicriviroc, Sch-D) incorporates a symmetrical pyrimidine, which halves the number of atropisomers. Relative to **46**, vicriviroc is more potent and has less activity against the hERG ion channel. It is currently in Phase 3 trials and shows an impressive human PK profile ($t_{1/2} = 28$–33 h).[93] Takeda's first agent **48** (E921, Tak-779) was a quaternary salt, which precluded oral delivery. Its development as a subcutaneous agent was also terminated as a result of irritation around the injection site.[94] However, the series was further developed, leading to the identification of **49** (Tak-652). Clinical data show the compound to be well-tolerated, with good bioavailability and low clearance ($t_{1/2} = 9$ h) following a 25 mg dose.[95] A second series yielded Takeda a third clinical candidate, **50** (Tak-220), which shows good *in vitro* activity ($EC_{50} = 0.6$ nM)[96] and an encouraging PK profile in monkey (bioavailability 29%, $t_{1/2} = 6$ h).[97] It is in Phase 1 trials with Tobira Therapeutics.

9.16 Conclusions

A decade of research into CCR5 antagonists for HIV has delivered a diverse array of chemotypes with a wide range of physicochemical properties. These compounds span a range of compound classes, from quaternary salts to bases, di-bases and zwitterions. All are high in molecular weight (496–700) and the majority have a clog $P > 3$. The need for potency has driven compound design into this challenging region of chemical space; but, despite that, it has still been possible to identify compounds worthy of progression. This success has undoubtedly been supported by the development of robust and meaningful assays that aid decision making and compound progression.

The ability to inhibit protein–protein interactions with small molecules is a significant achievement. The success observed with CCR5 must be put into context with the clinical failures seen when developing antagonists for other chemokine receptors.[16,98] For example, CCR1 and CCR2 antagonists for rheumatoid arthritis and multiple sclerosis have failed to show efficacy in Phase 2 trials (CP-481715, BX-471, MK-812). An antagonist (AMD3100) for CXCR4 was discontinued for HIV following leukocytosis, despite showing antiviral efficacy. Notably, this agent has recently been approved as a tool to harvest STEM cells.[99] The lack, or indeed unexpected efficacy, of these agents when targeting immune responses demonstrates the complexity of the immune system and our relatively weak understanding of the complex interplay between 40 + known chemokines and 19 receptors.

However, this same complex interplay between receptors and immune responses also offers significant future opportunities for safe and efficacious CCR5 agents (such as maraviroc) as our understanding develops. There is growing evidence that CCR5 antagonists could have applications in a wide range of alternative indications, from inflammation[100] and multiple sclerosis,[101] through to obesity,[102] cancer[103] and pain.[104] Given the costs of clinical trials, key to the future exploitation of agents such as maraviroc will be the development of more predictive animal models of human disease states.

Acknowledgements

Marcel de Groot is thanked for molecular modelling graphics.

References

1. H. Masur, M. A. Michelis, J. B. Greene, I. Onorato, R. A. Stouwe, R. S. Holzman, G. Wormser, L. Brettman, M. Lange, H. W. Murray and S. Cunningham-Rundles, *New Engl. J. Med.*, 1981, **305**, 1431–1438.
2. E. P. Gelmann, M. Popovic, D. Blayney, H. Masur, G. Sidhu, R. E. Stahl and R. C. Gallo, *Science*, 1983, **220**, 862–865.
3. R. C. Gallo, P. S. Sarin, E. P. Gelmann, M. Robert-Guroff, E. Richardson, V. S. Kalyanaraman, D. Mann, G. D. Sidhu, R. E. Stahl, S. Zolla-Pazner, J. Leibowitch and M. Popovic, *Science*, 1983, **220**, 865–867.
4. Y. Mehellou and E. De Clercq, *J. Med. Chem.*, 2010, **53**, 521–538.
5. S. J. Little, S. Holte, J.-P. Routy, E. S. Daar, M. Markowitz, A. C. Collier, R. A. Koup, J. W. Mellors, E. Connick, B. Conway, M. Kilby, L. Wang, J. M. Whitcomb, N. S. Hellmann and D. D. Richman, *New Engl. J. Med.*, 2002, **347**, 385–394.
6. J. P. Moore and R. W. Doms, *Proc. Natl. Acad. Sci. U.S.A.*, 2003, **100**, 10598–10602.
7. R. A. Weiss, *Nature*, 2001, **410**, 963–967.
8. J. P. Moore and M. Stevenson, *Nat. Rev. Mol. Cell Biol.*, 2000, **1**, 40–49.
9. U. Weiss, *Nature*, 2001, **410**, 961.
10. J. C. Stephens, D. E. Reich, D. B. Goldstein, H. D. Shin, M. W. Smith, M. Carrington, C. Winkler, G. A. Huttley, R. Allikmets, L. Schriml, B. Gerrard, M. Malasky, M. D. Ramos, S. Morlot, M. Tzetis, C. Oddoux, F. S. di Giovine, G. Nasioulas, D. Chandler, M. Aseev, M. Hanson, L. Kalaydjieva, D. Glavac, P. Gasparini, E. Kanavakis, M. Claustres, M. Kambouris, H. Ostrer, G. Duff, V. Baranov, H. Sibul, A. Metspalu, D. Goldman, N. Martin, D. Duffy, J. Schmidtke, X. Estivill, S. J. O'Brien and M. Dean, *Am. J. Hum. Genet.*, 1998, **62**, 1507–1515.

11. A. P. Galvani and M. Slatkin, *Proc. Natl. Acad. Sci. U.S.A.*, 2003, **100**, 15276–15279.
12. M. Perros, *Adv. Antiviral Drug Des.*, 2007, **5**, 185–212.
13. W. G. Glass, D. H. McDermott, J. K. Lim, S. Lekhong, S. F. Yu, W. A. Frank, J. Pape, R. C. Cheshier and P. M. Murphy, *J. Exp. Med.*, 2006, **203**, 35–40.
14. Y. Saita, E. Kodama, M. Orita, M. Kondo, T. Miyazaki, K. Sudo, K. Kajiwara, M. Matsuoka and Y. Shimizu, *J. Immunol.*, 2006, **177**, 3116–3122.
15. Y. Landry and J.-P. Gies, *Fundam. Clin. Pharmacol.*, 2008, **22**, 1–18.
16. R. Horuk, *Nat. Rev. Drug Discovery*, 2009, **8**, 23–33.
17. C. Blanpain, I. Migeotte, B. Lee, J. Vakili, B. J. Doranz, C. Govaerts, G. Vassart, R. W. Doms and M. Parmentier, *Blood*, 1999, **94**, 1899–1905.
18. M. Thelen, *Nat. Immunol.*, 2001, **2**, 129–134.
19. Y. Xu, H. Liu, C. Niu, C. Luo, X. Luo, J. Shen, K. Chen and H. Jiang, *Bioorg. Med. Chem.*, 2004, **12**, 6193–6208.
20. J.-M. Navenot, Z. Wang, J. O. Trent, J. L. Murray, Q. Hu, L. DeLeeuw, P. S. Moore, Y. Chang and S. C. Peiper, *J. Mol. Biol.*, 2001, **313**, 1181–1193.
21. J. J. Onuffer and R. Horuk, *Trends Pharmacol. Sci.*, 2002, **23**, 459–467.
22. C. Combadiere, S. K. Ahuja, H. L. Tiffany and P. M. Murphy, *J. Leukoc. Biol.*, 1996, **60**, 147–152.
23. A. L. Hopkins, C. R. Groom and A. Alex, *Drug Discovery Today*, 2004, **9**, 430–431.
24. P. D. Leeson and B. Springthorpe, *Nat. Rev. Drug Discovery*, 2007, **6**, 881–890.
25. T. Ryckmans, M. P. Edwards, V. A. Horne, A. M. Correia, D. R. Owen, L. R. Thompson, I. Tran, M. F. Tutt and T. Young, *Bioorg. Med. Chem. Lett.*, 2009, **19**, 4406–4409.
26. C. H. Reynolds, B. A. Tounge and S. D. Bembenek, *J. Med. Chem.*, 2008, **51**, 2432–2438.
27. D. Armour, M. J. de Groot, M. Edwards, M. Perros, D. A. Price, B. L. Stammen and A. Wood, *ChemMedChem*, 2006, **1**, 706–709.
28. H. Habashita, M. Kokubo, S.-I. Hamano, N. Hamanaka, M. Toda, S. Shibayama, H. Tada, K. Sagawa, D. Fukushima, K. Maeda and H. Mitsuya, *J. Med. Chem.*, 2006, **49**, 4140–4152.
29. G. Thoma, F. Nuninger, M. Schaefer, G. Akyel Kayhan, R. Albert, C. Beerli, C. Bruns, E. Francotte, M. Luyten, D. MacKenzie, L. Oberer, B. Streiff Markus, T. Wagner, H. Walter, G. Weckbecker and H.-G. Zerwes, *J. Med. Chem.*, 2004, **47**, 1939–1955.
30. C. P. Dorn, P. E. Finke, B. Oates, R. J. Budhu, S. G. Mills, M. MacCoss, L. Malkowitz, M. S. Springer, B. L. Daugherty, S. L. Gould, J. A. DeMartino, S. J. Siciliano, A. Carella, G. Carver, K. Holmes, R. Danzeisen, D. Hazuda, J. Kessler, J. Lineberger, M. Miller, W. A. Schleif and E. A. Emini, *Bioorg. Med. Chem. Lett.*, 2001, **11**, 259–264.

31. M. Shiraishi, Y. Aramaki, M. Seto, H. Imoto, Y. Nishikawa, N. Kanzaki, M. Okamoto, H. Sawada, O. Nishimura, M. Baba and M. Fujino, *J. Med. Chem.*, 2000, **43**, 2049–2063.
32. A. Palani, S. Shapiro, J. W. Clader, W. J. Greenlee, K. Cox, J. Strizki, M. Endres and B. M. Baroudy, *J. Med. Chem.*, 2001, **44**, 3339–3342.
33. P. Dorr, M. Westby, S. Dobbs, P. Griffin, B. Irvine, M. Macartney, J. Mori, G. Rickett, C. Smith-Burchnell, C. Napier, R. Webster, D. Armour, D. Price, B. Stammen, A. Wood and M. Perros, *Antimicrob. Agents Chemother.*, 2005, **49**, 4721–4732.
34. R. J. Temple and M. H. Himmel, *J. Am. Med. Assoc.*, 2002, **287**, 2273–2275.
35. M. J. de Groot, M. J. Ackland, V. A. Horne, A. A. Alex and B. C. Jones, *J. Med. Chem.*, 1999, **42**, 1515–1524.
36. D. Armour and A. Wood, *Prog. Med. Chem.*, 2005, **43**, 239–271.
37. S. W. Ellis, G. P. Hayhurst, G. Smith, T. Lightfoot, M. M. S. Wong, A. P. Simula, M. J. Ackland, M. J. E. Sternberg, M. S. Lennard, G. T. Tucker and C. R. Wolf, *J. Biol. Chem.*, 1995, **270**, 29055–29058.
38. M. Seto, N. Miyamoto, K. Aikawa, Y. Aramaki, N. Kanzaki, Y. Iizawa, M. Baba and M. Shiraishi, *Bioorg. Med. Chem.*, 2005, **13**, 363–386.
39. M. Perry, M. J. de Groot, R. Helliwell, D. Leishman, M. Tristani-Firouzi, M. C. Sanguinetti and J. Mitcheson, *Mol. Pharmacol.*, 2004, **66**, 240–249.
40. J. M. Strizki, C. Tremblay, S. Xu, L. Wojcik, N. Wagner, W. Gonsiorek, R. W. Hipkin, C.-C. Chou, C. Pugliese-Sivo, Y. Xiao, J. R. Tagat, K. Cox, T. Priestley, S. Sorota, W. Huang, M. Hirsch, G. R. Reyes and B. M. Baroudy, *Antimicrob. Agents Chemother.*, 2005, **49**, 4911–4919.
41. D. Kim, L. Wang, J. J. Hale, C. L. Lynch, R. J. Budhu, M. MacCoss, S. G. Mills, L. Malkowitz, S. L. Gould, J. A. DeMartino, M. S. Springer, D. Hazuda, M. Miller, J. Kessler, R. C. Hrin, G. Carver, A. Carella, K. Henry, J. Lineberger, W. A. Schleif and E. A. Emini, *Bioorg. Med. Chem. Lett.*, 2005, **15**, 2129–2134.
42. D. A. Price, D. Armour, M. De Groot, D. Leishman, C. Napier, M. Perros, B. L. Stammen and A. Wood, *Bioorg. Med. Chem. Lett.*, 2006, **16**, 4633–4637.
43. D. A. Price, D. Armour, M. de Groot, D. Leishman, C. Napier, M. Perros, B. L. Stammen and A. Wood, *Curr. Top. Med. Chem.*, 2008, **8**, 1140–1151.
44. K. Finlayson, L. Turnbull, C. T. January, J. Sharkey and J. S. Kelly, *Eur. J. Pharmacol.*, 2001, **430**, 147–148.
45. P. K. Dorr, S. Dobbs, G. Rickett, B. Lewis, M. Macartney, M. Westby and M. Perros, presented at the Interscience Conference on Antimicrobial Agents and Chemotherapy, Chicago, 2003, abstr. F1462.
46. C. A. Lipinski, F. Lombardo, B. W. Dominy and P. J. Feeney, *Adv. Drug Delivery Rev.*, 1997, **23**, 3–25.
47. K. L. Kirk, *Org. Process Res. Dev.*, 2008, **12**, 305–321.
48. W. K. Hagmann, *J. Med. Chem.*, 2008, **51**, 4359–4369.
49. Pfizer Ltd., *Pat. Appl.*, WO 2001/090106, 2001.

50. S. C. Piscitelli and K. D. Gallicano, *New Engl. J. Med.*, 2001, **344**, 984–996.

51. P. Dorr, M. Macartney, G. Rickett, S. Smith-Burchnell, S. Dobbs, J. Mori, P. Griffin, J. Lok, R. Irvine, M. Westby, C. Hitchcock, B. Stammen, D. Price, D. Armour, A. Wood and M. Perros, presented at the 10th Conference on Retroviruses and Opportunistic Infections, Boston, 2003.

52. D. A. Price, S. Gayton, M. D. Selby, J. Ahman, S. Haycock-Lewandowski, B. L. Stammen and A. Warren, *Tetrahedron Lett.*, 2005, **46**, 5005–5007.

53. D. A. Price, S. Gayton, M. D. Selby, J. Ahman and S. Haycock-Lewandowski, *Synlett*, 2005, 1133–1134.

54. J. E. Burks, L. Espinosa, E. S. LaBell, J. M. McGill, A. R. Ritter, J. L. Speakman, M. Williams, D. A. Bradley, M. G. Haehl and C. R. Schmid, *Org. Process Res. Dev.*, 1997, **1**, 198–210.

55. S. Abel, E. van der Ryst, M. C. Rosario, C. E. Ridgway, C. G. Medhurst, R. J. Taylor-Worth and G. J. Muirhead, *Br. J. Clin. Pharmacol.*, 2008, **65**, 5–18.

56. G. Fatkenheuer, A. L. Pozniak, M. A. Johnson, A. Plettenberg, S. Staszewski, A. I. M. Hoepelman, M. S. Saag, F. D. Goebel, J. K. Rockstroh, B. J. Dezube, T. M. Jenkins, C. Medhurst, J. F. Sullivan, C. Ridgway, S. Abel, I. T. James, M. Youle and E. van der Ryst, *Nat. Med.*, 2005, **11**, 1170–1172.

57. J. Lalezari, presented at the 14th Conference on Retroviruses and Opportunistic Infections, Los Angeles, 2007.

58. M. Nelson, presented at the 14th Conference on Retroviruses and Opportunistic Infections, Los Angeles, 2007.

59. M. Saag, presented at the 4th International AIDS Society Conference, Sydney, 2007.

60. D. Hardy, presented at the 15th Conference on Retroviruses and Opportunistic Infections, Boston, 2008.

61. S. Lieberman-Blum, *Clin. Ther.*, 2008, **30**, 1228–1250.

62. J. M. Whitcomb, *Antimicrob. Agents Chemother.*, 2007, **51**, 566–575.

63. C. Petropoulos, K. Limoli, J. M. Whitcomb, *et al.*, presented at the Fifth European HIV Drug Resistance Workshop, Cascais, Portugal, 2007.

64. M. Saag, J. Heera, J. Goodrich, *et al.*, presented at the 48th Interscience Conference on Antimicrobial Agents and Chemotherapy and the Infectious Diseases Society of America 46th Annual Meeting, Washington, 2008.

65. Pfizer Ltd., Selzentry: full prescribing information, http://media.pfizer.com/files/products/uspi_maraviroc.pdf.

66. R. Kondru, J. Zhang, C. Ji, T. Mirzadegan, D. Rotstein, S. Sankuratri and M. Dioszegi, *Mol. Pharmacol.*, 2008, **73**, 789–800.

67. T. Dragic, A. Trkola, D. A. Thompson, E. G. Cormier, F. A. Kajumo, E. Maxwell, S. W. Lin, W. Ying, S. O. Smith, T. P. Sakmar and J. P. Moore, *Proc. Natl. Acad. Sci. U.S.A.*, 2000, **97**, 5639–5644.

68. P. K. Dorr, K. Todd, R. Irvine, N. Robas, A. Thomas, M. Fidock, H. Sultan, J. Mills, F. Perrucio, K. Burt, G. Rickett, H. Perkins, P. Griffin, M. Macartney, D. Hamilton, M. Westby and M. Perros, presented at the 45th Interscience Conference on Antimicrobial Agents and Chemotherapy, Washington, 2005.
69. M. Westby, C. Smith-Burchnell, J. Mori, M. Lewis, J. Whitcomb, C. Petropoulos and M. Perros, presented at the XIII International HIV Drug Resistance Workshop, Tenerife, 2004.
70. M. Westby, C. Smith-Burchnell, J. Mori, M. Lewis, M. Mosley, M. Stockdale, P. Dorr, G. Ciaramella and M. Perros, *J. Virol.*, 2007, **81**, 2359–2371.
71. M. Westby, M. Lewis, J. Whitcomb, M. Youle, A. L. Pozniak, I. T. James, T. M. Jenkins, M. Perros and E. van der Ryst, *J. Virol.*, 2006, **80**, 4909–4920.
72. M. Westby, C. Smith-Burchnell, D. Hamilton, N. Robas, B. Irvine, M. Fidock, J. Mills, F. Perruccio, J. Mori, M. Macartney, C. Barber, P. Dorr and M. Perros, presented at the 12th Conference on Retroviruses and Opportunistic Infections, Boston, 2005.
73. J. A. Kramer, *Expert Opin. Drug Discovery*, 2008, **3**, 707–713.
74. Pfizer Ltd., *Pat. Appl.*, WO 2005/033107, 2005.
75. P. Dorr, M. Westby, L. McFadyen, J. Mori, J. Davis, F. Perruccio, R. Jones, P. Stupple, D. Middleton and M. Perros, presented at the 15th Conference on Retroviruses and Opportunistic Infections, Boston, 2008.
76. A. Hall, *Future Med. Chem.*, 2009, **1**, 431–434.
77. P. Dorr, M. Westby, L. McFadyen, J. Mori, J. Davis, F. Perruccio, R. Jones, P. Stupple, D. Middleton and M. Perros, presented at the 15th Conference on Retroviruses and Opportunistic Infections, Boston, 2007.
78. C. G. Barber, D. C. Blakemore, J.-Y. Chiva, R. L. Eastwood, D. S. Middleton and K. A. Paradowski, *Bioorg. Med. Chem. Lett.*, 2009, **19**, 1075–1079.
79. C. G. Barber, D. C. Blakemore, J.-Y. Chiva, R. L. Eastwood, D. S. Middleton and K. A. Paradowski, *Bioorg. Med. Chem. Lett.*, 2009, **19**, 1499–1503.
80. A. Wood, D. Armour, F. D. King and G. Lawton, *Prog. Med. Chem.*, 2005, **43**, 239–271.
81. R. J. Riley, A. J. Parker, S. Trigg and C. N. Manners, *Pharm. Res.*, 2001, **18**, 652–655.
82. M. Cereijido, J. Ehrenfeld, I. Meza and A. Martínez-Palomo, *J. Membr. Biol.*, 1980, **52**, 147–159.
83. K. Maeda, K. Yoshimura, S. Shibayama, H. Habashita, H. Tada, K. Sagawa, T. Miyakawa, M. Aoki, D. Fukushima and H. Mitsuya, *J. Biol. Chem.*, 2001, **276**, 35194–35200.
84. M. Baba, *Proc. Natl. Acad. Sci. U. S. A.*, 1999, **96**, 5698.

85. D. C. Pryde, M. Corless, D. R. Fenwick, H. J. Mason, B. C. Stammen, P. T. Stephenson, D. Ellis, D. Bachelor, D. Gordon, C. G. Barber, A. Wood, D. S. Middleton, D. C. Blakemore, G. C. Parsons, R. Eastwood, M. Y. Platts, K. Statham, K. A. Paradowski, C. Burt and W. Klute, *Bioorg. Med. Chem. Lett.*, 2009, **19**, 1084–1088.

86. D. C. Pryde, presented at the 16th Camerino-Noordwijkerhout Symposium, An Overview of Receptor Chemistry, Camerino, Italy, 2007.

87. G. Thoma, C. Beerli, M. Bigaud, C. Bruns, N. G. Cooke, M. B. Streiff and H.-G. Zerwes, *Bioorg. Med. Chem. Lett.*, 2008, **18**, 2000–2005.

88. P. D. Cornwell and R. G. Ulrich, *Toxicol. Pathol.*, 2007, **35**, 576–588.

89. B. M. Johnson, I. H. Song, K. K. Adkison, J. Borland, L. Fang, Y. Lou, M. M. Berrey, A. N. Nafziger, S. C. Piscitelli and J. S. Bertino Jr, *J. Clin. Pharmacol.*, 2006, **46**, 577–587.

90. W. G. Nichols, H. M. Steel, T. Bonny, K. Adkison, L. Curtis, J. Millard, K. Kabeya and N. Clumeck, *Antimicrob. Agents Chemother.*, 2008, **52**, 858–865.

91. *Drug Data Rep.*, 2007, **29**, 840.

92. A. Palani, S. Shapiro, J. W. Clader, W. J. Greenlee, D. Blythin, K. Cox, N. E. Wagner, J. Strizki, B. M. Baroudy and N. Dan, *Bioorg. Med. Chem. Lett.*, 2003, **13**, 705–708.

93. D. Schurmann, G. Fatkenheuer, J. Reynes, C. Michelet, F. Raffi, J. van Lier, M. Caceres, A. Keung, A. Sansone-Parsons, L. M. Dunkle and C. Hoffmann, *AIDS*, 2007, **21**, 1293–1299.

94. A. Mastrolorenzo, A. Scozzafava and C. T. Supuran, *Expert Opin. Ther. Pat.*, 2001, **11**, 1245–1252.

95. M. Baba, K. Takashima, H. Miyake, N. Kanzaki, K. Teshima, X. Wang, M. Shiraishi and Y. Iizawa, *Antimicrob. Agents Chemother.*, 2005, **49**, 4584–4591.

96. S. Imamura, Y. Nishikawa, T. Ichikawa, T. Hattori, Y. Matsushita, S. Hashiguchi, N. Kanzaki, Y. Iizawa, M. Baba and Y. Sugihara, *Bioorg. Med. Chem.*, 2005, **13**, 397–416.

97. S. Imamura, T. Ichikawa, Y. Nishikawa, N. Kanzaki, K. Takashima, S. Niwa, Y. Iizawa, M. Baba and Y. Sugihara, *J. Med. Chem.*, 2006, **49**, 2784–2793.

98. S. Ribeiro and R. Horuk, *Expert Opin. Drug Discovery*, 2009, **4**, 1017–1034.

99. A. F. Cashen, B. Nervi and J. DiPersio, *Future Oncol.*, 2007, **3**, 19–27.

100. C. Schroder, R. N. Pierson III, B.-N. H. Nguyen, D. W. Kawka, L. B. Peterson, G. Wu, T. Zhang, M. S. Springer, S. J. Siciliano, S. Iliff, J. M. Ayala, M. Lu, J. S. Mudgett, K. Lyons, S. G. Mills, G. G. Miller, I. I. Singer, A. M. Azimzadeh and J. A. DeMartino, *J. Immunol.*, 2007, **179**, 2289–2299.

101. D. Otaegui, J. Ruíz-Martínez, J. Olaskoaga, J. Emparanza and A. de Munain, *neurogenetics*, 2007, **8**, 201–205.

102. J. Huber, F. W. Kiefer, M. Zeyda, B. Ludvik, G. R. Silberhumer, G. Prager, G. J. Zlabinger and T. M. Stulnig, *J. Clin. Endocrinol. Metab.*, 2008, **93**, 3215–3221.
103. Y. Wu, Y.-Y. Li, K. Matsushima, T. Baba and N. Mukaida, *J. Immunol.*, 2008, **181**, 6384–6393.
104. S. Bhangoo, D. Ren, R. Miller, K. Henry, J. Lineswala, C. Hamdouchi, B. Li, P. Monahan, D. Chan, M. Ripsch and F. White, *Mol. Pain*, 2007, **3**, 38.

CHAPTER 10

The Discovery of GS-9131, an Amidate Prodrug of a Novel Nucleoside Phosphonate HIV Reverse Transcriptase Inhibitor

RICHARD MACKMAN

Gilead Sciences, Inc., 333 Lakeside Drive, Foster City, CA 94404, USA

10.1 Introduction

10.1.1 Human Immunodeficiency Virus Reverse Transcriptase

In 2006, the worldwide burden of human immunodeficiency virus (HIV) was estimated to affect a population of between 30.6 and 36.1 million.[1] Whilst no vaccine or cure is available, tremendous advances have been made in the development of drugs to suppress virus replication.[2] These drugs target several different viral proteins, including the crucial viral polymerase. HIV polymerase is a RNA-dependent DNA polymerase or reverse transcriptase (RT) whose function is to transcribe the HIV viral RNA genome into double-stranded proviral DNA for integration into the host cell genome.[3] Host enzymes and other proteins encoded by HIV generate the new viral proteins, which together with copied HIV RNA can be packaged and released from an infected cell as new virions. The approved RT inhibitors can be divided into two general classes, nucleoside and nucleotide inhibitors [NRTIs and N(t)RTIs,

RSC Drug Discovery Series No. 4
Accounts in Drug Discovery: Case Studies in Medicinal Chemistry
Edited by Joel C. Barrish, Percy H. Carter, Peter T. W. Cheng and Robert Zahler
© Royal Society of Chemistry 2011
Published by the Royal Society of Chemistry, www.rsc.org

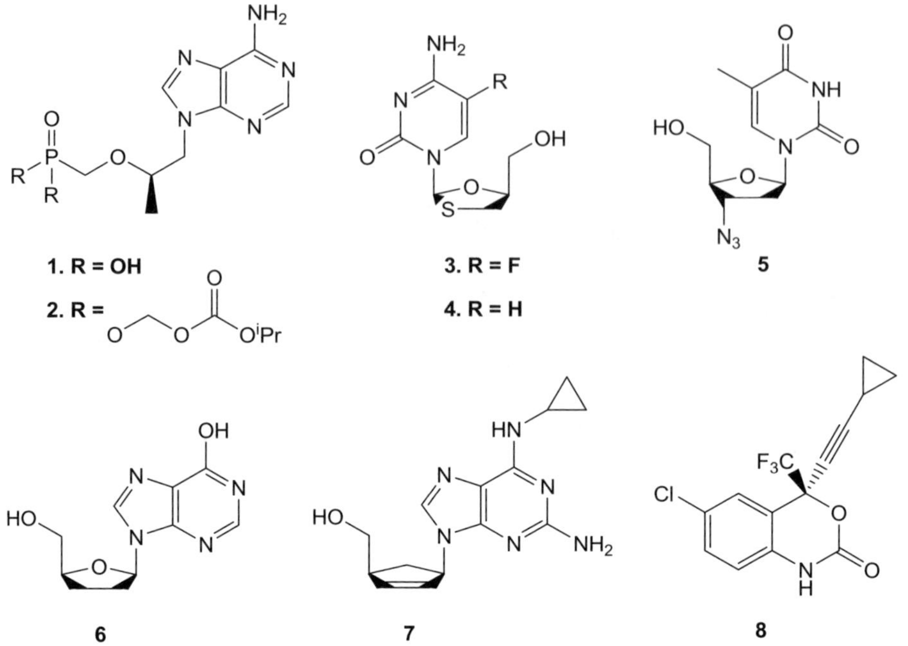

Figure 10.1   Structures of FDA approved N(t)RTI and NNRTIs.

respectively], and non-nucleoside inhibitors (NNRTIs). Examples of each class are shown in Figure 10.1.

Tenofovir disoproxil fumarate (TDF, **2**), a prodrug of tenofovir (TFV, **1**), is the only nucleotide analog approved for HIV treatment. Examples of NRTIs include emtricitabine (FTC, **3**), lamivudine (3TC, **4**), zidovudine (AZT, **5**), didanosine (ddI, **6**), and abacavir (ABC, **7**), whilst efavirenz (EFV, **8**) is an example of a NNRTI. Collectively the N(t)RTIs inhibit RT by competing with the natural deoxynucleotide triphosphate substrates at the active site, whereas the NNRTIs bind to an allosteric site on RT and block conformational changes required for efficient polymerization to occur. In 1995–1996, the first HIV protease inhibitors were approved and thus began the era of combination treatment now known as highly active antiretroviral therapy (HAART). The early HAART regimens required the administration of drugs according to complex dosing schedules and inevitably led to poor patient compliance, which in turn resulted in the emergence of resistance.[4] Therefore, major efforts across the pharmaceutical industry have been invested into the development of simplified HIV regimens, culminating in the approval of Atripla™, the first single-pill, once-daily (*qd*) complete regimen for HIV treatment. Atripla™ combines three *qd* drugs, **2**, **3**, and **8**, all of which target RT, emphasizing the critical importance of RT inhibitors in HIV therapy. Indeed the combination of two N(t)RTIs has become the backbone of choice for many HAART regimens.[5]

10.1.2 Target Profile for a Novel RT Inhibitor

The success of HAART therapy has increased life expectancy among infected patients to >20 years post diagnosis, in effect transforming HIV into a manageable life-long disease. In this context, the durability of therapy in an aging population can be limited by drug resistance, side effects due to chronic use, and drug–drug interactions. Therefore, more effective and convenient fixed-dose *qd* regimens are highly desirable in continuing the fight against HIV. New fixed-dose *qd* regimens combining **2** and **3** with HIV protease and HIV integrase inhibitors are being evaluated clinically.[6] The established role of N(t)RTIs as the backbone of choice in HAART therapy supports the development of new N(t)RTIs that can be combined with the existing treatment options or newer drugs in development. With this aim in mind, a program was initiated toward identifying a novel N(t)RTI that would have (i) an oral *qd* dosing schedule; (ii) an excellent resistance profile effective against clinically observed RT mutants, especially those that reduce susceptibility to **2** and **3**; (iii) a low potential for long-term toxicities; and (iv) the ability to be combined with other N(t)RTIs in HAART therapy.

10.1.3 Mechanism of Action of N(t)RTIs

A detailed understanding of the mechanism by which N(t)RTIs inhibit HIV RT is necessary to fully appreciate the strategy employed toward designing a new N(t)RTI drug. Consider first the NRTIs, which are neutral species at physiological pH and cross into cells passively or *via* nucleoside transporters (Figure 10.2).

The target cells for HIV inhibitors are cells in which the virus is replicating, primarily CD4+ memory T-cells and macrophages in lymph nodes and in peripheral blood. Inside target cells, kinases anabolise the nucleoside to its active triphosphate metabolite, for example **3**-TP, whereas phosphorylases catabolise degradation back to the nucleoside **3**. The metabolite **3**-TP is the active species and competes with the natural intracellular pools of deoxynucleotide substrate, in this case dCTP (**10**), for incorporation into the growing DNA polymerase chain. Inspection of the chemical structures of the N(t)RTIs **2–7** highlights the absence of a 3′- hydroxyl group on the deoxyribose ring or equivalent group of each inhibitor, unlike the natural substrates **9** or **10**, for example. This is an essential feature of all the N(t)RTI drugs, such that after a single incorporation of the inhibitor into the growing DNA, the inhibitor causes chain termination, that is, unless the blocking nucleotide can be excised from the end of the growing chain. Thus, the mechanism of action for the nucleoside inhibitors is passive or active transport into cells, activation by three intracellular kinases to the triphosphate, incorporation by HIV RT into the growing DNA, and chain termination of the DNA polymerization process.

In contrast, the N(t)RTIs, exemplified by nucleoside phosphonate **1**, present some unique features that set them apart. Phosphonate **1**, bearing an adenine nucleobase, is essentially a bioisostere of deoxyadenosine monophosphate, the

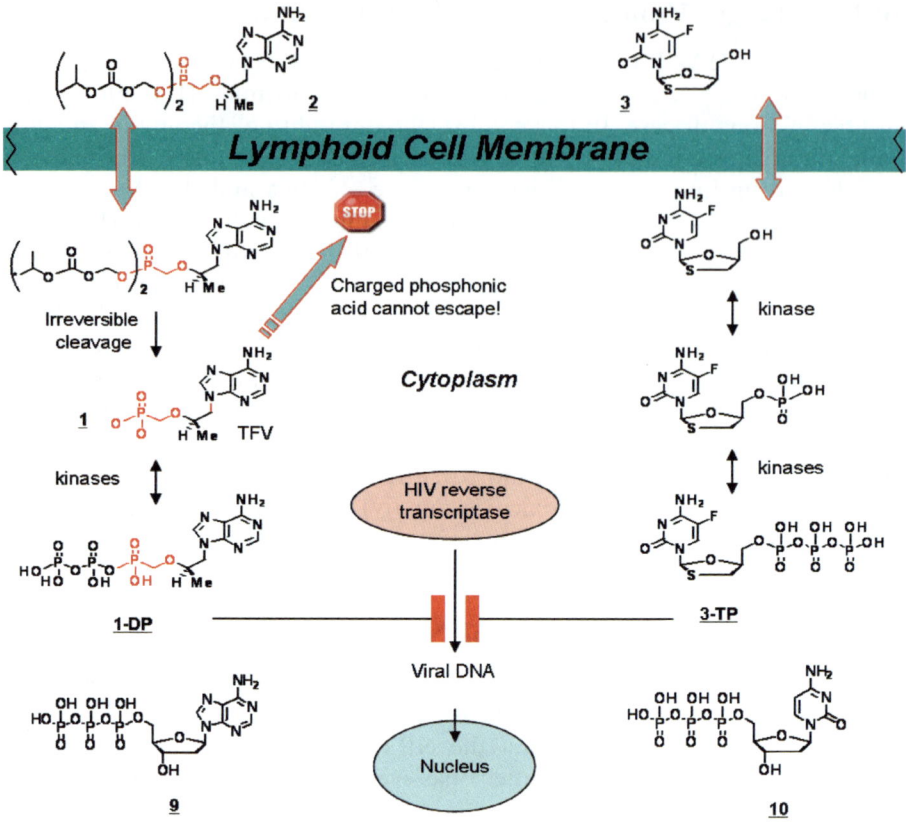

Figure 10.2   Intracellular metabolic metabolic pathway of N(t)RTI **2** compared to NRTI **3**.

first metabolite in the activation pathway of deoxyadenosine. Therefore, only two kinases are required to generate the active diphosphophosphonate (**1-DP**) metabolite which is equivalent to the deoxynucleoside triphosphate. Often the first metabolism step is rate limiting in the activation of nucleosides to their monophosphates, so the phosphonate effectively bypasses this step. A second distinction of the nucleoside phosphonates is that they are charged at physiological pH and therefore require prodrugs, such as the disoproxil ester groups found in **2**, to effectively cross the cell membrane. However, once the prodrug penetrates the cell, and is cleaved by intracellular esterases to release the phosphonic acid, the stability of the P–C bond to further cleavage by enzymes prevents the generation of a neutral species, effectively trapping the charged diacid in the cell. The combination of enhanced phosphonate stability and inefficient removal from the cell leads to prolonged intracellular levels of the diphosphate metabolite **1-DP**. Clinically, the intracellular half life of **1-DP** has been measured to exceed 150 h in activated lymphoid cells, a property that clearly favors administration of prodrug **2** on a *qd* regimen.[7] A final point worth noting is that the antiviral potency of the N(t)RTIs or NRTIs is reflected by the levels of diphosphophosphonate or triphosphate that can be maintained over time inside the target cells, and the ability of the metabolite to compete with deoxynucleotides and incorporate into the nascent DNA (determined *in vitro* as IC_{50} for RT). This can be contrasted with the other classes of HIV inhibitors, including NNRTI **8**, where the free fraction of drug in plasma is often used as a predictor for *in vivo* antiviral efficacy. For N(t)RTIs, antiviral activity results from an intracellular, charged metabolite species and is, therefore, not subject to these same free fraction correlations.

10.2   Discovery of the Novel Nucleoside Phosphonate GS-9148

10.2.1   Medicinal Strategy

The research program focused exclusively on the discovery of a novel N(t)RTI rather than NRTI for several reasons: (i) the expected long intracellular half-lives of the active metabolites would favor *qd* administration; (ii) the freedom to operate in the N(t)RTI space was much greater than NRTIs; (iii) prodrugs that target the N(t)RTI to the desired lymphoid cells could be developed to reduce therapeutic dose and *in vivo* toxicites; (iv) only two intracellular metabolism steps would be required, allowing the first and often rate-limiting step of NRTIs to be bypassed; and (v) cross resistance to a novel phosphonate may be less likely since only one N(t)RTI, prodrug **2**, is currently in clinical use for HIV.

There are many N(t)RTI-resistant mutations observed clinically, but initial evaluation of the new inhibitors focused primarily on the RT mutants that reduce susceptibility to **1**, **3**, **4**, and **7** due to their more abundant clinical use. Therefore, new analogs were profiled toward wild-type (wt) HIV-1 in cell culture, and also three recombinant RT mutants harboring K65R, M184V, and multiple thymidine analog mutations (6TAMs), respectively. The design of the novel N(t)RTIs was also structure based, using binding models of proposed compounds in the active site of wt RT. These models were based on the reported X-ray structure of the wt RT complex.[8] Particular attention was also paid to the characterization of potential N(t)RTI-related toxicities, such as mitochondrial DNA depletion and renal accumulation. Once a suitable N(t)RTI was identified, the focus would shift to prodrug development for efficient oral delivery. Although the disoproxil prodrug **2** was effective in the delivery of **1**, it was envisaged that a different class of prodrugs could be developed based on amidate **11** (GS-7340), shown in Figure 10.3.

During the early stages of the program, **11** was dosed in dogs and the $[AUC]_{0-24h}$ levels of **1** inside peripheral blood mononuclear cells (PBMCs) and the plasma compartment evaluated. Compared to the disoproxil prodrug **2**, the exposure ratio inside PBMCs of **1** following oral dosing was improved by 35-fold for an equivalent molar dose.[9] Furthermore, in proof-of-concept Phase 1 clinical studies, dosing of **11** at lower equivalent doses than **2** resulted in an

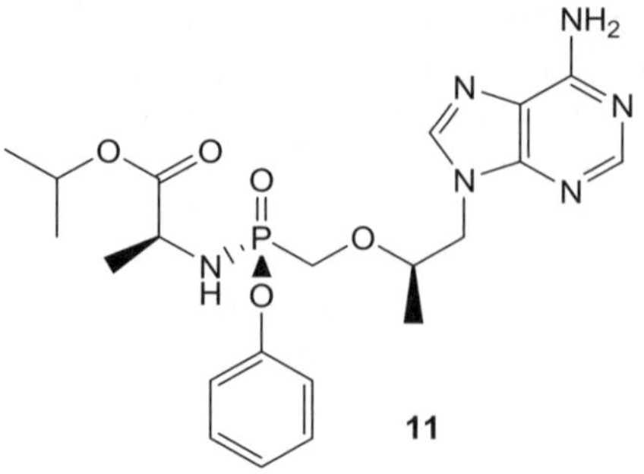

Figure 10.3   Monoamidate prodrug of **1** (GS-7340).

improved antiviral response over 14 days dosing.[10] The precedent established by **11** was of great significance since it was expected that a similar prodrug approach could be applied to a second-generation nucleoside phosphonate, leading to the tantalizing prospect of a new N(t)RTI preferentially targeted to PBMCs.

10.2.2   Design, Evaluation, and Modeling of Novel N(t)RTIs

Lead compounds for the new program were identified by examining the literature for nucleoside phosphonate analogs.[11–17] Initially, a number of new analogs of **1** were made, including modified nucleobases, extending the acyclic two carbon chain between the nucleobase and the oxygen atom, and changing the substituent on the two-carbon chain. Unfortunately, although many analogs were effective at inhibiting HIV in the antiviral cell assay, no new analog demonstrated a significantly improved resistance profile and cell-based toxicity profile compared to **1**. Focus then shifted to the more synthetically demanding cyclic analogs using the reported 2′,3′-dideoxydidehydro (d4) analogs, d4TP (**12**) and d4AP (**15**), as leads.[17] Selected analogs and their antiviral activities are shown in Table 10.1.

Replacing the thymine base of **12** with alternative pyrimidine bases, such as uridine analog **13** or 5-fluorouridine analog **14**, clearly reduced antiviral activity.[18,19] It was important to understand the reason for the reduced cell-based antiviral activity of the pyrimidines **12–14** compared to adenine **15**, since this could be due to (i) differences in cell penetration of the diacids, (ii) differences in intracellular metabolism to the active diphosphates, or (iii) differences in the ability of the active diphosphates to inhibit RT. Therefore, for many analogs the diphosphate metabolites were prepared and their inhibition of RT determined *in vitro*. This strategy provided valuable information concerning the structure–activity relationship (SAR) of the N(t)RTIs and allowed more rational design of target molecules. For example, thymidine **12-DP** had a greater RT potency than adenine **15-DP** despite its > 10-fold reduced antiviral potency, suggesting that **12** lacked antiviral activity due to deficiencies in cell

Table 10.1   Antiviral activity of N(t)RTIs and comparision of RT inhibition by N(t)RTI and NRTI

12. R= Me
13. R = H
14. R = F

15

16

Compound	wt HIV^a MT-2 EC_{50} (μM)	MT-2 cell CC_{50} (μM)	wt RT^b IC_{50} (μM)	Nucleoside^c wt RT IC_{50} (μM)
12	26 (8.8)	>1000	0.39 (0.08)	0.06 (0.03)
13	>200	>1000	5.7	0.55 (0.15)
14	>200	–	–	–
15	2.1 (1.0)	>1000	0.60 (0.16)	0.14 (0.07)
16	12 (2.5)	>1000	4.1 (1.8)	0.26 (0.14)

^{a}Values are results of at least two experiments; standard deviation is given in parentheses.
^{b}Data for the diphosphophosphonate metabolites.
^{c}Data for the equivalent nucleoside triphosphate analog with the same nucleoside core.

penetration and/or metabolism.[18,19] In general, adenine nucleoside phosphonates often show antiviral activity and appear to be more efficient at generating the intracellular diphosphates than the pyrimidines, so further SAR was focused toward adenine analogs. A second example is the saturated dideoxy (dd) analog **16**, which demonstrated reduced antiviral activity but on preparation of the active diphosphate was shown to also have reduced potency toward RT. Active site modeling in HIV RT provided a rationale for the reduced potency since the double bond of the ribose ring in the d4 analogs is positioned above the aryl ring of residue Tyr115, and likely gains binding potency through favorable π–π stacking interactions.[19] Amongst several pairs of dd and d4 N(t)RTI analogs with the same nucleobase, the unsaturated d4 analogs were always the more potent toward RT, a trend that was also consistent in the nucleoside analogs also shown in Table 10.1. These observations benefitted the program by providing greater confidence in the active-site models for the N(t)RTIs, and also the ability to apply information from the extensive nucleoside literature to future target design. A nice illustration of this came during the course of the program when a literature report established that the 4′-ethynyl-modified analog **17**, with the desirable d4 ribose core, had improved antiviral activity compared to the 4′-unsubstituted analog, including K65R hypersensitivity, one of the RT mutants of particular interest in the screening panel (Table 10.2).[20]

   The RT active-site model indicated that a small pocket existed to accommodate the 4′-ethynyl group, leading to additional binding interactions. In

Table 10.2 Antiviral activity, RT inhibition, and resistance profile of 4′-modified carbocyclic N(t)RTIs

Compound	wt HIV[a] MT-2 EC$_{50}$ (μM)	MT-2 cell CC$_{50}$ (μM)	wt RT[b] IC$_{50}$ (μM)	M184V[c] fold resistance	K65R[c] fold resistance
1	3.6 (1.5)	>1000	0.38 (0.20)	0.7 (0.2)	4.3 (1.5)
15	2.1 (1.0)	>1000	0.60 (0.16)	0.9 (0.6)	2.9 (1.1)
17	0.81 (0.4)	–	–	4.8 (0.1)	0.4 (0.2)
18	52 (13)	–	–	–	–
19	16 (9.6)	1450	0.13 (0.10)	>30	2.2 (0.5)
20	244	–	–	–	–

[a]Values are results of at least two experiments; standard deviation is given in parentheses.
[b]Data for the diphosphophosphonate metabolites.
[c]Resistance determined in MT-2 cell lines.

addition, carbocyclic d4 analog **7** and the N(t)RTI **18** are also known to be potent RT inhibitors.[21] Combining these SAR observations resulted in the proposed target **19**, a 4′-ethynyl-modified carbocyclic d4 adenine analog that was expected to have a good K65R resistance profile. The carbocyclic d4 core was employed since it was more synthetically accessible than the d4 ribose core and allowed for additional analogs, such as the 4′-methyl analog **20**, to be readily prepared.[21] The antiviral activity is shown in Table 10.2 and indicates that **19** has greater antiviral activity than **18**, and the profile toward K65R was superior to **1**. Methyl analog **20** was weakly active, suggesting the methyl group was too big for the pocket. Unfortunately, **19** demonstrated a very poor resistance profile toward M184V (>30-fold loss), far inferior to **1** and the cyclic analog **15**. Further active-site modeling, in which the wt residue 184 was mutated to valine, suggested this was probably due to unfavorable steric interactions with residue Y115 in the M184V RT mutant site.[21] Overall, despite much investigation of N(t)RTI SAR, compound **15** remained the most promising lead based on antiviral potency, RT inhibition, and resistance profile, but could not be pursued because of poor selectivity.

10.2.3 Improving Selectivity Toward HIV RT

Analog **15** has been reported to be an inhibitor of the host mitochondrial DNA polymerase γ and this leads to significant toxicity effects (MTC$_{50}$ = 3.6 μM)

through reduction of host mitochondrial DNA synthesis.[22] This raised the question as to whether selectivity toward HIV RT can be improved. Most nucleoside analogs have D-configurations like the natural substrates, but HIV RT can accommodate L-configured analogs as exemplified by **3** and **4**. Typically the L-analogs demonstrate good selectivity for HIV RT compared to mitochondrial polymerases, which suggested that the L-isomer **21** shown in Table 10.3 may be a promising solution to the selectivity issue.[23] A second potential solution also shown in Table 10.3 is fluoro analog **22**, rationally designed based on comparing the active site models of **15-DP** bound in HIV RT and DNA polymerase γ (Figure 10.4).[24,25]

Based on differences in the active site close to the 2′-position of the **15-DP** inhibitor, introduction of a 2′-fluorine atom was proposed to reduce binding to DNA polymerase γ yet retain good binding for HIV RT. Literature reports concerning 2′-β-F substitution onto saturated dd ribose nucleosides also described reduced inhibition of mitochondrial DNA polymerase γ and provided the final impetus to pursue this rationally designed target despite a challenging synthesis.[26,27] Both analogs **21** and **22** (GS-9148) were devoid of mitochondrial cell toxicity up to concentrations of 300 μM (Table 10.3).[28] The 2′-F analog diphosphate metabolite **22-DP** demonstrated weak inhibition of the host mitochondrial polymerases and was only 3- and 5-fold less potent than **15-DP** and **1-DP**, respectively, toward HIV RT. Consistent with this reduced potency toward RT, the antiviral activity of **22** was also marginally weaker than **15** and **1**. The unnatural L-isomer **21** was poorly active toward the M184V and 6TAMs RT mutations, whilst the 2′-F analog **22** was essentially identical to **15**. Overall, analog **22** demonstrated a very promising *in vitro* profile,

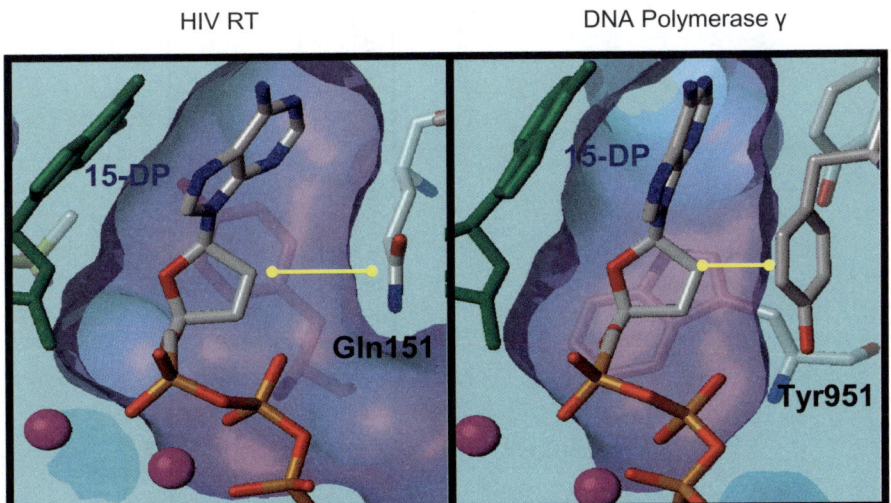

Figure 10.4   Binding models of **15-DP** in the active site of HIV RT and mitochondrial polymerase γ.

Table 10.3 Antiviral activity, mitochondrial toxicity, and resistance profile of novel N(t)RTIs

Cpd	wt RT IC$_{50}$ (μM)	Host DNA Polymerase IC$_{50}$ (μM)			wt HIV[a] EC$_{50}$ (μM)	MTC$_{50}$[b] CC$_{50}$ (μM)	6TAMs[c] fold resistance	M184V[c] fold resistance	K65R[c] fold resistance
		α	β	γ					
1	0.38	61	51	>300	3.6 (1.5)	>300	8.8 (3.7)	0.7 (0.2)	4.3 (1.5)
15	0.60	–	–	–	2.1 (1.0)	3.6 (1.5)	2.9 (1.0)	0.9 (0.6)	2.9 (1.1)
21	–	–	–	–	5.9 (0.3)	>300	14 (2.3)	11 (4.8)	3.6 (2.0
22	1.9	44	>175	>300	10.6 (2.4)	>300	4.3 (0.4)	0.8 (0.1)	1.1 (0.2)

[a]Values are results of at least two experiments; standard deviation is given in parentheses.
[b]Concentration required to reduce mtDNA by 50%.
[c]Resistance determined in MT-2 cell lines.

comparable to the approved N(t)RTI **1** toward wt virus and host polymerases, but with an improved resistance profile.

10.2.4 Clinical Isolate Resistance Profiling

The encouraging resistance profile of **22** justified more extensive profiling toward a panel of N(t)RTI-resistant patient isolates.[29] The Manhattan plot shown in Figure 10.5 compares the fold resistance of **22** and several marketed N(t)RTIs toward a panel of isolates containing the major known RT resistance mutations (M184V, K65R, multiple TAMs, L74V, and others) and their clinically relevant combinations.

Compared to all the N(t)RTIs in clinical use, **22** has an improved resistance profile with <3-fold change in susceptibility to all but one isolate. In the HIV RT model, the inhibitor binds in a similar orientation to dATP. Since **22**-DP is similar in size and shape to the natural substrate, it would be expected that RT mutations in the active site that discriminate against **22**-DP, whilst still retaining efficient incorporation of the natural dATP substrate, may be difficult to obtain. Efficient utilization of the natural substrates is of course necessary for the virus to remain fit and viable. The multi-drug-resistant mutation complex containing Q151M demonstrated the most significant reduction in susceptibility toward **22**. This was slightly worse than **1** but not as poor as most of the other NRTIs shown in Figure 10.5. The 2′-F group of **22** lies in close proximity to the side-chain of residue Q151 of RT shown in Figure 10.4 and provides a rationale for how mutations at residue Q151 could lead to resistance. Subsequent to the completion of this research program, the X-ray structure of

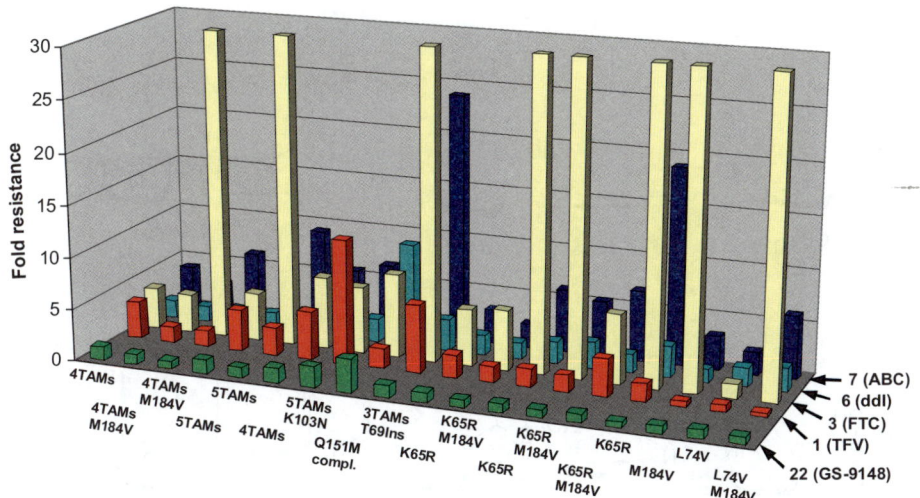

Figure 10.5 Resistance profile of **22** and other N(t)RTI toward patient isolates (Monogram Biosciences).

the diphosphate metabolite of **22** bound in the active site of **HIV RT** was solved and confirmed the binding orientation from the original modeling studies.[30] Overall, the unique and promising resistance profile of **22**, coupled with the potency and selectivity of the inhibitor toward RT, met our stringent target profile criteria and the program moved into the prodrug development phase.

10.2.5 Synthesis

In order to pursue a prodrug strategy on the target compound, an efficient route to multigram quantities was required. Compound **22** is a complex molecule, containing two chiral centers, a unique acetal-aminal configuration, and a fluorine, so the development of a viable stereocontrolled synthesis warrants some mention.[25] The most successful route, shown in Figure 10.6, utilized the readily available fluoro sugar **23**, which avoided the need to perform fluorine chemistry.

Stereochemical control of the nucleobase stereocenter was performed *via* S$_N$2 displacement of the anomeric α-bromide readily formed from **23**. The next key intermediate was glycal **26**, which was prepared through decarboxylative dehydration of acid **25**. Without the 2'-β-F present, the glycal intermediate was sufficiently stable to withstand high-temperature methods for effecting decarboxylation.[17] However, in the presence of 2'-β-F the glycal was prone to aromatization and so an alternative lower temperate method using Mitsunobu conditions had to be developed. The glycal **26** was formed *in situ* and preferentially activated on the opposite face to the purine ring by sterically large activating groups such as iodine or phenylselanyl.[17] This facial activation, in turn, directed nucleophilic attack of the phosphonate alcohol onto the same face as the purine ring to generate **27** with regio- and stereoselective control.

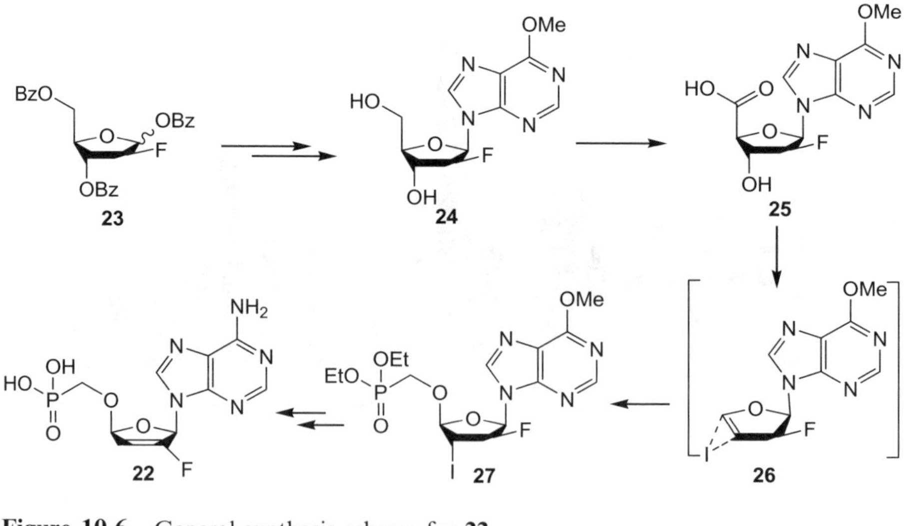

Figure 10.6   General synthesis scheme for **22**.

The final steps included formation of the double bond by oxidation and elimination of iodine, conversion of the purine methoxy group to an amino group, and dealkylation of the phosphonate esters to provide target **22** over eight steps with an overall yield of ~10%. Multigram quantities were prepared using this route, enabling prodrug development and 5-day toxicology.

10.3   Amidate Prodrug Development to Identify GS-9131

10.3.1   Designing Prodrugs that Target Lymphoid Cells

The disoproxil prodrug **2** effectively improved the cell permeability properties of **1** for oral delivery. However, the ideal prodrug not only improves cell permeability but also survives hepatic and plasma enzymes to enable sufficient plasma levels of prodrug to diffuse into lymphoid cells and then be preferentially cleaved within the lymphoid cells. Monoamidate **11** was developed with these goals in mind and succeeded in improving the delivery of **1** to PBMCs compared to prodrug **2**, both in beagle dogs and initial proof-of-concept clinical trials.[9,10] The intracellular cleavage enzyme inside PBMCs was later discovered to be lysosomal cathepsin A (Cat A), a protease highly expressed in PBMCs.[31] The proposed breakdown of the prodrug begins with Cat A-mediated cleavage of the amino acid ester, followed by intramolecular cyclization to **28**, and ejection of phenol as shown in Figure 10.7.

Hydrolysis of the P–O bond of cyclic intermediate **28** followed by chemical or enzymatic cleavage of the P–N bond then releases **1** into the cytosol. The strategy adopted for **22** was to design a range of amidate prodrugs similar to **11**, and profile the prodrugs for cleavage by recombinant Cat A, antiviral activity, and stability toward hepatic, plasma, and intestinal enzymes as shown in Table 10.4.[25] Amidate **11** provided a useful guidepost for prodrug optimization, since it was anticipated that the best prodrugs of **22** would equal or exceed the stability properties of **11**.

Initially, a range of bisamidate prodrugs bearing natural and unnatural amino acids were explored to avoid the generation of a phosphorus chiral

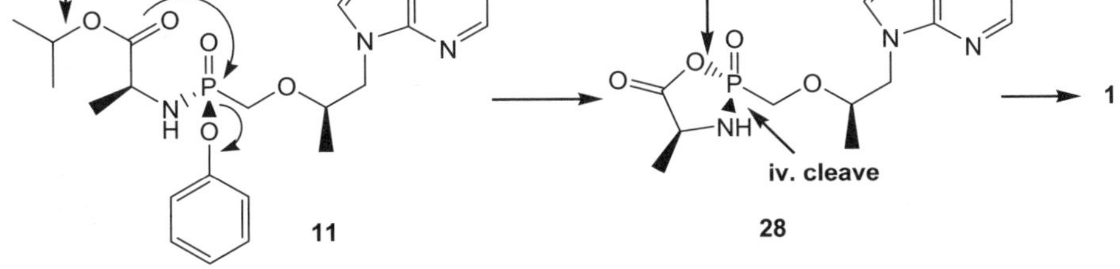

Figure 10.7   Proposed four step (i–iv) breakdown pathway for amidate prodrugs.

Table 10.4 Antiviral activity, Cat A cleavage, and stabilities of disamidate and monoamidate prodrugs of **22**

29 $R_1 = R_2$
30 $R_1 \neq R_2$

Cpd	$R_1{}^c$	$R_2{}^c$	wt HIV[a] EC_{50} (nM)	MT-2 CC_{50} (μM)	clog D^e	Human Cat A^b	Plasma (% remain)[f] Dog	Hum.	Hepatocytes $t_{1/2}$ (min) Dog	Hum.	Intestinal S9 $t_{1/2}$ (min) Dog	Hum.
11	Ala-Pr^i (B)^d	OPh	9.2	139	2.16	35 000	100	—	103	159	15	106
29a	POC	POC	105	—	1.80	—	—	—	—	—	—	—
29b	Phe-Bu^n	Phe-Bu^n	8.4	4.8	6.49	66 680	96	99	10	—	—	—
29c	Phe-Et	Phe-Et	8.9	—	4.45	41 020	95	100	—	—	—	—
29d	Ala-Bu^n	Ala-Bu^n	68	51	3.28	17 355	100	96	48	—	—	—
29e	Ala-Bu^i	Ala-Bu^i	119	>100	2.85	1770	100	100	193	—	—	—

30a	Phe-Bu^i	OCH$_2$CF$_3$	119	>100	4.05	30 107	84	91	27	—	—	—
30b	Phe-Bu^i	OPh	6.0	38	4.86	29 655	84	52	28	—	8.3	—
30c	Phe-Et	OPh	33	25	4.05	39 246	78	74	37	78	—	—
30d	Ala-Bu^n	OPh	92	91	3.47	6183	100	89	63	—	—	—
30e	Ala-Bu^i	OPh	58	>100	3.25	7916	93	81	38	86	—	—
30f	Ala-Pr^n	OPh	52	31	2.96	19 103	100	96	88	—	—	—
30g	Ala-Et	OPh	96	74	2.45	7280	84	88	165	189	146	318
30h	Ala-cBu	OPh	36	57	3.05	18 150	87	84	100	98	25	122
30i	Ala-Pr^i	OPh	728	—	2.98	216	78	97	88	127	—	—

^{a}Values for active compounds are results of at least two experiments in MT-2 cells; standard deviation is given in parentheses. ^{b}Activity is represented as pmol compound cleaved min^{-1} μg^{-1} of enzyme or extract. ^{c}Disoproxil (POC), phenylalanine (Phe), alanine (Ala), n-butyl (Bu^n), ethyl (Et), n-propyl (Pr^n), isobutyl (Bu^i), isopropyl (Pr^i), cyclobutyl (cBu), phenol (OPh). ^{d}A = first eluting isomer; B = second eluting isomer from chiral chromatography. ^{e}clog D was calculated using Pallas 3.0 from CompuDrug International, Sedona, AZ. ^{f}% prodrug remaining after 60 min incubation.

center. Overall, the best prodrugs for Cat A cleavage, **29b–e**, tended to be the ones containing natural amino acids. The more lipophilic phenylalanine bis-amidate prodrugs, **29b** and **29c**, were the best substrates for Cat A, superior even to **11**, and demonstrated potent antiviral activity below 10 nM. The less lipophilic alanine prodrugs **29d** and **29e** were generally poorer Cat A substrates, with glycine being the least efficient. In the monoamidate series of prodrugs, **30a–i**, one of the amino acid groups was replaced with phenol or tri-fluoroethoxy, leading to diastereomeric mixtures. Similar Cat A cleavage trends were observed amongst the linear alkyl and branched alkyl esters, with the more lipophilic phenylalanine prodrugs such as **30a–c** demonstrating good Cat A cleavage rates followed by the less lipophilic alanines **30d–g**. Interestingly, the replacement of phenol by trifluoroethoxy, example **30a**, did not affect the Cat A-mediated cleavage compared to prodrug **30b**, but antiviral activity dropped by almost 20-fold! This result likely reflected differences in the ability to cyclize following Cat A cleavage of the ester group, as proposed in the breakdown scheme described in Figure 10.7. The prodrugs **30h** and **30i** are also distinct in that the ester group on the amino acid is proximally branched. The rate of Cat A turnover of the isopropyl analog **30i** was surprisingly slow when considering this is the same prodrug as **11**, albeit as a diastereomeric mixture. The parent nucleoside phosphonate is clearly an important recognition motif for Cat A in addition to the phosphonate prodrug groups. Constraining the proximal branched ester into a ring, for example cyclobutyl prodrug **30h**, elevated the Cat A activity substantially, leading to good antiviral activity. In general, provided the ester group was linear, branched non-proximally, or proximally branched in a small ring, the natural L-alanine and L-phenylalanine amino acids provided sufficient Cat A cleavage rates to result in antiviral activities below 100 nM, more than two orders of magnitude lower than that of the diacid **22**. A final point of note, but of great significance, is that the Cat A cleavage rates correlated well with the cleavage rates observed in human PBMC extracts, providing evidence that Cat A is the dominant enzyme in the initial breakdown step, at least in the target PBMCs.

10.3.2 Optimization of *In vitro* ADME Properties

In parallel to the substrate SAR for Cat A, the ability of the prodrugs to survive intestinal enzymes, first-pass liver metabolism, and plasma exposure was evaluated (Table 10.4).[25] The prodrugs were profiled in both dog and human hepatocytes, intestinal S9 fraction, and plasma. Rodents were avoided since it is well known that they have high esterase activities compared to humans and would likely provide misleading results *in vivo*. Hepatocytes rather than microsomes were also used to more closely profile *in vivo* properties, where cell permeability and non-oxidative metabolism could have an important impact. All the amidate prodrugs of **22** demonstrated favorable stability in both dog and human plasma, but hepatocyte stability proved to be more challenging (Table 10.4). Although some prodrugs, especially the bisamidates **29b** and **29c**, were exceptionally good substrates for Cat A, they were also unstable when incubated with hepatocytes. This was not unexpected since hepatocytes also contain Cat A, in addition to many other esterases, leading to the conclusion that the optimal prodrug would not necessarily be the best Cat A substrate. Rather, the best prodrug would finely balance lability toward Cat A (and other esterases) to improve intestinal and hepatic stability whilst also allowing efficient PBMC loading. The role of other esterases was suspected because some prodrugs, such as **30i**, were very poor Cat A substrates, yet hepatic stability was the same as the much superior Cat A substrate **30f**. Overall, it was found that the monoamidate alanine prodrugs, especially n-propyl ester **30f** and ethyl ester **30g**, demonstrated the appropriate balance of Cat A lability in combination with hepatic and intestinal stability. The generally improved hepatic stability for the alanine prodrugs over the phenylalanine-based prodrugs was interpreted to be due to their lower lipophilicity. Therefore, a focused effort toward modifying the alanine ester groups was the next logical step in the optimization process, with the aim of maintaining low lipoholicity. A systematic exploration of hindered, proximally branched esters, such as *sec*-butyl, isopropyl, and small cycloalkyl rings, was moderately successful and resulted in the identification of alanine cyclobutyl ester **30h**, which demonstrated both greater hepatic stablility and promising Cat A substrate properties. Indeed, compound **30h** provided a balance of hepatic stability and Cat A lability very close to that of the targeted profile exemplified by compound **11**.

10.3.3 *In vivo* Evaluation of Prodrugs

An evaluation method using intravenous (iv) administration in beagle dogs was developed to identify the best prodrugs for oral studies and also reduced the burden for large quantities of prodrugs during the discovery process.[29] The method involved 30 min iv infusion of prodrug, followed by measuring the plasma prodrug over 24 h, and measuring the PBMC levels of **22** and its metabolites over the same time period. The ratio of **22** in PBMCs ([AUC] over 24 h) compared to plasma prodrug ([AUC] over 24 h) exposure was defined as the loading efficiency of the prodrug. Clearly, high levels of **22** inside PBMCs from relatively low exposures of prodrug would indicate an efficient prodrug for loading the PBMCs. Using this method, amidate **11** demonstrated a loading efficiency of 0.58 and a moderate clearance rate, as shown in Table 10.5.

Both bisamidate **29d** and monoamidate **30d** had high clearance rates, which was expected given their lower hepatic stabilities, but resulted in markedly different loading efficiencies! The difference can be attributed to the Cat A cleavage rates, which in the case of **30d** was sufficiently high to load PBMCs despite the rapid clearance. Thus, the ability to effectively load the PBMCs by iv was reflected by the balance between the rate of Cat A cleavage and clearance of prodrug. Unfortunately upon oral administration, **30d** had only 2.6% *F* due to a significant first-pass effect, and minimal intracellular PBMC levels of **22** and diphosphate **22-DP** were observed. The alanine prodrugs proved to be more effective due to their improved hepatocyte stabilities. Cyclobutyl prodrug

Table 10.5 The iv and oral pharmacokinetics of selected prodrugs and their separated diastereoisomers

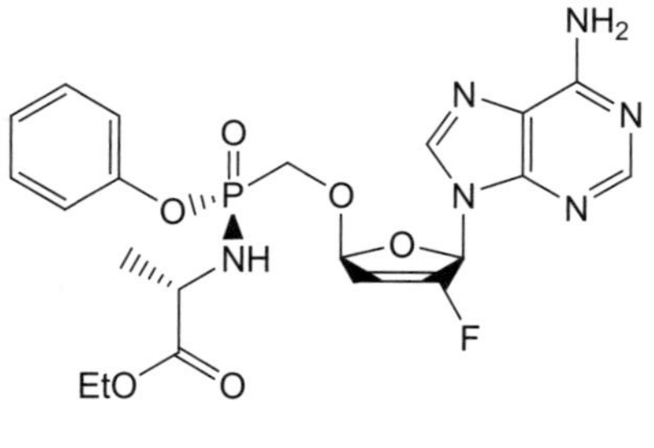

31

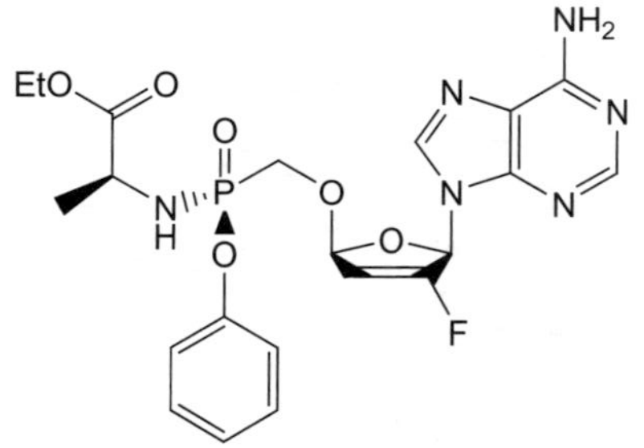

32

Cpd	Human Cat A	IV clearance ($L\,h^{-1}\,kg^{-1}$)	IV PBMC C_{max} 22[a] (μM)	Loading efficiency[b]	Oral plasma prodrug %F	Oral PBMC C_{max} 22-DP (μM)
11	35 000	2.23 (0.25)	–	0.58	21 (3)	–
29d	17 355	3.22 (1.17)	0.94 (0.26)	0.05	–	–
30d	29 655	3.70 (0.30)	22 (7.3)	1.94	2.6 (0.9)	–
30h	18 150	2.76 (0.88)	21 (5.3)	0.62	–	–
30g	7280	1.71 (0.41)	17 (1.5)	0.47	12 (3.5)	7.8 (3.0)
31	645	–	–	–	9.8 (6.8)[c]	5.2 (1.0)
32	12 371	–	–	–	18 (13)[c]	9.0 (2.3)

[a]Quantity of diacid **22** measured after dephosphorylation of samples with calf intestinal phosphatase.
[b]Dose normalized.
[c]Estimated based on iv parameters for isomeric mixture.

30h was cleared at a rate similar to **11**, whilst ethylalanine **30g** had a lower clearance rate, consistent with this prodrug demonstrating high hepatic stability. Although both of these prodrugs had lower Cat A cleavage rates than **11**, their loading efficiencies were comparable to **11** as a result of the more favorable plasma exposures. Initial oral studies on **30g** as a diastereomeric mixture demonstrated modest oral bioavailability for the prodrug (12% *F*) and excellent loading of the PBMCs with the diphosphate **22-DP**. Both the ethyl ester **30g** and cyclobutyl ester **30h** prodrugs were separated using chiral chromatography into their individual diastereoisomers for further oral evaluation. Although **30h** had a higher Cat A cleavage rate and corresponding loading efficiency, the lower hepatic and intestinal stability resulted in a lower oral bioavailability of the seperated prodrugs. Combined with potential toxicity concerns due to the release of cyclobutanol, this prodrug was abandoned in favor of the ethyl ester isomers **31** and **32**. Oral administration of the diastereomerically pure isomers **31** and **32** was undertaken at 3 mg kg^{-1} in beagle dogs, and established that the second eluting isomer **32** was the most promising prodrug isomer, resulting in PBMC levels of the active diphosphate **22-DP** that exceeded 9 µM from a single dose.[29] The full pharmacokinetic profiles of **32** and its parent are shown in Figure 10.8 and established that the prodrug is cleared from circulation within the first 1–2 h and has a half-life of less than 20 min.

Levels of diacid **22** in the plasma stay below 1 µM and peak in the first 2 h, consistent with the elimination of the prodrug. This suggests the cleavage of the prodrug in the liver and other sites is generating the diacid in the plasma. It is important to note that iv dosing of diacid **22** does not lead to significant levels of **22-DP** inside PBMCs. Intracellular PBMC levels of **22** peaked early at ∼10 µM, more than 10-fold higher than the plasma levels. This early C_{max} is consistent with the prodrug clearance from plasma and supports the prodrug being the dominant species for loading the PBMCs. Intracellular metabolism of

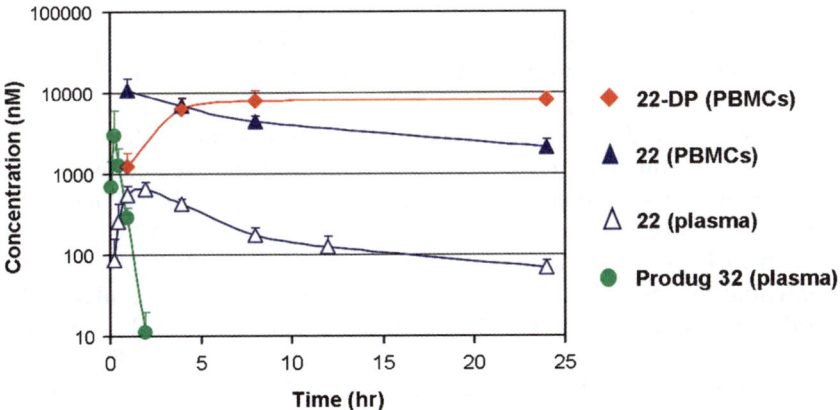

Figure 10.8   Plasma and PBMC pharmacokinetics of **32** delivered orally (3 mg kg^{-1}) in beagle dogs (*n* = 3).

Table 10.6 Activities of **22** and **32** against different subtypes of HIV-1 clinical
isolates in PBMCs.

	EC_{50} $(\mu M)^a$				
Compound	*UG-92-031 subtype A*	*B940374 subtype B*	*LJM subtype B*	*BR-92-025 subtype C*	*UG-92-024 subtype D*
5 (AZT)	0.16	0.31	0.33	0.26	0.13
22 (GS-9148)	12.5	17.6	8.2	5.1	12.6
32 (GS-9131)	0.037	0.030	0.027	0.068	0.023

^{a}Values are results of at least three experiments.

22 to its active diphosphate metabolite **22**-DP peaked at $\sim 5\,h$ with a concentration exceeding $9.0\,\mu M$, a level well above the RT IC_{50} of $1.9\,\mu M$. Furthermore, and very importantly, metabolite **22**-DP persists inside the PBMC for periods in excess of $24\,h$, thereby favoring a potential *qd* dosing regimen for the prodrug fulfilling a critical aspect of the target profile. Based on these data, prodrug **32** was chosen as the most promising candidate for preclinical toxicological evaluation. The prodrug **32** was evaluated against many subtypes of HIV, and the antiviral activity was significantly improved over that of the diacid **22** and more potent on average than **5** (Table 10.6).[29]

The absolute configuration of prodrug isomer **32** was established by crystallography and was the same stereochemistry at phosphorus reported for monoamidate **11**, supporting that Cat A preferentially recognizes a specific phosphorus isomer of this class of prodrugs.[32]

10.3.4   Toxicological Evaluation

Selection of the ideal prodrug **32** was followed by extensive evaluation with respect to potential toxicities and drug–drug interactions that could limit its clinical utility. N(t)RTIs are known to be renally eliminated by a combination of glomerular filtration and active tubular secretion.[33] Furthermore, accumulation of this class of compounds in kidney tissue has been observed in several animal species, and drug-related adverse events as a result of changes in renal proximal tubule function have been found in patients treated with certain nucleotide regimens. Studies carried out *in vitro* established that **22** has reduced efficiency for uptake *via* renal organic anion transporters hOAT1 and hOAT3 compared to **1**, but is still a substrate for the efflux pump MRP4, indicating a low potential for renal accumulation. Consistent with these molecular findings, the oral administration of $[^{14}C]$-**32** to dogs demonstrated limited accumulation of diacid **22** in kidney tissue. Extended 28-day oral dosing studies were performed in Sprague-Dawley rats, beagle dogs, and cynomolgus monkeys and established no significant renal findings.[33] The potential also exists for drug–drug interactions when dosing adenine analog **22** with other adenine-based nucleos(t)ide inhibitors as part of a combination regimen for HIV patients, since both compounds would presumably be dependent on the same intracellular phosphorylation enzymes. However, no antagonism of the intracellular pharmacology was found when combining **22** with **1** at suprapharmacologic concentrations or co-incubation of prodrugs **2** and **32** at 1 µM in activated PBMCs.[34] It should be noted that, under these incubation conditions, the concentration of metabolites of **2** exceed those observed in patient PBMCs by more than 100-fold. This result suggests that the adenosine phosphorylating enzymes have a high capacity and can accommodate co-administration of adenosine analogs, a result that is supported by the additive to slightly synergistic antiviral activity observed when combining **1** and **22**.[29] Additionally, metabolites of **1** have been reported to antagonize the catabolism of ddI (**5**) *via* inhibition of purine nucleoside phosphorylase. Studies have shown that **22** and its metabolites show a low inhibition for purine nucleoside phosphorylase and therefore a low potential to impact the pharmacokinetics of **5**. In summary, the combined data for **22** and its prodrug **32**, regarding interactions with renal transporters and intracellular pharmacokinetics when co-administered with other agents, supported a promising short- and long-term *in vivo* toxicity profile and enabled the compound to be selected for clinical development.

10.4 Conclusion

The discovery program for **22** and its amidate prodrug **32** was ultimately successful. The target profile established at the initiation of the program, although stringent, was met on all counts. The novel N(t)RTI **22** demonstrated good potency toward HIV RT, and the rational design of a 2'-F group enabled selectivity to be significantly improved. The resistance profile of **22** was established from broad-based patient isolate screening to be very favorable compared to all the approved drugs in the N(t)RTI class. Application of amidate prodrugs cleaved by Cat A allowed for **22** to be efficiently targeted to the desired lymphatic cells *in vivo*. In beagle dogs, a single oral dose of prodrug **32** delivered **22** and its active diphosphate metabolite at levels exceeding the RT IC_{50} and the active metabolite levels persisted for >24 h, favoring the potential for a *qd* therapeutic. Prodrug **32** therefore proceeded unabated into clinical development until its development was put on hold in early 2009. Over the course of the program, Atripla™ established itself as the dominant first-line therapy for HIV patients given the increased convenience of the once-daily fixed-dose option. This convenient regimen also led to changes in the patterns of N(t)RTI patient resistance. For example, the clinical manifestation of K65R-resistant strains, that were of concern in the initial years following the approval of **2**, have not materialized. For this, and other reasons, more focus has now been placed on the development of fixed-dose *qd* therapies that contain drugs targeted toward HIV integrase or other new antiviral enzyme targets. It is anticipated that this strategy may provide the next significant leap forward in HIV treatment options. Clinical trails are already ongoing to address these questions and as the results emerge, the suitability and need to develop **32** may well change. The final chapter on **32** may not be written.

References

1. http://www.who.int/hiv/data/2008_global_summary_AIDS_ep.png.
2. E. De Clercq, *Int. J. Antimicrob. Agents*, 2009, **33**, 307–320.
3. E. De Clercq, in *Comprehensive Medicinal Chemistry II*, ed. D. J. Triggle and J. B. Taylor, Elsevier, Oxford, 2006, **vol. 7**, pp. 253–293.
4. D. L. Paterson, S. Swindells, J. Mohr, M. Brester, E. N. Vergis, C. Squier, M. M. Wagener and N. Singh, *Ann. Intern. Med.*, 2000, **133**, 21–30.
5. S. M. Hammer, M. S. Saag, M. Schechter, J. S. G. Montaner, R. T. Schooley, D. M. Jacobsen, M. A. Thompson, C. C. J. Carpenter, M. A. Fischl, B. A. Gazzard, J. M. Gatell, M. S. Hirsch, D. A. Katzenstein, D. D. Richman, S. Vella, P. G. Yeni and P. A. Volberding, *J. Am. Med. Assoc.*, 2006, **296**, 827–843.
6. http://www.clinicaltrials.gov.
7. T. Hawkins, W. A. C. Veikley, R. L. Claire III, B. Guyer, N. Clark and B. Kearney, *JAIDS*, 2005, **39**, 406–411.
8. H. Huang, R. Chopra, G. L. Verdine and S. C. Harrison, *Science*, 1998, **282**, 1669–1675.
9. W. A. Lee, G.-X. He, E. Eisenberg, T. Cihlar, S. Swaminathan, M. Mulato and K. C. Cundy, *Antimicrob. Agents Chemother.*, 2005, **49**, 1898–1906.
10. W. A. Lee, presented at the XVII International Roundtable on Nucleosides, Nucelotides and Nucleic acids, Bern, Switzerland, September 2006, abstr. PL-04.
11. C. U. Kim, J. C. Martin, P. F. Misco and B. Y. Luh, *US Pat.*, 5 688 778, 1997.
12. C. U. Kim, B. Y. Luh, P. F. Misco, J. J. Bronson, M. J. M. Hitchcock, I. Ghazzouli and J. C. Martin, *J. Med. Chem.*, 1990, **33**, 1207–1213.
13. G. A. Freeman, J. L. Rideout, W. H. Miller and J. E. Reardon, *J. Med. Chem.*, 1992, **35**, 3192–3196.
14. D. M. Coe, D. C. Orr, S. M. Roberts and R. Storer, *J. Chem. Soc., Perkin Trans. I*, 1991, 3378–3379.
15. J. J. Bronson, L. M. Ferrara, J. C. Martin and M. M. Mansuri, *Bioorg. Med. Chem. Lett.*, 1992, **2**, 685–690.
16. A. L. Khandazhinskaya, E. A. Shirokova, Y. S. Skoblov, L. S. Victorova, L. Y. Goryunova, R. S. Beabealashvilli, T. R. Pronaeva, N. V. Fedyuk, V. V. Zolin, A. G. Pokrovsky and M. K. Kukhanova, *J. Med. Chem.*, 2002, **45**, 1284–1291.
17. C. U. Kim, B. Y. Luh and J. C. Martin, *J. Org. Chem.*, 1991, **56**, 2642–2647.
18. R. L. Mackman, L. Zhang, V. Prasad, C. G. Boojamra, J. Douglas, D. Grant, D. H. Hui, C. U. Kim, G. Laflamme, J. Parrish, A. D. Stoycheva, S. Swaminathan, K. Wang and T. Cihlar, *Bioorg. Med. Chem.*, 2007, **15**, 5519–5528.
19. R. L. Mackman, C. G. Boojamra, V. Prasad, L. Zhang, K.-Y. Lin, O. Petrakovsky, D. Babusis, J. Chen, J. Douglas, D. Grant, H. C. Hui, C. U. Kim, D. Y. Markevitch, J. Vela, A. Ray and T. Cihlar, *Bioorg. Med. Chem. Lett.*, 2007, **17**, 6785–6789.

20. K. Haraguchi, S. Takeda, H. Tanaka, T. Nitanda, M. Baab, G. E. Dutschman and C.-Y. Cheng, *Bioorg. Med. Chem. Lett.*, 2003, **13**, 3775–3777.

21. C. G. Boojamra, J. P. Parrish, D. Sperandio, Y. Gao, O. V. Petrakovsky, S. Lee, D. Y. Markevitch, J. E. Vela, G. Laflamme, J. M. Chen, A. S. Ray, A. C. Barron, M. L. Sparacino, M. C. Desai, C. U. Kim, T. Cihlar and R. L. Mackman, *Bioorg. Med. Chem. Lett.*, 2009, **17**, 1739–1746.

22. T. Cihlar and M. S. Chen, *Antiviral Chem. Chemother.*, 1997, **8**, 187–195.

23. A. A. Johnson, A. S. Ray, J. Hanes, Z. Suo, J. M. Colacino, K. S. Anderson and K. A. Johnson, *J. Biol. Chem.*, 2001, **276**, 40847–40857.

24. S. Doublié, S. Tabor, A. M. Long, C. C. Richardson and T. Ellenberger, *Nature*, 1998, **391**, 251–258.

25. R. L. Mackman. A. S. Ray, H. C. Hui, L. Zhang, G. Birkus, C. G. Boojamra, M. C. Desai, J. L. Douglas, Y. Gao, D. Grant, G. Laflamme, K.-Y. Lin, D. Y. Markevitch, R. Mishra, M. McDermott, R. Pakdaman, O. V. Petrakovsky, J. E. Vela and T. Cihlar, *Bioorg. Med. Chem.*, 2010, in press.

26. C. H. Tsai, S. L. Doong, D. G. Johns, J. S. Driscoll and Y. C. Cheng, *Biochem. Pharmacol.*, 1994, **48**, 1477–1481.

27. P. Wong-Kai-In, K. E. B. Parkes, D. Kinchington, S. Galpin, A. L. Hope, N. A. Roberts, J. A. Martin, J. H. Merrett, P. Machin and G. Thomas, *Nucleosides Nucleotides Nucleic Acids*, 1991, **10**, 401–404.

28. C. G. Boojamra, R. L. Mackman, D. Y. Markevitch, V. Prasad, A. S. Ray, J. Douglas, D. Grant, C. U. Kim and T. Cihlar, *Bioorg. Med. Chem. Lett.*, 2008, **18**, 1120–1123.

29. T. Cihlar, A. S. Ray, C. G. Boojamra, L. Zhang, H. Hui, G. Laflamme, J. E. Vela, D. Grant, J. Chen, F. Myrick, K. L. White, Y. Gao, K.-Y. Lin, J. L. Douglas, N. T. Parkin, A. Carey, R. Pakdaman and R. L. Mackman, *Antimicrob. Agents Chemother.*, 2008, **52**, 655–665.

30. E. B. Lansdon, D. Samuel, L. Lagpacan, K. L. White, C. G. Boojamra, R. L. Mackman, T. Cihlar, A. S. Ray, M. E. McGrath and S. Swaminathan, presented at the 16th Conference on Retroviruses and Opportunistic Infections, Montreal, February 2009.

31. G. Birkus, R. Wang, X. Liu, N. Kutty, H. MacArthur, T. Cihlar, C. Gibbs, S. Swaminathan, W. Lee and M. McDermott, *Antimicrob. Agents Chemother.*, 2007, **51**, 542–550.

32. H. Chapman, M. Kernan, E. Prisbe, J. Rohloff, M. Sparacino, T. Terhorst and R. Yu, *Nucleosides Nucleotides Nucleic Acids*, 2001, **20**, 621–628.

33. T. Cihlar, G. Laflamme, R. Fisher, A. C. Carey, J. E. Vela, R. Mackman and A. S. Ray, *Antimicrob. Agents Chemother.*, 2009, **53**, 150–156.

34. A. S. Ray, J. E. Vela, C. G. Boojamra, L. Zhang, H. Hui, C. Callebaut, K. Stray, K.-Y. Lin, Y. Gao, R. L. Mackman and T. Cihlar, *Antimicrob. Agents Chemother.*, 2008, **52**, 648–654.

CHAPTER 11

2'-F-2'-C-Methyl Nucleosides and Nucleotides for the Treatment of Hepatitis C Virus: from Discovery to the Clinic

MICHAEL J. SOFIA, PHILLIP A. FURMAN AND
WILLIAM T. SYMONDS

Pharmasset, Inc., 303A College Road East, Princeton, NJ 08540, USA

11.1   Introduction

The hepatitis C virus (HCV) infects approximately 170 million individuals worldwide, and ∼80% of those infected will develop chronic liver infection.[1] Of those chronically infected with HCV, ∼25% will develop liver cirrhosis and approximately 10% will go on to develop hepatocellular carcinoma.[2] The current treatment regimen for HCV-infected patients consists of regular injections of pegylated interferon (IFN) in combination with oral ribavirin (RBV) administration. However, this combination therapy is only effective in producing a sustained virological response in 40–60% of treated patients. Moreover, severe side effects, including flu-like symptoms, fatigue, hemolytic anemia, and depression, result in a high rate of drug discontinuations.[3,4] Consequently, due to the limited efficacy and intolerable side effects of the current standard of care (SOC), the need for the development of effective small-molecule, direct-acting antiviral agents is great.

RSC Drug Discovery Series No. 4
Accounts in Drug Discovery: Case Studies in Medicinal Chemistry
Edited by Joel C. Barrish, Percy H. Carter, Peter T. W. Cheng and Robert Zahler
© Royal Society of Chemistry 2011
Published by the Royal Society of Chemistry, www.rsc.org

HCV is a single-stranded, positive sense RNA virus in the *Flaviviridea* virus family.[5,6] There are six major viral genotypes and over 50 viral subtypes, demonstrating significant genomic sequence diversity.[5,6] The broad genotype diversity results from what is believed to be poor replicative fidelity associated with the HCV RNA-dependent RNA polymerase. In the Western world, genotypes 1a and 1b are the most common genotypes, with genotypes 2 and 3 comprising 20–30% of the total reported cases. In the developing world, genotypes 2–6 predominate. The genotype diversity and poor replicative fidelity associated with HCV has made the identification of a vaccine difficult. Consequently, efforts to identify anti-HCV agents have focused to a large extent on identification of small molecules that act directly on the virus and inhibit its ability to replicate.[6]

The hepatitis C virus is composed of a 9.6 kb plus strand RNA genome that encodes 10 proteins, three structural proteins, and seven non-structural proteins.[7] The non-structural proteins have been the focus of intensive efforts to identify novel and effective anti-HCV agents. One of these non-structural proteins, the NS5B RNA-dependent RNA polymerase (RdRp), is particularly attractive for the development of a direct-acting antiviral agent because it is required for HCV replication.[3] The HCV polymerase is part of a membrane-associated replication complex that includes other viral proteins, viral RNA, and altered cellular membranes. In concert with other non-structural proteins, NS5B is responsible for replicating the viral RNA chain by catalyzing the addition of nucleoside monophosphate building blocks to a growing RNA chain that is a complementary copy of the existing RNA template strand. Since NS5B is known to be devoid of a "proofreading" mechanism, HCV RNA replication is assumed to be a highly error-prone process, resulting in a high mutation rate. This high mutation rate can result in rapid emergence of drug-resistant variants.

The HCV NS5B polymerase has the classic palm–finger–thumb domain motif reminiscent of other viral polymerases.[8] As with the development of inhibitors of other viral polymerases, two approaches have been pursued to identify small-molecule HCV NS5B polymerase inhibitors. The first approach entails the identification of nucleoside derivatives that bind to the active site of the polymerase and get incorporated into the growing RNA chain, thereby inducing a chain termination event. The second approach leverages non-nucleoside inhibitors that function by binding to allosteric sites on the NS5B protein, thus preventing the polymerase from functioning effectively. In our efforts to discover and develop an HCV inhibitor, we chose to pursue the nucleoside strategy. Our choice was based on the long history of nucleosides as the backbone therapy for the treatment of viral diseases, and the demonstrated effectiveness of nucleosides as direct-acting antiviral agents for diseases such as human immunodeficiency virus (HIV), hepatitis B virus (HBV), herpes virus, and other viral diseases. In addition, nucleosides are attractive because of their well-established development path and the wealth of knowledge available pertaining to associated toxicological signals.[9]

Although nucleosides possess many positive attributes, they have their own challenges related to the discovery of therapeutically useful agents. For a nucleoside analog, it is the triphosphate metabolite that is the pharmacologically active agent. The triphosphate is the substrate for the viral polymerase ultimately incorporating the nucleoside monophosphate into the growing RNA chain and effecting a chain termination event (Figure 11.1). Therefore, in order for a nucleoside to be metabolized to the active 5′-triphosphate, it must be a substrate for at least three different kinases and undergo three phosphorylation steps. Consequently, for nucleosides it is difficult to predict if a particular nucleoside will be active against a suspect viral target using standard medicinal chemistry structure–activity relationship (SAR) principles, since any particular nucleoside must be a substrate for at least three kinases and one polymerase, thus potentially demonstrating competing SAR trends.

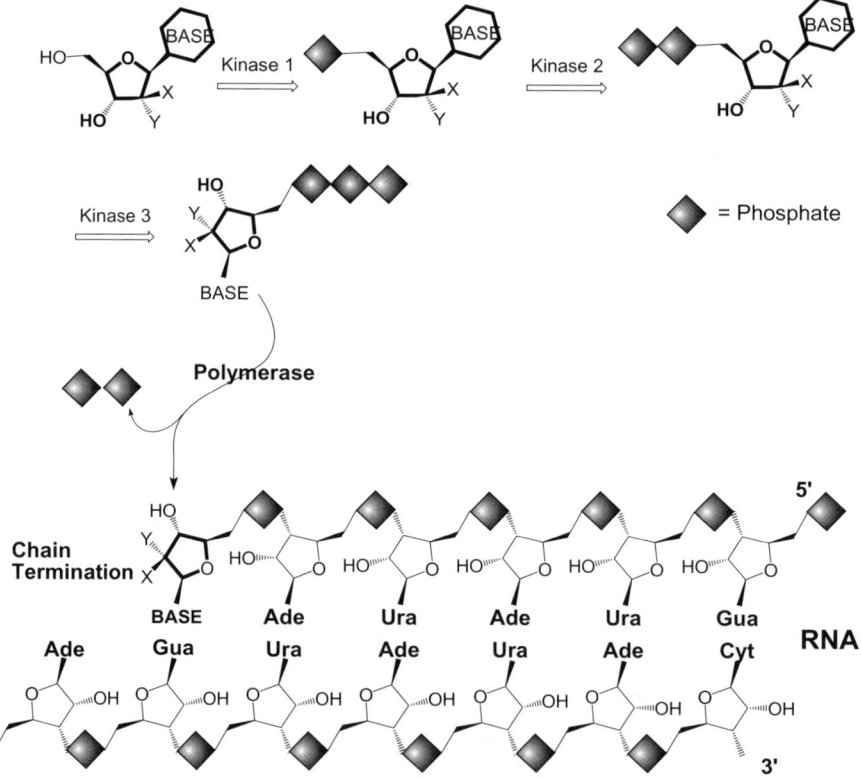

Figure 11.1   In order for a nucleoside to inhibit a viral polymerase, it must first be phosphorylated to its active triphosphate derivative and then be incorporated into the growing RNA chain to affect a chain termination event.

11.2 The Discovery of PSI-6130

11.2.1 PSI-6130: Rationale and Medicinal Chemistry

For the HCV NS5B RNA-dependent RNA polymerase, there was very little information available on ribonucleoside analog SAR that would guide a nucleoside discovery effort. However, it was known that attaching a methyl group to the 2'-β position of a ribonucleoside, as demonstrated by NM107 (**1**), produced an active inhibitor of a related RNA virus, the bovine virus diarrhea virus (BVDV), which was used as a surrogate for HCV.[10] Also, 2'-deoxy-2'-fluorocytidine (**2**) was shown to be a potent HCV inhibitor; however, its therapeutic potential was suspect because it lacked selectivity for the viral target over host cells. 2'-deoxy-2'-fluorocytidine triphosphate showed activity against both RNA and DNA polymerases. With this limited information in hand, the 2'-deoxy-2'-fluoro-2'-C-methylcytidine nucleoside PSI-6130 was prepared and evaluated as an inhibitor of HCV using the subgenomic HCV replicon assay.[11] PSI-6130 showed potent anti-HCV activity in this assay ($EC_{90} = 4.5\,\mu M$). In fact, had the commonly used BVDV surrogate assay been used, the activity of PSI-6130 (**3**) would not have been identified, because PSI-6130 was shown to be highly specific for HCV over other RNA or DNA viruses, and was therefore not active against BVDV.[12]

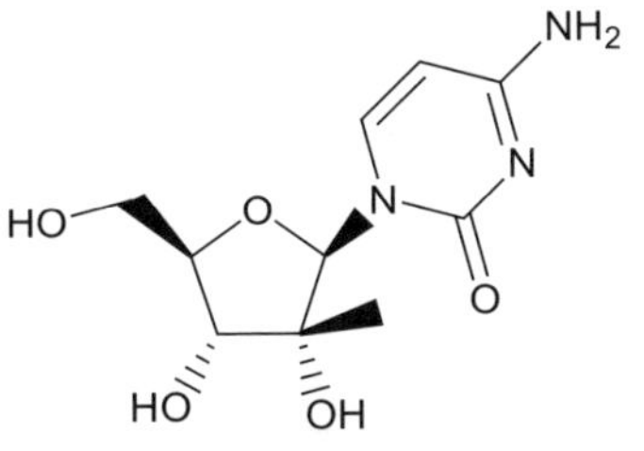

1

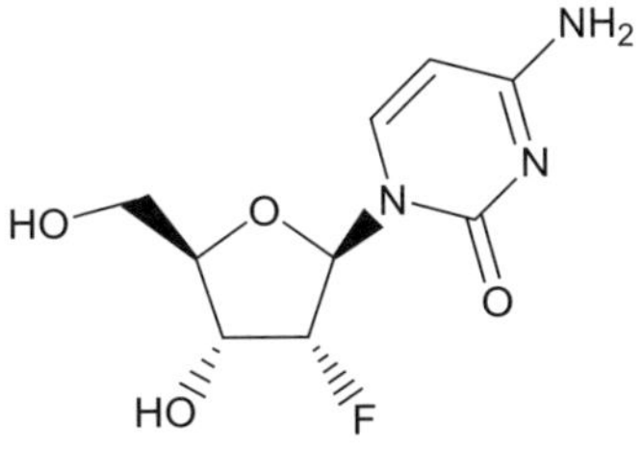

2

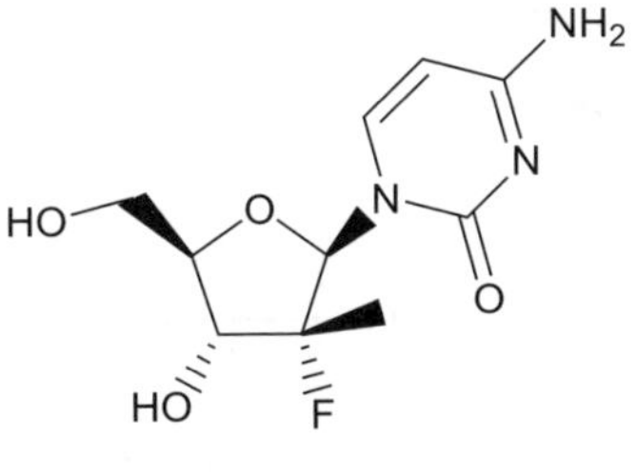

3
PSI-6130

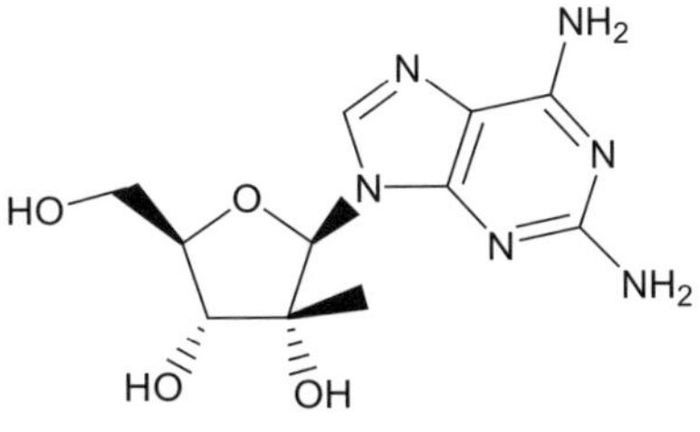

4

In further assessing the 2'-F-2'-C-methyl class of nucleosides, it became quite clear that this nucleoside class exhibited some very unique characteristics and that the combination of a 2'-C-β-methyl and 2'-α-F substitution imparted both potency and safety advantages. As previously mentioned, when PSI-6130 was tested against other members of the *Flaviviridae* family, including BVDV, little or no antiviral activity was detected.[12] PSI-6130 was also shown not to be active against HIV or HBV.[12] This was in contrast to the previously reported 2'-*C*-methylcytidine (**1**) and 2'-*C*-methyladenosine (**4**) analogs, which showed activity against other viruses. The 2'-F-2'-methyl combination also had unanticipated affects on the safety characteristics of PSI-6130. We observed that 2'-deoxycytidine nucleosides mono- or disubstituted with a fluorine atom at the 2'-position showed non-specific inhibitory activity against HCV in the subgenomic replicon assay, and were also shown to be substantially cytotoxic when assessed against a panel of cell lines (Table 11.1).[13] It was also observed that 2'-deoxy-2'-β-methylcytidine (**7**) showed activity against HCV but only at levels at which it was cytotoxic. So, it was surprising when a cytidine nucleoside analog (PSI-6130) that combined a 2'-β-methyl group and a 2'-α-F group exhibited exceptional potency, selectivity, and lack of both cytotoxicity and mitochondrial toxicity.

In efforts to assess the SAR around the 2'-F-2'-*C*-methyl nucleoside class of HCV inhibitors and potentially identify compounds with improved characteristics, we undertook an exhaustive analog development effort. This effort studied substitution variations at each of the positions on the ribose ring, including base modifications (Figure 11.2). One line of investigation maintained the 2'-F-2'-*C*-methyl substitution intact and varied the 3',4'-substitution and also varied the nature of the base. In addition, a carbocyclic analog was prepared in which the ring oxygen of PSI-6130 was replaced with a carbon

Table 11.1   2'-Substitution: HCV activity *vs.* cytotoxicity

Compound	X	Y	HCV activity replicon EC_{90} (μM)	Cytotoxicity (CC_{50}, μM)			
				Clone A	Hep G2	BxPC3	CEM
5	F	F	<1	<0.1	<1	<1	<1
2[a]	H	F	5.66	>100	400	10	6
6	F	H	<1	<50	200	5	5
7	CH₃	H	9.73	10.47	40	<1	<1
3 (PSI-6130)	CH₃	F	4.5	>100	>1000	>1000	>1000

[a]Demonstrates cytostatic effects.

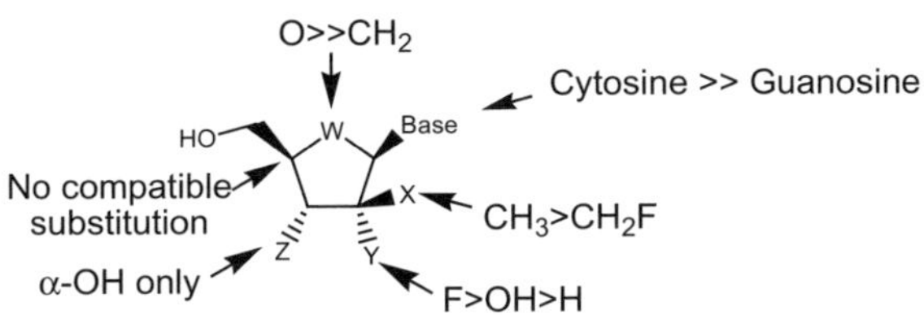

Figure 11.2 SAR overview of anti-HCV activity for the 2'-F-2'-*C*-methyl class of nucleosides is shown, thus demonstrating the unique specificity of this class of nucleosides.

atom. Overall, this study demonstrated that the 3'-α-hydroxyl group was critical for activity, no substitution at 4' was tolerated, the carbocyclic analog was inactive, and the only natural base or base analog that afforded workable activity was the cytosine base. The closest 2'-F-2'-*C*-methyl nucleoside analog that showed any activity was the guanosine analog, for which the activity was ∼ 10-fold less than PSI-6130.[14–16] The second line of investigation maintained the cytosine base and the 3'- and 4'-substitutions as in PSI-6130, but varied the 2'-α- and 2'-β-substitution. Again, PSI-6130 was the most potent analog by far. Although the β-CH$_2$F derivative showed some activity, no other substitution at the 2'-α- or β-positions provided active HCV inhibitors.[14–16] Based on these SAR studies, the 2'-F-2'-*C*-methyl class of nucleosides afforded limited opportunity to identify more effective nucleosides beyond PSI-6130.

11.2.2 Mechanism of Action of PSI-6130

With PSI-6130 identified as the lead development candidate, a complete understanding of the metabolism of this nucleoside was required to determine the mechanism of action. For inhibition of HCV RNA replication to occur, metabolism of the nucleoside to the corresponding active 5'-triphosphate form was required. After incubating either replicon cells or primary human hepatocytes with radiolabeled PSI-6130, the expected 5'-mono-, di-, and triphosphate derivatives of PSI-6130 were identified.[17,18] As shown in Figure 11.3, the deaminated metabolite of PSI-6130, β-D-2'-F-2'-methyluridine, PSI-6206 (**8**), and its corresponding 5'-mono-, di-, and triphosphate derivatives, were identified.[17,18] Phosphorylation of PSI-6130 to its corresponding 5'-monophosphate (PSI-6130-MP) was shown to be catalyzed by human deoxycytidine kinase ($k_{cat}/K_m = 1.9 \times 10^{-4}\,\mu M^{-1}\,s^{-1}$).[18] PSI-6130-MP was subsequently phosphorylated to the corresponding 5'-diphosphate (PSI-6130-DP) and 5'-triphosphate (PSI-6130-TP) by UMP-CMP kinase ($k_{cat}/K_m = 0.012\,\mu M^{-1}\,s^{-1}$) and nucleoside diphosphate kinase ($k_{cat}/K_m = 0.015\,\mu M^{-1}\,s^{-1}$), respectively.[18] To demonstrate that PSI-6130-TP was an inhibitor of the HCV RNA-dependent RNA polymerase (RdRp), it was tested using membrane-associated subcellular fractions containing HCV replication complexes (replicase assay) or

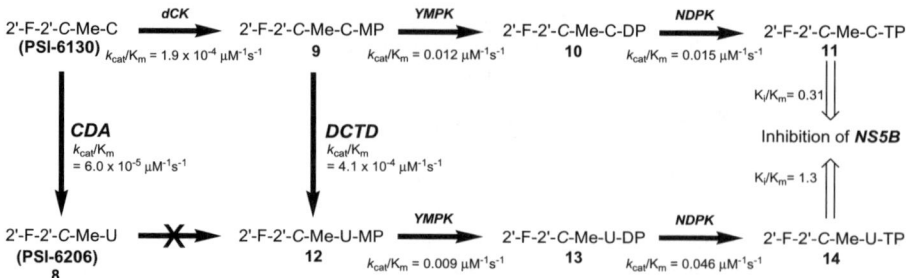

Figure 11.3   PSI-6130 is anabolized to its active triphosphate by the kinases dCK, YMPK, and NDPK. The monophosphate of PSI-6130 is also metabolized to the uridine monophosphate; the latter is further phosphorylated to the uridine triphosphate of PSI-6206, which is also an inhibitor of HCV polymerase. However, the uridine nucleoside, PSI-6206, is not a substrate for cellular nucleoside kinases and therefore is a dead-end metabolite of PSI-6130.

purified recombinant HCV NS5B RdRp. In the replicase assay, PSI-6130-TP gave a mean IC_{50} of 0.34 μM.[17] The inhibition constants (K_i) were determined in two separate studies using purified recombinant RdRp. When the 3′-end of the minus strand of the HCV genome was used as template, a K_i value of 0.023 μM was determined for PSI-6130-TP, and a K_i value of 0.059 μM was obtained when the minus strand of the HCV internal ribosomal entry site was used as template.[17,18] These results confirmed that PSI-6130 was a potent inhibitor of the HCV RdRp, and that it was a good substrate for cellular kinases that were important in metabolizing it to the active triphosphate form (Figure 11.3).

Since metabolism studies also revealed the presence of the uridine congener PSI-6206 (**8**) and its phosphorylated derivatives in both replicon cells and primary human hepatocytes, these compounds were further studied to understand their role in HCV inhibition.[17,19] Inhibition studies with PSI-6206-TP using the replicase assay and purified recombinant RdRp surprisingly demonstrated that PSI-6206-TP was also a potent inhibitor of HCV RNA synthesis.[17,19] This was the first time that a uridine analog was shown to be an inhibitor of HCV polymerase. The mean IC_{50} value for PSI-6206-TP in the replicase assay was 1.19 μM and the K_i values determined in two separate studies were 0.141 μM and 0.42 μM, respectively.[17,19] However, our earlier SAR work showed that the nucleoside PSI-6206 was inactive in the replicon assay. Enzyme studies indicated that the lack of activity in the replicon assay was due to the inability of PSI-6206 to be phosphorylated by cellular nucleoside kinases to PSI-6206-MP.[19] Instead, the formation of PSI-6206-TP was found to involve the deamination of PSI-6130-MP to PSI-6206-MP by deoxycytidylate deaminase ($k_{cat}/K_m = 4.1 \times 10^{-4}\,\mu M^{-1}\,s^{-1}$).[19] Phosphorylation to PSI-6206-DP and PSI-6206-TP was catalyzed by UMP-CMP kinase ($k_{cat}/K_m = 0.0091\,\mu M^{-1}\,s^{-1}$) and nucleoside diphosphate kinase ($k_{cat}/K_m = 0.046\,\mu M^{-1}\,s^{-1}$), respectively

(Figure 11.3).[19] The existence of the uridine metabolite and the discovery of the metabolic pathway for the formation of the uridine triphosphate (an active inhibitor of the HCV polymerase) added a unique dimension to the understanding of how PSI-6130 functioned as an HCV inhibitor. However, it also raised an issue regarding how PSI-6130 would be metabolized in humans and how the extent of formation of either the inactive unphosphorylated uridine metabolite (PSI-6206) or the active triphosphate uridine metabolite (PSI-6206-TP) was going to influence the pharmacokinetics (PK) and efficacy. As described later, the extent of PSI-6206 metabolite formation would become an issue in the clinic.

11.2.3   Resistance Profile of PSI-6130

Because nucleoside analogs that possess the 2′-C modification (**1** and **4**) were known to select for the S282T mutation (which resides in the active site of the HCV RdRp), PSI-6130 was tested for cross resistance using a HCV genotype 1b replicon that contained this mutation.[12] In general, this mutation caused a 20- to 100-fold reduction in sensitivity to 2′-C-methyl-substituted compounds such as NM107 (**1**), and 2′-C-methyladenosine (**4**).[20–22] In contrast, only a 2.4- to 6-fold reduction in activity was observed for PSI-6130 with the S282T mutant replicon.[12,23] Enzyme assays were performed using a RdRp that contained the S282T mutation to assess the ability of PSI-6130-TP and PSI-6206-TP to inhibit this mutant enzyme.[19,23] The presence of the S282T mutation in the RdRp resulted in a 7.5-fold and 23.7-fold reduction in the ability of PSI-6130-TP and PSI-6206-TP to inhibit the enzyme, respectively.[19] Further *in vitro* selection studies with PSI-6130 performed by Ali and colleagues demonstrated that long-term culturing of replicon cells in the presence of compound resulted in the emergence of the S282T mutation in a complex pattern with other amino acid changes.[23] However, the S282T mutation appeared to be the only mutation that provoked reduced sensitivity to PSI-6130. The presence of the S282T mutation led to a 15% decrease in the replication fitness of the replicon. Pairwise combinations with S282T resulted in a further decrease in replication fitness; however, certain multiple mutations combined with S282T resulted in enhanced replication fitness.

The modest impact of the S282T mutation on PSI-6130 activity and the challenge observed in obtaining this mutation was somewhat surprising. Previous mutation studies with 2′-methylcytidine (**1**, NM107) and 2′-methyladenosine (**4**) nucleosides showed that the S282T mutation arose relatively quickly, and that these 2′-C-Me nucleosides were quite sensitive to the mutation. Based on these results, it was speculated that the 2′-methyl group was responsible for the mutational sensitivity; therefore, one would have expected PSI-6130 to show similar sensitivity.[20] The relatively low fold-change in sensitivity to the S282T mutation again demonstrates that the 2′-F-2′-C-methyl combination imparts differentiating characteristics relative to the unfluorinated analog.

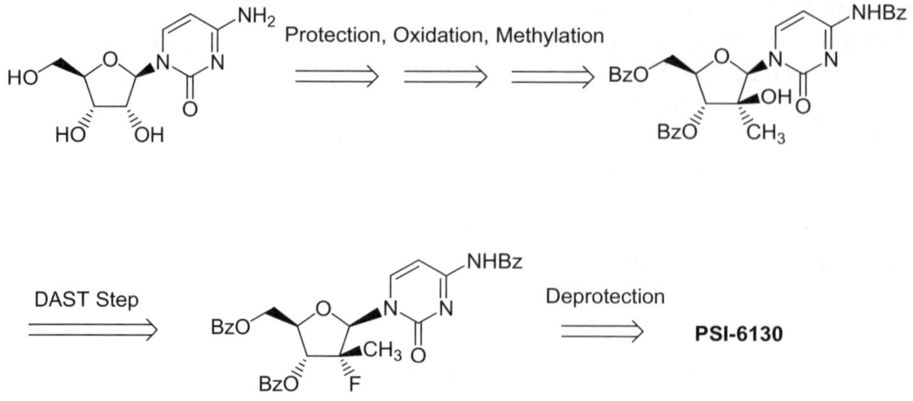

Scheme 11.1

11.2.4   The Synthesis of PSI-6130

With PSI-6130 selected as a preclinical development candidate, a significant chemistry challenge existed in order to progress this nucleoside into development. The challenge involved development of an efficient and scalable asymmetric synthesis of this unusual 2'-α-F-2'-*C*-β-methyl-substituted nucleoside. The original medicinal chemistry route began with cytidine and introduced the 2'-functional array by oxidation of the 2'-hydroxyl group to the ketone, followed by installation of the methyl group using an organometallic methylating reagent (Scheme 11.1).[11] A diethylaminosulfur trifluoride (DAST) fluorination was then used to introduce the 2'-F substituent in the correct 2'-β configuration.[11] However, this route posed some significant problems. The DAST fluorination proceeded in low yield and gave many side products. In addition, the route required many chromatographic steps and gave an overall yield of only 3%. Consequently, a more effective synthetic route was needed.

The synthetic route that was ultimately developed started with a readily available chiral starting material, D-glyceraldehyde, and constructed a chiral lactone to which was attached the cytosine base unit using a stereoselective Verbruggen-like coupling (Scheme 11.2).[24] The key chiral lactone intermediate was constructed by converting the isopropylidene D-glyceraldehyde **14** to a chiral α,β-unsaturated ester. This ester was stereoselectively dihydroxylated to provide the *cis*-diol **15**. The stereoselective introduction of the key fluorine atom was accomplished *via* a nucleophilic fluorination reaction on the cyclic sulfate **16** of the diol **15**. Acid treatment gave the crystalline lactone **18**. The chiral lactone was then reduced to the lactol and acetylated. The *O*-acetate was conjugated with silylated *N*⁴-benzoylcytosine in the presence of tin(IV) chloride, giving a 4:1 β/α anomeric ratio. Homochiral PSI-6130 was obtained after crystallization. This improved route gave an overall yield of

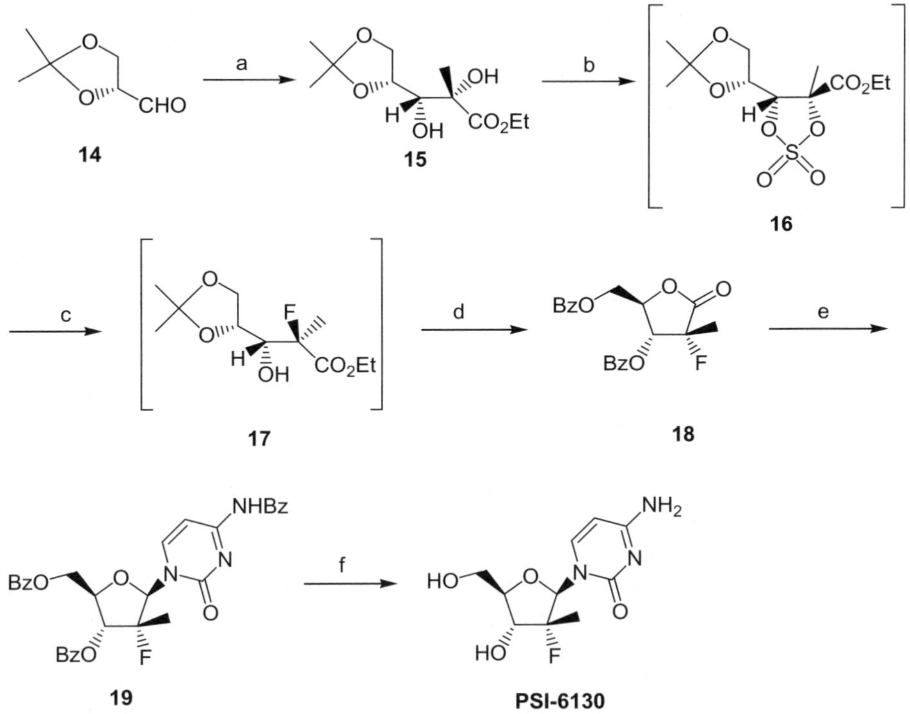

Scheme 11.2 Reagents: (a) Ph₃PC(Me)CO₂Et, DCM, –40 °C; then KMnO₄, acetone, 0 °C; (b) SOCl₂, TEA, DCM, 0 °C; then aq. NaOCl, TEMPO, NaHCO₃, MeCN, 0 °C; (c) TEAF, dioxane, 100 °C; then (MeO)₂CMe₂, conc. HCl, dioxane, RT; (d) EtOH, conc. HCl, RT; then BzCl, pyridine, RT; (e) Li(OBuᵗ)₃AlH, THF, –20 °C; then Ac₂O, DMAP, –20 °C; then silylayed N⁴-Bz-cytosine (prepared from N-benzoylcytosine, NH₄SO₄, hexamethyldisilazane, reflux), SnCl₄, PhCl, RT to 65 °C; (f) NH₃, MeOH, RT.

>20% and was subsequently used to produce multikilogram quantities of PSI-6130.[24]

11.3 The Prodrug RG7128

11.3.1 From PSI-6130 to RG7128

PSI-6130 was progressed into clinical studies after demonstrating an exceptional safety profile in preclinical toxicology assessments. In the Phase I single ascending dosing study to assess safety, tolerability, and PK, PSI-6130 was given orally at doses of 500, 1500, and 3000 mg *QD* as a solution in grape juice. The compound was well tolerated at all doses, and no adverse events were observed. In fact, no maximum tolerated dose was achieved. However, in assessing the human PK, it was determined that PSI-6130 was significantly metabolized to the inactive uridine metabolite, PSI-6206, confirming what was observed in biochemical studies.[17–19] Further complicating the PK profile, PSI-6130 was determined to have an approximate oral bioavailability of 25% in humans. The presence of a significant inactive metabolite in conjunction with the low bioavailability raised questions about the development path for this compound. To address the bioavailability and metabolite issues, a prodrug strategy was proposed. It was hypothesized that a prodrug approach would deliver enhanced exposure of the parent nucleoside and might protect against metabolism to the uridine congener.

To obtain an acceptable prodrug of PSI-6130, the prodrug would have to maintain stability in the gastrointestinal tract, be efficiently absorbed, and release the parent nucleoside once it reached the systemic circulation. A number of prodrug analogs were assessed where prodrug moieties were attached to the 3'- and/or 5'-hydroxyl groups or the N4 amino group of the cytidine base. The prodrug derivatives included esters, carbonates, carbamates at the 3'- and 5'- hydroxyl groups and amides, carbamates, and ureas at the cytidine N4 amino group. The prodrugs were first triaged based on activity in the HCV subgenomic replicon assay, stability characteristics in simulated gastric and intestinal fluids, Caco cell permeability profile, and liver S9 stability. The prodrugs that demonstrated acceptable characteristics in these *in vitro* studies were evaluated *in vivo*. Rat and monkey studies were performed to assess plasma levels of parent nucleoside subsequent to oral administration of the prodrug. Of all the prodrug derivatives investigated, the 3'- and/or 5'-esters and carbonates provided the best profile and, ultimately, the 3',5'-diisobutryrate ester prodrug RG7128 (**20**) was selected for clinical evaluation for the treatment of HCV (Table 11.2).[25]

11.3.2   RG7128 Clinical Studies

RG7128 was initially studied in a single protocol which included three distinct components: a single ascending dose escalation conducted in healthy volunteers, a 14-day multiple ascending dose monotherapy evaluation in HCV-infected patients who had previously failed standard therapy, and a 28-day combination with pegylated IFN (Pegasys®) and RBV (Copegus®) in both treatment-naïve patients infected with HCV genotype 1 and treatment-experienced patients with HCV genotypes 2 or 3. This approach allowed for very rapid early clinical development by realizing the synergies inherent in a single protocol. To date, RG7128 has been generally well tolerated and is in Phase 2b development as of January 2010.

11.3.2.1   Single Ascending Dose Study in Healthy Volunteers

Single oral doses of RG7128 from 500 mg up to 9000 mg were administered to 46 healthy subjects in an escalating fashion, with PK and safety assessments

Table 11.2 Comparison of PSI-6130 to prodrug RG7128

20
(RG7128)

Compound	EC_{90} (μM) Clone A	CC_{50} (μM)	Stability SGF^a (pH 1.2, 37°C) $t_{1/2}$ (h)	Stability SIF^a (pH 7.4, 37°C) $t_{1/2}$ (h)	Caco2 Papp ($\times 10^{-6}$ cm s^{-1})	Rat PK 10 mg kg^{-1} AUC_{0-24} (μg h mL^{-1}) C_{max} (μg mL^{-1})
PSI-6130	3.03	>100	>20	>20	0.21	AUC (parent) = 2.97 C_{max} = 0.6
RG7128	2.5	>100	25	36	6.4	AUC (parent) = 16.17 C_{max} = 1.86

^{a}SGF = simulated gastric fluid; SIF = simulated intestinal fluid.

conducted throughout the study periods. The effect of food was also assessed in this trial in order to provide guidance with regard to food for the latter portions of this study. Few adverse events were reported, with headache being the most frequent, as is commonly observed in healthy volunteer studies. In general, the compound was well tolerated across the dose range with no safety issues identified and no maximum tolerated dose noted. Plasma exposure to RG7128 was negligible, as would be expected with the prodrug. Systemic exposure to PSI-6130 and PSI-6206, the uridine metabolite of PSI-6130, increased with increasing doses of RG7128, with PSI-6206 accounting for 14–19% of the total drug exposure over this dose range. Terminal half-life was approximately 5 h for PSI-6130 and 19 h for PSI-6206. Food increased exposure of PSI-6130 by approximately 20%. Overall, the PK profile indicated good exposure to the active moiety, PSI-6130, and no dose-related adverse events or laboratory abnormalities were observed. Based upon these results, the 14-day multiple-dose study of RG7128 was initiated in HCV-infected genotype 1 subjects who previously had failed IFN therapy.[26]

11.3.2.2   Multiple Ascending Dose Studies in HCV Genotype 1 Infected Subjects

A 14-day monotherapy study was conducted in 40 HCV genotype 1-infected patients who had previously failed an IFN-based therapeutic regimen. Single daily (*QD*) doses of 750 mg or 1500 mg, and twice daily (*BID*) doses of 750 mg or 1500 mg, were tested. All regimens resulted in significant reductions in HCV RNA in a dose-dependent manner. Table 11.3 shows the mean $\log_{10}$ decline in plasma HCV RNA in all RG7128 treatment groups. The decline occurred in a dose-dependent manner and reached nadir values at day 15, suggesting the potential for further viral suppression with continued therapy, and a low potential for escape mutations during early dosing. The decrease in HCV RNA from baseline in the highest cohort, 1500 mg *BID*, ranged from –1.2 to –4.2 $\log_{10}$ IU mL^{-1}. One subject's HCV RNA was suppressed below the limit of detection, 15 IU mL^{-1}. An important conclusion to be drawn from this investigation was the apparent benefit of *BID* dosing over *QD* dosing due to higher

Table 11.3   Mean plasma HCV RNA ($\log_{10}$ copies mL^{-1}) change from baseline (MAD)

	Placebo	*750 mg QD*	*1500 mg QD*	*750 mg BID*	*1500 mg BID*
Number of subjects	8	8	8	9	7
Mean $\log_{10}$ change on day 15	–0.19	–0.90	–1.48	–2.12	–2.67
Range	0.1 to –0.4	–0.3 to –1.4	–0.8 to –2.3	–1.7 to –3.5	–1.2 to –4.2

C_{min} values following *BID* dosing, which appeared to correlate with the observed antiviral activity. By including two cohorts with the same total daily dose of 1500 mg, but with one administered on a *BID* schedule, we were able to determine that the *BID* regimen resulted in an additional 0.6 $\log_{10}$ IU mL^{-1} decrease in HCV RNA.[27]

Sequence analysis of patient samples was performed using both population sequencing and clonal analysis. No mutations were identified that could be associated with a resistant phenotype. In particular, no NS5B S282T variants–which have been previously identified by *in vitro* replicon selection to have reduced susceptibility to PSI-6130 and RG7128 (see above)–were detected.[12,23]

11.3.2.3 28-Day Combination Study with RG7128 Added to Pegasys® and Copegus® in HCV Genotype 1-Infected Treatment-Naïve Subjects

A 28-day study comparing SOC (Pegasys® and Copegus®) versus three dose levels of RG7128 (500, 1000, and 1500 mg, *BID*) added onto SOC was conducted in treatment-naïve, HCV genotype 1 subjects. All doses were generally well-tolerated. Adverse events at all doses of RG7128 were similar to those reported for the SOC alone cohort. During the course of this investigation, the 500 mg and 1500 mg *BID* cohorts were studied initially and then a PK/pharmacodynamic analysis was conducted to characterize the emerging dose–response relationship. Using an E_{max} model with change in HCV RNA from baseline *versus* plasma trough drug concentrations, it appeared that the maximum antiviral effect had been achieved at the 1500 mg *BID* dose and that a modeled 1000 mg *BID* dose would result in a similar level of antiviral activity. Thus, a third cohort utilizing 1000 mg *BID* was conducted to test this premise. The effects on mean HCV RNA levels are shown in Table 11.4. As predicted, the antiviral effects were essentially identical at the two higher doses and substantially less at the 500 mg dose level. Maximum reductions were observed at both the 1000 mg and 1500 mg dose levels, with mean HCV RNA dropping by approximately 5 $\log_{10}$ IU mL^{-1}. More importantly, after 28 days of RG7128 plus SOC, 30%, 88%, and 85% in the 500 mg *BID*, 1000 mg *BID*, and 1500 mg bid groups, respectively, had reached RVR (Rapid Virologic Response),

Table 11.4 Antiviral effects of R7128 after 28 days of therapy

	SOC^a	500 mg BID + SOC	1000 mg BID + SOC	1500 mg BID + SOC
Number of subjects	15	20	25	20
Mean $\log_{10}$ change on day 29	−2.85	−3.86	−5.05	−5.09
Percent subjects with HCV RNA < 15 IU	20	30	88	85

^{a}SOC = Pegasys® + Copegus®.

defined as the proportion of patients with HCV RNA below the limit of detection after 28 days of treatment, versus 20% of the SOC patients.[28,29]

As part of this study, phenotypic and sequence analyses were performed on clinical isolates to monitor for the development of viral resistance *in vivo*. Sequence analysis of the entire NS5B coding region from all baseline samples showed no pre-existing S282T mutation and no common amino acid substitution across the patients treated with RG7128 that could be a predictor of lack of response to treatment. Phenotypic characterization of the NS5B clinical isolates obtained from these patients showed that susceptibility to RG7128 and to IFN was similar between baseline and end-of-treatment samples. These data show the lack of selection of resistance to RG7128 in HCV genotype 1 after four weeks of treatment with RG7128 in combination with SOC.[30]

11.3.2.4 28-Day Combination Study with RG7128 Added to SOC in HCV Genotype 2 or 3 Infected Subjects Who Had Failed Previous Interferon-Based Treatment

Based upon the preclinical data demonstrating similar antiviral potency across multiple viral genotypes and the excellent antiviral activity observed in HCV genotype 1-infected patients, a final cohort was studied to demonstrate the utility of RG7128 for patients infected with HCV genotypes 2 or 3. Thus, a cohort of HCV genotype 2 or 3 subjects who were non-responders to prior IFN-based treatment were recruited to receive a 28-day treatment with SOC alone or 1500 mg *BID* of RG7128 added onto SOC. As was observed in genotype 1 subjects, 1500 mg *BID* plus SOC was generally well-tolerated, with adverse events similar to those commonly reported for IFN and RBV. Following 28 days of therapy, the combination of RG7128 plus SOC resulted in a 5.03 $\log_{10}$ decrease from baseline in plasma HCV RNA, *versus* a 3.26 $\log_{10}$ decrease from baseline in patients who received SOC alone. Higher rates of RVR were observed in the RG7128 1500 mg *BID* plus SOC treatment group *versus* the SOC only group, with 40% (2/5) of SOC patients achieving RVR versus 90% (18/20) in the RG7128 1500 mg *BID* plus SOC cohort.[31]

As with the HCV genotype 1-infected subjects, phenotypic and sequence assessments of resistance selection to RG7128 were performed as part of this combination study in HCV genotypes 2- or 3-infected subjects. Sequence analysis of the entire NS5B coding region from all baseline samples showed no pre-existing S282T mutation and no common amino acid substitution across the patients treated with RG7128 that could be a predictor of lack of response to treatment. Phenotypic characterization of the NS5B clinical isolates obtained from these patients showed that susceptibility to RG7128 and to IFN was similar before and at the end of treatment. These data showed the lack of selection of resistance to RG7128 in HCV genotypes 2 or 3 after four weeks of treatment with RG7128 in combination with SOC.[30] This success of RG7128 against genotype 2 and 3 infection represented the first time that a direct-acting antiviral showed efficacy against non-genotype 1-infected patients, thus translating the broad genotype coverage shown in the laboratory into the clinical setting.

11.3.2.5 INFORM 1

As has been the case in the treatment of HIV infection, the combination of potent direct-acting antivirals (DAAs) targeting different viral enzymes may offer advantages over single DAA strategies, including: enhanced potency, slower emergence of drug resistance, and possible elimination of the need for PEG-IFN $+/-$ ribavirin. The combination of RG7128 and RG7227, an HCV protease inhibitor, offers the potential for a highly potent regimen with a high genetic barrier to resistance. To this end, the results of the INFORM-1 study presented in early 2009 had a significant impact on the outlook for the potential development of an interferon-free combination regimen for the treatment of chronic HCV infection.[32,33] INFORM 1 was a randomized, double-blind, ascending dose Phase I trial. HCV-infected adults (genotype 1) were enrolled to receive up to 14 days of oral combination therapy with RG7128 and RG7227. Study participants were either treatment-naïve or treatment-experienced, including some who had a null response to previous interferon-based therapy (HCV RNA decrease <1 $\log_{10}$ in 4 weeks or <2 $\log_{10}$ in 12 weeks). The first two study groups (A and B) received low-dose monotherapy with either 500 mg RG7128 twice-daily or 100 mg RG7227 three-times-daily on days 1–3. Both cohorts then received a combination of both drugs on days 4–7. Additional cohorts received escalating doses of RG7128 (500 or 1000 mg twice-daily) plus RG7227 (600 or 900 mg twice-daily, or 100 or 200 mg three-times-daily) for 14 days (Table 11.5). Safety, viral kinetics, resistance, and pharmacokinetics of RG7128/RG7227 were evaluated in all cohorts. All dosing groups completed the study with no treatment-related significant adverse events, dose modifications, or discontinuations. Pharmacokinetic analysis in Groups A and B confirmed no drug/drug interaction between RG7128 and RG7227. The HCV RNA change from baseline for all dose groups is presented in Table 11.5.

The orally administered combination of RG7128 and RG7227 provided significant antiviral potency in both treatment-naïve and treatment-experienced patients, suppressed viral rebound, and appeared safe and well-tolerated for 14 days. These results demonstrated that a HCV protease and nucleoside polymerase inhibitor can be combined *in vivo*, and are a promising combination for future treatment.[32,33]

11.4 PSI-7851: a Nucleotide Prodrug of the 2'-F-2'-C-Methyl Class of Nucleosides

11.4.1 PSI-7851: Rationale and Medicinal Chemistry

RG7128 clearly demonstrated that a nucleoside has the potential to become an important addition to anti-HCV therapy. In addition, with its unique overall

Table 11.5 INFORM-1 antiviral activity for cohorts B–G

Regimen (RG7128 mg/ RG7227 mg)	Median baseline HCV RNA (log_{10} IU mL^{-1})	N	Patient population	HCV RNA median change from baseline (log_{10} IU mL^{-1}) (range)	HCV RNA <LLOQ^a (<43 IU mL^{-1}) N (%)	HCV RNA <LLOD^b (<15 IU mL^{-1}) N (%)
B 500 BID/100 TID^c	6.5	8	naive	−3.9 (−5.0 to −2.9)	1/8 (13)	1/8 (13)
C$_1$ 500 BID/200 TID	6.9	8	naive	−5.2 (−5.5 to −3.1)	5/8 (63)	2/8 (25)
C$_2$ 1000 BID/100 TID	6.4	7	naive	−4.8 (−5.7 to −4.5)	5/7 (71)	2/7 (29)
D 1000 BID/200 TID	6.3	8	naive	−4.8 (−5.5 to −2.7)	5/8 (63)	2/8 (25)
E 1000 BID/600 BID	6.2	8	TF^d (non-null)	−4.0 (−6.0 to −2.5)	4/8 (50)	1/8 (13)
F 1000 BID/900 BID	6.5	8	TF (null)	−4.9 (−5.3 to −3.5)	4/8 (50)	2/8 (25)
G 1000 BID/900 BID	6.5	8	naive	−5.1 (−5.9 to -3.0)	7/8 (88)	5/8 (63)

^{a}LLOQ = lower limit of quantification by Roche TaqMan Assay (<43 IU mL^{-1}).
^{b}LLOD = lower limit of detection by Roche TaqMan Assay (<15 IU mL^{-1}).
^{c}TID = three times daily.
^{d}TF = treatment failures.

characteristics, RG7128 showed that nucleosides are positioned to become the cornerstone of any HCV combination therapy, whether they are added to SOC or combined with other direct-acting antiviral agents. The unique efficacy and safety profile, demonstrated high barrier to resistance, and broad genotype coverage clearly positioned the 2'-F-2'-C-methyl class of nucleosides as a platform to further extend the nucleoside strategy and provide new compounds with improved overall profiles. Consequently, even with the clinical success of RG7128, we were interested in developing a second-generation compound that improved on several of the perceived limitations of RG7128. This desire was driven by the belief that combinations of direct-acting antiviral agents would be dictating therapeutic regimens in the future, thus potentially displacing the need for IFN and RBV. To optimally position a nucleoside to take advantage of this future paradigm, a compound which demonstrated *QD* dosing and a low pill burden was desired. Since RG7128 required a 1000 mg *BID* dose to achieve optimal viral load reduction, investigation of second-generation compounds was undertaken.

To achieve our objective of identifying a compound that reduced the pill burden and provided improved PK, we wanted to identify a nucleos(t)ide that had increased potency over PSI-6130 and provided a high liver/plasma ratio. The ability to target the liver would reduce the circulating level of nucleos(t)ide and therefore reduce systemic exposure. At the same time, we wanted to leverage the exceptional characteristics we had observed with the 2'-F-2'-C-methyl class of nucleosides. To achieve this goal, we decided to take advantage of what was known about the metabolism of PSI-6130. As detailed earlier in this chapter (Figure 11.3), PSI-6130-MP is metabolized to PSI-6206-MP, the uridine 5'-monophosphate metabolite, which is subsequently metabolized to the triphosphate of PSI-6206, a potent inhibitor of the HCV RdRp. In addition, this uridine triphosphate has a significantly long half-life (38 h) in hepatocytes. The problem with PSI-6206 is its inability to be phosphorylated to the monophosphate precursor of the active triphosphate. Therefore, we speculated that if we could bypass the nonproductive monophosphorylation step and deliver a monophosphate to the cell, we could achieve sustained inhibition of HCV because of the long half-life of the triphosphate. In addition, this long intracellular half-life might achieve the desired effect of less frequent dosing *in vivo*. However, the challenge was to find a way to deliver the monophosphate intracellularly such that subsequent phosphorylation steps could proceed.

To accomplish the liver targeting aspect of our strategy, we wanted to take advantage of first-pass metabolism. Since the liver would be the first organ to which the compound would be exposed after absorption, we wanted to identify a nucleoside monophosphate prodrug in which the monophosphate of the parent nucleoside would be revealed upon exposure to liver enzymes. The phosphoramidate nucleotide prodrug approach described by McGuigan appeared to have the attributes we desired.[34-36] We speculated that liver esterases would cleave the terminal carboxylate ester to the acid, thereby triggering the subsequent chain of decomposition events that would ultimately reveal the desired nucleoside monophosphate (Figure 11.4). However, although

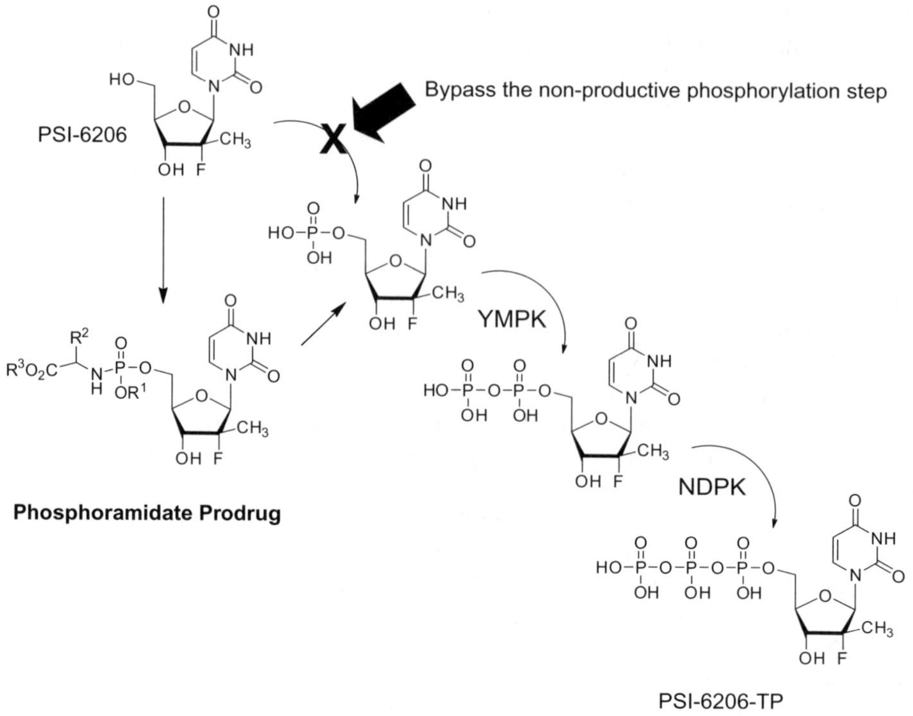

Figure 11.4   To produce the active triphosphate metabolite of PSI-6206, a phosphoramidate prodrug approach can be used to bypass the phosphorylation step where the kinase does not accept PSI-6206 as a substrate.

the phosphoramidate prodrug strategy had been known for a number of years and was demonstrated to work *in vitro*, it had never been demonstrated to be effective *in vivo*; presumably, the prodrug moiety decomposed before the intact prodrug could get to the target cell type or organ. Nevertheless, it seemed to us that applying this prodrug strategy to treat HCV would be an ideal application of this methodology because the liver would be the first organ to which the prodrug would be exposed after absorption. Consequently, we needed to identify a phosphoramidate prodrug of PSI-6206 that could survive exposure to the gastrointestinal tract and subsequently undergo the desired decomposition pathway to reveal the desired uridine monophosphate in hepatocytes.

   With our strategy and approach formulated, we prepared an extensive series of phosphoramidate prodrugs of the 2′-F-2′-*C*-methyluridine nucleoside PSI-6206. Our SAR strategy focused on varying the nature of the carboxylate ester, phosphate ester, and amino acid side-chain moieties of the phosphoramidate functionality.[37–39] To assess which of these prodrug candidates possessed the appropriate properties to progress into *in vivo* studies, we established an *in vitro* testing paradigm that would assess potency, *in vitro* safety, and stability characteristics   when   exposed   to   environments   that   simulated   the gastrointestinal tract, blood (or circulation), and the liver (Figure 11.5). Our desire was to identify a prodrug that was stable in simulated gastric fluid, simulated intestinal fluid, and blood, but that would decompose readily to the monophosphate derivative when exposed to liver S9 fractions. At the same time, we evaluated the safety profile of our compounds by assessing cytotoxicity against a panel of cell lines and by assessing mitochondrial toxicity.[39] This initial assessment phase led to a selection of a set of compounds that we progressed into PK evaluation in rats, where we looked not only for circulating levels of prodrug and key metabolites, but also at liver levels of prodrug and key metabolites (particularly the pharmacologically active PSI-6202 triphosphate metabolite).[40] Simultaneously, we evaluated the potential for these prodrugs to generate the active PSI-6206 triphosphate metabolite in isolated primary rat, human, dog, and monkey hepatocytes.

After the initial screening phase, a selected set of compounds met the criteria of improved potency over PSI-6130 in the HCV replicon assay, lack of cytotoxicity in a panel of different cell lines, lack of mitochondrial toxicity, adequate gastrointestinal stability, and a half-life when incubated with a liver S9 fraction that supported rapid conversion in the liver by first-pass metabolism (see Figure 11.6). Rats were dosed orally with each of these compounds to compare their metabolic fate and to assess if they generated appropriate liver levels of the desired triphosphate-active metabolite. In fact, each compound demonstrated high levels of the triphosphate in rat livers, thus supporting the concept that a phosphoramidate prodrug of PSI-6206 could be delivered effectively to the liver

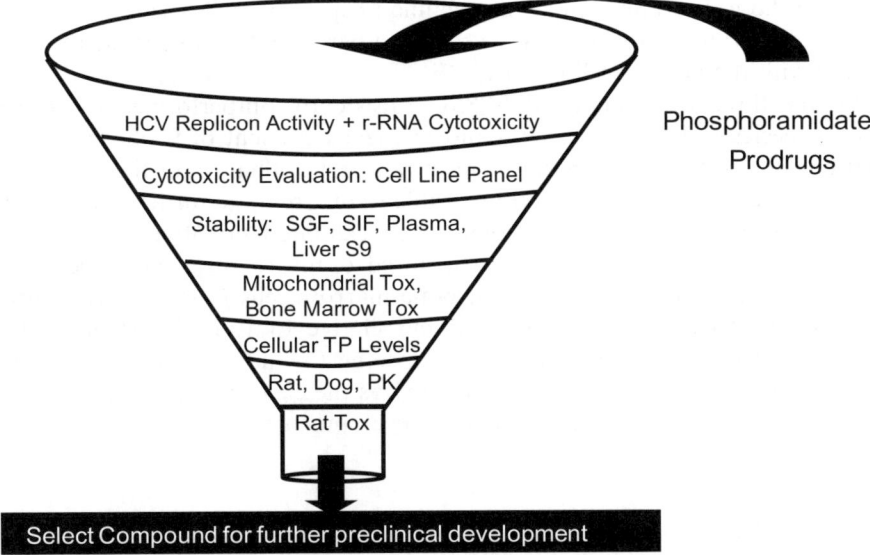

Figure 11.5 To select a preclinical development candidate, the phosphoramidate prodrugs of PSI-6206 were progressed through a screening process assessing potency, *in vitro* and *in vivo* toxicity, and PK characteristics.

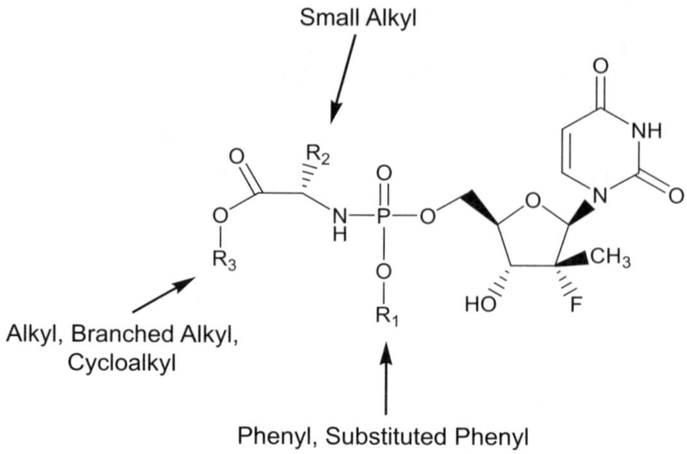

Small Alkyl

R₂

Alkyl, Branched Alkyl,
Cycloalkyl

Phenyl, Substituted Phenyl

Figure 11.6   SAR overview of HCV replicon activity for the PSI-6206 5′-phosphor-
amidate derivatives is shown.

and produce the desired pharmacological effect. Three compounds, PSI-7672, PSI-7851, and PSI-8118, were chosen for progression into additional studies that would result in selection of a preclinical development candidate (see Figure 11.7). We were particularly interested in further evaluating the PK characteristics of these compounds in dog and monkey in order to determine which compound behaved best across species. In addition, we wanted to assess *in vitro* triphosphate levels in primary hepatocytes of rat, dog, and monkey, and compare these levels to those determined in primary human hepatocytes. These *in vitro* and *in vivo* studies would provide the necessary comparative data to choose the compound with the best overall profile.

Of the three compounds, PSI-7851 consistently outperformed the others across species in both the *in vitro* and *in vivo* assessment of triphosphate formation. *In vivo* toxicology assessment in rats and evaluation of their ability to induce bone marrow toxicity were not able to differentiate further between the three candidates. Each of the compounds exhibited no observable adverse effects levels of $>1800\,\mathrm{mg\,kg^{-1}}$ when dosed orally to rats, and none of the compounds demonstrated effects on bone marrow cells *in vitro* at concentrations up to $50\,\mu\mathrm{M}$. Its superior efficacy relative to RG7128, strong safety profile, and demonstrated superior PK profile (including formation of high levels of triphosphate) positioned PSI-7851 as the clear winner. Consequently, PSI-7851 was selected as the preclinical development candidate.

11.4.2   PSI-7851 Metabolic Characterization

In deciding to take PSI-7851 into development, a question remained on whether we would get therapeutic levels of the active PSI-6206-TP in human liver.

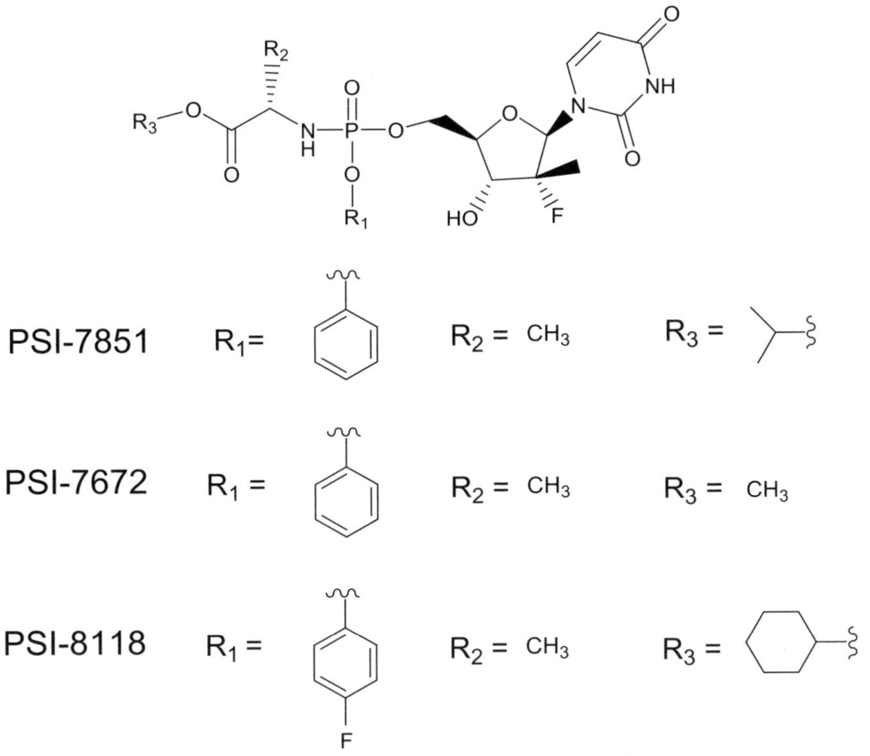

Figure 11.7 The three lead PSI-6206 phosphoramidate derivatives, PSI-7851, 7976, and 8118, are shown. They provided the best overall profile after lead optimization assessment.

We had *in vivo* animal data to support our hypothesis, but similar experiments in humans to assess exposure were not practical. So we needed to determine the potential for PSI-7851 to generate PSI-6206-TP in human liver cells. Although the mechanism by which phosphoramidate prodrugs are metabolized to the active triphosphate has been reported, to our knowledge no one has demonstrated that a phosphoramidate nucleotide prodrug could be converted to a nucleoside triphosphate in human liver.[34–36] We speculated that the first step in the activation pathway of PSI-7851 involved the hydrolysis of the carboxyester moiety and that this hydrolytic step was catalyzed by human cathepsin A and carboxyesterase 1 (Figure 11.8). Hydrolysis of the ester was then likely followed by a putative nucleophilic attack of the phosphorus by the carboxyl group, resulting in the spontaneous elimination of phenol and producing an intermediate diacid metabolite. We were able to substantiate the above mechanistic hypothesis both by demonstrating through Western blot analysis that significant levels of both cathepsin A and carboxyesterase 1 were expressed in primary human hepatocytes, and through identifying the intermediate diacid metabolite as PSI-352707 (Figure 11.8).

This diacid metabolite, PSI-352707, was deaminated to the nucleoside monophosphate by human histidine triad nucleotide binding protein 1. Subsequent phosphorylation to the corresponding di- and triphosphate metabolites was catalyzed by UMP-CMP kinase 1 and nucleoside diphosphate kinase,

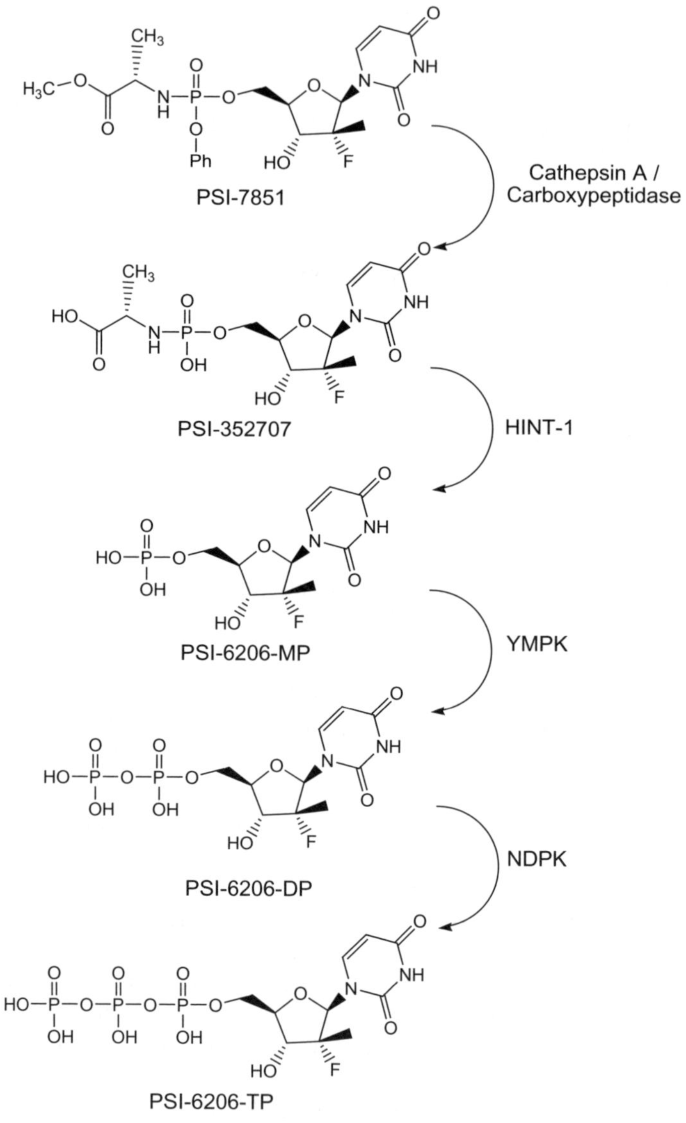

Figure 11.8 The nucleoside phosphoramidate PSI-7851 is metabolized to the monophosphate of PSI-6206 by the action of the enzymes cathepsin A/carboxypeptidase and HINT-1. This monophosphate is then anabolized to the active triphosphate by the kinases YMPK and NDPK.

respectively (Figure 11.8). Cell-based metabolism studies showed that high levels of PSI-6206-TP were formed when replicon cells and primary human hepatocytes were incubated with [³H]-PSI-7851. In primary human hepatocytes incubated with 5 μM [³H]-PSI-7851, the level of PSI-6206-TP reached an intracellular concentration of approximately 100 μM. Studies to determine the intracellular half-life of PSI-6206-TP revealed that after reaching steady-state levels, the concentration of PSI-6206-TP decreased with a mean half-life of 38.1 ± 16.1 h. This exceptionally high level of intracellular triphosphate formation gave us confidence that if PSI-7851 could make it to the liver we should see *in vivo* efficacy in humans. In fact, the levels of triphosphate metabolite generated by delivering PSI-7851 were much higher than what was observed with PSI-6130/RG7128. Since RG7128 shows exceptional clinical efficacy, we expected to see efficacy in humans with PSI-7851 if it could reach the liver.

11.4.3 PSI-7851 Clinical Characterization and Studies

As noted above, PSI-7851 has a number of favorable preclinical characteristics which made it an attractive development candidate for the treatment of chronic HCV infection. We speculated that the long intracellular half-life of PSI-6206-TP and the demonstrated improvements in *in vitro* antiviral potency, coupled with the improved liver/plasma ratio observed in preclinical studies, would translate into a potent antiviral which would potentially be administered *QD* at a lower daily dose than RG7128. As of January 2010, PSI-7851 has been evaluated following single ascending doses in healthy volunteers and in a three-day monotherapy study in patients infected with genotype 1 HCV.

11.4.3.1 Single Ascending Dose Study in Healthy Volunteers

Single oral doses of PSI-7851 or matching placebo were administered to 42 healthy subjects enrolled in four alternating cohorts to receive either placebo or PSI-7851 at doses of 25, 50, 100, 200, 400, and 800 mg. A solution formulation at a dose of 50 mg was also compared to the 50 mg capsule dose. Safety assessments were conducted throughout the study and plasma exposure to PSI-7851 and its metabolites (PSI-352707 and the uridine nucleoside metabolite, PSI-6206) were evaluated. Single ascending doses of PSI-7851 were generally safe and well-tolerated in this study, with no dose-limiting toxicities or a maximum tolerated dose identified. The PK results demonstrated a systemic exposure profile consistent with rapid uptake of the drug by the liver with low plasma exposure to PSI-7851 and PSI-352707, and relatively higher plasma PSI-6206 exposure. The 50 mg solution dose resulted in plasma exposures 15–16% higher than the capsule formulation.

11.4.3.2   Multiple Ascending Dose Study Over Three Days in Patients Infected with Genotype 1 HCV

PSI-7851 was evaluated following multiple oral doses of 50, 100, 200, and 400 mg or placebo administered *QD* for three days to 40 treatment-naïve, HCV-infected patients. Extensive safety, PK, and virology assessments were conducted throughout the study in order to guide the dose escalations and to assess the acute antiviral activity of PSI-7851. In this study, PSI-7851 was generally well-tolerated, with no dose-limiting toxicities identified. Plasma PK exposure to the prodrug, PSI-7851, was negligible, as observed in the preclinical studies. The nucleoside metabolite (PSI-6206) exhibited low but consistent plasma exposure, which increased with dose, a profile consistent with a liver-targeted prodrug. Following three days of treatment with PSI-7851, HCV RNA declined in a dose-dependent manner, with mean changes from baseline of − 0.49, −0.61, −1.01, and −1.95 $\log_{10}$ IU mL^{-1} in the 50, 100, 200, and 400 mg *QD* cohorts, respectively. No substantial decline in HCV RNA was observed in the placebo group.[41] This level of antiviral activity with PSI-7851 exceeded that estimated at day 3 from the RG7128 monotherapy trial of −1.07 $\log_{10}$ IU mL following 1500 mg *BID*, which went on to achieve an 85% RVR in combination with IFN and RBV.[27,42] In addition, there was no pre-existing or treatment-emergent S282T mutations detected, nor was there evidence of viral resistance following three days of monotherapy based upon population sequencing of the NS5B region. These results support the continued development of PSI-7851 for the treatment of chronic HCV infection in combination with IFN and RBV, or as a component of small-molecule combination regimens.

11.5   Summary

The current SOC has provided limited and sometimes intolerable options for those suffering from HCV infection. The drive to find new medicines to treat patients with HCV has focused to a large extent on identifying small-molecule direct-acting antiviral agents. With the HCV NS5B RdRp recognized as an attractive target for drug therapy, the opportunity to develop nucleoside-based agents became a viable strategy. The 2′-F-2′-C-methyl nucleoside class represented by PSI-6130 was discovered to have a unique efficacy, safety, selectivity, resistance, and genotype profile not seen with other small-molecule HCV antivirals. The subsequent development of the PSI-6130 prodrug, RG7128, demonstrated clinically that a nucleoside for HCV could provide exceptional clinical efficacy; and, because of its broad genotype coverage and compatibility with other small-molecule HCV agents, RG7128 showed that a nucleoside could be combined with other small-molecule anti-HCV drugs to serve as the cornerstone of future HCV combination therapies. The discovery of PSI-7851 has taken the development of novel HCV nucleos(t)ide therapies to the next level by providing a method to more efficiently deliver the active triphosphate to the liver, and thus further extending the potential of the 2′-F-2′-C-methyl class of nucleoside anti-HCV agents.

Acknowledgements

We would like to acknowledge all those whose contributions from 1999 to 2009 were important to the success of these programs.

Pharmasset (PSI-6130, RG7128, PSI-7851): Albanis, E., Bansal. S., Bao, D., Bao, H., Beard, A., Bennett, M., Berrey, M., Biroc, S., Cartee, L., Chang, W., Chun, B.-K., Clark, J., Cleary, D., Denning, J., DeSilva, L., Du, J., Espiritu, C., Grier, J., Hobbs, A., Hollecker, L., Julander, J., Keilman, M., Kupsch, L., Lam, A., Li, D., Liu, J., Lostia, S., Mason, J., McBrayer, T., Morrey, J., Murakami, E., Nachman, T., Nagarathnam, D., Niu, C., Otto, M., Pankie-wicz, K., Patterson, S., Rachakonda, S., Ramesh, M., Reddy, G., Shi, J., Stec, W., Steuer, H., Stuyver, L., Tharnish, P., Tolstykh, T., Wang, P., Watanabe, K., Whitaker, T., Xie, M.-Y., Zhang, H.-R.

Roche (RG7128): Alfredson, T., Blue, D., Brandl, M., Cammack, N., Chow, J., Dei Rossi, D., Gordon, D., Hegde, S., Hill, G., Holper, M., Ipe, D., Jiang, W.-R., Lakatos, I., Larrabee, S., LePogam, S., Lin, F., Ma, H., Najera, I., Olocco, K., Rubas, W., Sarma, K., Symons, J., Van Natta, K., Washington, C., Wong, P., Wu, X., Yeo, H., Zhu, J.

References

1. P. Marcellin and N. Boyer, *Best Pract. Res. Clin. Gastroenterol.*, 2003, **17**, 259–275.
2. H. J. Alter and L. B. Seeff, *Semin. Liver Dis.*, 2000, **20**, 17–35.
3. M. P. Manns, G. R. Foster, J. K. Rockstroh, S. Zeuzem, F. Zoulim and M. Houghton, *Nat. Rev. Drug Discovery*, 2007, **6**, 991–1000.
4. M. W. Fried, M. L. Shiffman, K. R. Reddy, C. Smith, G. Marinos, F. L. Goncales, D. Haussinger, M. Diango, G. Carosi, D. Dhumeaux, A. Craxi, A. Lin, J. Hoffman and J. Yu, *New Engl. J. Med.*, 2002, **347**, 975–982.
5. P. Simmonds, *J. Gen. Virol.*, 2004, **85**, 3173–3188.
6. H. Le Guillou-Guillemette, S. Vallet, C. Gaudy-Graffin, C. Payan, A. Pivet, A. Goudeau and F. Lunel-Fabiani, *World J. Gastroenterol.*, 2007, **13**, 2416–2426.
7. B. D. Linddenbach and C. M. Rice, *Nature*, 2005, **436**, 933–938.
8. C. A. Lesburg, M. B. Cable, E. Ferrari, Z. Hong, A. F. Mannarino and P. C. Weber, *Nat. Struct. Biol.*, 1999, **6**, 937–943.
9. E. De Clerq, *Curr. Opin. Microbiol.*, 2005, **8**, 552–560.
10. C. Pierra, A. Amador, S. Benzaria, E. Cretton-Scott, M. D'Amours, J. Mao, S. Mathieu, A. Moussa, E. G. Bridges, D. N. Standring, J.-P. Sammadossi, R. Storer and G. Gosselin, *J. Med. Chem.*, 2006, **49**, 6614–6620.
11. J. L. Clark, L. Hollecker, J. C. Mason, L. J. Stuyver, P. M. Tharnish, S. Lostia, T. R. McBrayer, R. Schinazi, K. A. Watanabe, M. J. Otto, P. A. Furman, W. J. Stec, S. E. Patterson and K. W. Pankiewicz, *J. Med. Chem.*, 2005, **48**, 5504–5508.

12. L. J. Stuyver, T. R. McBrayer, P. M. Tharnish, J. L. Clark, L. Hollecher, S. Lostia, T. Nachman, J. Grier, M. A. Bennett, M.-Y. Xie, R. F. Schinazi, J. D. Morrey, J. L. Julander, P. A. Furman and M. J. Otto, *Antiviral Chem. Chemother.*, 2006, **17**, 79–87.

13. L. J. Stuyver, T. R. McBrayer, T. Whitaker, P. M. Tharnish, M. Ramesh, S. Lostia, L. Cartee, J. Shi, A. Hobbs, R. F. Schinazi, K. A. Watanabe and M. J. Otto, *Antimicrob. Agents Chemother.*, 2004, **48**, 651–654.

14. M. J. Sofia, J. Du, P. Wang, K. Chun, S. Rachakonda, B. Ross, H. Steuer, C. Espiritu, E. Murakami, H. Bao, M. J. Otto and P. A. Furman, presented at the 238th ACS National Meeting, Washington, August 2009, MEDI-101.

15. J. Liu, J. Du, P. Wang, D. Nagarathnam, R. Mosley, C. Espiritu, H. Bao, E. Murakami, P.A. Furman, M. J. Otto and M. J. Sofia, presented at the 238th ACS National Meeting, Washington, August 2009, MEDI-107.

16. P. Wang, M. J. Sofia, B.-K. Chun, J. Du, S. Rachakonda, H. Steuer, E. Murakami, H. Bao, D. Nagarathnam, M. J. Otto and P. A. Furman, presented at the 238th ACS National Meeting, Washington, August 2009, MEDI-100.

17. H. Ma, W. R. Jiang, N. Robledo, V. Leveque, S. Ali, T. Lara-Jaime, M. Masjedizadeh, D. B. Smith, N. Cammack, K. Klumpp and J. Symons, *J. Biol. Chem.*, 2007, **282**, 29812–29820.

18. E. Murakami, H. Bao, M. Ramesh, T. R. McBrayer, T. Whitaker, H. M. Micolochick Steuer, R. F. Schinazi, L. J. Stuyver, A. Obikhod, M. J. Otto and P. A. Furman, *Antimicrob. Agents Chemother.*, 2007, **51**, 503–509.

19. E. Murakami, C. Niu, H. Bao, H. M. Micolochick Steuer, T. Whitaker, T. Nachman, M. J. Sofia, P. Wang, M. J. Otto and P. A. Furman, *Antimicrob. Agents. Chemother.*, 2008, **52**, 458–464.

20. G. Migliaccio, J. E. Tomassini, S. S. Carroll, L. Tomei, S. Altamura, B. Bhat, L. Bartholomew, M. R. Bosserman, A. Ceccacci, L. F. Colwell, R. Cortese, F. R. De, A. B. Eldrup, K. L. Getty, X. S. Hou, R. L. LaFemina, S. W. Ludmerer, M. MacCoss, D. R. McMasters, M. W. Stahlhut, D. B. Olsen, D. J. Hazuda and O. A. Flores, *J. Biol. Chem.*, 2003, **278**, 49164–49170.

21. D. B. Olsen, A. B. Eldrup, L. Bartholomew, B. Bhat, M. R. Bosserman, A. Ceccacci, L. F. Colwell, J. F. Fay, O. A. Flores, K. L. Getty, J. A. Grobler, R. L. LaFemina, E. J. Markel, G. Migliaccio, M. Prhavc, M. W. Stahlhut, J. E. Tomassini, M. MacCoss, D. J. Hazuda and S. S. Carroll, *Antimicrob. Agents Chemother.*, 2004, **48**, 3944–3953.

22. S. Le Pogam, W. R. Jiang, V. Leveque, S. Rajyaguru, H. Ma, K. Kang, S. Jiang, M. Singer, S. Ali, K. Klumpp, D. Smith, J. Symons, N. Cammack and I. Najera, *Virology*, 2006, **351**, 349–359.

23. S. Ali, V. Leveque, S. Le Pogam, H. Ma, F. Philipp, N. Inocencio, M. Smith, A. Alker, H. Kang, I. Najera, K. Klumpp, J. Symons, N. Cammack and W. R. Jiang, *Antimicrob. Agents Chemother.*, 2008, **52**, 4356–4369.

24. P. Wang, B.-K. Chun, S. Rachakonda, J. Du, N. Khan, J. Shi, W. Stec, D. Cleary, B. Ross and M. J. Sofia, *J. Org. Chem.*, 2009, **74**, 6819–6824.

25. T. Alfredson, K. Sarma, M. Brandl, M. Sofia, P. Furman, P. Wang, J. Zu, F. Li, J. Chow, M. Holper, S. Larrabee, X. Wu, K. Olocco, S. Hedge, H. Yeo, K. Van Natta, W. Rubas, H. Maag, D. Smith, C. Washington, G. Hill, R. Robson, P. Wong, J. Symons, W.-R. Jiang, H. Ma, N. Cammack, D. Ipe, I. Lakatos, D. Gordon, D. Dei Rossi and D. Blue, presented at the American Association of Pharmaceutical Sciences Meeting, Los Angeles, November 2009, abstr. no. 3957.

26. M. J. Otto, R. Robson, C. Rodriguez, A. Beard, W. Symonds, G. Hill and M. Berrey, presented at the 14th International Symposium on Hepatitis C Virus and Related Viruses, Glasgow, September 2007, abstr. no. 268.

27. R. Reddy, M. Rodriguez-Torres, E. Gane, R. Robson, J. Lalezari, G. Everson, E. DeJesus, J. McHutchison and H. Vargas, presented at the 58th Annual Meeting of the American Association for the Study of Liver Diseases (AASLD), Boston, November 2007, abstr. no. LB9.

28. J. McHutchison, *et al.*, presented at the 43rd Annual Meeting of the European Association for the Study of the Liver (EASL), Milan, April 2008.

29. M. Rodriguez-Torres, J. Lalezari, E. Gane, E. DeJesus, D. Nelson, G. Everson, I. Jacobsen, R. Reddy, J. McHutchison, A. Beard, S. Walker, W. Symonds and M. Berrey, presented at the 59th Annual Meeting of the American Association for the Study of Liver Diseases (AASLD), San Francisco, November 2008, abstr. no. 1899.

30. S. Le Pogam, A. Seshaadri, A. Kosaka, S. Hu, A. Beard, J. Symons, N. Cammack and I. Najera, presented at the 44th Annual Meeting of the European Association for the Study of the Liver (EASL), Copenhagen, April 2009.

31. E. Gane, M. Rodriquez-Torres, D. Nelson, I. Jacobsen, J. McHutchison, L. Jeffers, A. Beard, S. Walker, N. Schulman, W. Symonds, E. Albanis and M. Berrey, presented at the 59th Annual Meeting of the American Association for the Study of Liver Diseases (AASLD), San Francisco, November 2008, LB10.

32. E. Gane, S. Roberts, C. Stedman, P. Angus, R. Ritchie, R. Elston, D. Ipe, L. Baher, P. Morcos, I. Nagera, M. Mannino, B. Brennan, M. Berrey, W. Bradford, E. Yetzer, N. Shulman and P. Smith, presented at the 44th Annual Meeting of the European Association for the Study of the Liver (EASL), Copenhagen, April 2009.

33. E. Gane *et al.*, presented at the 60th Annual Meeting of the American Association for the Study of Liver Diseases (AASLD), Boston, November 2009.

34. C. McGuigan, D. Cahard, H. M. Sheeka, E. De Clercq and J. Balzarini, *J. Med. Chem.*, 1996, **39**, 1748–1753.

35. C. McGuigan, P. W. Sutton, D. Cahard, K. Turner, G. O'Leary, Y. Wang, M. Gumbleton, E. De Clercq and J. Balzarini, *Antiviral Chem. Chemother.*, 1998, **9**, 473–479.

36. D. Saboulard, L. Naesens, D. Cahard, A. Salgado, R. Pathirana, S. Velazquez, C. McGuigan, E. De Clercq and J. Balzarini, *Mol. Pharmacol.*, 1999, **56**, 693–704.

37. M. J. Sofia, P. Wang, J. Du, H. M. Steuer, C. Niu, P. Furman and M. J. Otto, presented at the 14th International Symposium on Hepatitis C Virus and Related Viruses, Glasgow, September 2007, P-259.

38. M. J. Sofia, P. Wang, J. Du, H. M. Steuer, C. Niu, P. Furman and M. J. Otto, presented at the 2nd International Workshop on Hepatitis C Resistance and New Compounds, Boston, October 2007, abstr. no. 7.

39. M. J. Sofia, P. Wang, J. Du, H. M. Steuer, C. Niu, B. S. Ross, S. Rachakonda, D. Bao, B. Symonds, P. A. Furman, M. J. Otto and D. Nagarathnam, presented at the 236th ACS National Meeting, Philadelphia, August 2008, MEDI-330.

40. P. Furman, D. Bao, W. Chang, J. Du, H. M. Steuer, D. Nagarathnam, C. Niu, S. Rachakonda, B. Ross, W. Symonds, P. Wang, M. J. Otto and M. J. Sofia, presented at the 15th International Symposium on Hepatitis C Virus and Related Viruses, San Antonio, October 2008, abstr. no. 275.

41. M. Rodriguez-Torres, E. Lawitz, S. Flach, J. Denning, E. Albanis, W. Symonds and M. Berrey, presented at the 60th Annual Meeting of the American Association for the Study of Liver Diseases (AASLD), Boston, November 2009, LB17.

42. J. Lalezari, E. Gane, M. Rodriguez-Torres, E. DeJesus, D. Nelson, G. Everson, I. Jacobson, R. Reddy, G. Hill, A. Beard, W. Symonds and M. Berrey, presented at the 43th Annual Meeting of the European Association for the Study of the Liver (EASL), Milan, April 2008, LB16.

CHAPTER 12

A Case History on the Challenges of Central Nervous System and Dual Pharmacology Drug Discovery

PAUL V. FISH, ANTHONY HARRISON, FLORIAN WAKENHUT AND GAVIN A. WHITLOCK*

Pfizer Global Research and Development, Ramsgate Road, Sandwich, Kent CT13 9NJ, UK

12.1 Introduction

Dual serotonin (5-HT) and noradrenaline (NA) reuptake inhibitors (SNRI) have proven to be an effective treatment for a number of indications, including depression, anxiety disorders, fibromyalgia, painful peripheral neuropathy and stress urinary incontinence.[1–5] The search for potent and selective SNRIs has identified venlafaxine **1**, desvenlafaxine **2**, milnacipran **3** and duloxetine **4** (Figure 12.1), which are all marketed drugs.[6–10] Furthermore, several small-molecule SNRIs have been reported to be in early clinical development or undergoing preclinical optimization and evaluation.[11–14] Hence, the discovery of new SNRIs with enhanced properties continues to be an attractive target for the pharmaceutical industry.

RSC Drug Discovery Series No. 4
Accounts in Drug Discovery: Case Studies in Medicinal Chemistry
Edited by Joel C. Barrish, Percy H. Carter, Peter T. W. Cheng and Robert Zahler
© Royal Society of Chemistry 2011
Published by the Royal Society of Chemistry, www.rsc.org

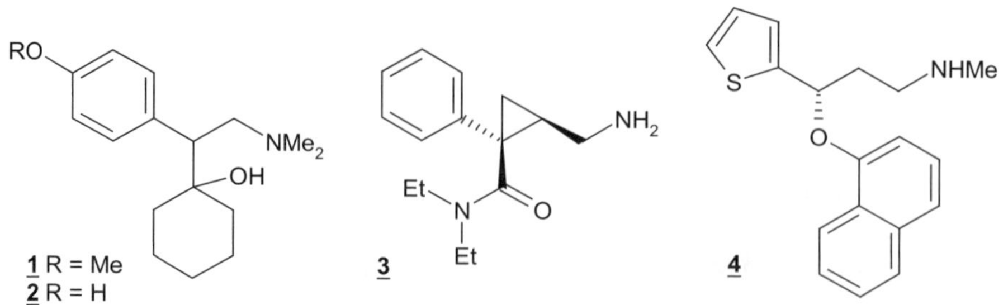

Figure 12.1 Structures of marketed SNRIs.

12.1.1 Disease Background

Stress urinary incontinence (SUI) is the loss of urine coincident with an increase in intra-abdominal pressure on the bladder that exceeds the urethral resistance. Examples of triggers of involuntary leakage of urine are coughing, sneezing, laughing and physical exercise. The dominant risk factor for SUI is gender, with women representing over 90% of sufferers. Women are vulnerable to SUI due to experiencing changes in their lower urinary tract anatomy during pregnancy, childbirth and menopause. Additional risk factors are age and obesity. Urinary incontinence tends to be regarded as an embarrassing illness and is accepted as part of motherhood and/or aging. There are believed to be 13 million sufferers of SUI in the US and very few seek treatment.[15] People afflicted with SUI typically experience physical disability, emotional stress and disruption to their daily life. The most widely adopted approaches to SUI are maintenance of the condition through incontinence pads and rehabilitation through training of the pelvic floor muscles. Surgical procedures are an option once other approaches have failed and involve elevation of the bladder above the pelvic floor or introduction of slings to provide additional support for the urethra. Until recently, there were no pharmacotherapies licensed for SUI and treatment was limited to the off-label use of non-selective α-adrenergic agonists (improve urethral tone), tricyclic antidepressants (improve urethral tone), β-adrenergic agonists (bladder relaxant) and estrogens (reverse atrophy).[16] In 2004, Eli Lilly introduced the dual SNRI duloxetine as the first licensed drug therapy specifically for the treatment of moderate to severe SUI.[5]

12.1.2 Target Background

The neurotransmitter 5-HT mediates its effects through at least 14 different receptor subtypes that have been classified into seven major families, 5-HT_{1-7}.[17] In a similar manner, the neurotransmitter NA acts through the α_1-, α_2- and β-adrenoceptors.[18] The release of 5-HT and NA into the synaptic cleft results in receptor activation followed by reuptake of the neurotransmitters by their respective transporter proteins. Inhibition of these transporters with a dual SNRI prolongs the action of the 5-HT and NA in the synapse. Selectivity over the dopamine (DA) reuptake transporter (DRI) would be desirable to avoid any dopaminergic effects.

In the context of treating SUI, SNRIs are believed to work by blocking reuptake of 5-HT and NA in Onuf's nucleus in the sacral spinal cord.[16] Pudendal motor neurons located in Onuf's nucleus regulate the urethral striated muscle sphincter and Onuf's nucleus has a high density of 5-HT and NA receptors.[16] Furthermore, it has been reported that the prevention of urine leakage could be mediated by stimulation of central 5-HT$_2$ and α_1 adrenergic receptors, resulting in more effective closure of the urethral sphincter.[19] In addition, comparison of the effects of an SRI, an NRI, co-administration of an SRI with an NRI, and a dual SNRI in lower urinary tract function in cats has shown that improving bladder storage is unique to the dual SNRI mechanism.[20]

12.1.3 Compounds in Class

There are currently four SNRIs approved for clinical use. Venlafaxine is an orally active, selective dual SNRI developed and launched by Wyeth for the treatment of depression.[6] Venlafaxine was launched in 1994 as Effexor™ as "first in class" of the second-generation of antidepressants with dual SNRI activity and has a much improved side-effect profile compared with the classical tricyclic antidepressants. In a recent small clinical study, venlafaxine has shown potential utility as a treatment for SUI.[21] In 2008, Wyeth then launched desvenlafaxine (Pristiq™) for the treatment of major depressive disorder (MDD). Desvenlafaxine is the major active metabolite of venlafaxine and offers a reduced risk of drug–drug interactions as a potential advantage over other SNRIs.[10] Milnacipran is also a dual SNRI and was introduced in Europe by Pierre Fabre in 1997 as Ixel™. Milnacipran is approved for the treatment of depression[7] and is currently in phase 3 trials for the treatment of fibromyalgia.[22]

Duloxetine was first launched in the US in 2004 as Cymbalta™ by Eli Lilly for the treatment of major depressive disorder, generalised anxiety disorder and diabetic peripheral neuropathic pain.[8] Duloxetine has also been approved for the treatment of SUI in Europe and is marketed as Yentreve™.[5,14,16,19] (+)-(S)-Duloxetine is a combined SNRI with potent inhibitory activity at both the human 5-HT ($K_i = 0.8$ nM) and NA ($K_i = 7.5$ nM) transporters.[23] Duloxetine weakly inhibits DA reuptake ($K_i = 240$ nM) with no significant activity at aminergic receptors (M, H$_1$, α_1, α_2, D).[24] Further pharmacological evaluation *in vivo*, in microdialysis experiments, showed that duloxetine dose-dependently increased extracellular 5-HT and NA levels in interstitial fluid of the frontal cortex and hypothalamus of conscious rats.[24] Duloxetine is well absorbed after oral administration, with bioavailability of 32–80% and a mean plasma half-life of 12 h, consistent with the recommendation of a twice-daily dosing regime. The primary route of duloxetine clearance is through hepatic oxidative metabolism to a number of metabolites and *in vitro* studies have shown that duloxetine is predominantly metabolised by the enzymes CYP1A2 and CYP2D6. Duloxetine plasma exposure is increased 3- to 6-fold when co-dosed with inhibitors of CYP2D6 and/or CYP1A2.[25] Duloxetine is a moderate CYP2D6 inhibitor (*in vitro* $IC_{50} = 7\,\mu M$)[26] and increases the exposure of drugs metabolised by CYP2D6.[25] The clinical efficacy of duloxetine (40 mg bid) in the treatment of SUI has been established in nine double-blind, placebo-controlled studies that randomised adult women ($n = 3063$) with symptoms of SUI.[5] In clinical trials across multiple indications at clinically effective doses, duloxetine has also been shown to increase systolic and diastolic blood pressure by up to 2 mmHg.[25] In a clinical pharmacology study designed to evaluate the effects of duloxetine on various parameters, at the highest 200 mg twice-daily dose, increases in mean blood pressure were 4.7–6.8 mmHg (systolic) and 4.5–7 mmHg (diastolic) up to 12 hours after dosing.[25]

12.1.4   Program Objectives and General Medicinal Chemistry Strategy

At the outset of our research program, duloxetine was already in advanced clinical studies for the treatment of SUI and our realistic object was to identify a "best in class" agent.

Hence our program goal was to identify a chronic use, oral therapy with once-daily dosing suitable for the treatment of SUI. Our laboratory objectives were to identify a potent, centrally penetrant, dual SNRI ($K_i < 10\,nM$) with a balance of SRI and NRI activity similar to that of duloxetine, and having excellent selectivity over DRI and other receptors, ion channels and enzymes (> 100-fold). Preclinical differentiation from duloxetine would be evaluated and achieved through:

- Similar or better efficacy in preclinical models of SUI
- Similar or better cardiovascular profile in preclinical models
- Weak CYP2D6 inhibition
- No significant metabolism by polymorphically expressed cytochrome P450 (CYP) enzymes
- Potential for once-a-day dosing in humans

A new SNRI that satisfied these criteria would not only be a potential "best in class" treatment for SUI but would also have a profile suitable for clinical evaluation as a treatment for other indications where SNRIs have been shown to have a beneficial effect.

As part of our research efforts to identify potential new SNRI drug candidates, we adopted a strategy of exploring multiple chemical templates in parallel in order to increase our chances of having compounds survive to become advanced clinical candidates.

A key component of our screen sequence was the evaluation of test compounds in a preclinical *in vivo* canine model of SUI.[27] Our screening funnel was designed to identify quality compounds for assessment in this model and several *in vitro* assays were run in parallel to facilitate rapid decision-making for

Tier 1 *in vitro* screens:

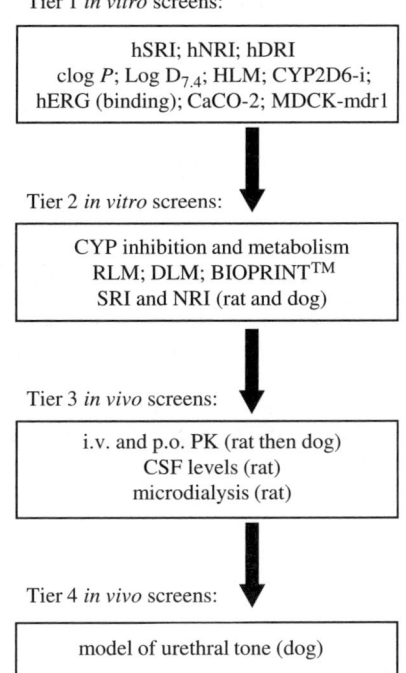

Figure 12.2 Project screening sequence.

each series (Figure 12.2). In the canine model of SUI, test compounds were administered by *i.v.* infusion to anaesthetized female dogs and a pressure transducer inserted into the urethra as far as the bladder neck. The transducer would then be slowly withdrawn whilst recording the pressure, and a measurement of peak urethral pressure (PUP) would be taken: the percentage difference in PUP over vehicle was plotted against steady-state free plasma concentrations of drug. The percentage increase in PUP was used as a measure of efficacy of the drug in improving urethral tone. This model could also be used for a preliminary measurement of cardiovascular safety data such as mean arterial pressure (MAP) and heart rate (HR).

Evaluation of duloxetine in this model showed a drug-related, dose-dependent increase in (PUP) with a maximal increase of 35% at free plasma levels of ≤ 15 nM ($n = 5$) (Figure 12.3). An increase of 26% in PUP was observed at a free plasma concentration of 4 nM, which we believed to be close to the free plasma levels from a therapeutic dose of duloxetine for SUI (40 mg bid). The increase in PUP was also accompanied by a small dose-dependent increase in MAP relative to control, which is consistent with the clinical findings.[25] Evolution of the methodology of this model by means of the application in conscious dogs was thought to translate through to the clinical setting and allow a direct comparison of efficacy between dog, healthy volunteers and patients, *i.e.* we felt the model would serve as a translatable clinical biomarker.[27]

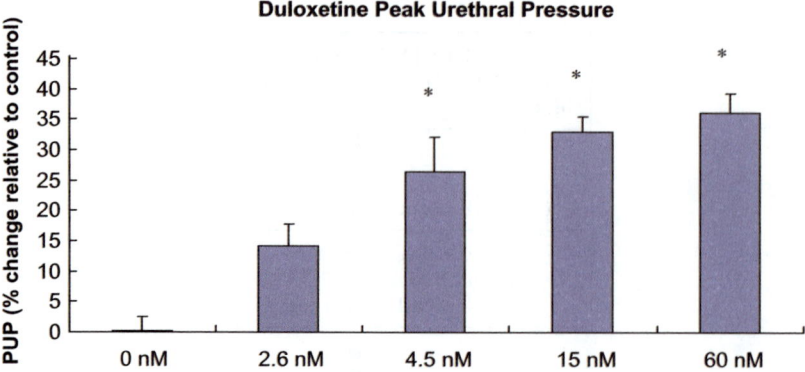

Figure 12.3 Increases in PUP *versus* free plasma drug concentration for duloxetine in an anaesthetized dog model of SUI. Significance relative to control noted thus *; $P < 0.05$.

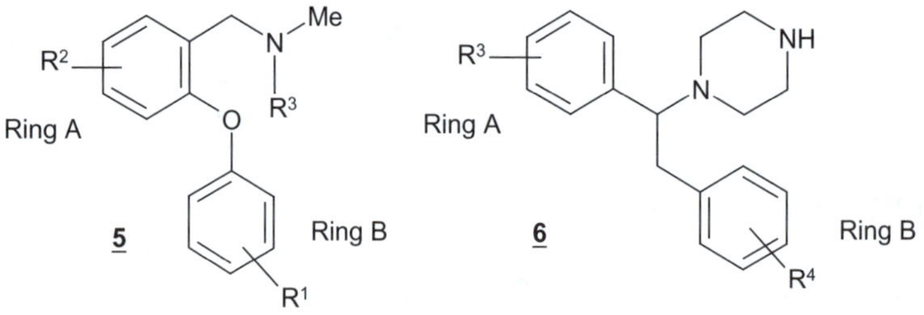

Figure 12.4 Structures of lead series.

12.2 Medicinal Chemistry: Hit, Lead and Candidate Generation

As discussed in the introduction, we chose to prosecute this medicinal chemistry program with multiple lead-series in parallel. However, for clarity these will be discussed in sequence in this section. One of our two main leads consisted of a proprietary series of diphenyl ethers **5** (Figure 12.4), which had initially been prosecuted in the discovery of novel selective serotonin reuptake inhibitor (SSRI) development candidates.[28]

Our second main lead resulted from *in silico* virtual screening of the Pfizer compound file, which identified biarylpiperazines **6** as a structure which possessed interesting levels of SNRI pharmacology and attractive lead-like properties (Figure 12.4).

Structure–activity relationships (SAR) within the B-ring of diphenyl ether **5** quickly identified a 2,4-disubstitution pattern that was important for dual SNRI potency. Further optimisation to a 2-OMe + 4-Cl substitution alongside

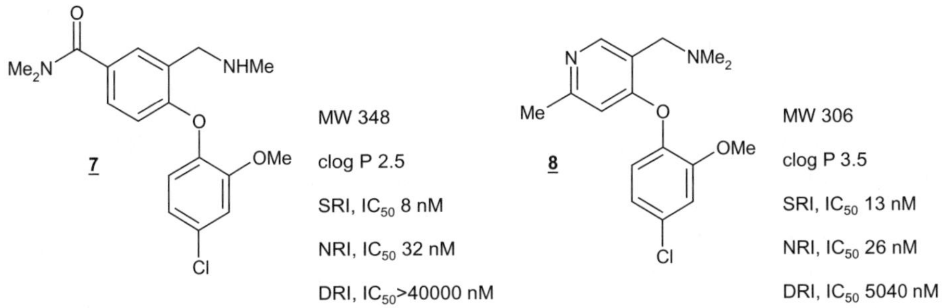

Figure 12.5 *In vitro* profile of compounds **7** and **8**. DRI = dopamine reuptake inhibitor; NRI = noradrenaline reuptake inhibitor; SRI = serotonin reuptake inhibitor.

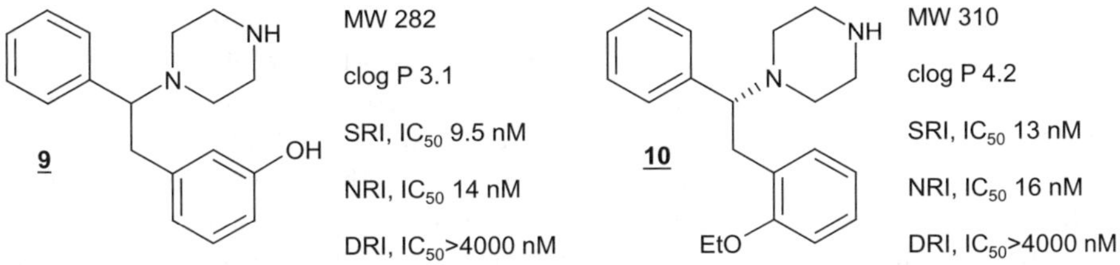

Figure 12.6 *In vitro* profile of compounds **9** and **10**.

incorporation of a polar amide R^2 group to reduce lipophilicity gave **7** as an early lead (Figure 12.5).[29]

Although **7** possessed good microsomal stability, rapid amide hydrolysis was detected in human hepatocytes, and therefore a structural change to the amide was required. This was achieved with the pyridyl phenyl ether **8**, which had improved microsomal and hepatocyte stability whilst retaining the target *in vitro* pharmacological profile.[30] However, during the development of SAR in this pyridyl phenyl ether series it became apparent that NA reuptake activity was highly dependent on lipophilicity, with a clog *P* of >3.5 needed to achieve NA reuptake potency of <25 nM. The relationship between high clog *P* and increased risk of off-target pharmacology and adverse toxicological outcomes has recently been reported.[31] This relationship was reflected in the poor off-target selectivity of compound **8**, which was active against a large number of GPCR receptor targets when screened at CEREP. Based on the poor selectivity profile of **8** and other pyridyl phenyl ether analogues, this series was not progressed for dual SNRI activity any further.

In parallel to investigations on the above compounds **7** and **8**, we had also mined the Pfizer compound file for new analogues which possessed some of the key elements of the 3-aryloxypropanamine series, namely two aromatic rings suitably spaced from a basic amine. This work highlighted the piperazine **9** as a potentially interesting lead (Figure 12.6). This analogue had originally been made as an intermediate for a δ-opioid ligand program, but possessed interesting levels of monoamine reuptake activity.[32]

During SAR investigations of this series, piperazine 10 was discovered which possessed excellent *in vitro* pharmacology[32] and, despite its high clog *P*, it was highly selective against other GPCR targets when screened at CEREP. However, piperazine 9 did possess potent activity against a range of ion channels, in particular Na channels (site 2, $IC_{50} = 410$ nM) and Ca channels (L-type, diltiazem site, $IC_{50} = 730$ nM). To determine the impact of this *in vitro* ion channel activity, compound 10 was assessed in a conscious dog cardiovascular study. A complex haemodynamic profile emerged for compound 10, primarily involving changes in heart rate and contractility with only a moderate therapeutic index between projected efficacious free drug levels in human and cardiovascular side effects and so further development of this compound was halted. We felt that the high clog *P* of 10 was a major contributing factor to its ion channel activity and therefore sought to reduce clog *P* in this series whilst maintaining SNRI activity. Unfortunately, as with the pyridyl phenyl ethers above, there was a clear relationship between clog *P* and NA activity and none of the more polar piperazine analogues that were synthesised possessed the correct balance of SNRI potency and other key properties such as weak P450 inhibition.

At this stage we chose to investigate the role of the piperazine ring and a scaffold-hopping exercise to an aminopyrrolidine template was carried out.[33] This was guided by synthetic expedience rather than any specific SAR hypothesis, as suitably protected enantiomerically pure 3-aminopyrrolidine starting materials were commercially available. This small change in structure brought about a significant change in the potency/lipophilicity relationship, which can be seen graphically in Figure 12.7. Now, a significant number of analogues had combined clog *P* values <3.5 and NA potency <50 nM, whereas very few piperazine analogues combined those properties.

Among the compounds synthesised in the aminopyrrolidine series, the tetrahydropyranyl analogue 11 was profiled extensively. The combination of SNRI potency with significantly reduced ion channel activity, when compared to the piperazine analogue 10, made 11 an exciting new lead (Figure 12.8).

However, when the metabolic profile of 11 was investigated in *in vitro* microsomal preparations, it was discovered that 11 was predominantly metabolized by CYP2D6 (>80%) *via* oxidation of the benzylic ring and to a lesser degree by CYP3A4 *via* dealkylation (Figure 12.8). Predominant metabolism of a drug via CYP2D6 can result in either significantly increased or decreased exposure levels in different patient groups. For example, there is a risk of significant over-exposure of 11 in poor metabolisers (patients who have mutations in their CYP2D6 genes which result in reduced CYP2D6 activity) or underexposure in ultra-rapid metabolisers (patients who have multiple copies of the CYP2D6 gene which results in increased CYP2D6 activity), relative to exposure observed in extensive metabolisers (patients with normal levels of CYP2D6 activity).[34] Based on this finding, it was decided to halt the progression of compound 11. Blocking the CYP2D6 oxidation pathway of 11 through judicious modification of the benzylic substituent was successful, as illustrated by

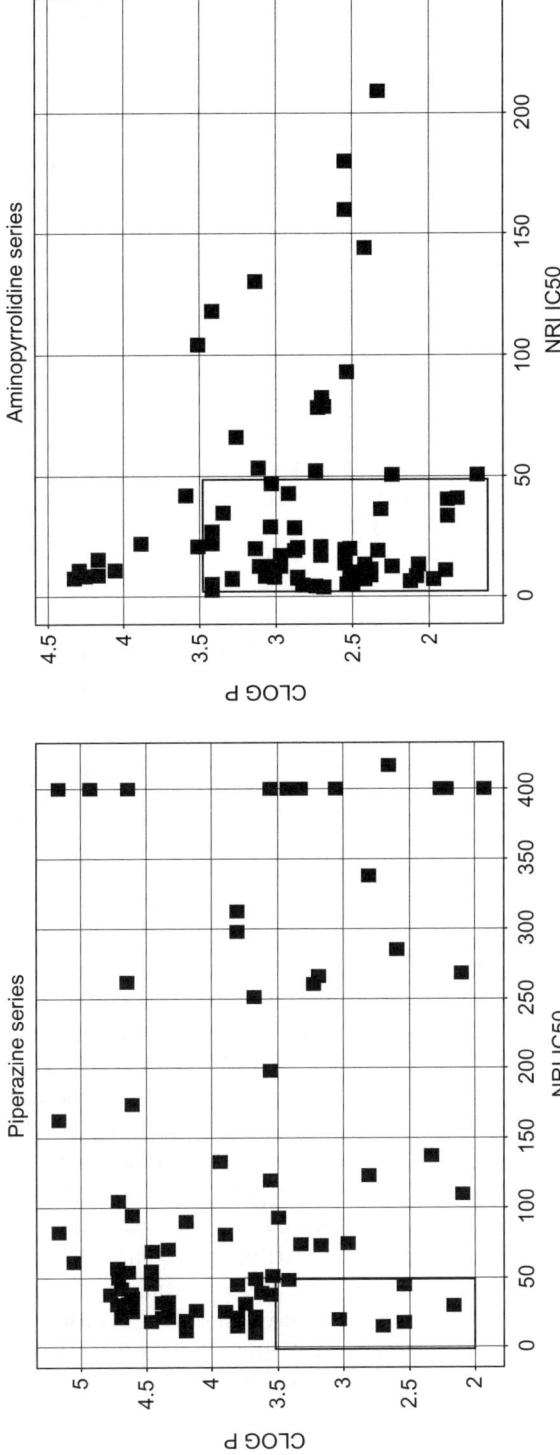

Figure 12.7 NA potency and clog *P* relationship for piperazine and aminopyrrolidine series.

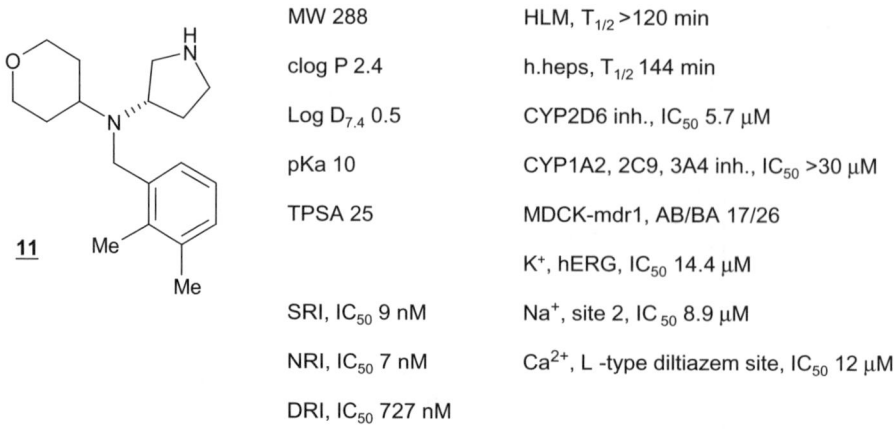

MW 288   HLM, $T_{1/2}$ >120 min

clog P 2.4   h.heps, $T_{1/2}$ 144 min

Log $D_{7.4}$ 0.5   CYP2D6 inh., IC_{50} 5.7 µM

pKa 10   CYP1A2, 2C9, 3A4 inh., IC_{50} >30 µM

TPSA 25   MDCK-mdr1, AB/BA 17/26

                         K^+, hERG, IC_{50} 14.4 µM

SRI, IC_{50} 9 nM Na^+, site 2, IC_{50} 8.9 µM

NRI, IC_{50} 7 nM Ca^{2+}, L -type diltiazem site, IC_{50} 12 µM

DRI, IC_{50} 727 nM

Figure 12.8 Profile of compound **10**. hERG = human Ether-a-go-go Related Gene; h.heps = human hepatocytes; HLM = human liver microsomes; MDCK mdr-1 = Madin–Darby Canine Kidney cell line overexpressing the mdr-1 gene which encodes P-glycoprotein.

compound **12** for which dealkylation was found to be the primary route of metabolism (>80%). Unfortunately, the primary metabolizing enzyme for **12** was found to be CYP2D6, whereas for compound **11** it was CYP3A4, illustrating the impact small structural modifications can have on metabolic pathways. We then elected to block the *N*-dealkylation of **12** by replacing the 4-tetrahydropyranyl substituent by an amide group (Figure 12.9).

SAR investigations in this novel carboxamide template have been reported and were greatly facilitated by the simplicity of the chemistry route.[35] From these analyses, compound **13** emerged as combining potent dual SNRI activity with selectivity against the dopamine transporter. Compound **13** also exhibited good membrane permeability in the CaCO-2 cell line and good metabolic stability both in human liver microsomes and human hepatocytes, predictive of a low *in vivo* clearance. Compound **13** had no significant inhibition of all major CYP450 enzymes, modest affinity for the hERG channel (Figure 12.10) and no significant off-target pharmacology when assessed against a panel of 150 receptors, enzymes and ion channels (200-fold selectivity for SNRI activity, CEREP/Bioprint™). When the metabolic profile of compound **13** was investigated in an *in vitro* human microsomal preparation, the compound was found to be so metabolically stable that no measurable turnover could be detected. As a result, assessment of the CYP2D6 contribution to metabolism was not possible. Should **13** have progressed further, the potential for increased exposures in poor metabolisers would therefore have had to be monitored in clinical studies. Finally, pharmacokinetic data in the dog showed that compound **13** had low clearance, a low volume of distribution, a long half-life and high bioavailability, all predictive of a human pharmacokinetics profile consistent with our project objectives.

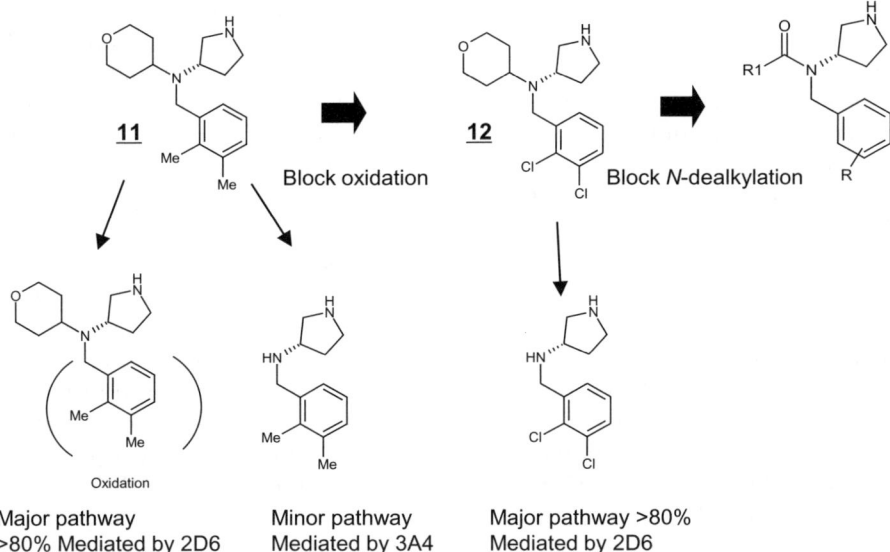

Figure 12.9 Metabolic pathway of compounds **10** and **11** and strategy to design out CYP2D6 metabolism.

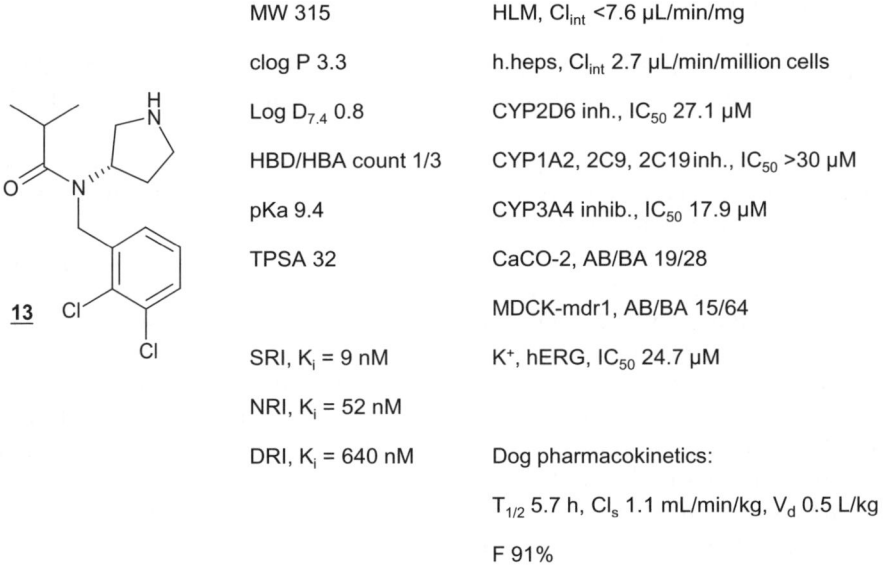

MW 315

clog P 3.3

Log $D_{7.4}$ 0.8

HBD/HBA count 1/3

pKa 9.4

TPSA 32

SRI, K_i = 9 nM

NRI, K_i = 52 nM

DRI, K_i = 640 nM

HLM, Cl_{int} <7.6 µL/min/mg

h.heps, Cl_{int} 2.7 µL/min/million cells

CYP2D6 inh., IC_{50} 27.1 µM

CYP1A2, 2C9, 2C19 inh., IC_{50} >30 µM

CYP3A4 inhib., IC_{50} 17.9 µM

CaCO-2, AB/BA 19/28

MDCK-mdr1, AB/BA 15/64

K$^+$, hERG, IC_{50} 24.7 µM

Dog pharmacokinetics:

$T_{1/2}$ 5.7 h, Cl_s 1.1 mL/min/kg, V_d 0.5 L/kg

F 91%

Figure 12.10 *In vitro* and pharmacokinetic profile of compound **13**. CaCO-2 = colorectal adenocarcinoma cells.

Based on these data, compound **13** was selected for evaluation in the *in vivo* pre-clinical efficacy canine model for SUI described earlier. Unfortunately, **13** did not produce a response consistent with its primary pharmacology, as statistically significant increases in PUP were only observed at very high free plasma concentrations ($>1\,\mu M$). Although unexpected at first, this experimental finding was subsequently rationalized as a consequence of poor blood–brain barrier penetration for **13** as illustrated by (i) the very low cerebrospinal fluid (CSF) to free plasma concentration ratio observed for **13** in both rat and dog at steady state (<0.1), and (ii) pharmacological evaluation *in vivo* in microdialysis experiments demonstrating that **13** significantly increased striatal 5-HT levels after systemic administration only at free plasma concentrations much higher than expected based on the increase observed following intrastriatal perfusion of **13**. As compound **13** exhibited physicochemical properties consistent with central nervous system (CNS) target space,[36] *i.e.* low molecular weight (315), moderate lipophilicity (clog $P = 3.3$), low topological polar surface area (TPSA $= 32\,\text{Å}^2$), H-bond donor and acceptor counts (HBD 1, HBA 3), we postulated that the poor blood–brain barrier penetration was probably the consequence of recognition by the efflux P-glycoprotein (P-gp) transporter. This hypothesis was further supported by the significant efflux ratio observed for **13** in the MDCK-mdr1 cell line (15/64, efflux ratio $= 4.3$).[37] In order to overcome this issue, we decided to focus our medicinal chemistry design on disrupting the hydrogen bonding ability of **13**, as it is well established to be one of the key recognition elements for P-gp transport.[38] Since we had already demonstrated through preliminary SAR investigations that the pyrrolidine N–H was paramount for SNRI potency, our options were somewhat limited and our main strategy relied on isomerising the amide carbonyl group into the benzylic position, postulating that in doing so we might disrupt the possible P-gp recognition motif while retaining the very attractive *in vitro* profile of **13**. This modification led to the discovery of the novel benzamide **14**, aka PF-184298.[39]

Compound **14** was found to be a potent SNRI, exhibiting selectivity over DRI, metabolic stability both in human liver microsomes and human hepatocytes, and selectivity over ion channel activity as measured by binding to the hERG channel (K^+, hERG), consistent with our project objectives (Figure 12.11). More excitingly, **14** exhibited a reduced efflux ratio compared to **13** when evaluated in the MDCK-mdr1 cell line. Compound **14** also demonstrated good membrane permeability in the CaCO-2 cell line, suggesting the potential for good oral absorption. Additional attractive features included weak inhibition of all major CYP enzymes and no significant off-target pharmacology when evaluated in a panel of receptors, ion channels and enzymes (Cerep, BIOPRINT^{TM}), exhibiting $>50\%$ inhibition at $10\,\mu M$ only for the Na channel (site 2; $IC_{50} = 3.6\,\mu M$), the sigma receptor ($IC_{50} = 5.1\,\mu M$) and the kappa opioid receptor ($IC_{50} = 5.8\,\mu M$). To better understand the impact of the reduction in efflux ratio and therefore P-gp recognition observed in the MDCK-mdr1 cell line, the CSF to unbound plasma concentration ratio for **14** was measured in the rat at steady state following an intravenous infusion

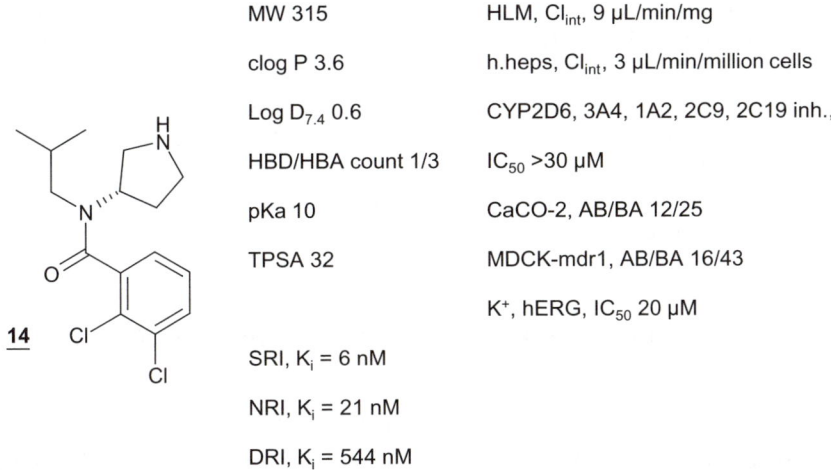

MW 315

clog P 3.6

Log $D_{7.4}$ 0.6

HBD/HBA count 1/3

pKa 10

TPSA 32

HLM, Cl_{int}, 9 µL/min/mg

h.heps, Cl_{int}, 3 µL/min/million cells

CYP2D6, 3A4, 1A2, 2C9, 2C19 inh.,

IC_{50} >30 µM

CaCO-2, AB/BA 12/25

MDCK-mdr1, AB/BA 16/43

K^+, hERG, IC_{50} 20 µM

14

SRI, K_i = 6 nM

NRI, K_i = 21 nM

DRI, K_i = 544 nM

Figure 12.11 *In vitro* profile of compound **14**.

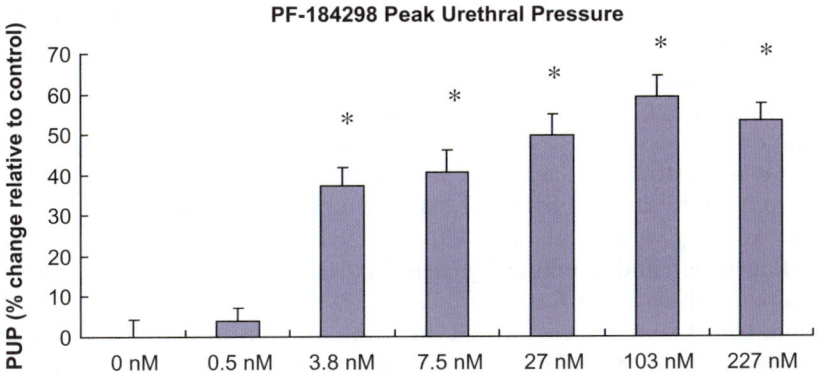

Figure 12.12 Increases in PUP *versus* free plasma concentration of **14** in an anaesthetized dog model (*i.v.* infusion; $n = 4$). Significance relative to control noted thus *; $P < 0.05$.

regimen. The ratio obtained of 0.45 for **14** was clearly superior to the ratio of 0.1 observed for **13**. Encouraged by these data and the overall promising *in vitro* profile of **14**, its performance was evaluated in the dog SUI efficacy model (Figure 12.12). We were pleased to find that **14** elicited a robust dose-dependent increase in urethral pressure in the model, confirming good blood–brain barrier penetration. Additionally, at the estimated human therapeutic free plasma concentration of duloxetine for SUI, *i.e.* 4 nM (corresponding to a dose of 40 mg bid), **14** elicited a superior response to duloxetine, producing a 35% increase in PUP compared to only 26% for duloxetine. Compound **14** also had a superior haemodynamic profile. A slight increase in MAP of 2.5%, consistent with the mechanism of action, was also observed, which was smaller than a 9.4% increase in MAP observed for duloxetine at a similar free plasma concentration.

Finally, compound **14** did not produce dose-limiting toxicity in sub-chronic (14-day) oral rat (up to $30 \, \text{mg kg}^{-1}$) and dog toxicology studies (up to $7.5 \, \text{mg kg}^{-1}$). It was also clean in *in vitro* and *in vivo* genetic toxicology screens and did not produce significant effects on spontaneous locomotor activity compared to vehicle control when administered orally up to $15 \, \text{mg kg}^{-1}$ in rats.

12.3 Pre-clinical and Clinical Pharmacokinetics of PF-184298

Preclinical pharmacokinetic studies with PF-184298 following intravenous administration to male Sprague Dawley rats and male beagle dogs showed the compound to have moderate to high plasma clearance ($38 \, \text{mL min}^{-1} \, \text{kg}^{-1}$ and $40 \, \text{mL min}^{-1} \, \text{kg}^{-1}$, respectively) and volumes of distribution of $3.4 \, \text{L kg}^{-1}$ and $7.1 \, \text{L kg}^{-1}$, respectively. These properties resulted in terminal elimination half-life values of 1.2 h in the rat and 2.2 h in the dog (Table 12.1). The high volumes of distribution were in keeping with the physicochemistry of PF-184298 (log $D_{7.4} = 0.6$, $pK_a = 10.0$). Following oral administration, PF-184298 was rapidly absorbed in rat and dog, and showed moderate oral bioavailability (50% and 28%, respectively), despite liver blood flow-limited clearance in the dog (dog liver blood flow assumed to be $40 \, \text{mL min}^{-1} \, \text{kg}^{-1}$) (Table 12.2). Bioavailability in the rat suggested complete absorption from the gut, based on a clearance of $38 \, \text{mL min}^{-1} \, \text{kg}^{-1}$ and an assumed liver blood flow of $70 \, \text{mL min}^{-1} \, \text{kg}^{-1}$. This was in keeping with high solubility of $> 12 \, \text{mg mL}^{-1}$ in aqueous media (including water and simulated gastric fluids), and high intrinsic membrane permeability observed in CaCO-2 cells.

A number of studies were carried out in an attempt to understand the likely clearance mechanism of PF-184298 in man and to aid in the clearance prediction. *In vitro* metabolism studies showed that hepatic metabolism occurred in rat and dog *via* oxidative ("Phase 1") processes (Figure 12.13). Furthermore, metabolites detected *in vitro* in rat hepatocytes were also present *in vivo* (excreted in bile). Administration of [^{3}H]-PF-184298 to dogs resulted in

Table 12.1 Pharmacokinetic parameters of PF-184298 in plasma following intravenous administration to rats and dogs.

	Rat (n = 2)[a]	*Dog (n = 2)*[a]
Dose (mg kg^{-1})	2	0.0125
Cl ($\text{mL min}^{-1} \, \text{kg}^{-1}$)	38 (29–48)	40 (25–54)
$\text{AUC}_{0-\infty}$ (ng h mL^{-1})	933 (694–1171)	6.1 (3.8–8.3)
V_D (L kg^{-1})	3.4 (2.7–4.2)	7.1 (5.6–8.6)
$T_{1/2}$ (h)	1.2 (0.7–1.7)	2.2 (1.8–2.6)

[a]Data presented as arithmetic mean; figures in parentheses represent range.

Table 12.2 Pharmacokinetic parameters of PF-184298 in plasma following oral administration to rats and dogs.

	Rat (n=2)[a]	*Dog (n=2)[a]*
Dose $(mg\,kg^{-1})$	1.7	0.05
C_{max} $(ng\,mL^{-1})$	81 (74–88)	2.2 (1.2–3.2)
T_{max} (h)	0.4 (0.25–0.5)	0.6 (0.5–0.75)
$t_{1/2}$ (h)	1.1 (1.0–1.3)	2.6 (1.8–3.3)
$AUC_{0-\infty}$ $(ng\,h\,mL^{-1})$	394 (328–459)	6.8 (3.6–10)
F (%)	50	28 (24–30)

[a]Data presented as arithmetic mean; figures in parentheses represent range.

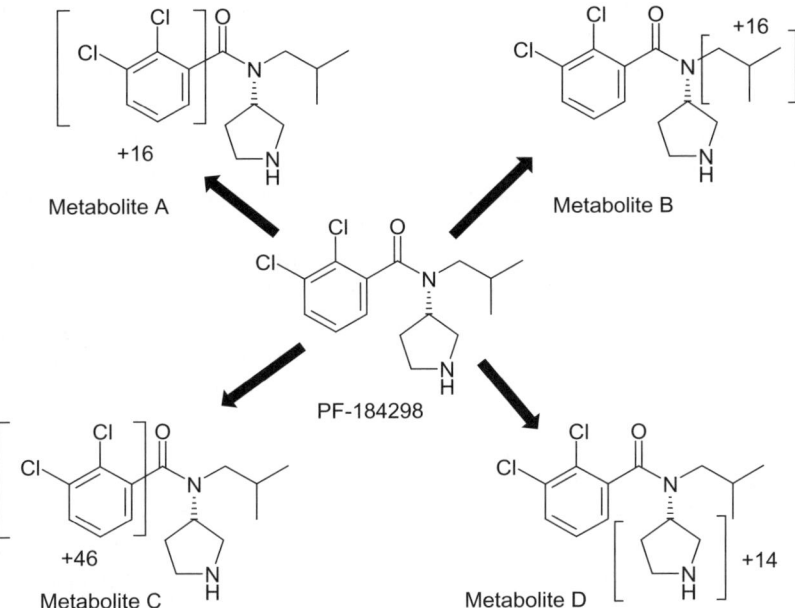

Figure 12.13 Metabolism of PF-184298 in preclinical studies.

excretion of a large number of metabolites, the majority of which were excreted in urine (metabolite B) and were also formed in dog hepatocytes, providing further evidence that hepatic metabolism plays an important role in the clearance of PF-184298. There was no evidence for significant renal or hepatobiliary elimination in the dog or rat; in total, unchanged PF-184298 in dog accounted for no more than 15% of total radioactivity excreted (with only 2% of the dose eliminated unchanged in urine), and elimination in rat bile accounted for <1% of the administered dose. Metabolism studies in human *in vitro* systems were hindered by the slow turnover of PF-184298 $(9\,\mu L\,min^{-1}\,mg^{-1}$ in human liver microsomes). Furthermore, this prevented a quantitative assessment of the contribution of CYP-mediated oxidative

processes to the metabolism of the compound. A mixture of human recombinant CYPs was used to provide sufficient quantities of metabolites for characterisation by mass spectrometry, and demonstrated a similar metabolism
pathway to that observed in dog (metabolite D, Figure 12.13). Whilst not
conclusive, the package of data from preclinical species provided evidence that
Phase 1 (cytochrome P450) mediated metabolism was a significant mechanism
in the clearance of PF-184298.

Despite the evidence around the clearance mechanism, confidence in a
quantitative prediction of clearance using human liver microsomes was
impacted by lack of *in vitro/in vivo* correlation in preclinical pharmacokinetic
studies in different species. Whilst high intrinsic clearance (Cl_{int}) in rat liver
microsomes ($407 \, \mu L \, min^{-1} \, mg^{-1}$) resulted in high clearance of unbound drug
in vivo ($576 \, mL \, min^{-1} \, kg^{-1}$, calculated from systemic clearance of $38 \, mL$
$min^{-1} \, kg^{-1}$ and a free fraction in plasma of 0.066), PF-184298 clearance *in vivo*
in the dog was limited by liver blood flow despite a lack of measurable turnover
in dog liver microsomes ($Cl_{int} < 8.5 \, \mu L \, min^{-1} \, mg^{-1}$). The reasons for this
remain unexplained, but nevertheless impacted the confidence in decision
making for progression of the compound towards clinical studies.

Using the above preclinical information, pharmacokinetic parameters in
humans were estimated. Clearance predictions were made using single species
scaling (SSS) and direct scaling from human liver microsomal intrinsic
clearance methodologies.[40] Applying these approaches resulted in a range of
outcome scenarios in human that covered low systemic clearance ($2.6 \, mL$
$min^{-1} \, kg^{-1}$ predicted from human liver microsomes; equivalent to an oral
clearance of $3.4 \, mL \, min^{-1} \, kg^{-1}$) to clearance limited by liver blood flow
(assumed to be $20 \, mL \, min^{-1} \, kg^{-1}$ in human) predicted from rat and dog SSS.
These extremes of clearance prediction represented outcomes in humans that
could be either supportive of continued progression of the compound or
indicate that PF-184298 was not suitable to take into clinical studies.

Although exhibiting a discrepancy in the clearance prediction between *in
vitro* and SSS methods, PF-184298 demonstrated many other properties that
made it attractive to take into human clinical trials. Exploratory human
pharmacokinetic studies have been increasingly used over recent years to
address pharmacokinetic risk whilst minimising investment to reach a decision
point in humans.[41] Due to the uncertainties in the prediction of the human
pharmacokinetic profile of PF-184298, a preliminary human study was conducted with oral doses of 0.25–2.5 mg that would allow an accurate estimate of
pharmacokinetics in healthy male volunteers (Study 1). This study was
underwritten by non-clinical safety pharmacology and toxicology studies
designed to ensure adequate safety coverage of 25-fold over the maximum
predicted C_{max} in humans, and which allowed human volunteers to receive a
single dose of the compound. This approach, using male animals only in preclinical regulatory studies, enabled a pharmacokinetic readout in man using less
than 100 g of bulk material. Furthermore, this safety multiple approach
allowed assessment of pharmacokinetics at pharmacologically relevant exposures in man, thereby removing the risk of non-linear pharmacokinetics at very

low doses that can be associated with traditional microdosing studies (typically limited to 1/100th of the anticipated clinically efficacious dose). Study 1 demonstrated that PF-184298 had pharmacokinetics consistent with once-a-day administration (Figure 12.14), with an estimated terminal elimination half-life of 29 h and free exposures at 24 h ($\sim$2 nM) close to that considered to be required for efficacy.

This outcome provided confidence for investment in bulk and toxicology studies to underwrite a more traditional single-dose study in man to fully characterise safety, toleration and pharmacokinetics (Study 2, Table 12.3).

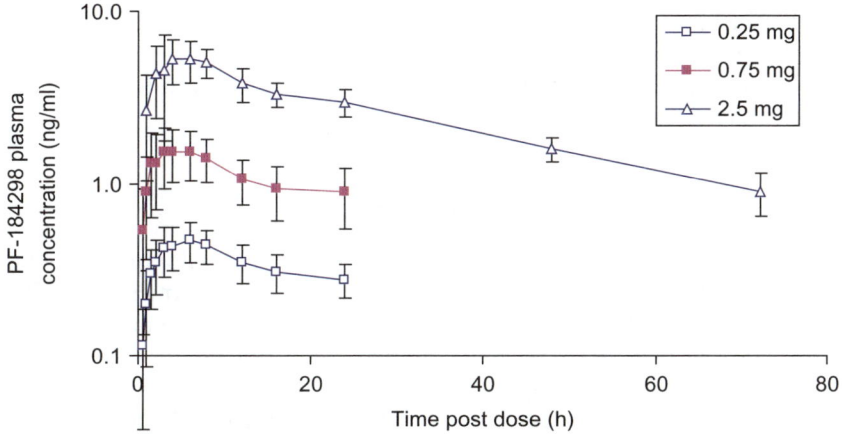

Figure 12.14 Plasma pharmacokinetic profiles of PF-184298 following oral administration to healthy volunteers in Study 1.

Table 12.3 Pharmacokinetic parameters of PF-184298 in plasma following single oral administration to healthy volunteers (Study 2).

Parameter[a]	Dose (mg)			
	1	*3*	*10*	*20*
n	6	6	6	6
C_{max} (ng mL^{-1})	2.3 (0.9)	6.9 (1.8)	25 (9.6)	48 (16)
T_{max} (h)	5.0 (5-8)	3.5 (3–4)	4.0 (3–5)	3.0 (2–4)
$AUC_{0-\infty}$ (ng h mL^{-1})	81 (14)	261 (42)	930 (191)	1741 (314)
$t_{1/2}$ (h)	31 (10)	30 (8.3)	29 (4.0)	31 (8.7)
Cl/F (mL min^{-1} kg^{-1})	2.8 (0.4)	2.2 (0.4)	2.4 (0.4)	2.5 (0.5)
V_D/F (L kg^{-1})	5.8 (2.2)	5.0 (1.4)	5.5 (1.4)	5.8 (2.1)
Ae_{48} (% dose)	23 (3.9)	12 (1.0)	26 (6.1)	18 (3.4)

[a]C_{max}, maximum observed concentration [geometric mean, standard deviation (SD)]; T_{max}, time at which maximum concentration observed (median, range); $AUC_{0-\infty}$, area under the plasma concentration–time curve to infinity (geometric mean, SD); $t_{1/2}$, terminal elimination half-life (arithmetic mean, SD); Cl/F, oral clearance (geometric mean, SD); V_D/F, oral volume of distribution (geometric mean, SD); Ae_{48}, amount excreted unchanged in urine (arithmetic mean, SD).

PF-184298 exhibited linear pharmacokinetics in human subjects over the dose range studied (1–20 mg), with a mean terminal elimination half-life of 30 h. The specific CYP contribution to the clearance of PF-184298 could not be determined from *in vitro* data due to the very low turnover of the compound. However, CYP2D6 genotyping of healthy volunteers in Study 1 demonstrated that this polymorphic CYP is unlikely to play a major role in the clearance of PF-184298, since an ultra-rapid metaboliser and poor metabolisers had similar exposures compared to extensive metabolisers. This was a significant obser-vation since CYP2D6 metabolism was a key issue in the research programme that led to the discovery of PF-184298 and demonstrated a successful outcome of the efforts to reduce CYP2D6 contribution in the series which were described earlier in this chapter.

Further insight into the contribution of individual CYPs came from studies in healthy volunteers with the CYP3A4 inhibitor ketoconazole, which resulted in a modest increase in exposures of PF-184298, demonstrating a role of this enzyme in the clearance of the compound.

12.4   Summary and Conclusions

This drug discovery program successfully delivered a development candidate with the potential to be a "best in class" agent for the treatment of SUI and other SNRI-mediated diseases. Key elements for the successful advancement of PF-184298 to the clinic were (i) high confidence in rationale for the drug target, (ii) a high-quality *in vivo* model which could also serve as a translatable bio-marker to the clinic, (iii) a package of data from preclinical species provided evidence that Phase 1 (cytochrome P450) mediated metabolism was a sig-nificant mechanism in the clearance of PF-184298 and that this compound had very low turnover in human *in vitro* metabolic systems and (iv) use of a bulk-sparing pre-clinical toxicology package which allowed a rapid readout of human pharmacokinetics. This project also highlighted to us the complex nature of working in dual pharmacology and CNS target space. The interplay of many activities, for example potency, selectivity, off-target pharmacology, routes of metabolism and blood–brain barrier penetration proved to be extremely challenging to optimise into a single molecule.

Acknowledgements

The authors would like to thank the following Pfizer colleagues for their valuable contributions to this project:

Ian Gurrell, Gill Allan, Lynn Purkins, Jonathan Fray, Gerwyn Bish, Alan Stobie, Stephen Phillips, Donald Newgreen, Miles Tackett, Arnaud Lemaitre, Julian Blagg, Liz Hopkins, David Winpenny, Alison Bridgeland, Debbie Lovering, Carol Bains, Kerry Paradowski, Iain Gardner, Peter Bungay, Jackie Kendal, Edel Evrard, Nicola Lindsay, Edward Pegden, Paul Blackwell, Tim Buxton, Russell Cave, Yann Lecouturier, Malcolm Mackenny, Melanie Skerten, Hans Rollema,

Fidelma Atkinson, Robert Webster, Neil Attkins, Christopher Kohl, Helena Barker, Thomas Ryckmans, James Mills, Mark Savage, Jackie Luckwell, Nick Clarke, Simon Westbrook, Dominique Westbrook, Rachel McCoy, Gillian Burgess.

References

1. S. Montgomery, *Int. J. Psych. Clin. Pract.*, 2006, **10**(S2), 5.
2. D. S. Baldwin, *Int. J. Psych. Clin. Pract.*, 2006, **10**(S2), 12.
3. N. Ueceyler, W. Haeuser and C. Sommer, *Arthritis Rheum.*, 2008, **58**, 1279.
4. J. F. Wernicke, S. Iyengar and M. D. Ferrer-Garcia, *Curr. Drug Therapy*, 2007, **2**, 161.
5. P. Mariappan, A. Alhasso, Z. Ballantyne, A. Grant and J. N'Dow, *Eur. Urol.*, 2007, **51**, 67.
6. S. M. Holliday and P. Benfield, *Drugs*, 1995, **49**, 280.
7. C. M. Spencer and M. I. Wilde, *Drugs*, 1998, **56**, 405.
8. X. Rabasseda, *Drugs Today*, 2004, **40**, 773.
9. F. Artigas, *CNS Drugs*, 1995, **4**, 79.
10. M. T. C. Lourenco and S. H. Kennedy, *Neuropsychiatr. Dis. Treat.*, 2009, **5**, 127.
11. M. W. Walter, *Drug Dev. Res.*, 2005, **65**, 97.
12. Y. Huang and W. A. Williams, *Expert Opin. Ther. Pat.*, 2007, **17**, 889.
13. S. Liu, Shuang and B. F. Molino, *Annu. Rep. Med. Chem.*, 2007, **42**, 13.
14. G. A. Whitlock, M. D. Andrews, A. D. Brown, P. V. Fish, A. Stobie and F. Wakenhut, in *Topics in Medicinal Chemistry*, ed. S. Napier, M. Bingham, Springer, Berlin, 2009, vol. 4, chap. 3, pp. 53–94.
15. K. B. Thor and I. Yalcin, in *Comprehensive Medicinal Chemistry II*, ed. D. J. Triggle and J. B. Taylor, Elsevier, Oxford, 2006, **vol. 8**, p. 123.
16. N. R. Zinner, *Expert Opin. Investig. Drugs*, 2003, **12**, 1559.
17. S. P. H. Alexander, A. Mathie and J. A. Peters, *Br. J. Pharmacol.*, 2006, **147**(S3), S6.
18. S. P. H. Alexander, A. Mathie and J. A. Peters, *Br. J. Pharmacol.*, 2006, **147**(S3), S11.
19. K. B. Thor, M. Kirby and L. Viktrup, *Int. J. Clin. Pract.*, 2007, **61**, 1349.
20. M. A. Katofiasc, J. Nissen, J. E. Audia and K. B. Thor, *Life Sci.*, 2002, **71**, 1227.
21. A. Erdinc, B. Gurates, H. Celik, A. Polat, S. Kumru and M. Simsek, *Arch. Gynecol. Obstet.*, 2009, **279**, 343.
22. R. T. Owen, *Drugs Today*, 2008, **44**, 653.
23. F. P. Bymaster, E. E. Beedle, J. Findlay, P. T. Gallagher, J. H. Krushinski, S. Mitchell, D. W. Robertson, D. C. Thompson, L. Wallace and D. T. Wong, *Bioorg. Med. Chem. Lett.*, 2003, **13**, 4477.
24. D. T. Wong, *Expert Opin. Investig. Drugs*, 1998, **7**, 1691.
25. For a detailed overview of the pre-clinical and clinical profile of duloxetine, see label information at: www.accessdata.fda.gov/scripts/cder/drugsatfda/index.cfm.

26. B. L. Paris, B. W. Ogilvie, J. A. Scheinkoenig, F. Ndikim-Moffor, R. Gibson and A. Parkinson, *Drug Metab. Dispos.*, 2009, **37**, 2045.
27. K. Conlon, C. Christy, S. Westbrook, G. Whitlock, L. Roberts, A. Stobie and G. McMurray, *J. Pharmacol. Exp. Ther.*, 2009, **330**, 892.
28. D. S. Middleton, M. A. Andrews, P. Glossop, G. Gymer, D. Hepworth, A. Jessiman, P. S. Johnson, M. MacKenny, A. Stobie, K. Tang, P. Morgan and B. Jones, *Bioorg. Med. Chem. Lett.*, 2008, **18**, 5303.
29. G. A. Whitlock, J. Blagg and P. V. Fish, *Bioorg. Med. Chem. Lett.*, 2008, **18**, 596.
30. G. A. Whitlock, P. V. Fish, M. J. Fray, A. Stobie and F. Wakenhut, *Bioorg. Med. Chem. Lett.*, 2008, **18**, 2896.
31. (a) P. D. Leeson and B. Springthorpe, *Nat. Rev. Drug Discovery*, 2007, **6**, 881; (b) D. A. Price, J. Blagg, L. Jones, N. Greene and T. Wager, *Expert Opin. Drug Metab. Toxicol.*, 2009, **5**, 921; (c) J. D. Hughes, J. Blagg, D. A. Price, S. Bailey, G. A. DeCrescenzo, R. V. Devraj, E. Ellsworth, Y. M. Fobian, M. E. Gibbs, R. W. Gilles, N. Greene, E. Huang, T. Krieger-Burke, J. Loesel, T. Wager, L. Whiteley and Y. Zhang, *Bioorg. Med. Chem. Lett.*, 2008, **18**, 4872.
32. (a) M. J. Fray, G. Bish, A. D. Brown, P. V. Fish, A. Stobie, F. Wakenhut and G. A. Whitlock, *Bioorg. Med. Chem. Lett.*, 2006, **16**, 4345; (b) M. J. Fray, G. Bish, A. D. Brown, P. V. Fish, A. Stobie, F. Wakenhut and G. A. Whitlock, *Bioorg. Med. Chem. Lett.*, 2006, **16**, 4349.
33. P. V. Fish, M. J. Fray, A. Stobie, F. Wakenhut and G. A. Whitlock, *Bioorg. Med. Chem. Lett.*, 2007, **17**, 2022.
34. P. A. H. M. Wijnen, R. A. M. Op den Buijsch, M. Drent, P. M. J. C. Kuipers, C. Neef, A. Bast, O. Bekers and G. H. Koek, *Aliment. Pharmacol. Ther.*, 2007, **26**(S2), 211.
35. F. Wakenhut, P. V. Fish, M. J. Fray, I. Gurrell, J. E. Mills, A. Stobie and G. A. Whitlock, *Bioorg. Med. Chem. Lett.*, 2008, **18**, 4308.
36. K. M. Mahar Doan, J. E. Humphreys, L. O. Webster, S. A. Wring, L. J. Shampine, C. J. Serabjit-Singh, K. K. Adkison and J. W. Polli, *J. Pharmacol. Exp. Ther.*, 2002, **303**, 1029.
37. The MDCK-mdr1 cell line is commonly used to assess P-gp recognition; see M. Hammarlund-Udenaes, U. Bredberg and M. Friden, *Curr. Top. Med. Chem.*, 2009, **9**, 148.
38. For a substrate model of the P-gp transporter, see: A. Seelig, *Eur. J. Biochem.*, 1998, **251**, 252.
39. F. Wakenhut, G. A. Allan, P. V. Fish, M. J. Fray, A. C. Harrison, R. McCoy, S. C. Phillips, A. Stobie, D. Westbrook, S. L. Westbrook and G. A. Whitlock, *Bioorg. Med. Chem. Lett.*, 2009, **19**, 5078.
40. N. A. Hosea, W. T. Collard, S. Cole, T. S. Maurer, R. X. Fang, H. Jones, S. M. Kakar, Y. Nakai, W. J. Smith, R. Webster and K. Beaumont, *Clin. Pharmacol.*, 2009, **49**, 513.
41. R. A. Boyd and R. L. Lalonde, *Clin. Pharmacol. Ther.*, 2007, **81**, 24.

CHAPTER 13

The Discovery of TRPV1 Antagonists: Turning up the Heat

MARK H. NORMAN

Director of Medicinal Chemistry, Amgen, Inc., One Amgen Center Drive, Thousand Oaks, CA 91320, USA

13.1 Introduction

Pain is a complex perceptual experience that can have a profound impact on the quality of a person's life. From the beginning of human existence, man has endured pain in all its many forms, and for centuries has sought methods to relieve it. For example, pharmaceutical intervention dates as far back as 5000 years ago, when Sumerian texts reported the effect of the poppy plant (opium) for treating pain and Egyptians used willow extracts (salicylic acid) to reduce the redness and pain of inflamed joints.[1–3] In the intervening years, physicians and scientists have discovered and developed many analgesic agents that have dramatically improved people's lives.[4] However, despite many years of effort to effectively manage pain, individuals still suffer from its physical and emotional consequences and improved treatments are still needed today.[5]

Ironically, one of the most recently identified pain targets has also been modulated unknowingly for centuries.[6] This pain target is linked to the natural product capsaicin (1, Figure 13.1), the pungent component of chili peppers.[7] For example, early European folk medicine employed hot pepper extracts (hence capsaicin) to relieve toothaches and Native Americans were also known

RSC Drug Discovery Series No. 4
Accounts in Drug Discovery: Case Studies in Medicinal Chemistry
Edited by Joel C. Barrish, Percy H. Carter, Peter T. W. Cheng and Robert Zahler
© Royal Society of Chemistry 2011
Published by the Royal Society of Chemistry, www.rsc.org

to have massaged chili pods onto their inflamed gums to ease dental pain.[8,9] The analgesic effects of capsaicin was found to act through the ion channel known as the vanilloid receptor 1 (VR1 or TRPV1). TRPV1 (transient receptor potential vanilloid 1) is a polymodal nociceptor, which belongs to the transient receptor potential (TRP) ion channels family and was first cloned and characterized by Caterina and Julius at the University of California, San Francisco.[10] Subsequently, an explosion of research from pharmaceutical and biotech companies ensued that positioned TRPV1 as a key target for developing novel pain therapeutics.[11–14] As a result of these efforts, TRPV1 as a potential therapeutic target,[15–17] as well as the progress made towards identifying selective antagonists of TRPV1, has been extensively reviewed.[18–22] This highly Ca^{2+} permeable receptor is predominantly expressed in sensory neurons[23–25] and is involved in the detection of painful stimuli. The endogenous activators of TRPV1 are generated as a result of tissue injury and inflammatory conditions in research animals[26–28] and in humans,[29,30] and include heat ($>42\,^{\circ}C$), protons (pH 5),[31] and ligands such as the endocannabinoid anandamide[32] and lipoxygenase metabolites.[33]

When delivered topically, the exogenous ligand capsaicin selectively activates TRPV1 in a certain population of unmyelinated primary sensory neurons. Although painful initially, TRPV1 hyperactivation has an analgesic effect since it leads to long-term desensitization of the sensory neurons to additional agonist challenges. Capsaicin is still in wide use today and is the active pharmacological ingredient of several topical analgesic creams used to relieve minor aches and pains of muscles and joints associated with arthritis.[34] The clinical uses of TRPV1 agonists such as capsaicin, however, are limited due to side effects of burning sensation and irritation, and neurotoxicity[3] resulting from the continuous influx of Ca^{2+} ions into the cells. On the other hand, blockade of the pain-signaling pathway with a TRPV1 antagonist represented a promising new strategy for the development of novel analgesics[35] with potentially fewer side effects. The concept is that TRPV1 antagonists would be able to produce analgesia without the associated nerve damage seen with capsaicin. Early support of this approach came from the study of capsaicin derivatives, for example capsazepine (**2**), the first reported competitive antagonist of TRPV1 (Figure 13.1). It was demonstrated that capsazepine blocks the capsaicin-induced uptake of Ca^{2+} in neonatal rat dorsal root ganglia, and shows species-dependent efficacies in various *in vivo* models of inflammatory

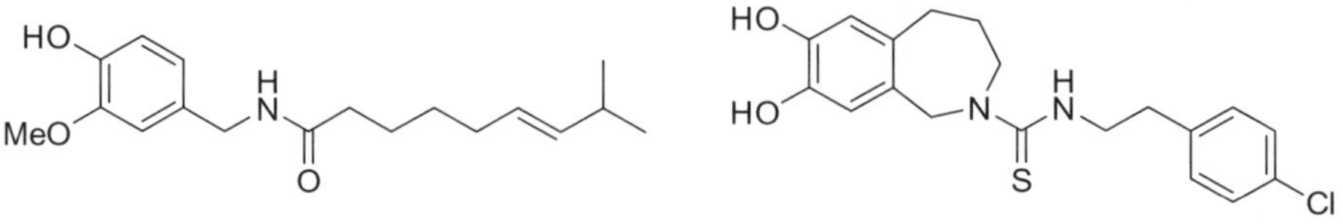

Figure 13.1   The TRPV1 agonist capsaicin (**1**) and the TRPV1 antagonist capsazepine (**2**).

hyperalgesia and chronic pain.[36–38] However, it was also found that capsazepine blocks receptors other than TRPV1, such as voltage-gated Ca^{2+} channels[39] and nicotinic acetylcholine receptors,[40] and does not act as an antagonist when the TRPV1 channel is activated by heat or acid. Other evidence for the role of TRPV1 in pain sensation has come from studies with TRPV1 knock-out mice.[41,42] These TRPV1-deficient mice have a decreased sensitivity to painful heat in inflammatory pain models.

Based on these considerations, we set out to discover and develop TRPV1 antagonists that block all modes of activation. We hypothesized that this type of agent may offer a rapid onset of analgesic action by blockade of the pain-signaling pathway with potentially fewer side effects.[43,44]

13.2 Lead Identification

To generate and identify new leads we conducted a high-throughput screen (HTS) of our corporate library consisting of approximately 1 million compounds. For the purposes of our study, we employed rat or human TRPV1 channel recombinantly expressed in Chinese hamster ovary (CHO) cells in a cell-based assay, where an increase in intracellular calcium resulted in light emission through transfected aequorin. We then confirmed antagonism of the primary hits in assays measuring both capsaicin- and pH-mediated influx of $^{45}Ca^{2+}$.[45] As a result of the screen we identified a number of hits from various structural classes (*e.g.*, compounds **3–5**; Figure 13.2).

Through traditional medicinal chemistry investigations, several potent benzimidazole,[46] thiazole amide,[47] imidazole,[48] and cinnamide[49] TRPV1 antagonists were generated from these original hits and their structure–activity

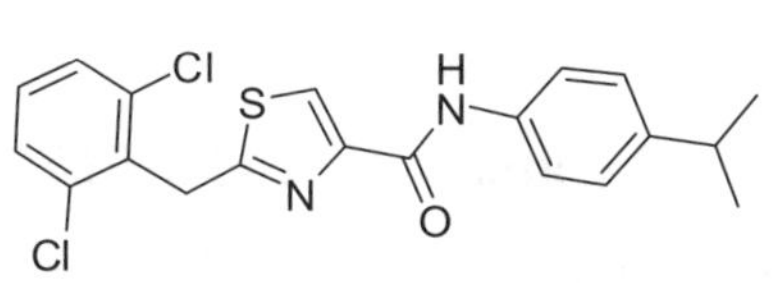

3

hTRPV1(Cap) IC_{50} = 370 nM

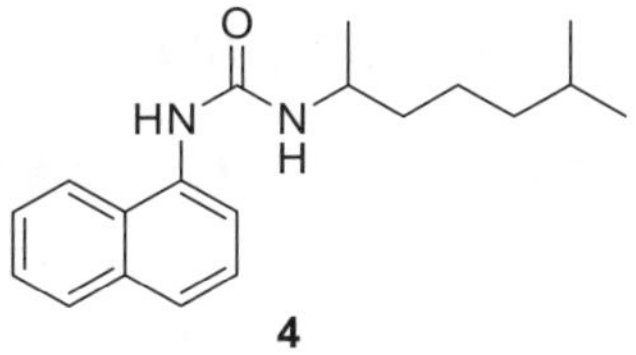

4

hTRPV1(Cap) IC_{50} = 5,800 nM

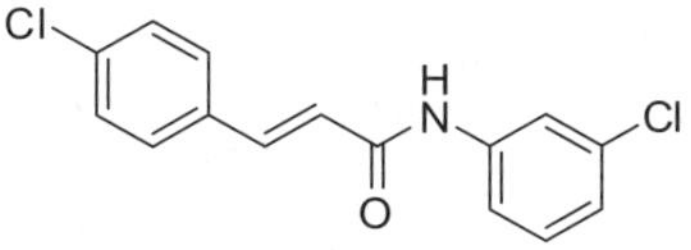

5

hTRPV1(Cap) IC_{50} = 1,200 nM

Figure 13.2 Examples of initial hits obtained from TRPV1 high-throughput screening efforts.

relationship (SAR) development has been reported. However, the journey leading to the clinical candidates described herein started with the cinnamide hit **5**. The following discussion will summarize the challenges encountered with the discovery of AMG 517 as well as the second-generation compound, AMG 628. The key advances in the SAR investigations will be highlighted to illustrate project milestones that were achieved and selected data will be shown for each compound. For full details and additional information regarding the complete SAR development of the compounds in this series, the reader is referred to the original journal articles referenced herein.

13.3   Identification of a Proof-of-concept Molecule

The initial goal of the project was to identify a proof-of-concept molecule for evaluation in *in vivo* models. This was rapidly accomplished by the initial screening of commercially available cinnamides based on the HTS hit **5**. From this effort we identified *N*-arylcinnamide **6** as a potent, competitive antagonist that blocked the capsaicin-, heat-, and pH-induced uptake of $^{45}Ca^{2+}$ in hTRPV1-expressing CHO cells (Figure 13.3).[49]

Schild analysis and electrophysiology experiments indicated that cinnamide **6** was competitive with capsaicin and bound reversibly. Furthermore, compound **6** was shown to be selective against a panel of over 80 other targets, including various voltage- and ligand-gated ion channels, G-protein-coupled receptors, and transporters. *In vivo*, we demonstrated that cinnamide **6** was effective at preventing a capsaicin-induced eye wiping response in a dose-dependent manner, as well as reversing thermal hyperalgesia in a model of inflammatory pain induced by intraplantar injection of complete Freund's adjuvant (CFA) in rats (Figures 13.4A and 13.4B, respectively).[45]

13.4   Lead Optimization

13.4.1   Identification of the Clinical Candidate AMG 517

Although cinnamide **6** served well as a proof-of-concept molecule to support pursuing TRPV1 antagonists as potential pain therapeutics, it suffered from

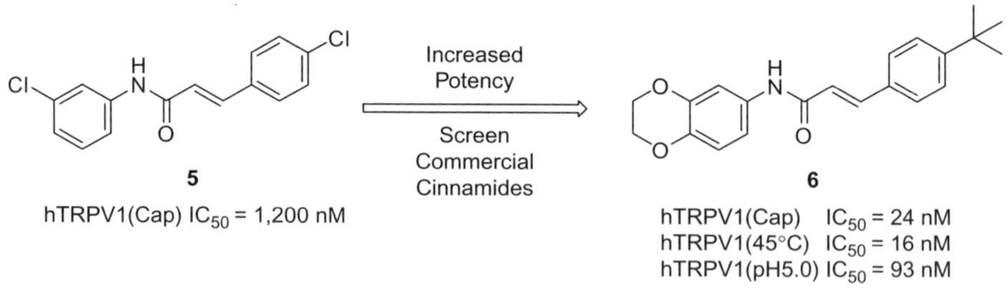

5
hTRPV1(Cap) IC$_{50}$ = 1,200 nM

Increased Potency
Screen Commercial Cinnamides

6
hTRPV1(Cap)   IC$_{50}$ = 24 nM
hTRPV1(45°C)   IC$_{50}$ = 16 nM
hTRPV1(pH5.0) IC$_{50}$ = 93 nM

Figure 13.3   Identification of the proof-of-concept molecule, *N*-arylcinnamide (**6**).

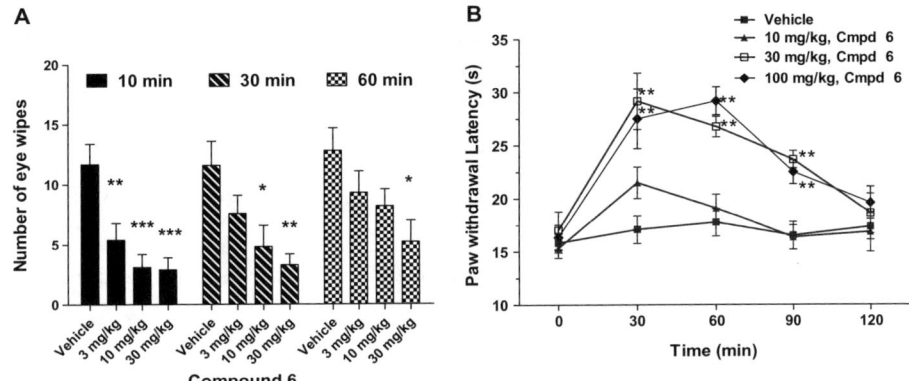

Figure 13.4 (**A**) Pretreatment with cinnamide **6** dose-dependently inhibited capsaicin-induced eye-wiping behavior in rats (i.p. administration). (**B**) Effect of **6** on CFA-induced thermal hyperalgesia in rats. Significant reversal of CFA-induced thermal hyperalgesia was observed with 30 and 100 mg kg^{-1} (i.p.) doses of cinnamide **6**. Reproduced with permission of the American Society for Pharmacology and Experimental Therapeutics from *J. Pharmacol. Exp. Ther.*, 2005, **313**, 474.

high first-pass metabolism and poor oral absorption in rats. Therefore, we undertook an extensive SAR investigation to improve both the potency and pharmacokinetic properties of compound **6**. Early metabolism identification studies showed that the *para tert*-butyl group was extensively oxidized in human liver microsomes (HLM $Cl_{in\ vitro} = 574\,\mu$L min^{-1} mg^{-1}).[50] In this preliminary study we found that a trifluoromethyl group could replace the *para tert*-butyl group and served to give a compound with increased metabolic stability (HLM $Cl_{in\ vitro} = 72\,\mu$L min^{-1} mg^{-1}). In addition, we found that the 7-quinolinyl group was an acceptable substitution for the benzodioxan-7-yl moiety and gave analogues with increased TRPV1 potency (*e.g.*, compound **7**; Figure 13.5). Unfortunately, cinnamide **7** still suffered from poor oral bioavailability in rats ($F_{oral} = 5\%$).

To further advance our understanding of this series, we also conducted a conformational analysis of several *N*-arylcinnamides to gain insights into the optimum orientation of the pharmacophoric elements required for TRPV1 inhibitory activity.[51,52] In these studies, we examined the conformational preferences of several *N*-arylcinnamides using Monte Carlo searching and *ab initio* quantum mechanical calculations at the 6-31G* level.[53] We found that the s-*cis* conformation of compound **6** was preferred over the s-*trans* conformation by 2.6 kcal mol^{-1} and that analogues that were better able to adopt the s-*cis* conformation were more potent TRPV1 antagonists (Figure 13.6). Based on this conformational analysis, we proposed that the bioactive conformer of the *N*-arylcinnamide **6** was s-*cis*, with a coplanar arrangement of the amide carbonyl, the cinnamide double bond, and the β-substituted aryl group being optimal.

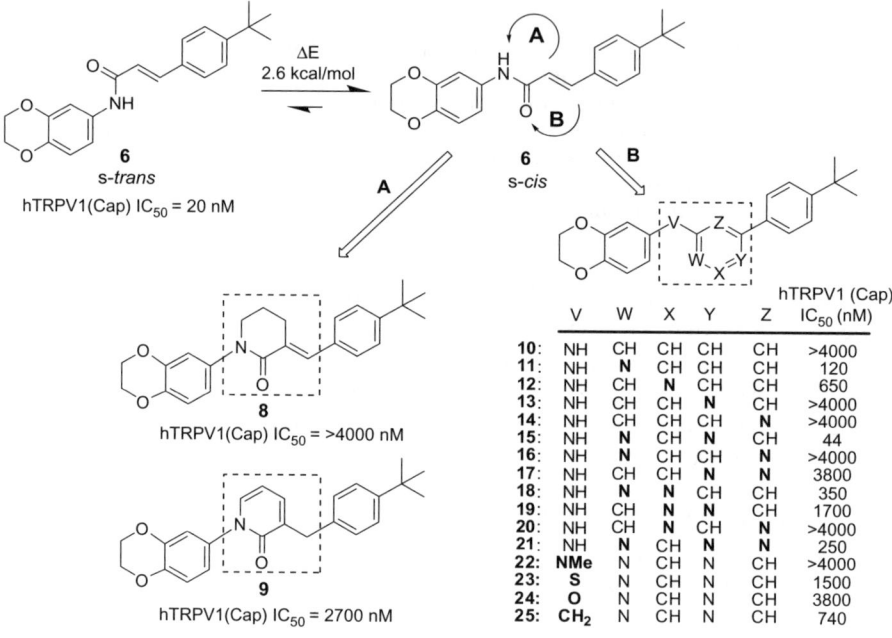

Figure 13.5 Initial cinnamide SAR development to block metabolism and increase potency.

	V	W	X	Y	Z	hTRPV1 (Cap) IC50 (nM)
10:	NH	CH	CH	CH	CH	>4000
11:	NH	N	CH	CH	CH	120
12:	NH	CH	N	CH	CH	650
13:	NH	CH	CH	N	CH	>4000
14:	NH	CH	CH	CH	N	>4000
15:	NH	N	CH	N	CH	44
16:	NH	N	CH	CH	N	>4000
17:	NH	CH	CH	N	N	3800
18:	NH	N	N	CH	CH	350
19:	NH	CH	N	N	CH	1700
20:	NH	CH	N	CH	N	>4000
21:	NH	N	CH	N	N	250
22:	NMe	N	CH	N	CH	>4000
23:	S	N	CH	N	CH	1500
24:	O	N	CH	N	CH	3800
25:	CH2	N	CH	N	CH	740

Figure 13.6 Conformationally restricted analogs of the proposed bioactive *s-cis* conformer of cinnamide **6**.

To test this hypothesis we investigated a variety of replacements of the acrylamide core (**8–21**; Figure 13.6).[54,55] First we examined the effect of restricting the *s-cis* conformer of cinnamide **6** in two general ways: (1) by connecting the amide nitrogen to the α-position of the cinnamide to form lactam **8** and the corresponding dihydro isomer, pyridone **9** (cyclization A; Figure 13.6); and (2) by incorporating the β-position of the cinnamide and the amide carbonyl into various six-membered aromatic rings to give compounds **10–21** (cyclization B; Figure 13.6). Constraining cinnamide **6** in these two ways allowed for the thorough evaluation of 14 different core replacements in which the position of the ring heteroatoms were systematically varied (W, X, Y, and Z; Figure 13.6), while the position of the *tert*-butylphenyl and the benzodioxane groups remained constant. As a second phase of this study, we modified the nature of the "linker" group (V; Figure 13.6) between the pyrimidine heterocyclic core and the benzodioxane moiety to give compounds **22–25**.

In this study of conformationally constrained analogues of the s-*cis* conformer of cinnamide **6** we found that cyclization via "Route A" was detrimental to TRPV1 activity, while the 4-aminopyrimidine core was found to be a suitable isosteric replacement for the acrylamide moiety [compound **15**; hTRPV1 (Cap) $IC_{50} = 44$ nM]. This conformational constraint allowed for a favorable planar alignment of the key pharmacophoric elements found in cinnamide **6** (*e.g.*, the benzodioxane and *tert*-butylphenyl) and also established that the nitrogen adjacent to the aminobenzodioxane group in pyrimidine **15** (*i.e.*, W = N) could serve as a replacement for the cinnamide carbonyl oxygen.

In the next phase of our investigation we made changes to the linking group between the pyrimidine heterocycle and the benzodioxane moiety. We determined that the NH linker was not required for activity in this new series of antagonists and found that it could be replaced with an oxygen atom and still maintain potency (*e.g.*, compound **24**). Finally, we demonstrated that potent, orally available TRPV1 antagonists could be prepared by combining the 7-substituted quinoline and the (trifluoromethyl)phenyl groups (found in our initial cinnamide SAR investigation)[49] with the new 4-oxypyrimidine core. In fact, the new 4-oxypyrimidine derivative **26** was equipotent to cinnamide **7**, but it also had significantly improved pharmacokinetic properties (*e.g.*, lower clearance: rat $Cl_{in\ vivo} = 1.2$ *vs.* 3.9 L h^{-1} kg^{-1}, and higher oral bioavailability: $F_{oral} = 31\%$ *vs.* 5%, respectively) (Figure 13.7).

Compound **26** was shown to be effective *in vivo* at blocking capsaicin-induced hypothermia in rats; it did not, however, show significant activity in the CFA-induced pain model.[54] We postulated that the minimal effect observed in the pain model may be due to a combination of insufficient intrinsic TRPV1 potency and inadequate exposure *in vivo*. Therefore, at this stage of our SAR investigation we sought to improve not only the potency of the 4-oxypyrimidine TRPV1 antagonists, but also to enhance their pharmacokinetic properties *in vivo*.[56]

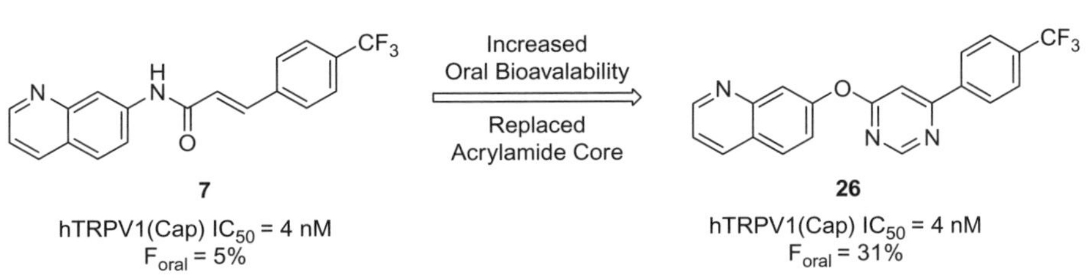

7
hTRPV1(Cap) IC_{50} = 4 nM
F_{oral} = 5%

Increased
Oral Bioavalability
⇒
Replaced
Acrylamide Core

26
hTRPV1(Cap) IC_{50} = 4 nM
F_{oral} = 31%

Figure 13.7 Optimization of the central core leading to enhanced oral bioavailability.

As an initial step towards improving the overall profile of these antagonists, we examined compound **26** in more detail. While compound **26** demonstrated acceptable metabolic stability in rat liver microsomes (RLM $Cl_{in\ vitro} = 111\ \mu L\ min^{-1}\ mg^{-1}$) and good *in vivo* pharmacokinetic properties in rats ($t_{1/2} = 2.8\ h$, $Cl_{in\ vivo} = 1.2\ L\ h^{-1}\ kg^{-1}$, $F_{oral} = 31\%$ at $5\ mg\ kg^{-1}$), measurements of stability in human liver microsomes (HLM $Cl_{in\ vitro} = 250\ \mu L\ min^{-1}$ mg_{-1}) suggested that this compound would be extensively metabolized in humans. Biotransformation studies of compound **26** revealed the region of the molecule that was most susceptible to metabolism. Upon incubation of compound **26** with human liver microsomes in the presence of NADPH, only two oxidative metabolites were found. Although the exact sites of oxidation were not ascertained, analysis by MS/MS indicated that both metabolites were the result of mono-oxidation on the quinoline ring, while the center pyrimidine core and the (trifluoromethyl)phenyl moieties were unaffected. Therefore, our strategy to increase metabolic stability, as well as enhance potency in this series, was to systematically explore alternative heterocycles as replacements for the quinoline ring found in compound **26** (Figure 13.8).

The initial set of quinoline and isoquinoline derivatives examined revealed that the 8-quinolinyl isomer was slightly less potent in the TRPV1 capsaicin-mediated assay [hTRPV1 (Cap), $IC_{50} = 18\ nM$]; however, it was more stable than compound **26** upon incubation in rat or human liver microsomes (HLM $Cl_{in\ vitro} = 135\ \mu L\ min^{-1}\ mg^{-1}$). Furthermore, we found that additional TRPV1

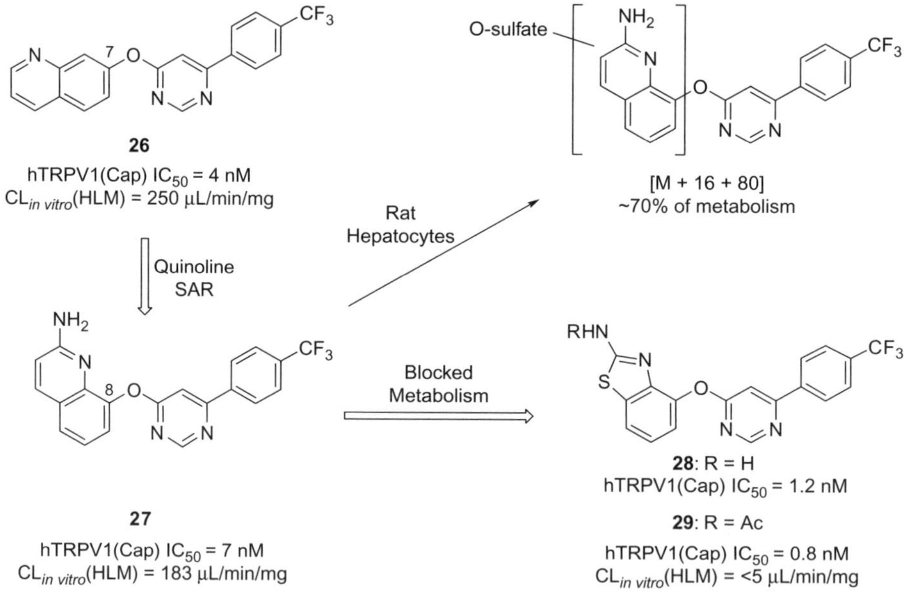

Figure 13.8   Optimization of the heterocyclic ring led to improved potency and reduced metabolism.

potency in the acid-mediated assay could be obtained by substituting the 2-position of the quinoline ring with an amino group [compound **27**; hTRPV1 (pH 5), $IC_{50} = 73$ nM]. However, metabolite identification studies in rat hepatocytes showed that compound **27** was still extensively metabolized on the 2-aminoquinoline moiety (Figure 13.8).

With this information in hand, we prepared a series of analogs designed to block the proposed sites of metabolism, focusing on the 2-aminopyridyl portion of **27**. The 2-aminobenzothiazole analogue **28** was designed to block two of the proposed metabolic hot spots on the quinoline ring of **27** by replacing the C3–C4 atoms with sulfur. This modification resulted in a dramatic improvement in the metabolic stability in human liver micosomes ($Cl_{in\ vitro} < 5\,\mu L\,min^{-1}\,mg^{-1}$). In addition, compound **28** showed a 6-fold improvement in potency over **27** in the capsaicin-mediated assay ($IC_{50} = 1.2$ nM *vs.* 7 nM); however, activity in the acid-mediated assay was sacrificed ($IC_{50} > 4000$ nM).

It had been our experience that subtle modifications to this region of the heterocycle can impact activity in the TRPV1 acid-mediated assay. Therefore, a series of compounds was prepared to investigate the effect of an additional substitution on the nitrogen of compound **28**. We found that the *N*-acylated analogue **29** had enhanced potency in both assays ($IC_{50} = 0.76$ and 0.62 nM in the capsaicin- and acid-mediated assays, respectively), while also maintaining excellent metabolic stability in human liver micosomes ($Cl_{in\ vitro} < 5\,\mu L\,min^{-1}\,mg^{-1}$).

With compound **29** selected as the lead candidate, its pharmacokinetic properties were examined in rats, dogs, and monkeys (Table 13.1). In all species examined, compound **29** exhibited low clearance and moderately high volumes of distribution, with consequently long half-lives. Oral bioavailability among the three species tested ranged from 23 to 52%.

Table 13.1 Mean pharmacokinetic parameters for compound **29** following i.v. and p.o. administration to fasted rat, dog, and monkey, with projected values for human.

| Species | Intravenous dosing[a] | | | | Oral dosing[b] |
	AUC_{0-inf} $(ng\,h\,mL^{-1})$	Cl $(mL\,h^{-1}\,kg^{-1})$	V_{ss} $(mL\,kg^{-1})$	$t_{1/2}$ (h)	F (%)
Rat	8800	120	4000	31	51
Dog	7400	140	7000	41	23
Monkey	37000	30	2300	62	52
Human (projected)	–	50	4500	60–120	–

[a] 1 mg kg^{-1} in 80% PEG-400/H$_2$O; $n = 3$ animals per group. Variability for AUC_{0-inf}, Cl, V_{ss}, and $t_{1/2}$ values ranged from 7 to 38% for all species.
[b] 1 mg kg^{-1} suspension in 10% Pluronic/Oraplus; $n = 3$ animals per group. Variability for F_{oral} ranged from 26 to 50% for all species.

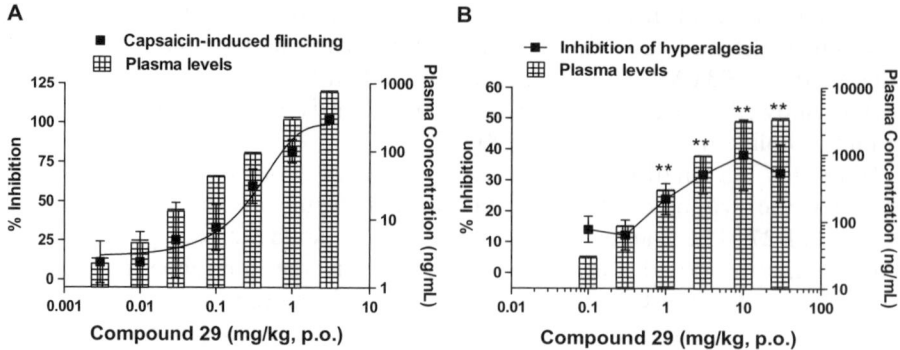

Figure 13.9 (A) Compound **29** blocked capsaicin-induced flinch in rats. Compound **29** was administered p.o. to rats 60 min prior to capsaicin challenge. The number of flinches in the first 1 min were counted and plotted against the dose of compound **29**. (B) AMG 517 blocked thermal hyperalgesia in CFA model of pain. Significant reversal of CFA-induced thermal hyperalgesia was observed at doses $\geq 1 \, mg \, kg^{-1}$ of compound **29**. Reproduced with permission of the American Society for Pharmacology and Experimental Therapeutics from *J. Pharmacol. Exp. Ther.*, 2007, **323**, 128.

Human pharmacokinetic parameters were projected based on allometric scaling and are presented in Table 13.1. Based on the allometric projection, compound **29** was predicted to have a long half-life in humans (60–120 h). The projected half-life in humans was expected to result in a 4-fold accumulation at steady state, based on a QD dosing regimen, with minimal peak-to-trough fluctuations and low variability.

In vivo evaluation of compound **29** showed that it was effective in a rodent on-target biochemical challenge model (capsaicin-induced flinch, $ED_{50} = 0.33 \, mg \, kg^{-1}$ p.o.) (Figure 13.9A). In the CFA pain model in rats, compound **29** showed a dose-dependent inhibition of thermal hyperalgesia ($MED = 0.3 \, mg \, kg^{-1}$, p.o.) (Figure 13.9B).

In addition to reversing inflammation-induced pain behavior in rats, we found that compound **29** also caused hyperthermia in rodents, dogs, and monkeys (1–1.5 °C) but not in TRPV1 knockout mice, suggesting that TRPV1 is tonically activated *in vivo* and involved in body temperature regulation (rat data shown in Figure 13.10A).[57,58] Interestingly, this hyperthermia was attenuated with repeated dosing of antagonists to rats, dogs, and monkeys (monkey data shown in Figure 13.10B).[59] Other than the hyperthermic effect of compound **29**, it was well tolerated in preclinical safety studies, with safety margins of 78-fold in 14-day rat toxicology studies (based on C_{max} compared to rat flinch EC_{50} of $96 \, ng \, mL^{-1}$). Because of its excellent pharmacokinetic properties, *in vivo* efficacy, and good safety profile, compound **29** was selected for further evaluation in human clinical trails for the treatment of inflammatory pain and was redesignated AMG 517.

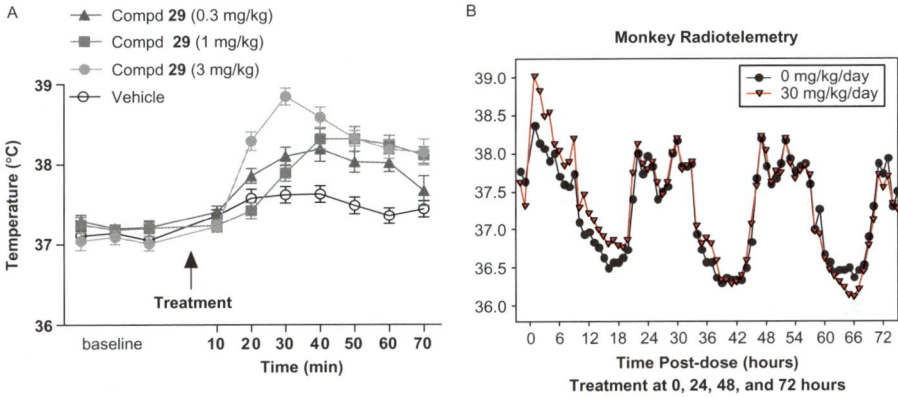

Figure 13.10 Compound **29** effect on body temperature. (A) Body temperature of rats administered with either vehicle or different doses of compound **29**. A significant increase in body temperature was seen 30–40 min after administration of compound **29**. (B) Compound **29**-induced hyperthermia attenuates after repeat dosing in female monkeys. Compound **29** caused an increase of approximately 1.1 °C on day 1 at 3 h post administration but not on days 2–4. Reproduced with permission of the American Society for Pharmacology and Experimental Therapeutics from *J. Pharmacol. Exp. Ther.*, 2007, **323**, 128.

13.4.2 Identification of the Second-generation Clinical Candidate AMG 628

As AMG 517 progressed into clinical development, we sought to identify a second-generation TRPV1 antagonist with an improved profile as a potential backup compound.[60] The two areas that we thought AMG 517 could be improved upon were its half-life and solubility. In preclinical studies, AMG 517 was found to have very long half-lives in multiple species (Table 13.1). In addition, AMG 517 has low aqueous solubility ($<1\,\mu g\,mL^{-1}$ in PBS or 0.01 N HCl).[61] These two factors, although not critical for further progression of AMG 517, presented challenges to the development of this candidate. Therefore, our goal was to identify a novel second-generation clinical candidate with a similar pharmacological profile to AMG 517 but with increased aqueous solubility and a reduced half-life. Towards this end we examined four different approaches, as illustrated in Figure 13.11: (1) alternatives to the benzothiazole moiety,[62] (2) substitutions on the 2-position of the pyrimidine core,[63] (3) substitutions at the *ortho*-position of the 4-(trifluoromethyl)phenyl ring,[64] and (4) replacements of the 4-(trifluoromethyl)phenyl group with various saturated aza heterocycles (*e.g.*, compound **33**, where $Y = N$ and $Z = CH$, O, or N) to create the potential for salt formation.[60] Each of these approaches met with some success (*e.g.*, compounds **30–32** had good potency, improved solubility, and decreased half-lives); however, it was the fourth approach of replacing of the 4-(trifluoromethyl)phenyl group with saturated aza heterocycles which proved to be the most successful and led to the backup candidate, AMG 628.

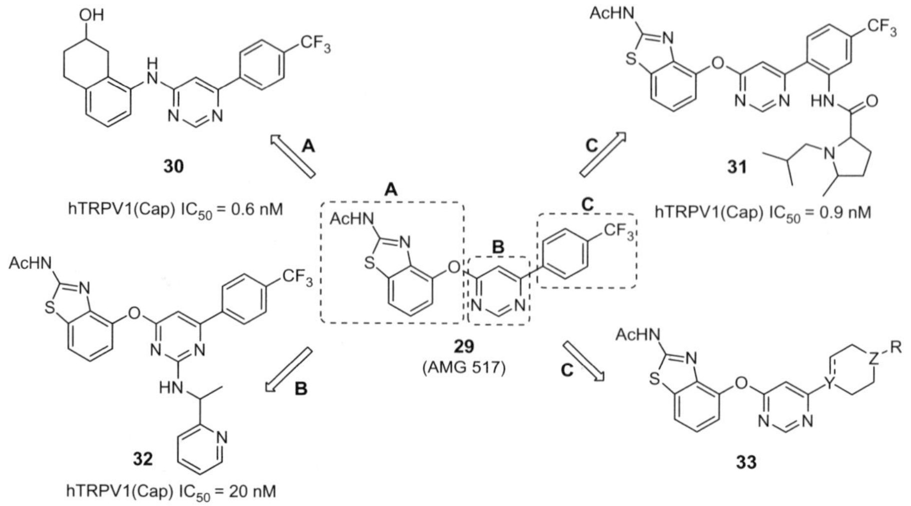

Figure 13.11   Four approaches to reduce half-life and increase solubility of AMG 517.

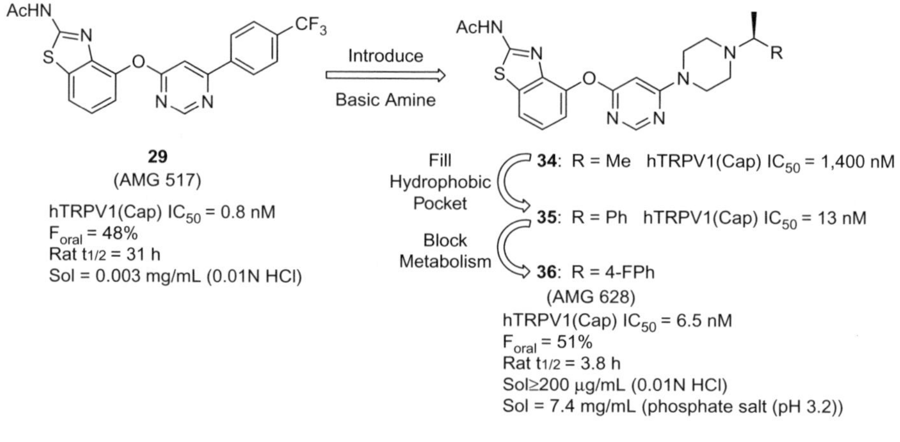

Figure 13.12   Key SAR leading to AMG 628, a second-generation TRPV1 antagonist with increased solubility and reduce half-life.

In this investigation we demonstrated that replacement of the 4-(trifluoromethyl)phenyl moiety with heterocycles was effective at increasing aqueous solubility and modifying the pharmacokinetic properties of AMG 517. Some of the key SAR milestones are illustrated in Figure 13.12. Initial replacement of the 4-(trifluoromethyl)phenyl group with isopropylpiperazine resulted in a compound with significantly increased solubility; however, TRPV1 potency was compromised (compound **34**). We found that potency could be

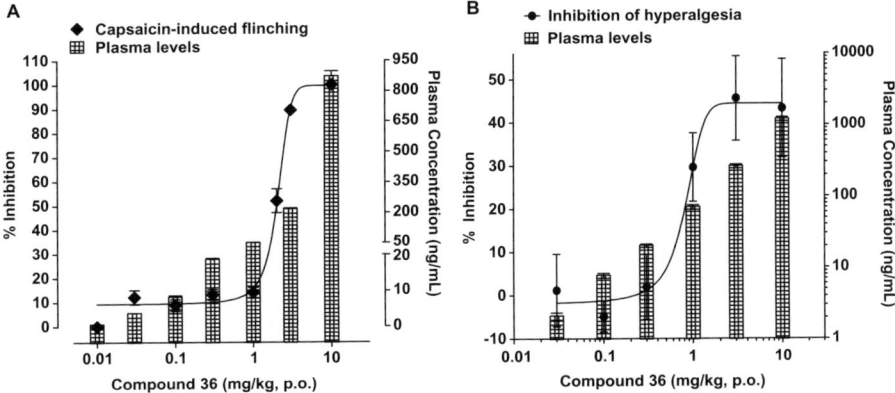

Figure 13.13 (A) Effect of compound **36** on capsaicin-induced flinching in rats and corresponding plasma levels. (B) Dose–response curve with compound **36** in CFA-induced thermal hyperalgesia with corresponding plasma levels. Reproduced with permission of the American Chemical Society from *J. Med. Chem.*, 2007, **50**, 3528.

regained by increasing the size of the hydrophobic group in the 4-position of the piperazine ring with a 1-(1-phenylethyl)piperazine group (compound **35**). While this analog showed excellent potency in both assays ($IC_{50} = 13$ nM and 17 nM in the capsaicin- and acid-mediated assays, respectively) and good solubility ($\geq 200\,\mu g\,mL^{-1}$ in 0.01 N HCl), the compound suffered from high *in vivo* clearance ($Cl_{in\ vivo} = 3.9\,L\,h^{-1}\,kg^{-1}$). Therefore, we prepared a series of substituted analogs of compound **35** with the intention of blocking putative metabolism on the aromatic ring and thus reducing clearance. The 4-fluoro derivative **36** provided the best results.

Compound **36** demonstrated excellent *in vitro* potencies [hTRPV1(Cap) $IC_{50} = 6.5$ nM], *in vivo* efficacy (Figure 13.13; capsaicin-induced flinch $ED_{50} = 1.9$ mg kg^{-1} and CFA-induced thermal hyperalgesia MED = 1 mg kg^{-1}), and had increased solubility (≥ 200g mL^{-1} in 0.01 N HCl) and a reduced half-life (rat $t_{1/2} = 3.8$ h, dog $t_{1/2} = 2.7$ h, monkey $t_{1/2} = 3.2$ h) as compared to AMG 517. Based on its enhanced overall profile, compound **36** (AMG 628) was chosen as our second-generation TRPV1 clinical candidate for evaluation as a potential new treatment for inflammatory pain.

13.5 Phase I Clinical Data for AMG 517

Three Phase I studies were conducted with AMG 517: (1) a single-dose escalation study, (2) a multi-dose study, and (3) a study in subjects having undergone third-molar extraction.[65] The results of the three studies are summarized below.

The first study involved oral administration of increasing doses (1–25 mg) of AMG 517 to healthy volunteers. In this study, AMG 517 demonstrated rapid absorption ($T_{max} = 1$–2 h) and linear pharmacokinetics (C_{max} and AUC_{0-inf}

increased in an approximately dose-proportional manner; Figure 13.14A). Consistent with the long half-life of this molecule in rats, dogs, and monkeys, a long mean $t_{1/2}$ range of 13–23 days was observed in humans across the dose range evaluated. No indications of safety issues were observed upon AMG 517 administration except moderate, generally dose- and plasma concentration-dependent, increases in body temperature (Figure 13.14B). An increase in body temperature typically occurred between 1 and 4 h after AMG 517 administration, and returned to baseline values within 24 h, suggesting the transient nature of hyperthermia induced by AMG 517. In addition, the C_{max} was observed at 1–2 h post-dosing, usually earlier than the peak increase in body temperature, suggesting that the development of hyperthermia after exposure to AMG 517 in healthy subjects was the result of a physiologic response to TRPV1 blockade. During this study, an exception to plasma concentration-dependent hyperthermia was observed in one subject (in the 25 mg dose group) whose body temperature was increased to 39.9 °C, although this person did not have the highest C_{max} (Figure 13.14B). Thus, despite overall dose- and plasma concentration-dependent hyperthermia, the magnitude of temperature elevation after exposure to single doses of AMG 517 also reflected individual susceptibility to the hyperthermic effect of TRPV1 blockade.

As mentioned previously, when AMG 517 was administered daily to rats, dogs, or monkeys, hyperthermia was attenuated to values indistinguishable from the vehicle-treated groups. This habituation, or attenuation, to the hyperthermic effects of AMG 517 supported the initiation of a Phase I multiple-dose study in humans. When subjects were exposed to daily single oral doses of AMG 517 (2, 5, or 10 mg) for seven days, they had dose-dependent increases in body temperature, with the mean maximal temperature reaching 38.3 + 0.1 °C on the first day of drug administration (Figure 13.15A). There was a statistically significant

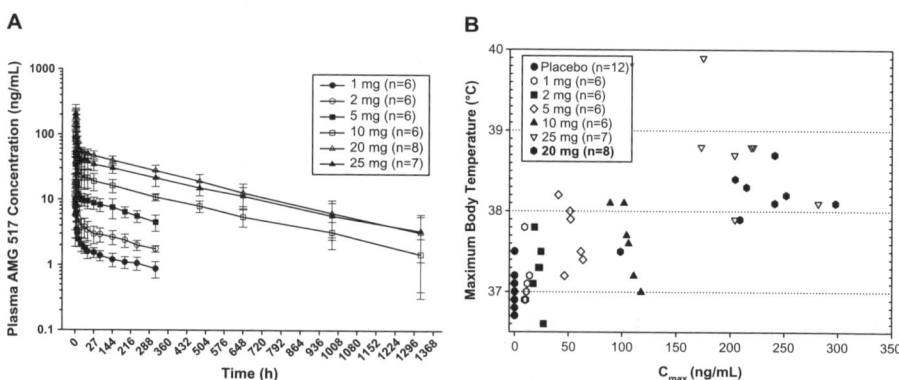

Figure 13.14   (A) Pharmacokinetic profile of several doses of AMG 517 in human volunteers (1–25 mg). (B) Mean of the maximum body temperature (tympanic) *vs.* maximum observed plasma concentration following oral administration of single doses of AMG 517 (1–25 mg). Reproduced with permission of the International Association for the Study of Pain.

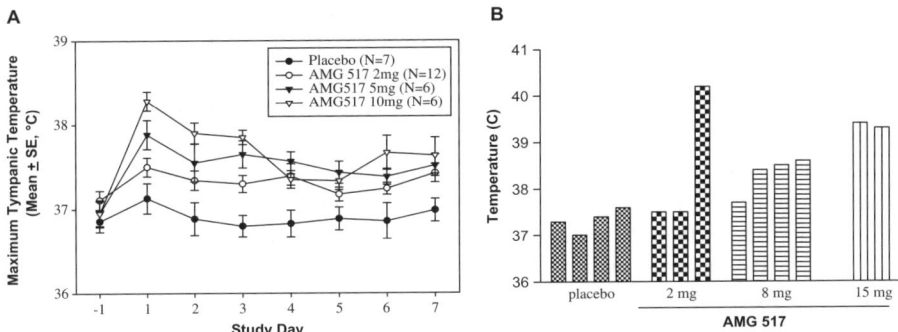

Figure 13.15 (A) Mean of the maximum body temperature (tympanic) after daily oral administration of multiple doses of AMG 517 (2, 5, 10 mg) for 7 days. Maximum tympanic temperature on day 1 for 10 mg AMG 517 dose group was significantly higher compared to days −1 and 2–7. Maximum tympanic temperature on day 1 for 5 mg AMG 517 dose group was significantly higher compared to days −1 and 2, 4–7. Mean maximum body temperatures of all doses of AMG 517 were significantly greater than placebo on days 1–7. There was no significant temperature change for placebo group. (B) Tympanic body temperature measurements of subjects exposed to placebo or different doses of AMG 517 after third-molar extraction. Reproduced with permission of the International Association for the Study of Pain.

attenuation of hyperthermia at the highest dose (10 mg dose of AMG 517) on days 2 through 7; however, the mean maximal body temperatures of all subjects administered with all doses of AMG 517 were still significantly higher compared to placebo-treated group over the 7-day dosing period. The reasons for unambiguous attenuation of hyperthermia after repeated exposure to all doses of AMG 517 in preclinical species but not in humans are unclear. One possibility is the differences in the doses used in preclinical species and in human studies. Rats and monkeys received 3 mg kg^{-1} and 30–500 mg kg^{-1}, respectively, whereas the maximum repeated dose for humans was 10 mg, or approximately 0.14 mg kg^{-1} (199 ± 36 ng mL^{-1} plasma concentration), which is about 20-fold lower than the lowest dose used in rats. Doses greater than 0.14 mg kg^{-1} were not used for repeated dosing studies in humans because of the undesirable hyperthermia observed at the higher doses in the first Phase I study, and because the long half-life of AMG 517 in humans was expected to cause an accumulation during repeated dosing. The plasma concentration of 10 mg dose of AMG 517 on day 8 was 199 ± 36 ng mL^{-1}, which was approximately equivalent to the C_{max} obtained at 20 and 25 mg doses used in the first Phase I study.

Although AMG 517 caused a generally plasma concentration-dependent hyperthermia, it was unknown what plasma concentration of AMG 517 would produce analgesia in humans and whether that concentration could be reached without triggering hyperthermia. Since TRPV1 is expressed in the dental pulp and is believed to contribute to pain after molar extraction, we decided to proceed to an efficacy study to treat acute pain after molar extraction. In this Phase Ib study, subjects were administered a single dose of placebo, 2, 8, or 15 mg of AMG 517 as soon as they experienced moderate to severe post-operative pain. Body temperature measurements obtained from this study are illustrated in Figure 13.15B. A 2 mg (or approximately $29 \, \mu g \, kg^{-1}$) dose of AMG 517 triggered hyperthermia that exceeded 40 °C in one subject, with hyperthermia of > 39 °C that persisted for three days despite multiple doses of antipyretic medication. Similar marked hyperthermia that persisted for two or more days was also observed in two subjects who received a 15 mg dose of AMG 517 after molar extraction. Hyperthermia of > 38 °C was observed in three out of four subjects who received 8 mg of AMG 517. Body temperatures of 39–40.2 °C persisted for 1–4 days in 33% of the subjects who received AMG 517 after molar extraction, with no other observed cause for this marked and persistent hyperthermia. These studies also indicated that the largest-magnitude hyperthermia (> 40 °C), which occurred in one out of three subjects who received 2 mg of AMG 517, appeared to be related to individual susceptibility. The emergence of this marked and persistent hyperthermia observed in subjects undergoing molar extraction presented a major hurdle for the clinical development of AMG 517. Although the plasma concentrations of AMG 517 observed in this Phase Ib study were similar to those obtained in the first pharmacokinetic and safety study, such large magnitude and persistent hyperthermia was not observed after exposure to AMG 517 in healthy subjects. It is possible that the surgical procedure and TRPV1 blockade may have acted additively or synergistically to produce an undesirable, marked, and persistent hyperthermia in susceptible individuals. Alternately, access of AMG 517 in solution to normal or sensitized TRPV1 receptors in the surgical wound may have contributed to the differences in hyperthermia produced by AMG 517 in these two populations.

Because of hyperthermia after exposure to single and multiple doses of AMG 517 in healthy volunteers, and due to marked and persistent hyperthermia after exposure to single doses of this molecule in subjects who underwent molar extraction, clinical studies of AMG 517 were discontinued. Unfortunately, too few subjects were evaluated to determine the potential analgesic effect of AMG 517 due to the early termination of this study and therefore the target coverage of AMG 517 required for analgesia in humans remains unknown. Because the second-generation backup compound, AMG 628, had a similar *in vitro* profile as AMG 517 and it also caused hyperthermia in preclinical models, it was not advanced into clinical trials.

13.6 Approaches to Address the Hyperthermic Response

Due to the undesirable hyperthermia observed for AMG 517 in humans, we examined the effect of TRPV1 antagonists on body temperature regulation more closely and sought possible ways to eliminate or minimize this on-target effect. Initially, we examined the effects of several TRPV1 antagonists represented by various chemotypes, including: cinnamides, ureas, amides, benzimidazoles, and piperazine carboxamides on rat body temperature.[57] We found that all chemotypes that we tested caused a 0.5–1.5 °C increase in body temperature, suggesting that antagonist-induced hyperthermia was not chemotype specific but rather occured because of TRPV1 blockade *in vivo*. Furthermore, we also found that a TRPV1-selective antagonist did not cause hyperthermia in TRPV1 knockout mice, demonstrating that the entire hyperthermic effect was TRPV1 mediated.[57]

We also took two approaches in attempts to eliminate or minimize the hyperthermic response: (1) we examined the effect of peripherally restricted TRPV1 antagonists,[66] and (2) we identified and evaluated compounds that display differential pharmacologies (*i.e.*, compounds that differentially modulate distinct modes of *in vitro* TRPV1 activation such as capsaicin, pH 5, and heat) *in vivo*.[67]

13.6.1 Peripherally Restricted TRPV1 Antagonists[68]

Because the hypothalamus is the key area of the brain known to be involved in thermoregulation,[69] we postulated that we may be able to separate the analgesic effect of our TRPV1 antagonists from their hyperthermic effect if they were excluded from the central nervous system (CNS). Unfortunately, all of the initial derivatives of AMG 517 we had prepared and evaluated at that time had significant brain penetration. For example, AMG 517 had exhibited significant CNS penetration with a brain-to-plasma ratio (B/P) of 1. Therefore, to test our hypothesis we needed to prepare and evaluate novel derivatives of AMG 517 that had low brain-to-plasma ratios while at the same time maintaining potent TRPV1 activities.

Our approach to the discovery of peripherally restricted derivatives of AMG 517 was based on the premise that passive diffusion is the primary process for translocation of these compounds from the bloodstream to the brain. CNS permeability (and in general transcellular permeability) is a complex function of the physicochemical properties of molecules such as size, lipophilicity, hydrogen-bonding potential, charge, and conformation.[70,71] In general, CNS-penetrant compounds are somewhat smaller than other biologically active molecules (90% of them have molecular weights of less than 500). They also have anywhere from 2 to 7 hydrogen-bonding groups, while the range for non-CNS-penetrant agents is 2–9. In addition, over 90% of the CNS-active drugs have 7 or fewer rotatable bonds, while this number is 10 for peripherally restricted compounds. Finally, compounds that access the CNS are in general more lipophilic (larger log *P*) than peripherally targeted compounds,[72,73] and they generally have polar surface area (PSA) values less than 90.[74] Therefore, with these guiding principles in mind, we sought to restrict the CNS penetration of our TRPV1 antagonists by: (1) increasing molecular weight; (2) increasing the number of hydrogen bond donors (HBD); (3) increasing the number of rotatable bonds; (4) decreasing lipopholicity; and (5) increasing the polar surface area.[75]

As we modified the compounds in this series to minimize CNS penetration, it was important to make specific structural changes that would maintain excellent TRPV1 antagonism. Therefore, as a basis for the design of new peripherally restricted derivatives, we were guided by the results obtained from our previous SAR investigations.[55,63,76] For example, we employed the 2-aminoquinoxalinone ring system that had previously demonstrated superior TRPV1 antagonism (Table 13.2; ring system B). This heterocycle provided compounds with higher PSA values and added two additional hydrogen-bond donors. Our earlier studies also revealed that substituents at the 2-position of the central pyrimidine ring (R^1) were, in general, well tolerated for TRPV1 potency.[61] Additionally, we found that we could introduce substituents at the *ortho*-position (R^2) of the phenyl ring (X = CH) or pyridine ring (X = N) and maintain good TRPV1 activities.[62] Therefore, we prepared a series of derivatives with additional heteroatoms and polar groups at R^1, R^2, and X and examined their effect on potency, CNS penetration,[77] efficacy (as determined by inhibition of *in vivo* capsaicin-induced flinching), and body core temperature. Modifying AMG 517 in this way led to the identification of five potent and peripherally restricted TRPV1 antagonists (Table 13.2) that showed good on-target *in vivo* efficacy in the capsaicin-induced flinch model (Figure 13.16A).

These derivatives were tested in a telemetry model to examine their effect on body core temperature. Rats implanted with radiotelemetry probes were dosed orally with a single dose of the TRPV1 antagonists; body core temperatures were recorded 30 min prior to drug administration and continued for 2 h post-dose. Figure 13.16B shows the increase in body core temperature over control animals at the 60-min time point, which usually corresponds with the maximum increase in body core temperature. Unfortunately, all five TRPV1 antagonists showed a significant increase in body core temperature ($> 0.5\,^{\circ}C$) compared to vehicle control. These results suggested that peripheral restriction alone was not sufficient to eliminate TRPV1 antagonist-induced hyperthermia and the site of action for the hyperthermic effect is predominantly outside of the blood–brain barrier. From this study we concluded that minimizing brain penetration was not sufficient to separate the analgesic properties of TRPV1 antagonists from the on-target hyperthermia observed in rodents.

13.6.2 TRPV1 Antagonists with Differential *In vivo* Pharmacologies

The final approach we took to eliminate TRPV1-induced hyperthermia was to evaluate compounds that display differential TRPV1 *in vitro* pharmacologies.[67] Until this point we had targeted compounds that blocked all modes of activation and had found that all compounds with this profile caused hyperthermia. However, since TRPV1 is a poly-modal sensor of lipids, protons, heat, and

Table 13.2 *In vitro* rat TRPV1 activities and calculated physicochemical properties for AMG 517 and several peripherally restricted derivatives.[a]

Compd	Het	R¹	R²	X	rTRPV1[b] (Cap)	MW	PSA	clog P	HBD	Rot	B/P[c]
AMG 517	A	H	H	CH	0.9	430	77	5.1	1	5	0.73
37	B	(structure)	H	CH	0.6	472	128	2.9	4	8	0.05
38	B	(structure)	H	CH	2.8	519	132	3.6	4	7	0.04
39	B	H	NH₂	N	0.8	415	146	3.6	5	4	0.11
40	B	H	(structure)	N	0.7	473	141	3.6	4	8	0.04
41	B	(structure)	NH₂	CH	3.8	501	154	2.9	6	7	0.07

[a]Physicochemical properties were calculated using Amgen proprietary software (ADAAPT).[75]
[b]IC$_{50}$ values based on inhibition of capsaicin (500 nM) of ^{45}Ca^{2+} into rat TRPV1 expressing CHO cells.
[c]Study in male Sprague-Dawley rats dosed (i.v.) at 5 mg kg^{-1} solution in DMSO with sampling at 0.5 h.

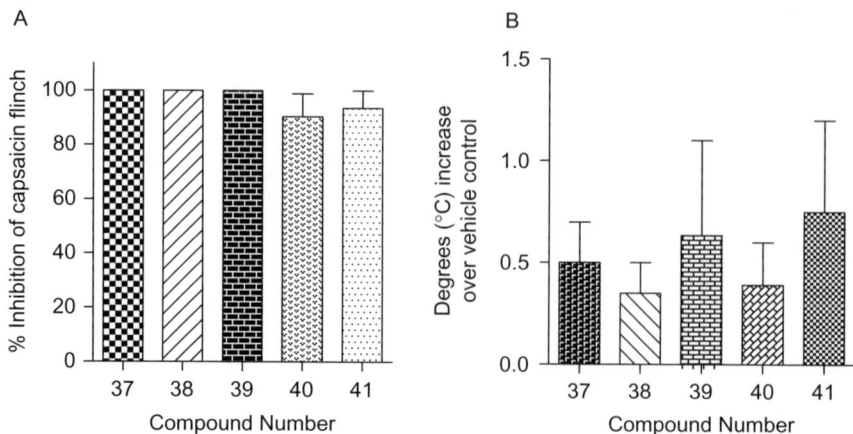

Figure 13.16 (A) Inhibition of capsaicin-induced flinching. Compounds dosed orally at $1\,mg\,kg^{-1}$ (**37**, **40**), $3\,mg\,kg^{-1}$ (**39**), or $10\,mg\,kg^{-1}$ (**38**, **41**) in 5% Tween 80/Oralplus 1 h prior to capsaicin challenge in male Sprague Dawley rats. (B) Effect of TRPV1 antagonists on body core temperature. Compounds dosed orally at $1\,mg\,kg^{-1}$ (**37**, **40**), $3\,mg\,kg^{-1}$ (**39**), or $10\,mg\,kg^{-1}$ (**38**, **41**) in 5% Tween 80/Oralplus to male Sprague Dawley rats implanted with radiotelemetry probes. Body core temperature baselines were recorded 30 min before dosing, and temperature recordings were continued for 2 h. The histogram shows body core temperatures at 60 min post-drug administration. Adapted with permission of the American Chemical Society from *J. Med. Chem.*, 2008, **51**, 2744.

other stimuli, we postulated that compounds with alternative profiles (depending on how the channel is activated) may also provide different pharmacological responses. To identify such compounds we mined our database and examined TRPV1 activities against rat capsaicin, pH 5, and heat. From this analysis we identified compounds with four distinct rat TRPV1 modulation profiles (Figure 13.17). Compound **42** (profile A) blocked all modes of activation (similar to AMG 517); compound **43** (profile B) blocked capsaicin and heat but not proton activation; compound **44** (profile C) blocked capsaicin but not heat activation and potentiated pH 5 activation; and compound **45** (profile D) blocked capsaicin activation and potentiated both heat and pH 5 activation of TRPV1. It should be noted that both compounds **44** and **45** by themselves did not induce $^{45}Ca^{2+}$ uptake into CHO cells expressing the rat TRPV1 channels at physiological pH (pH 7.2), indicating that they are not partial agonists. In addition to blocking capsaicin activation, all four compounds inhibited putative endogenous ligand (AEA and NADA) activation of TRPV1 (data not shown).

To evaluate the effect of the four unique TRPV1 modulation profiles *in vivo*, compounds **42–44** were administered to rats implanted with radiotelemetry probes to monitor body temperature. The effects of the different profiles on thermal regulation were striking. Both **42** and **43** (profiles A and B,

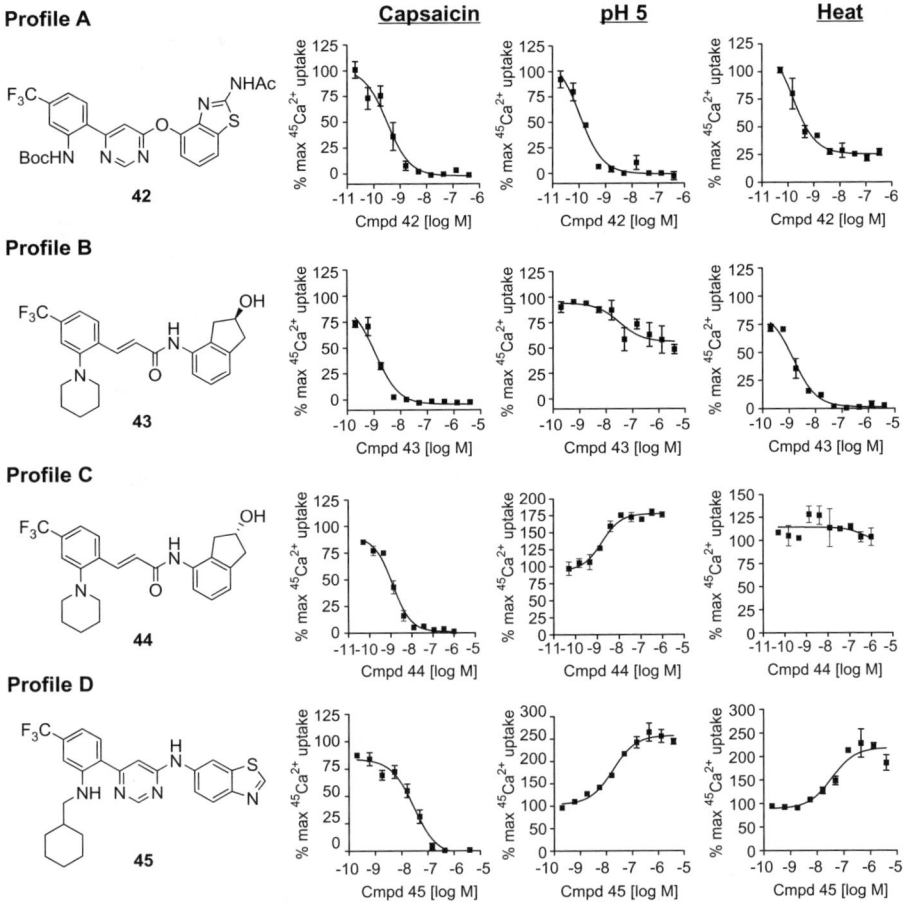

Figure 13.17 Characterization of *in vitro* profiles of TRPV1 modulators **42–45**. Effects on capsaicin (0.5 µM), pH 5, and heat (45 °C) activation of rat TRPV1. Profiles A–D: each point in the graph is an average + SD of an experiment conducted in triplicate. Reproduced with permission of the American Society for Pharmacology and Experimental Therapeutics from *J. Pharmacol. Exp. Ther.*, 2008, **326**, 218.

respectively) produced a significant 1.0–1.5 °C increase in body temperature (data not shown), while compound **44** (profile C) did not cause hyperthermia and, in fact, a modest decrease in temperature (− 0.6 °C) was observed (Figure 13.18A). Interestingly, compound **45** (profile D), which potentiated both heat and pH 5, caused a marked hypothermic response (− 2.9 °C) (Figure 13.18B).

With the identification of a selective modulator of rodent TRPV1 that did not cause hyperthermia, we evaluated compound **44** in a biochemical challenge model (capsaicin-induced flinch), and three pain models: CFA- and surgical incision-induced thermal hyperalgesia in rats, and acetic-acid induced writhing

A   **B**

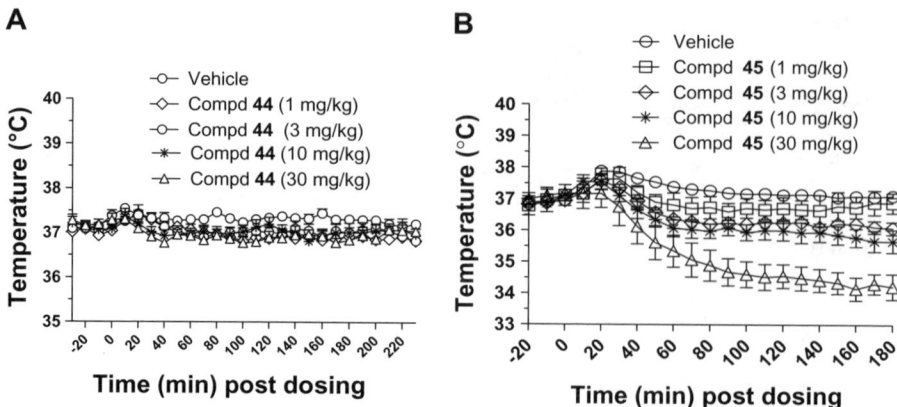

Figure 13.18 Effect on core body temperature of compounds **44** and **45**. (A) Compound **44** dosed at 1, 3, 10, 30 mg kg^{-1} p.o. slightly decreased body temperature with a maximum decrease of 0.6 °C occurring at 160 min post dosing in the 1 mg kg^{-1} dosing group. (B) Compound **45** dosed at 0.3, 1, 3, 10, 30 mg kg^{-1} p.o. significantly decreased body temperature with a maximum decrease of 2.9 °C occurring at 160 min post dosing. Reprinted with permission of the American Society for Pharmacology and Experimental Therapeutics from *J. Pharmacol. Exp. Ther.*, 2008, **326**, 218.

in mice (Figure 13.19). We found that compound **44** significantly blocked capsaicin-induced flinching behavior, and produced statistically significant efficacy in CFA- and skin incision-induced thermal hyperalgesia, and acetic acid-induced writhing models.

The unique profile of compound **44** indicates that it is feasible to eliminate hyperthermia while preserving antihyperalgesia by differential modulation of distinct modes of TRPV1 activation. However, when tested at the human TRPV1 receptor, compound **44** behaved as a "profile B" antagonist (blocking all modes of activation: Figure 13.20) and would be expected to cause hyperthermia in humans. Therefore, feasibility of "profile C" modulation of both rat and human TRPV1 by a single small molecule is yet to be demonstrated.

13.7 Summary

The information described above outlines the case study of the TRPV1 antagonist program conducted at Amgen, Inc., which started from an initial high-throughput screening hit and culminated in the identification of two clinical candidates, AMG 517 and AMG 628. Unfortunately, clinical studies of AMG 517 were discontinued due to the hyperthermia observed after exposure to single and multiple doses of this molecule in healthy volunteers, and due to marked and persistent hyperthermia after exposure to single doses of AMG 517

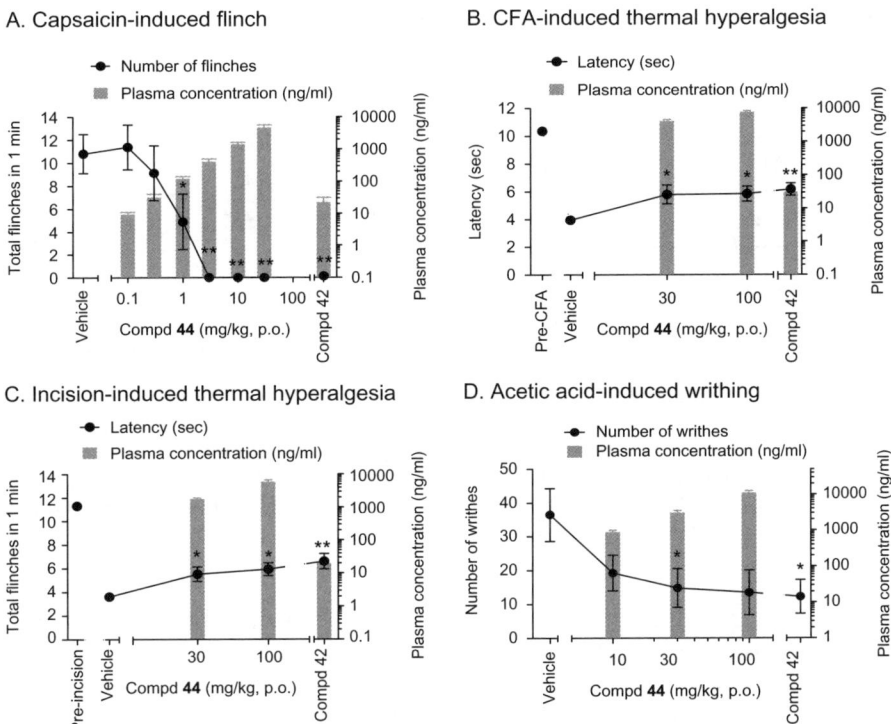

Figure 13.19 *In vivo* efficacy of compound **44** in biochemical challenge as well as inflammatory and surgical pain models. (A) Compound **44** significantly decreased capsaicin-induced flinching behavior. (B) Compound **44** significantly reversed CFA-induced thermal hyperalgesia. (C) Compound **44** significantly reversed skin incision-induced thermal hyperalgesia. (D) Compound **44** reduced acetic acid-induced writhing in mice. Note: compound **44** showed similar *in vitro* profiles at both rat and mouse TRPV1. Reproduced with permission of the American Society for Pharmacology and Experimental Therapeutics from *J. Pharmacol. Exp. Ther.*, 2008, **326**, 218.

in subjects who underwent third-molar extraction. Although AMG 517 did not progress to a successful registration, valuable lessons regarding hit identification, metabolism, pharmacokinetics, toxicology, and pharmaceutics were encountered in this "near miss" drug discovery effort. For example, in this study we used a unique high-throughput functional assay to identify several high-quality leads for medicinal chemistry optimization, and through extensive SAR investigations, a good understanding of the pharmacophores needed for potent TRPV1 activity was obtained. The lead optimization efforts also illustrated the effective utilization of metabolite identification studies for the design of new derivatives and resulted in the preparation of compounds with improved pharmacokinetic profiles. Furthermore, we demonstrated that we could increase solubility and reduced half-life by modulating the physiochemical properties of the compounds

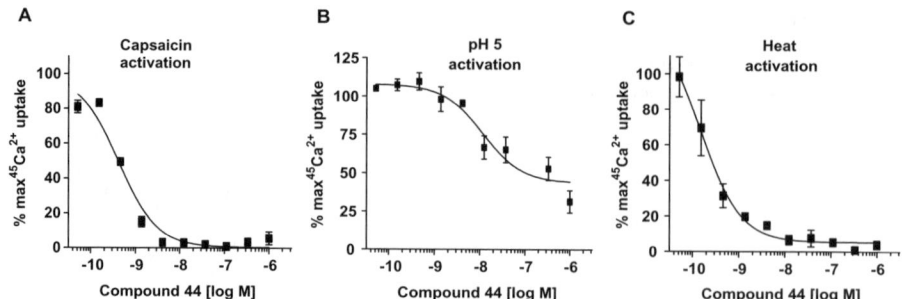

Figure 13.20   Effects of compound **44** on capsaicin (0.5 μM), pH 5, and heat (45 °C) activation of human TRPV1 (A to C); each point in the graph is an average ± S.D. of an experiment conducted in triplicate. Note that compound **44** blocked pH 5 activation partially and capsaicin and heat activation fully, making it a "profile B" compound at the human receptor. Reproduced with permission of the American Society for Pharmacology and Experimental Therapeutics from *J. Pharmacol. Exp. Ther.*, 2008, **326**, 218.

through the introduction of ring saturation and the incorporation of basic groups in this series of TRPV1 antagonists.

In addition to the identification of the two clinical candidates, subsequent efforts to better understand TRPV1 antagonist-induced hyperthermia lead to greater insights of the role of this channel in body-temperature maintenance. We found that peripheral restriction alone was not sufficient to eliminate TRPV1 antagonist-induced hyperthermia, suggesting that the site of action is predominantly peripherally mediated. Additional investigations with modality-specific TRPV1 antagonists also provided intriguing results. These studies led to the identification of a unique "profile C" type of TRPV1 modulator that blocks capsaicin activation, potentiates pH 5 activation, and does not impact heat activation while lacking the hyperthermia liability.

The recent results of TRPV1-induced hyperthermia reported by our laboratory and others have triggered a surge of interest in the role of TRPV1 in thermoregulation.[78,79] For example, Romanovsky *et al.* have critically dissected the thermoregulatory responses of TRPV1 agonist and antagonists and propose differential effects on two principal neuronal populations.[80] Studies such as these continue to increase our understanding of how TRPV1 functions in thermoregulation and help biologists gain further insights into the pharmacological role of this complex channel. As these preclinical studies continue, several other companies are also evaluating TRPV1 antagonists in the clinic (*e.g.*, AZD1386, GRC 6211, JTS 653, MK 2295, and SB-705498). Recent information from conference presentations and the clinical trial website, clinicaltials.gov, indicate that several of these trials have been completed, terminated, or suspended. Results from these clinical studies should provide valuable information about TRPV1 pharmacology and ultimately help afford resolution on the potential therapeutic utility of TRPV1 antagonists.

Acknowledgements

The author would like to thank Amgen's management for the support of this program as well as the many scientists that contributed to the TRPV1 research described in this chapter. Their tireless dedication, determination, and persistence to the project were inspirational. Special thanks go to Narender Gavva, James Treanor, Elizabeth Doherty, Anthony Bannon, Sekhar Surapaneni, Sonya Lehto and Nuria Tamayo, for their outstanding contributions to this research program. Finally, appreciation goes to Andrew Tasker and for critically proof reading this chapter.

References

1. *The History of Anesthesia*, ed. J. C. Diz, A. Franco, D. R. Bacon, J. Rupreht and J. Alvarez, Proceedings of the 5th International Symposium on the History of Anesthesia, Santiago, Spain, September 2001, Elsevier, Amsterdam, 2003.
2. C. R. McCurdy and S. S. Scully, *Life Sci.*, 2005, **78**, 476.
3. F. T. Verosick Jr, *Why We Hurt: The Natural History of Pain*, Harcourt, Orlando, FL, 2000, p. 141.
4. For example, several nonsteroidal anti-inflammatory drugs (also known as NSAIDs; *e.g.*, aspirin, ibuprofen, and naproxen) have been developed that help patients cope with many types of chronic pain. For the management of more severe pain, narcotics such as morphine, codeine, and methadone can be effective analgesic agents. More recent entries to the battery of pain therapeutics include drugs such as Lyrica, Neurontin, and Cymbalta that were originally developed for alternative indications. In addition to their anticonvulsant or antidepressant activities, these agents have also been found to be effective in treating various pain indications such as fibromyalgia, diabetic neuropathy, and post-herpetic neuralgia.
5. J. Rome, *in Mayo Clinic on Chronic Pain*, Mayo Clinic Health Information/Kensington Publishing Corporation, New York, 2002.
6. F. Lembeck, *Acta Physiol. Hung.*, 1987, **69**, 265.
7. T. Suzuki and K. Iwai, in *The Alkaloids: Chemistry and Pharmacology*, ed. A. Brossi, Academic Press, Orlando, FL, 1984, **vol. 23**, chap. 4.
8. A. Szallasi and P. M. Blumberg, *Pharmacol. Rev.*, 1999, **51**, 159.
9. A. Naj, *Peppers. A Story of Hot Pursuits*, Knopf, New York, 1992.
10. M. J. Caterina, M. A. Schumacher, M. Tominaga, T. A. Rosen, J. D. Levine and D. Julius, *Nature*, 1997, **389**, 816.
11. J. Lazar, L. Gharat, N. Khairathkar-Joshi, P. M. Blumberg and A. Szallasi, *Expert Opin. Drug Discovery*, 2009, **4**, 159.
12. G. Y. Wong and N. R. Gavva, *Brain Res. Rev.*, 2009, **60**, 267.
13. V. L. Korlipara, *Curr. Bioact. Compd.*, 2008, **4**, 110.
14. D. G. Lambert, *Br. Anaesth. J.*, 2009, **102**, 153.
15. L. S. Premkumar and P. Sikand, *Curr. Neuropharmacol.*, 2008, **6**, 151.

16. E. Palazzo, F. Rossi and S. Maione, *Mol. Cell. Endocrinol.*, 2008, **286**(S1), S79.
17. Y. Wang, *Neurochem. Res.*, 2008, **33**, 2008.
18. G. Appendino, E. Munoz and B. L. Fiebich, *Expert Opin. Ther. Pat.*, 2003, **13**, 1825.
19. K. H. Rami and M. J. Gunthorpe, *Drug Discovery Today*, 2004, **1**, 97.
20. G. Breitenbucher, S. R. Chaplan and N. I. Carruters, *Annu. Rep. Med. Chem.*, 2005, **40**, 185.
21. L. M. Broad, S. J. Keding and M. J. Blanco, *Curr. Top. Med. Chem.*, 2008, **8**, 1431.
22. L. Gharat and A. Szallasi, *Drug Dev. Res.*, 2007, **68**, 477.
23. A. Szallasi and P. M. Blumberg, *Pharmacol. Rev.*, 1999, **51**, 159.
24. M. J. Caterina and D. Julius, *Annu. Rev. Neurosci.*, 2001, **24**, 487.
25. A. Szallasi, *Am. J. Clin. Pathol.*, 2002, **118**, 110.
26. M. Tominaga, M. J. Caterina, A. B. Malmberg, T. A. Rosen, H. Gilbert, K. Skinner, B. E. Raumann, A. I. Basbaum and D. Julius, *Neuron*, 1998, **21**, 531.
27. R. R. Ji, T. A. Samad, S. X. Jin, R. Schmoll and C. J. Woolf, *Neuron*, 2002, **36**, 57.
28. L. J. Hudson, S. Bevan, G. Witherspoon, C. Gentry, A. Fox and J. Winter, *Eur. J. Neurosci.*, 2001, **13**, 2105.
29. P. Tympanidis, M. A. Casula, Y. Yiangou, G. Terenghi, P. Dowd and P. Anand, *Eur. J. Pain*, 2004, **8**, 129.
30. C. L. Chan, P. Facer, J. B. Davis, G. D. Smith, J. Egerton, C. Bountra, N. S. Williams and P. Anand, *Lancet*, 2003, **361**, 385.
31. M. Tominaga, M. J. Caterina, A. B. Malmberg, T. A. Rosen, H. Gilbert, K. Skinner, B. E. Raumann, A. I. Basbaum and D. Julius, *Neuron*, 1998, **21**, 531.
32. D. Smart, M. J. Gunthrope, J. C. Jerman, S. Nasir, J. Gray, A. I. Muir, J. K. Chambers, A. D. Randal and J. B. Davis, *Br. J. Pharmacol.*, 2000, **129**, 227.
33. S. W. Hwang, H. Cho, J. Kwak, S. Y. Lee, C. J. Kang, J. Jung, S. Cho, K. H. Min, Y. G. Suh, D. Kim and U. Oh, *Proc. Natl. Acad. Sci. USA.*, 2000, **97**, 6155.
34. Brand names of some topical analgesic creams containing capsaicin: Capsin, Capzasin-HP Arthritis Formula, Capzasin-P, Dolorac, Dr.s Cream, Menthac Arthritis Cream with Capsaicin, NO PAIN, RT Capsin, Salonpas Pain Patch with Capsaicin, Trixaicin, Trixaicin HP, Zostrix, Zostrix Sports, and Zostrix-HP.
35. G. Appendino, E. Munoz and B. L. Fiebich, *Expert Opin. Ther. Pat.*, 2003, **13**, 1825.
36. J. Y. Kwak, J. Y. Jung, S. W. Hwang, W. T. Lee and U. Oh, *Neuroscience*, 1998, **86**, 619.
37. A. R. S. Santos and J. B. Calixto, *Neurosci. Lett.*, 1997, **235**, 73.
38. K. M. Walker, L. Urban, S. J. Medhurst, S. Patel, M. Panesar, A. J. Fox and P. McIntyre, *J. Pharmacol. Exp. Ther.*, 2003, **304**, 56.

39. R. J. Docherty, J. C. Yeats and A. S. Piper, *Br. J. Pharmacol.*, 1997, **121**, 1461.
40. L. Liu and S. A. Simon, *Neurosci. Lett.*, 1997, **228**, 29.
41. M. J. Caterina, A. Leffler, A. B. Malmberg, W. J. Martin, J. Trafton, K. R. Petersen-Zeitz, M. Koltzenburg, A. I. Basbaum and D. Julius, *Science*, 2000, **288**, 306.
42. J. B. Davis, J. Gray, M. J. Gunthrope, J. P. Hatcher, P. T. Davey, P. Overend, M. H. Harries, J. Latcham, C. Clapham, K. Atkinson, S. A. Hughes, K. Rance, E. Grau, A. J. Harper, P. L. Pugh, D. C. Rogers, S. Bingham, A. Randall and S. A. Sheardown, *Nature*, 2000, **405**, 183.
43. G. Appendino, E. Munoz and B. L. Fiebich, *Exp. Opin. Ther. Pat.*, 2003, **13**, 1825.
44. Introduction from *J. Med. Chem.*, 2007, **50**, 3497; adapted with permission.
45. N. R. Gavva, R. Tamir, Y. Qu, L. Klionsky, T. J. Zhang, D. Immke, J. Wang, D. Zhu, T. W. Vanderah, F. Porreca, E. M. Doherty, M. H. Norman, K. D. Wild, A. W. Bannon, J. C. Louis and J. J. S. Treanor, *J. Pharmacol. Exp. Ther.*, 2005, **313**, 474.
46. V. I. Ognyanov, C. Balan, A. W. Bannon, Y. Bo, C. Dominguez, C. Fotsch, V. K. Gore, L. Klionsky, V. M. Ma, Y. X. Qian, R. Tamir, X. Wang, N. Xi, S. Xu, D. Zhu, N. R. Gavva, J. J. S. Treanor and M. H. Norman, *J. Med. Chem.*, 2006, **49**, 3719.
47. N. Xi, Y. Bo, E. M. Doherty, C. Fotsch, N. Gavva, N. Han, R. W. Hungate, L. Klionsky, Q. Liu, R. Tamir, S. Xu, J. J. S. Treanor and M. H. Norman, *Bioorg. Med. Chem. Lett.*, 2005, **15**, 5211.
48. V. K. Gore, V. V. Ma, R. Tamir, N. R. Gavva, J. J. S. Treanor and M. H. Norman, *Bioorg. Med. Chem. Lett.*, 2007, **17**, 5825.
49. E. M. Doherty, C. Fotsch, Y. Bo, P. Chakrabarti, N. Chen, N. R. Gavva, N. Han, M. G. Kelly, J. Kincaid, L. Klionsky, Q. Liu, V. I. Ognyanov, R. Tamir, Q. Wang, Z. Zhu, M. H. Norman and J. J. S. Treanor, *J. Med. Chem.*, 2005, **48**, 71.
50. Human and rat liver microsome incubation studies reported herein were conducted in a 1 M phosphate pH 7.4 buffer for a 10-min period with a final liver microsome concentration of $0.1 \, \text{mg} \, \text{mL}^{-1}$, a final compound concentration of $1 \, \mu\text{M}$, and a final NADPH concentration of 1 mM. Under these conditions, a cut off of $< 100 \, \mu\text{L} \, \text{min}^{-1} \, \text{mg}^{-1}$ was considered desirable.
51. J. Zhu, V. N. Viswanadhan, V. I. Ognyanov, Y. Bo, N. Chen, P. Chakrabarti, E. M. Doherty, C. Fotsch, N. R. Gavva, N. Han, L. Klionski, Q. Liu, R. Tamir, X. Wang, Y. Sun, J. J. S. Treanor and M. H. Norman, presented at the 229th ACS National Meeting, San Diego, March 2005, abstr. MEDI-101.
52. V. N. Viswanadhan, Y. Sun and M. H. Norman, *J. Med. Chem.*, 2007, **500**, 5608.
53. D. B. Boyd, *Reviews in Computational Chemistry*, ed. K. B. Lipkowitz and D. B. Boyd, VCH, Weinheim, 1990, **vol. 1**, p. 321.
54. M. H. Norman, J. Zhu, C. Fotsch, Y. Bo, N. Chen, P. Chakrabarti, E. M. Doherty, N. R. Gavva, N. Nishimura, T. Nixey, V. I. Ognyanov, R. M.

Rzasa, M. Stec, S. Surapaneni, R. Tamir, V. N. Viswanadhan and J. J. S. Treanor, *J. Med. Chem.*, 2007, **50**, 3497.

55. Similar isosteric replacements of *trans*-cinnamides have also been reported by G. T. Wang, S. Wang, R. Gentles, T. Sowin, S. Leitza, E. B. Reilly and T. W. von Geldern, *Bioorg. Med. Chem. Lett.*, 2005, **15**, 195.

56. E. M. Doherty, C. Fotsch, A. W. Bannon, Y. Bo, N. Chen, C. Dominguez, J. Falsey, N. R. Gavva, J. Katon, T. Nixey, V. I. Ognyanov, L. Pettus, R. M. Rzasa, M. Stec, S. Surapaneni, R. Tamir, J. Zhu, J. J. S. Treanor and M. H. Norman, *J. Med. Chem.*, 2007, **50**, 3515.

57. N. R. Gavva, A. W. Bannon, S. Surapaneni, D. N. Hovland Jr, S. G. Lehto, A. Gore, T. Juan, Todd, H. Deng, B. Han, L. Klionsky, R. Kuang, A. Le, R. Tamir, J. Wang, B. Youngblood, D. Zhu, M. H. Norman, E. Magal, J. J. S. Treanor and J. C. Louis, *J. Neurosci.*, 2007, **27**, 3366.

58. A. A. Steiner, V. F. Turek, M. C. Almeida, J. J. Burmeister, D. L. Oliveira, J. L. Roberts, A. W. Bannon, M. H. Norman, J. C. Louis, J. J. S. Treanor, N. R. Gavva and A. A. Romanovsky, *J. Neurosci.*, 2007, **27**, 7459.

59. N. R. Gavva, A. W. Bannon, D. Hovland Jr, S. Lehto, L. Klionsky, S. Surapaneni, D. C. Immke, C. Henley, L. Arik, A. Bak, J. Davis, N. Ernst, G. Hever, R. Kuang, L. Shi, R. Tamir, J. Wang, W. Wang, G. Zajic, D. Zhu, M. H. Norman, J. C. Louis, E. Magal and J. J. S. Treanor, *J. Pharmacol. Exp. Ther.*, 2007, **323**, 128.

60. H. L. Wang, J. Katon, C. Balan, A. W. Bannon, C. Bernard, E. M. Doherty, C. Dominguez, N. R. Gavva, V. Gore, V. Ma, N. Nishimura, S. Surapaneni, P. Tang, R. Tamir, O. Thiel, J. J. S. Treanor and M. H. Norman, *J. Med. Chem.*, 2007, **50**, 3528.

61. H. Tan, D. Semin, M. Wacker and J. Cheetham, *J. Assoc. Lab. Automation.*, 2005, **10**, 364.

62. E. M. Doherty, D. Retz, N. R. Gavva, R. Tamir, J. J. S. Treanor and M. H. Norman, *Bioorg. Med. Chem. Lett.*, 2008, **18**, 1830.

63. X. Wang, P. P. Chakrabarti, V. I. Ognyanov, L. H. Pettus, R. Tamir, H. Tan, P. Tang, J. J. S. Treanor, N. R. Gavva and M. H. Norman, *Bioorg. Med. Chem. Lett.*, 2007, **17**, 6539.

64. M. M. Stec, Y. Bo, P. P. Chakrabarti, L. Liao, M. Ncube, N. Tamayo, R. Tamir, N. R. Gavva, J. J. S. Treanor and M. H. Norman, *Bioorg. Med. Chem. Lett.*, 2008, **18**, 5118.

65. N. R. Gavva, A. Garami, L. Fang, S. Surapaneni, A. Akrami, F. Alvarez, A. Bak, M. Darling, A. Gore, G. R. Jang, P. Kesslak, L. Ni, M. H. Norman, G. Palluconi, M. J. Rose, M. Salfi, E. Tan, J. J. S. Treanor, A. Romanovsky, C. Banfield and G. Davar, *Pain*, 2008, **136**, 202.

66. N. Tamayo, H. Liao, M. M. Stec, X. Wang, P. Chakrabarti, D. Retz, E. M. Doherty, S. Surapaneni, R. Tamir, A. W. Bannon, N. R. Gavva and M. H. Norman, *J. Med. Chem.*, 2008, **51**, 2744.

67. S. G. Lehto, R. Tamir, H. Deng, L. Klionsky, R. Kuang, A. Le, D. Lee, J. C. Louis, E. Magal, B. H. Manning, J. Rubino, S. Surapaneni, N. Tamayo, T. Wang, J. Wang, J. Wang, W. Wang, B. Youngblood,

M. Zhang, D. Zhu, M. H. Norman and N. R. Gavva, *J. Pharmacol. Exp. Ther.*, 2008, **326**, 218.

68. Adapted in part with permission from *J. Med. Chem.*, 2008, **51**, 2744.
69. A. A. Romanovsky, *Am. J. Physiol. Regul. Integr. Comp. Physiol.*, 2007, **292**, R37.
70. P. S. Burton, R. A. Conradi, N. F. Ho, A. R. Hilgers and R. Borchardt, *J. Pharm. Sci.*, 1996, **85**, 1336.
71. S. A. Hitchcock and L. D. Pennington, *J. Med. Chem.*, 2006, **49**, 7559.
72. C. Hansch, J. P. Bjorkroth and A. Leo, *J. Pharm. Sci.*, 1987, **76**, 663.
73. W. R. Glave and C. J. Hansch, *J. Pharm. Sci.*, 1972, **61**, 589.
74. K. M. Mahar Doan, J. E. Humphreys, L. O. Webster, S. A. Wring, L. J. Shampine, C. J. Serabjit-Singh, K. K. Adkison and J. W. Polli, *J. Pharmacol. Exp. Ther.*, 2002, **303**, 1029.
75. S. J. Cho, X. Sun and W. Harte, *J. Comput. Aided Mol. Des.*, 2006, **20**, 249.
76. X. Wang, P. P. Chakrabarti, V. I. Ognyanov, L. H. Pettus, R. Tamir, H. Tan, P. Tang, J. J. S. Treanor, N. R. Gavva and M. H. Norman, *Bioorg. Med. Chem. Lett.*, 2007, **17**, 6539.
77. CNS penetration was evaluated by measuring brain-to-plasma ratios. Plasma and brains were collected 30 min after dosing and total concentration of compound was determined by HPLC. For the purpose of this study we measured compound concentrations in non-perfused whole-brain samples. Taking into account the residual concentration of drug present as a consequence of brain capillary blood volume, we considered compounds having a brain-to-plasma ratio of less than or equal to 0.1 as peripherally restricted.
78. M. J. Caterina, *Pain*, 2008, **136**, 3.
79. S. S. Ayoub, J. C. Hunter and D. L. Simmons, *Pharmacol. Rev.*, 2009, **61**, 225.
80. A. A. Romanovsky, M. C. Almeida, A. Garami, A. A. Steiner, M. H. Norman, S. F. Morrison, K. Nakamura, J. J. Burmeister and T. B. Nucci, *Pharmacol. Rev.*, 2009, **61**, 228.

CHAPTER 14

Uncoupling Neuroprotection from Immunosuppression: the Discovery of ILS-920

EDMUND I. GRAZIANI

Worldwide Medicinal Chemistry, Pfizer Global Research & Development, 455 Eastern Point Road, Groton, CT 06340, USA

14.1 Introduction

Ischemic stroke—injury due to an obstruction within a blood vessel that restricts blood supply to the brain—is the leading cause of neurological disability worldwide and the most common cause of adult long-term disability in the United States.[1] Existing approved therapies to treat ischemic stroke are limited to recombinant tissue plasminogen activator (rt-PA), which acts as a thrombolytic that targets the underlying vessel blockage without addressing stroke-induced neurological deficits. Moreover, rt-PA must be administered within the first six hours following an ischemic event. Since rt-PA also increases the likelihood of hemorrhagic stroke, a CT scan is required prior to administration to rule out intracranial bleeding. Thus, only 5% of stroke patients ever receive rt-PA treatment.[1]

There are no marketed neuroprotective agents for ischemic stroke, despite decades of clinical trials. The recent and highly publicized Phase 3 failure of NXY-059, a novel free-radical scavenger neuroprotectant, hastened the call for an end to neuroprotective therapy research for treating stroke.[2] Nonetheless, the Stroke Therapy Academic Industry Roundtable (STAIR) has recently

RSC Drug Discovery Series No. 4
Accounts in Drug Discovery: Case Studies in Medicinal Chemistry
Edited by Joel C. Barrish, Percy H. Carter, Peter T. W. Cheng and Robert Zahler
Published by the Royal Society of Chemistry, www.rsc.org

re-issued and updated[3] its recommendations for improving the quality of preclinical studies of stroke treatments with the hope of improving the probability of successful outcomes in the clinic, and so reports on the abandonment of ischemic stroke treatments by the pharmaceutical industry are perhaps premature.

Our own interest in beginning a medicinal chemistry program to identify candidate compounds for treating neurological deficits induced by ischemic stroke was therefore predicated on highly stringent requirements that would address the deficits of earlier, failed efforts. These requirements consisted chiefly in adopting animal models that were the most challenging in which to demonstrate efficacy, and, most importantly, to expand time-to-treat therapeutic windows to well beyond six hours. In setting challenging goals and empowering our team with strict "go/no go" decision points along the way, we hoped to rapidly identify whether our approach would yield a compound with robust efficacy, prior to subjecting it further to the newest STAIR requirements.

In mid-2003, a program was initiated to determine the utility of Wyeth's collection of analogs of the immunosuppressant rapamycin (sirolimus, Rapamune®) in treating neurodegenerative disorders. This initial evaluation identified a potential synthetic strategy to use rapamycin as a starting point to prepare new analogs with desirable properties for further evaluation. Specifically, the mTOR binding region of rapamycin was targeted with the aim of reducing immunosuppressive activity while maintaining or augmenting neuroprotective or neuroregenerative activities of the series. This account will aim to provide a brief introduction to this remarkable natural product, its protein targets (the immunophilins and the mammalian target of rapamycin), and describe the rationale, approach, and outcome of this effort that led to the potent, non-immunosuppressive, brain-penetrant and efficacious immunophilin ligand, ILS-920.

14.2   Rapamycin: Structure and Biochemical Mechanism of Action

Rapamycin (Figure 14.1) is a natural product produced by the bacterium *Streptomyces hygroscopicus* NRRL 5491.[4,5] The compound was discovered as part of an antifungal screening program at Ayerst Laboratories in Montreal, Canada, and was also determined to possess immunosuppressive and antiproliferative activities.[6–8]

The structure of rapamycin was unique when first determined; the related compounds FK-506[9] and FK-520 (ascomycin)[10] were discovered shortly thereafter. One of the largest macrolide rings (31 atoms) then known, the compound consists of an extended cyclohexane starter side-chain, elaborated with a polyketide chain and pipecolic acid (the six-membered ring analog of proline), before terminating in a macrolactone. Only three other members of this family of natural products have since been described, the meridamycins,[11,12] the antascomicins,[13] and the nocardiopsins,[14] all of which are

Immunosuppressive

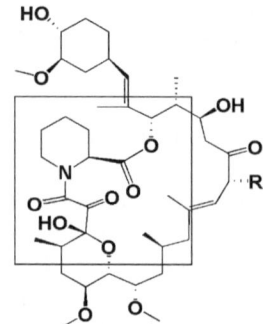

rapamycin

fujimycin (FK-506): R = ethyl
ascomycin (FK-520): R = allyl

Non-immunosuppressive

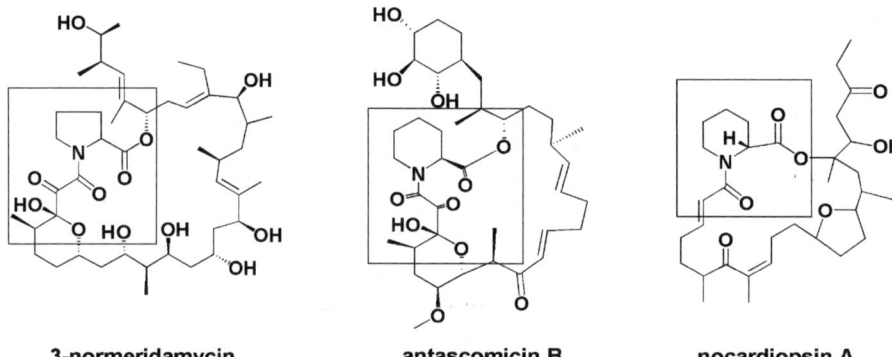

3-normeridamycin   **antascomicin B**   **nocardiopsin A**

Figure 14.1   The structures of natural product immunophilin ligands.

non-immunosuppressive (Figure 14.1). Affinity methods were used to identify putative protein-binding partners,[15,16] named FK-506 binding proteins (FKBPs), a subfamily of the immunophilins exemplified by the 12 kDa FKBP12. FKBPs possess peptidyl-prolyl *cis-trans* isomerase (PPIase) activities[17] and are known to form complexes with ion channels,[18–20] steroid receptor complexes,[21] transcription factors,[22] tubulin,[23] and B-cell CLL/lymphoma 2 (Bcl-2) upon conditional activation.[24] It soon became apparent that binding to FKBP12 did not on its own account for rapamycin's antifungal and immunosuppressive activities, but rather that they arose from the formation of a high affinity ternary complex between the rapamycin–FKBP12 complex and what was then a new protein of unknown function named target of rapamycin (TOR) in yeast.[25]

The highly conserved mammalian homolog of TOR (mTOR, also called FRAP, RAFT, and SEP), a large molecular weight (>250 kDa) member of the phosphoinositide-3-kinase-related kinase (PIKK) family, was identified soon after.[26–29] It was later discovered that mTOR associates with the regulatory associated protein of mTOR (raptor)[30] and mammalian counterpart of yeast Lst8 (MLST8) (also known as G protein β-subunit-like protein, GβL)[31] in a protein complex now referred to as mTOR complex 1 (mTORC1; see Figure 14.2). The central role of mTOR and its complexes in cell signaling (particularly in the context of cancer) has been reviewed extensively.[32] In addition, mTOR also exists in a second protein complex, mTORC2, consisting of mTOR, MLST8 (GβL), stress-activated protein kinase-interacting protein (sin1), and rapamycin-insensitive companion of mTOR (rictor),[33] but not raptor. As the name implies, this complex is not inhibited by rapamycin, except under

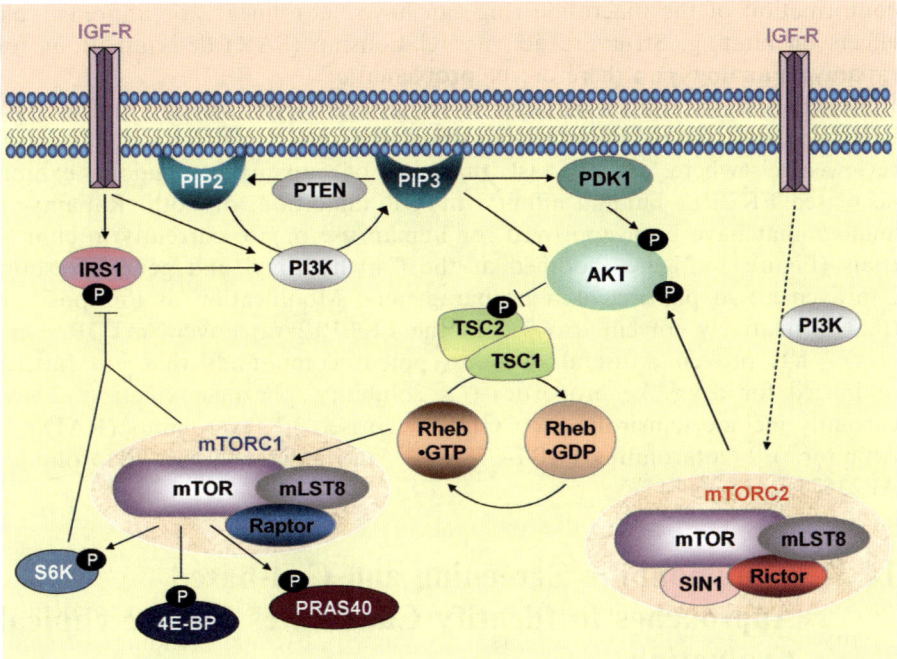

Figure 14.2 The protein kinase B–mammalian target of rapamycin signaling pathway. 4E-BP, eIF-4E-binding protein; Akt, protein kinase B; IGF-R, insulin-like growth factor receptor; IRS1, insulin receptor substrate 1; mLST8, mammalian counterpart of yeast Lst8; mTORC, mammalian target of rapamycin complex; PDK, phosphoinositol-dependent protein kinase; PI3K, phosphatidylinositol 3-kinase; PIP2, phosphatidylinositol (4,5)-bisphosphate; PIP3, phosphatidylinositol (3,4,5)-trisphosphate; PRAS-40, proline-rich Akt substrate of 40 kDa; PTEN, phosphatase and tensin homolog deleted on chromosome 10; REDD1, regulated in development and DNA damage responses; S6K, p70 S6 kinase; SIN1, stress-activated protein kinase interaction protein 1; TSC, tuberous sclerosis. Reproduced with permission from R.T. Abraham and C. H. Eng, *Exp. Opin. Ther. Targets*, 2008, **12**, 209–222.

long-term treatment in certain cell types.[34,35] The mechanisms whereby mTORC2 is generally unresponsive to rapamycin, either due to steric hindrance to binding at the FKBP12–rapamycin binding (FRB) domain or another mechanism, is unknown at this time.

Any structural changes to the rapamycin scaffold can affect binding to mTOR both directly and indirectly as a function of binding to FKBP12.[36,37] In fact, the interaction of the rapamycin–FKBP12 complex with mTOR correlates well with the conformational flexibility of the so-called effector domain of rapamycin. This domain consists of regions of the molecule that make important hydrophobic interactions with the FRB domain of mTOR, and consists principally of the triene region (C17 to C22) of rapamycin and a number of methyl groups, notably the OMe group at C16, and the methyls attached to C23, C29, and C31.[38,39] Thus, modifications that change the global conformation of the macrolide ring can have "non-linear" or unpredictable effects on binding. Structure–activity relationship (SAR) determinations for rapamycin analogs can therefore be problematic.

Nonetheless, a number of useful correlates between structure and activity have been observed by a number of groups working in this area and have been reviewed elsewhere.[40] The vast majority of rapamycin analogs exhibit decreased FKBP12 binding affinity, mTOR inhibition, or both. Rapamycin analogs that have been approved for human use or are currently in clinical trials (Figure 14.3) are modified at the C40 hydroxyl and generally show improvement in pharmacokinetic parameters. Modification at this position (that is relatively solvent exposed in the FKBP12–rapamycin–mTOR complex)[38] has proven a useful route to potent compounds that are further optimized for drug-like properties (*i.e.* solubility, pharmacokinetics). These currently include temsirolimus (CCI-779, Torisel®),[41] everolimus (RAD001, Affinitor®),[42] zotarolimus (ABT-578),[43–45] and ridaforolimus (deforolimus, AP23573, MK-8669).[46]

14.3 Immunophilin Screening and Cell-based Approaches to Identify Candidates for Pre-clinical Evaluation

Rapamycin and FK-506 have also been reported to exhibit neuroprotective activity *in vitro*[47] that has been attributed to FKBP binding[48] rather than mTOR inhibition. However, it had also been demonstrated that while the immunophilin ligand FK-506 showed efficacy in a rodent model of ischemic stroke,[49] rapamycin had no effect in this model, suggesting that immunophilin binding may only play a partial role in neuroprotection/neuroregeneration. Non-immunosuppressive immunophilin ligands—such as GPI-1046 (Figure 14.4), which contains the FKBP-binding moiety of rapamycin and FK-506 but lacks an mTOR or calcineurin binding domain—have been pursued as potential therapies for neurodegenerative disorders.[50] However, the neuroprotective activity of GPI-1046 is marginal in both neurite outgrowth studies in

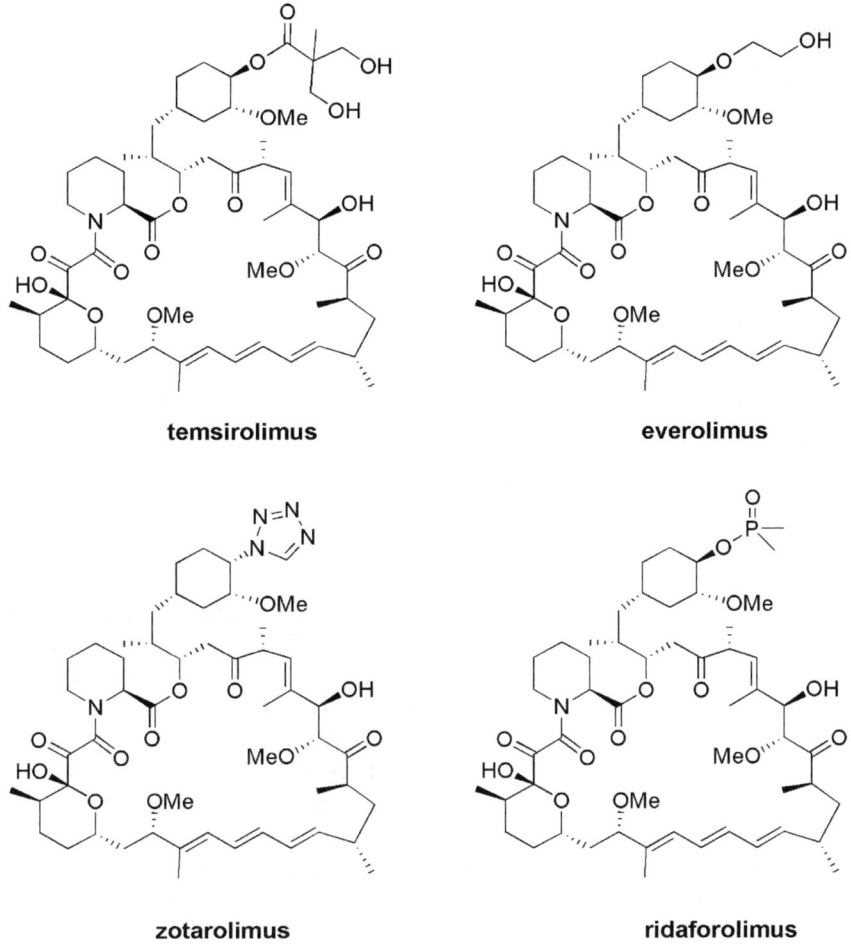

Figure 14.3   Structures of rapamycin analogs currently approved for human use or in clinical trials: temsirolimus, everolimus, zotarolimus, and ridaforolimus.

chicken dorsal root ganglia (DRG) explants, rat DRGs, and in a sciatic nerve injury model.[51] It has also been proposed that neuroprotection is mediated by immunophilins other than FKBP12, notably FKBP52 and FKBP38.[24,48] When we started our efforts, attempts to develop therapeutically useful compounds by uncoupling neuroprotection from immunosuppression in this class of compounds had met with limited success.[51]

We undertook a focused initiative to identify new, neuroprotective natural product immunophilin ligands based on multiple hypotheses: (i) that FKBP binding alone cannot account for the mechanism of neuroprotection; (ii) that the marginal neuroprotective effects of rapamycin, in contrast to FK506, suggest that mTOR binding is not required for this activity; and (iii) that therefore compounds with FKBP binding domains and modified "effector"

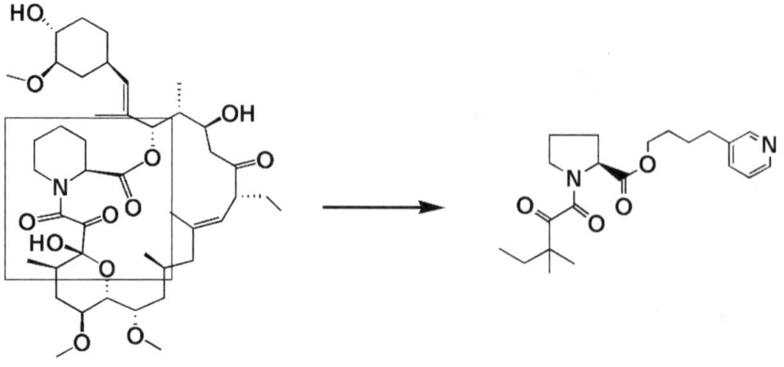

FK-506   GPI-1046

Figure 14.4 The structure of GPI-1046 showing its origin from the immunophilin binding domain of FK-506.

domains may be more potent neuroprotective agents, and may also be less immunosuppressive.

In the course of screening our library of microbial fermentation extracts for compounds that bind to immunophilins, we identified a culture (subsequently identified as *Streptomyces* sp. LL-C31037) that produced a previously unreported compound, 3-normeridamycin (Figure 14.1). This new natural product exhibited potent restoration of dopamine uptake in 1-methyl-4-phenylpyridinium (MPP^+) challenged neurons, with statistical significance down to low nanomolar concentrations. The biological activity of 3-normeridamycin stands in marked contrast to FK-506 and rapamycin (Table 14.1). The apparent lack of any correlation between neuroprotection and binding to FKBP12 is consistent with previous reports. The discovery of the neuroprotective activity of 3-normeridamycin supported our hypothesis that non-calcineurin/mTOR effector domains may be relevant for neuroprotection, and furthermore suggested that as-yet unreported protein partners may bind to the natural product–immunophilin complex.

Encouraged by the results of screening for immunophilin binders, we then reexamined a representative set of compounds developed in the course of earlier medicinal chemistry programs based on the rapamycin scaffold. Approximately 70 compounds were selected on the basis of structural diversity, purity/integrity, and availability. These compounds were tested for promotion of neuronal survival using a neurofilament (NF) ELISA assay (CTX) and neurite outgrowth (NRO) activity in cortical neurons using a Cellomics Arrayscan assay platform.[52] We found that the non-immunosuppressive rapamycin analog (rapalog) WAY-124466 (**1**) had activity ($EC_{50} = 0.8\,\mu M$) in promoting survival of cortical neurons in culture.[47] The initial observations from this analysis of library rapalogs are summarized in Figure 14.5.

Therefore, a medicinal chemistry effort was initiated to identify a series of rapalogs modified at the C19,C22 diene employing Diels–Alder chemistry with

Table 14.1 Biological activities and kinetic constants of immunophilin ligands.

Compounds	T-cell inhibition (IC_{50}, μM)	Neurite outgrowth (EC_{50}, μM)	Cortical neuron survival (EC_{50}, μM)	$[^3H]$-DA uptake (EC_{50}, μM)	FKBP12 binding (K_d, nM)
FK506	0.001	0.45	0.01	0.049	0.33 ± 0.03 (lit. 0.4)
ILS-920	>5	0.54	0.15	NT[a]	110 ± 11
WYE-592	0.15	0.42	0.7	NT	4.7 ± 0.4
Rapamycin	0.005	1.6	1.3	>10	0.33 ± 0.03
3-Normeridamycin	>10	200	8.8	0.11	50

[a]NT = not tested.

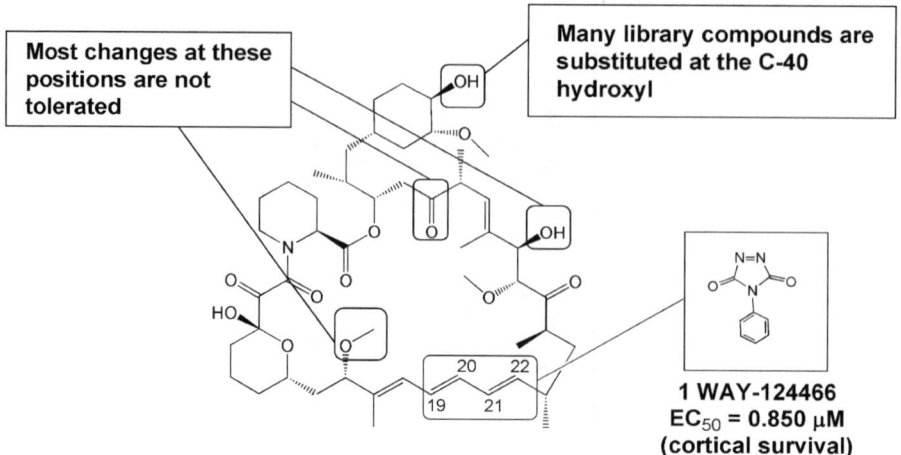

Most changes at these positions are not tolerated

Many library compounds are substituted at the C-40 hydroxyl

20 22
19 21

1 WAY-124466
$EC_{50} = 0.850 \, \mu M$
(cortical survival)

Figure 14.5 General trends observed in activity of library rapamycin analogs screened for neurite outgrowth inducing activity.

novel dienophiles. The initial goal of the program was to improve the potency of the WAY-124466 scaffold, and in this way identify a lead compound with acceptable brain penetration and pharamacokinetics for *in vivo* proof-of-concept. Since the C19,C22 diene cannot easily access the Diels–Alder transition state, presumably for reasons of geometry and electronics, only very reactive dienophiles were observed to yield [4 + 2] cycloadducts at this position. Figure 14.6 summarizes the preliminary structure–activity work that was done, with reference to various substituted dienophiles.

WYE-592 was synthesized via [4 + 2] cycloaddition of rapamycin with nitrosobenzene. WYE-592 exhibited both good brain penetration (brain/whole blood = 0.52) and increased potency in promoting neurite outgrowth and neuronal survival compared to both rapamycin and WAY-124466, but the compound inhibited IL-2-stimulated T-cell proliferation with an IC_{50} of 150 nM. This activity is consistent with the 3–5% conversion of WYE-592 to rapamycin starting material *via* a retro-Diels–Alder reaction that had been observed in physiological buffer. This hypothesis is further supported by the observation that WYE-592 does not bind *in vitro* to the FRB domain of mTOR in the presence of FKBP12 (as expected due to the modification of triene), whereas rapamycin itself exerts its immunosuppressive activity via formation of an inhibitory complex with FKBP12 and mTOR.

In order to prevent this back-conversion to rapamycin from occurring, the C20,C21 alkene in WYE-592 was reduced by catalytic hydrogenation over a Pd/C catalyst to yield ILS-920. In contrast to WYE-592, this compound did not inhibit T-cell proliferation up to 5 μM. ILS-920 was at least five times more potent than WYE-592, rapamycin, or GPI-1046 at promoting cortical neuronal survival (and therefore an order of magnitude more potent than rapamycin), but yet maintained good brain penetration and acceptable ADME properties.

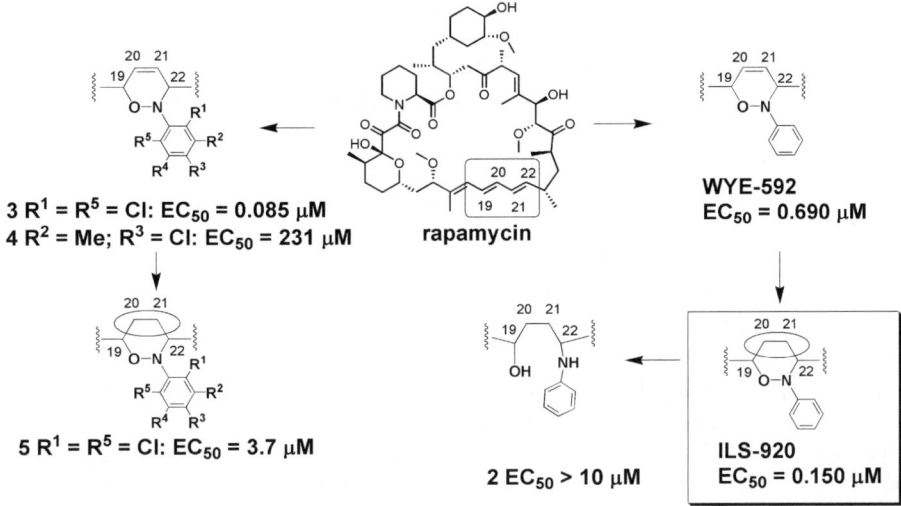

3 R^1 = R^5 = Cl: EC$_{50}$ = 0.085 μM
4 R^2 = Me; R^3 = Cl: EC$_{50}$ = 231 μM   **rapamycin**

WYE-592
EC$_{50}$ = 0.690 μM

5 R^1 = R^5 = Cl: EC$_{50}$ = 3.7 μM

2 EC$_{50}$ > 10 μM

ILS-920
EC$_{50}$ = 0.150 μM

Figure 14.6 Preliminary structure–activity relationships observed for rapamycin [4 + 2] cycloadducts. EC$_{50}$ values refer to activity measured in a cortical neuronal survival assay (*cf.* Liu *et al.*[52]).

We were initially curious to know whether modification at the triene of rapamycin with nitrosobenzene and subsequent hydrogenation to yield the stable analog ILS-920 was a path to revealing an inherent, mTOR-independent neurite outgrowth activity of the rapamycin scaffold (*via* steric perturbation of rapamycin-FKBP12 binding to mTORC1), or if the aromatic moiety so installed was also an important contributor towards activity. Interestingly, further reduction of ILS-920 yielded the ring-opened secondary amine (compound **2**, Figure 14.6) with an EC$_{50}$ > 10 μM in the cortical NF ELISA assay. This was a key finding and suggested that the presence and conformation of a phenyl ring at this position was important for activity either directly or because of induced changes to the macrocyclic ring conformation. A preliminary evaluation of substitution around the phenyl ring of WYE-592 and ILS-920 further confirmed the importance of this moiety for activity. The 2,6-dichloro analog of WYE-592, **3**, was significantly more potent in the NF ELISA assay than WYE-592 or the 3-methyl-4-chloro analog, **4**. Surprisingly, reduction of **3** to give the dichloro analog of ILS-920, **5**, yielded a compound with an EC$_{50}$ = 3.7 μM.

14.4 Chemical Biology of ILS-920: Towards an Understanding of the Molecular Mechanism of Action

With potent, non-immunosuppressive compounds in hand, we set out to better understand their molecular mechanism of action by a series of chemical biology experiments. We were particularly interested in testing the hypothesis, mentioned earlier, that immunophilin ligand-induced neurite outgrowth is mediated by immunophilins other than FKBP12, notably FKBP52 and/or FKBP38–Ca^{2+}–calmodulin (CaM).[24,48] Specifically, FKBP52 was proposed to mediate the neurite outgrowth action of FK506 through activation of steroid receptor complexes that mediate downstream responses to estrogen, androgen, and glucocorticoid hormones. However, rapamycin is a potent inhibitor of FKBP52, has some neuroprotective activity, but lacks efficacy in an animal model. Alternately, an FKBP38–calmodulin–Ca^{2+} complex formed at high Ca^{2+} concentrations was recently proposed to mediate Ca^{2+} overload-induced cell death, through its interaction as a negative effector of the anti-apoptotic Bcl-2 protein.

Affinity purification from F-11 cell lysates using resin with covalently bound WYE-592 and ILS-920 identified FKBP52 and the α1 subunit of L-type Ca^{2+} channels (CACNB1) as the major binding partners for the compounds. Moreover, ILS-920 demonstrated binding selectivity for FKBP52 over FKBP12. The observed magnitude difference in binding affinities for ILS-920 among homologous immunophilins appears consistent with the reported differences in binding affinities for FKBP25 between rapamycin and FK506,[53] and with previous observations that modification distant to the FKBP binding domain of rapamycin can affect immunophilin binding.

Having demonstrated that ILS-920 is a binding partner for FKBP52, we further investigated whether a functional correlation between FKBP52 inhibition and neuroprotection could be established. The observation that neurite outgrowth was essentially unchanged in CACNB1 siRNA-treated neurons, but significantly increased in FKBP52 siRNA-treated neurons, suggested that inhibition of FKBP52 was a mediator of neurite outgrowth of the compounds, given that neurite outgrowth of this magnitude is consistent with the reported activity of an FKBP52 antibody.[48] Preliminary transcriptional analysis of cortical neuronal cultures treated with WYE-592 or ILS-920 suggested up-regulation of genes involved in cholesterol biosynthesis, consistent with activation of glucocorticoid and other steroid receptors, an effect that can be attributed to the immunophilin ligand-induced dissociation of FKBP52 from steroid receptor complexes. Stimulation of cholesterol biosynthesis is a common feature of agents that promote neurite outgrowth, including FK506, sterol hormones, and geldanamycin.[48]

ILS-920 was also observed to bind the β1 subunit of L-type calcium channels and partially inhibit L-type voltage-gated calcium channel (VGCC) current. Interestingly, the *in vivo* efficacy, or lack thereof, of ILS-920, WYE-592, FK506, and rapamycin correlated with their capacity to inhibit L-type VGCC-mediated Ca^{2+} signaling pathways. Note that ILS-920 inhibits the L-type VGCC directly, whereas FK506 inhibits calcineurin, a key component of L-type VGCC signaling. We hypothesized that attenuation of Ca^{2+} influx might be mechanistically important, since Ca^{2+} overload is generally considered to be a critical event in excitotoxic mediated neuronal death,[54] although additional work to test this further is needed.

ILS-920 was selected as a suitable candidate for *in vivo* testing based upon its lack of immunosuppression, superior potency, and pharmacokinetic properties in addition to its improved stability relative to WYE-592. A rodent model of surgically-induced ischemic stroke was chosen for initial evaluation, with subsequent follow up in non-human primates, with a particular focus on extending the treatment window. Treatment with ILS-920 at 10 and 30 mg kg^{-1}, administered intravenously (iv) 4 h post-occlusion, significantly reduced infarct volume and improved neurological recovery in a transient mid-cerebral artery occlusion (tMCAO) rat model of ischemic stroke. Additionally, rats were given iv doses of 10 or 30 mg kg^{-1} of ILS-920 at 24 h post-induction of permanent MCAO, with four additional doses administered once daily, and neurological deficits assessed over 21 days post-stroke. The data show that rats treated with 30 mg kg^{-1} ILS-920 displayed a significantly improved slope of recovery and significantly improved neurological scores at 21 days post-MCAO relative to vehicle treated animals.[55]

The *in vivo* efficacy of ILS-920 is in marked contrast to rapamycin, which failed to reduce infarct volume in a rat tMCAO model.[49] GPI-1046 was also reported to have limited or no efficacy in comparable models of ischemic stroke.[51] *N*-(*N′*,*N′*-Dimethylcarboxamidomethyl)cycloheximide (DM-CHX) was reported to reduce infarct volume by up to 44% when delivered *via* intracerebroventricular (icv) application 2 h post-occlusion (but not 6 h) in a rat model of endothelin-induced transient focal ischemia.[24] Since icv application is impractical for human stroke patients, in the absence of data employing alternate routes of delivery it is unclear at present what the clinical utility of such a compound might be. In contrast, treatment with ILS-920 results in significant enhancement of neurological recovery in a tMCAO model of stroke with at least a 24 h therapeutic window (time elapsed between insult and initial dose) when delivered intravenously. Clinical evaluation of ILS-920 was initiated in 2009.[56]

14.5 Concluding Remarks

Current paradigms of drug discovery present an interesting, if vexing, conundrum. It is universally acknowledged that therapeutic indications having high unmet medical need represent the best areas to invest resources and effort in terms of both the potential impact to health and untapped market potential. However, these same areas of focus are also expected to demand innovative approaches outside those tried before (approaches that would presumably have yielded solutions long ago), which adds a considerable amount of risk to an already long, difficult, expensive, and attrition-prone R&D process. Ironically, an older paradigm from an earlier era of drug discovery, namely natural product research, represents a potential bridge to balancing risk and innovation that complements modern synthetic chemistry approaches. On the one hand, these privileged structures (that have withstood the selective pressures of eons of natural selection) have already proven their utility in human medicine.

Moreover, many natural product structures are vital tool compounds in deciphering signaling pathways within the cell. Our approach has therefore been to attempt to modulate the activity of a medically useful natural product in order to perturb biological activity towards a desired profile, while at the same time learning how structural changes affect mechanism. With such an approach, and using cell-based methods for screening, we hoped to obtain novel structures with a unique activity profile that could serve both as clinical candidates and tools to explore disease biology.[57]

Specifically, our work supports the hypothesis that modification of rapamycin at the mTOR binding region can provide non-immunosuppressive compounds with potent neuroprotective activity, and significant efficacy in an animal model of ischemic stroke. The role of natural products as privileged scaffolds for semi-synthesis has been under-explored in recent years, but remains a useful complement to modern approaches to medicinal chemistry. Combined with cell-based screening, the preparation of biologically active rapamycin analogs has yielded a clinical candidate, ILS-920. Preliminary explorations of the chemical biology of the compound suggest that the *in vivo* efficacy of ILS-920 derives from the compound's dual functions as a potential activator of glucocorticoid and other steroid receptors via dissociation of FKBP52 from the receptor complexes, and as an inhibitor of L-type voltage-gated Ca^{2+} channels via binding to the $\beta 1$ subunit.

References

1. G. A. Donnan, M. Fisher, M. Macleod and S. M. Davis, *Lancet*, 2008, **371**, 1612–1623.
2. S. I. Savitz, *Stroke*, 2009, **40**, S115–118.
3. M. Fisher, G. Feuerstein, D. W. Howells, P. D. Hurn, T. A. Kent, S. I. Savitz and E. H. Lo, *Stroke*, 2009, **40**, 2244–2250.
4. S. N. Sehgal, H. Baker and C. Vezina, *J. Antibiot. (Tokyo)*, 1975, **28**, 727–732.
5. C. Vezina, A. Kudelski and S. N. Sehgal, *J. Antibiot. (Tokyo)*, 1975, **28**, 721–726.
6. H. Baker, A. Sidorowicz, S. N. Sehgal and C. Vezina, *J. Antibiot. (Tokyo)*, 1978, **31**, 539–545.
7. K. Singh, S. Sun and C. Vezina, *J. Antibiot. (Tokyo)*, 1979, **32**, 630–645.
8. R. R. Martel, J. Klicius and S. Galet, *Can. J. Physiol. Pharmacol.*, 1977, **55**, 48–51.
9. T. Kino, H. Hatanaka, M. Hashimoto, M. Nishiyama, T. Goto, M. Okuhara, M. Kohsaka, H. Aoki and H. Imanaka, *J. Antibiot. (Tokyo)*, 1987, **40**, 1249–1255.
10. H. Hatanaka, T. Kino, S. Miyata, N. Inamura, A. Kuroda, T. Goto, H. Tanaka and M. Okuhara, *J. Antibiot. (Tokyo)*, 1988, **41**, 1592–1601.
11. G. M. Salituro, D. L. Zink, A. Dahl, J. Nielsen, E. Wu, L. Huang, C. Kastner and F. J. Dumont, *Tetrahedron Lett.*, 1995, **36**, 997–1000.

12. M. Y. Summers, M. Leighton, D. Liu, K. Pong and E. I. Graziani, *J. Antibiot. (Tokyo)*, 2006, **59**, 184–189.
13. T. Fehr, J. J. Sanglier, W. Schuler, L. Gschwind, M. Ponelle, W. Schilling and C. Wioland, *J. Antibiot. (Tokyo)*, 1996, **49**, 230–233.
14. R. Raju, A. M. Piggott, M. Conte, Z. Tnimov, K. Alexandrov and R. J. Capon, *Chem.– Eur. J.*, 2010, **16**, 3194–3200.
15. M. W. Harding, A. Galat, D. E. Uehling and S. L. Schreiber, *Nature*, 1989, **341**, 758–760.
16. H. Fretz, M. W. Albers, A. Galat, R. F. Standaert, W. S. Lane and S. J. Burakoff, *J. Am. Chem. Soc.*, 1991, **113**, 1409–1411.
17. F. Edlich and G. Fischer, *Handb. Exp. Pharmacol.*, 2006, 359–404.
18. E. Lam, M. M. Martin, A. P. Timerman, C. Sabers, S. Fleischer, T. Lukas, R. T. Abraham, S. J. O'Keefe, E. A. O'Neill and G. J. Wiederrecht, *J. Biol. Chem.*, 1995, **270**, 26511–26522.
19. W. G. Sinkins, M. Goel, M. Estacion and W. P. Schilling, *J. Biol. Chem.*, 2004, **279**, 34521–34529.
20. A. M. Cameron, J. P. Steiner, D. M. Sabatini, A. I. Kaplin, L. D. Walensky and S. H. Snyder, *Proc. Natl. Acad. Sci. U. S. A.*, 1995, **92**, 1784–1788.
21. T. Ratajczak, B. K. Ward and R. F. Minchin, *Curr. Top. Med. Chem.*, 2003, **3**, 1348–1357.
22. Y. J. Jin and S. J. Burakoff, *Proc. Natl. Acad. Sci. U. S. A.*, 1993, **90**, 7769–7773.
23. B. Chambraud, H. Belabes, V. Fontaine-Lenoir, A. Fellous and E. E. Baulieu, *FASEB J.*, 2007, **21**, 2787–2797.
24. F. Edlich, M. Weiwad, D. Wildemann, F. Jarczowski, S. Kilka, M. C. Moutty, G. Jahreis, C. Lucke, W. Schmidt, F. Striggow and G. Fischer, *J. Biol. Chem.*, 2006, **281**, 14961–14970.
25. J. Heitman, N. R. Movva and M. N. Hall, *Science*, 1991, **253**, 905–909.
26. E. J. Brown, M. W. Albers, T. B. Shin, K. Ichikawa, C. T. Keith, W. S. Lane and S. L. Schreiber, *Nature*, 1994, **369**, 756–758.
27. C. J. Sabers, M. M. Martin, G. J. Brunn, J. M. Williams, F. J. Dumont, G. Wiederrecht and R. T. Abraham, *J. Biol. Chem.*, 1995, **270**, 815–822.
28. Y. Chen, H. Chen, A. E. Rhoad, L. Warner, T. J. Caggiano, A. Failli, H. Zhang, C. L. Hsiao, K. Nakanishi and K. L. Molnar-Kimber, *Biochem. Biophys. Res. Commun.*, 1994, **203**, 1–7.
29. D. M. Sabatini, H. Erdjument-Bromage, M. Lui, P. Tempst and S. H. Snyder, *Cell*, 1994, **78**, 35–43.
30. D. H. Kim, D. D. Sarbassov, S. M. Ali, J. E. King, R. R. Latek, H. Erdjument-Bromage, P. Tempst and D. M. Sabatini, *Cell*, 2002, **110**, 163–175.
31. G. G. Chiang and R. T. Abraham, *Trends Mol. Med.*, 2007, **13**, 433–442.
32. D. A. Guertin and D. M. Sabatini, *Cancer Cell*, 2007, **12**, 9–22.
33. D. D. Sarbassov, S. M. Ali, D. H. Kim, D. A. Guertin, R. R. Latek, H. Erdjument-Bromage, P. Tempst and D. M. Sabatini, *Curr. Biol.*, 2004, **14**, 1296–1302.

34. D. D. Sarbassov, S. M. Ali, S. Sengupta, J. H. Sheen, P. P. Hsu, A. F. Bagley, A. L. Markhard and D. M. Sabatini, *Mol. Cell*, 2006, **22**, 159–168.

35. Z. Zeng, D. Sarbassov dos, I. J. Samudio, K. W. Yee, M. F. Munsell, C. Ellen Jackson, F. J. Giles, D. M. Sabatini, M. Andreeff and M. Konopleva, *Blood*, 2007, **109**, 3509–3512.

36. J. A. Kallen, R. Sedrani and S. Cottens, *J. Am. Chem. Soc.*, 1996, **118**, 5857–5861.

37. R. Sedrani, L. H. Jones, A. M. Jutzi-Eme, W. Schuler and S. Cottens, *Bioorg. Med. Chem. Lett.*, 1999, **9**, 459–462.

38. J. Choi, J. Chen, S. L. Schreiber and J. Clardy, *Science*, 1996, **273**, 239–242.

39. L. A. Banaszynski, C. W. Liu and T. J. Wandless, *J. Am. Chem. Soc.*, 2005, **127**, 4715–4721.

40. C. E. Caufield, *Curr. Pharm. Des.*, 1995, **1**, 145–160.

41. C. J. Punt, J. Boni, U. Bruntsch, M. Peters and C. Thielert, *Ann. Oncol.*, 2003, **14**, 931–937.

42. G. I. Kirchner, I. Meier-Wiedenbach and M. P. Manns, *Clin. Pharmacokinet.*, 2004, **43**, 83–95.

43. Y. W. Chen, M. L. Smith, M. Sheets, S. Ballaron, J. M. Trevillyan, S. E. Burke, T. Rosenberg, C. Henry, R. Wagner, J. Bauch, K. Marsh, T. A. Fey, G. Hsieh, D. Gauvin, K. W. Mollison, G. W. Carter and S. W. Djuric, *J. Cardiovasc. Pharmacol.*, 2007, **49**, 228–235.

44. R. Palaparthy, R. Pradhan, J. Chan, Q. Wang, Q. Ji, R. Achari, T. Chira, L. B. Schwartz and R. O'Dea, *Clin. Drug Investig.*, 2005, **25**, 491–498.

45. C. S. Karyekar, R. S. Pradhan, T. Freeney, Q. Ji, T. Edeki, W. Chiu, W. M. Awni, C. Locke, L. B. Schwartz, R. G. Granneman and R. O'Dea, *J. Clin. Pharmacol.*, 2005, **45**, 910–918.

46. M. M. Mita, A. C. Mita, Q. S. Chu, E. K. Rowinsky, G. J. Fetterly, M. Goldston, A. Patnaik, L. Mathews, A. D. Ricart, T. Mays, H. Knowles, V. M. Rivera, J. Kreisberg, C. L. Bedrosian and A. W. Tolcher, *J. Clin. Oncol.*, 2008, **26**, 361–367.

47. J. P. Steiner, M. A. Connolly, H. L. Valentine, G. S. Hamilton, T. M. Dawson, L. Hester and S. H. Snyder, *Nat. Med.*, 1997, **3**, 421–428.

48. B. G. Gold, V. Densmore, W. Shou, M. M. Matzuk and H. S. Gordon, *J. Pharmacol. Exp. Ther.*, 1999, **289**, 1202–1210.

49. J. Sharkey and S. P. Butcher, *Nature*, 1994, **371**, 336–339.

50. S. Harper, J. Bilsland, L. Young, L. Bristow, S. Boyce, G. Mason, M. Rigby, L. Hewson, D. Smith, R. O'Donnell, D. O'Connor, R. G. Hill, D. Evans, C. Swain, B. Williams and F. Hefti, *Neuroscience*, 1999, **88**, 257–267.

51. A. Bocquet, G. Lorent, B. Fuks, R. Grimee, P. Talaga, J. Daliers and H. Klitgaard, *Eur. J. Pharmacol.*, 2001, **415**, 173–180.

52. D. Liu, H. B. McIlvain, M. Fennell, J. Dunlop, A. Wood, M. M. Zaleska, E. I. Graziani and K. Pong, *J. Neurosci. Methods*, 2007, **163**, 310–320.

53. A. Galat, W. S. Lane, R. F. Standaert and S. L. Schreiber, *Biochemistry*, 1992, **31**, 2427–2434.
54. A. Ghosh and M. E. Greenberg, *Science*, 1995, **268**, 239–247.
55. S. Liang, D. Liu, Y. Chen, M. L. T. Mercado, E. I. Graziani, P. H. Reinhart, M. N. Pangalos, A. Wood, and M. M. Zaleska, presented at the Society for Neuroscience Annual Meeting, Chicago, 2009.
56. http://clinicaltrials.gov/ct2/show/NCT00827190
57. J. W. H. Li and J. C. Vederas, *Science*, 2009, **325**, 161–165.

CHAPTER 15

Identification of α7 Nicotinic Acetylcholine Receptor Agonists for their Assessment in Improving Cognition in Schizophrenia

BRUCE N. ROGERS*, E. JON JACOBSEN, CHRISTOPHER J. O'DONNELL, CHRISTOPHER L. SHAFFER, DANIEL P. WALKER AND DONN G. WISHKA

Pfizer Global Research and Development, Groton Laboratories, Eastern Point Road, Groton, CT 06340, USA

15.1 Introduction

Nicotinic acetylcholine receptors (nAChRs) are a well-defined subset of pentameric ligand-gated ion channels.[1,2] Neuronal nAChRs are found throughout the central and peripheral nervous systems, as well as in neuromuscular junctions,[3] and assemble into homo-pentamers (such as α7) or, more commonly, are formed by combinations of α and β subunits into hetero-pentamers (such as α4β2).[4] To date, 17 distinct nAChR isoforms presenting a vast diversity of possible combinations have been discovered, but only a small subset of these has been shown to give rise to functionally and physiologically relevant

RSC Drug Discovery Series No. 4
Accounts in Drug Discovery: Case Studies in Medicinal Chemistry
Edited by Joel C. Barrish, Percy H. Carter, Peter T. W. Cheng and Robert Zahler
© Royal Society of Chemistry 2011
Published by the Royal Society of Chemistry, www.rsc.org

channels. The homo-pentameric α7 nAChR consists of five α7 subunits, each providing an orthosteric binding site for its endogenous ligand, acetylcholine.[5]

Numerous studies have established the importance of this class of ion channels within the central nervous system (CNS), particularly its link to higher functions such as memory, cognition, and sensory processing.[6] Receptors in the homomeric α7 subtype are prevalent in the hippocampus and are thought to modulate a variety of attention and cognitive processes.[7] Deficits in auditory sensory processing are thought to lead to sensory overload and contribute to attention and cognitive problems in a variety of CNS diseases, especially schizophrenia.[8] Individuals with schizophrenia have decreased levels of α7 nAChRs as well as sensory gating deficits that can be corrected with nicotine; analogous deficits in rodent models can also be restored with the α7 nAChR partial agonist GTS-21.[9] Improvements in sensory gating have been shown to correlate with improvements in cognitive performance in animal models and schizophrenic patients,[10] suggesting that a selective α7 nAChR agonist may play a role in treating schizophrenia.[11,12] Interest in α7 nAChR agonists has greatly expanded during the past two decades, and many novel and selective chemical entities have been identified. An extensive review of the medicinal chemistry literature reveals that the majority of these ligands have been derived from various aminoquinuclidine-containing scaffolds,[13] including such structures as spiro-oxazolidinones (AR-R17779), carbamates, ethers, and amides. Our interest in this target began in 1998, when the few structures other than anabasine and GTS-21 described in the literature were fairly weak and non-selective for the α7 nAChR. Our goal was to discover and develop orally active α7 nAChR agonists to treat cognitive deficits in schizophrenia or Alzheimer's disease.

15.2 Identifying the Right Lead

15.2.1 High-throughput Screening to Identify High-quality Hits

When our program commenced, the short-term goals of the team focused on the development of the reagents, assays, and technologies required to use the α7 nAChR as a drug target for a high-throughput screen (HTS). Consistent with these goals, early synthetic efforts focused on the synthesis of known activators of the α7 nAChR, such as anabasine, GTS-21, AR-R17779, and galanthamine to aid in assay development and validation. These chemical tools were critical in validating a robust fluorometric imaging plate reader-based screen (FLIPR; Molecular Devices, Sunnyvale, CA) for an HTS using a chimeric surrogate for the α7 nAChR (α7/5-HT$_3$) that joined the extracellular portion of the α7 nAChR with the transmembrane portion of the 5-HT$_3$ receptor, a serotonin-gated ion channel closely related to nAChRs.[14] Quinuclidine amide **1** (Figure 15.1) was identified via this HTS as an agonist with weak ligand binding affinity (K_i) of the α7/5-HT$_3$ chimera, and was a known serotonin 5-HT$_3$ antagonist ($K_i = 1$ nM) typical of the quinuclidine amide 5-HT$_3$ inhibitor

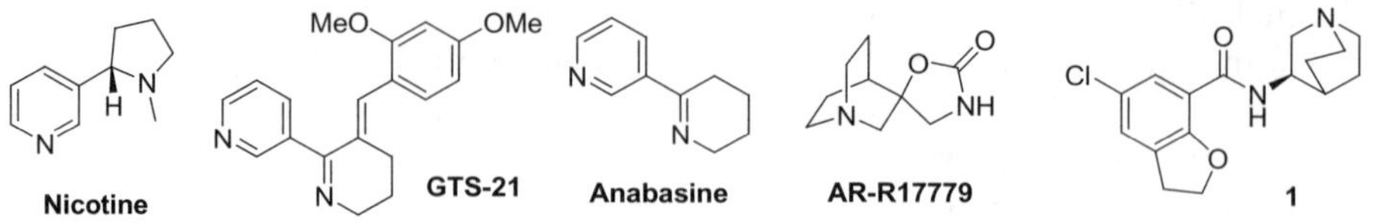

Figure 15.1   α7 nAChR agonists.

template.[15] More generally, **1** is a member of the CNS class of the 1,2-diamine-monoamides that have found utility as modulators of a variety of CNS receptors.

15.2.2   Using Parallel Medicinal Chemistry to Rapidly Identify Leads

Because few selective α7 nAChR agonists were described in the literature in 1998 on which to build a structure–activity relationship (SAR), parallel synthesis was used to explore the potential of this diamine-amide class of molecules as agonists of the α7 nAChR.[16] The FLIPR assay we had implemented was not only reliable but also provided our program with a robust and rapid measure of potency and functional activity in a single HTS. To accelerate hit follow-up and assess whether α7/5-HT$_3$ selectivity could be attained, we prepared and tested focused libraries of commercially available 1,3-diamino-amides,[17] a hit expansion strategy that allowed for swift SAR evaluation. In addition, we focused on a series with key lead attributes such as "drugability" (Lipinski rule of five), structures lacking toxic functionality, nAChR and 5-HT$_3$ selectivity, good pharmacokinetic (PK) properties, and excellent CNS penetration. The wide availability of amide bond-forming procedures allowed us to screen a diverse set of substrates (Figure 15.2), which were quickly evaluated for potency and functional activity in a FLIPR assay that used SHEP cells expressing the α7/5-HT$_3$ chimera (Figure 15.2). The resulting libraries showed a wide range of activity, indicating a general and useful SAR. For the amine it was clear that the only analogs in this series with potent activity contained the 3-aminoquinuclidine unit, wherein the (*R*) configuration was preferred over the (*S*). It was also apparent that 5-HT$_3$ selectivity could be attained by substitution in the 4-position of the aryl ring, whereas substitution in the 2-position resulted in potent 5-HT$_3$ activity. The most promising compound that emerged from the parallel synthesis was 4-chlorobenzamide (C7), PNU-282987 (Figure 15.2), which was potent at the α7/5-HT$_3$ chimera [half maximal effective concentration (EC$_{50}$) = 178 nM] and selective versus 5-HT$_3$.[17]

15.2.3   Developing a Robust Screening Strategy

With PNU-282987 in hand as a key tool compound, we could develop the full screening strategy for the program. To understand the *in vitro* α7 nAChR

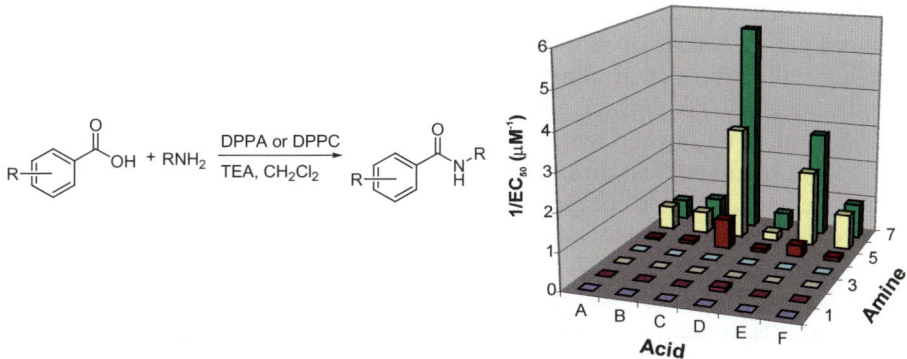

Figure 15.2 Coupling conditions for acid and amine fragments (*left*) and data from the evaluation of the compounds in the FLIPR assay (*right*). Adapted with permission of the American Chemical Society from *J. Med. Chem.*, 2005, **48**, 905.

profile fully, we used a binding assay and a hippocampal neuron assay to validate the chimera assay as a primary screen predictive of native activity. In the α7 nAChR binding assay, PNU-282987 displaced the α7 nAChR-selective antagonist methyllycaconitine (MLA) from rat brain homogenate ($K_i = 27$ nM), and when applied to cultured rat hippocampal neurons it evoked a rapidly desensitizing inward whole-cell current that was concentration dependent and blocked by MLA, consistent with the opening of the α7 nAChR.[18] Another critical aspect of the screen was selectivity, starting with 5-HT$_3$ selectivity. PNU-282987 was selective against the neuromuscular junction form of the receptor (α1β1γδ) and the predominant ganglionic nAChR (α3β4), which was critical because activation of these receptors causes many of the undesirable effects of nonspecific nAChR agonists such as epibatidine and nicotine.[19]

In addition to its selectivity profile, PNU-282987 had very desirable PK properties, with robust metabolic stability in rat and human liver microsomes (HLM), low *in vitro* clearance, excellent brain penetration, and high oral bioavailability in rats.[17,20] A key safety issue identified in the early profiling of this compound was hERG (human ether-à-go-go) potassium channel activity. Significant hERG activity with this compound was an obstacle for further development and would be a critical criterion in identifying additional chemical matter, but 14-day rodent *in vivo* toxicology studies strengthened our confidence in this template as a safe chemotype for further SAR studies.

One of the most significant milestones for PNU-282987 was its early *in vivo* proof of pharmacology, which was established through tests in a rat model of impaired sensory gating (P50) that was also validated with GTS-21.[8] P50 was measured in a rat model of auditory sensory gating in which normal auditory gating was disrupted by systemic administration of *d*-amphetamine.[21] Systemic administration of *d*-amphetamine (0.3 or 1 mg kg^{-1} iv) significantly disrupted auditory gating in anesthetized rats owing to simultaneous decreases of conditioning responses and increases in test responses.[22] Subsequent administration of PNU-282987 (1 or $3\,\text{mg}\,\text{kg}^{-1}$ iv) significantly reversed the *d*-amphetamine-induced gating deficit; in contrast, administration of vehicle to control rats did not normalize the deficit. Furthermore, PNU-282987 ($1\,\text{mg}\,\text{kg}^{-1}$ iv) had no significant effect on normal gating in anesthetized rats.[23] With this evidence as a starting point, a valid template for chemistry, and a deep screening strategy to serve in the identification of an *in vivo* active compound, we were poised to identify clinically viable compounds.

As described earlier, HTS and parallel synthetic chemistry identified quinuclidine amide PNU-282987 as a potent and selective agonist of the α7 nAChR.[17] PNU-282987 had been alleged to be a useful treatment for various CNS diseases; however,[24] preclinical cardiac safety studies revealed significant hERG potassium channel activity. Thus the program objective was to identify a novel quinuclidine aryl amide with *in vitro* and *in vivo* profiles equal to or better than that of PNU-282987.

15.3 Pursuing the First Clinical Candidate by Defining the Desired Properties to Deliver a High-Quality Molecule

Another potent α7 nAChR agonist identified during parallel synthesis was indole **2** (Figure 15.3). Although azabicyclic systems had been identified in the literature for the treatment of various CNS disorders,[25] the team believed that indole **2** provided an avenue to novel analogs with improved safety profiles. To this end, a large number of custom acid fragments were prepared *via* multistep syntheses to build the SAR systematically around 6,5-fused heterocyclic amides of the 3-aminoquinuclidine.[20,26,27] The first set of analogs focused on aryl rings with two heteroatoms in the five-membered ring (Figure 15.4). Benzoxazole **3** showed affinity and functional activity at the α7 receptor equal to that of PNU-282987 (α7 nAChR, $K_i = 32$ nM; α7/5-HT₃, $EC_{50} = 130\,\text{nM}$). Unfortunately, the compound showed no stability in rat liver microsomes (RLM) (4% remaining after 1 h). Regioisomeric benzoxazole **4** was significantly less potent than PNU-282987. Interestingly, benzothiazole **5**, a sulfur isostere of benzoxazole **3**, showed poor activity toward the α7 nAChR, but regioisomeric benzothiazole **6** demonstrated strong affinity and functional activity equal to that of PNU-282987. Whereas indazole **7** showed low affinity for the α7 nAChR, indazole **8** exhibited activity similar to that of PNU-282987. Benzimidazole **9**

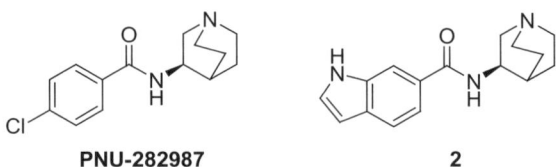

PNU-282987   **2**

Figure 15.3   6,5-Fused bicyclic heteroaryl amides of the 3-aminoquinuclidine prototype.

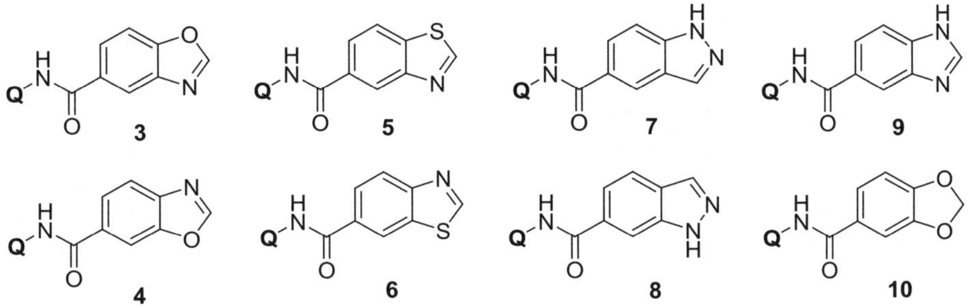

Q = (R)-quinuclidine-3-yl

Figure 15.4 Structures of novel heteroaryl amidoquinuclidines.

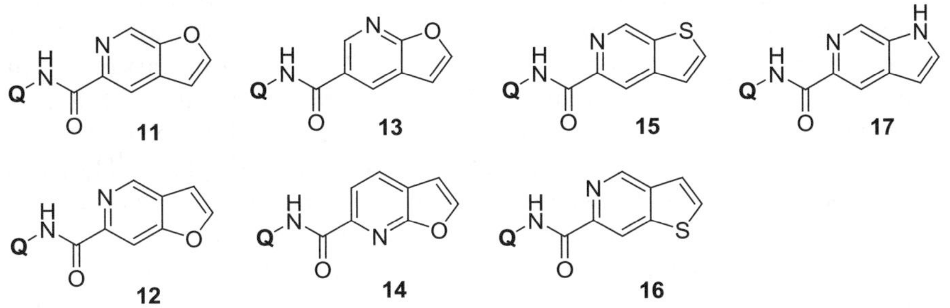

Q = (R)-quinuclidine-3-yl

Figure 15.5 Structures of novel heteroaryl amidoquinuclidines.

showed poor α7 activity, whereas benzodioxole **10** was one of the most potent compounds screened in our α7/5-HT$_3$ chimera assay (α7/5-HT$_3$, EC$_{50}$ = 37 nM). From a metabolic stability perspective, compounds **6** and **8** were stable, but compound **10** showed no stability in RLM.

Several 6,5-fused heteroaryl amidoquinuclidines with a nitrogen in the six-membered ring were also prepared (Figure 15.5).[20] One compound, furopyridine **11**, with an α7/5-HT$_3$ EC$_{50}$ of 65 nM, was two times more potent than PNU-282987, showed significant affinity for the native α7 nAChR (K_i = 8.8 nM), and had robust stability in RLM. Compound **12**, a regioisomer of compound **11**, was three times less potent in the α7/5-HT$_3$ functional assay than PNU-282987. Interestingly, moving the pyridyl nitrogen next to either bridgehead carbon, as in compounds **13** and **14**, resulted in complete loss of α7 activity (α7/5-HT$_3$, EC$_{50}$ > 10 000 nM). Replacing the furanyl oxygen atom of **11** and **12** with sulfur gave rise to thienopyridines **15** and **16**, respectively, both of which were equipotent with PNU-282987. Unfortunately, neither compound was stable in RLM. Pyrrolopyridine **17** was ten times less potent than PNU-282987 and was not considered further.

Based on their potency and RLM stability profiles (Figures 15.4 and 15.5), benzothiazole **6**, indazole **8**, and furopyridine **11** were selected for further profiling.[20] These three compounds were evaluated for functional selectivity against the 5-HT$_3$ receptor expressed in SHEP1 cells, against endogenous receptors of TE671 cells to evaluate muscle-like nAChRs (α1β1$\gamma$$\delta$), and against endogenous receptors of SH-SY5Y cells to evaluate ganglion-like nAChRs (α3β4). Given the high degree of homology between orthosteric sites in α7 nAChRs and 5-HT$_3$ receptors,[28] some degree of cross-reactivity was anticipated, and a goal was to identify compounds that afforded reasonable binding selectivity and functional antagonism at the 5-HT$_3$ receptor. The risk associated with 5-HT$_3$ receptor antagonism was assumed to be minimal and, in fact, could be beneficial because clinical studies suggest that the potent and selective 5-HT$_3$ receptor antagonist ondansetron may provide benefits to tardive dyskinesia and psychotic symptoms in patients with schizophrenia.[29] In contrast, compounds that displayed functional 5-HT$_3$ agonism were not of interest owing to a potential link to cardiovascular events.[30] Each of the three compounds demonstrated antagonist activity at the 5-HT$_3$ receptor and a 25-fold increase in selectivity compared to PNU-282987 in a binding assay for the α7 nAChR. All three compounds showed no detectable agonist activity ($>100\,\mu$M) and negligible antagonist activity at both TE671 and SH-SY5Y cell receptors. Further, the compounds did not significantly displace [^{3}H]cytisine from rat brain homogenates when examined at a compound concentration of 1 μM, suggesting a high selectivity over the α4β2 nAChR subtype.

One of the key requirements in identifying a compound for clinical development was an improved therapeutic index for hERG. *In vitro* cardiovascular safety for lead compounds was assessed by measuring their propensity to block the hERG channel during electrophysiology studies. Prolongation of the QT interval is believed to correlate with the risk of cardiac arrhythmia in humans and potentially contribute to ventricular fibrillation.[31] Measuring the effectiveness of a compound to block the hERG potassium channel is an important preclinical assessment of its proarrhythmic potential.[32] Compounds **6**, **8**, and **11** were evaluated in a patch-clamp hERG K$^+$ channel assay at concentrations of 2 and 20 μM,[33] and each showed significantly reduced hERG activity compared to PNU-282987. To evaluate the interactions with the hERG potassium channel further, we determined concentration–response profiles for PNU-282987 and compound **11**. In agreement with the screening data, compound **11** was less potent at inhibiting the hERG channel-mediated currents. Although **11** produces insufficient blockade at 20 μM, the highest tested concentration to establish the half-maximal inhibitory concentration (IC$_{50}$), extrapolating the blockade produced at this concentration (29% for **11**) to the fitted curve for PNU-282987 suggests that the potency for hERG blockade is reduced at least tenfold (Figure 15.6).

Compounds **6**, **8**, and **11** were evaluated in a constant-infusion rat *in vivo* PK model,[20,26] revealing moderate plasma clearance ranging from 34 mL min^{-1} kg^{-1} for furopyridine **11** to 68 mL min^{-1} kg^{-1} for benzothiazole **6**. Although *in vitro* RLM was an accurate predictor of *in vivo* clearance for compound **11**, RLM under-predicted the *in vivo* clearance of benzothiazole **6** and indazole **8**. Rat oral bioavailabilities ranged from 65% for furopyridine **11** to 3% for

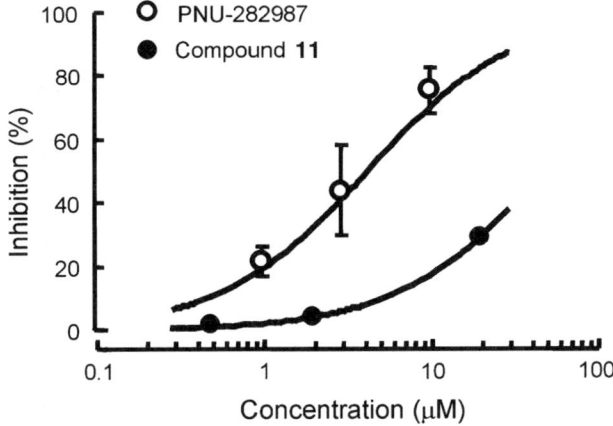

Figure 15.6 hERG potassium channel concentration response profiles of PNU-282987 and **11**. Reproduced with permission of the American Chemical Society from *J. Med. Chem.*, 2006, **49**, 4425.

indazole **8**. Compound metabolic stability was also assessed using HLM (half-life > 120 min) and revealed moderate stability for **6** and robust stability for **8** and **11**.[34]

To assess blood–brain barrier penetration, the three compounds were evaluated in a multidrug resistance (MDR) P-glycoprotein (PgP) assay.[35] Compounds that show PgP efflux, as evidenced by an MDR ratio > 3, tend to have hindered brain penetration. Benzothiazole **6** showed significant PgP efflux (MDR ratio $= 13$); indazole **8** and furopyridine **11** showed no significant PgP efflux. CNS penetration was further assessed in a mouse brain uptake assay.[36] Compounds **8** and **11** showed good to excellent brain penetration, but brain penetration was hindered in benzothiazole **6**, which was consistent with its high MDR efflux ratio. The PK and CNS penetration data clearly differentiated furopyridine **11** (PHA-543613) from benzothiazole **6** and indazole **8**, justifying further *in vivo* evaluation.[37]

The functional activity of PHA-543613 was confirmed using native α7 nAChRs of rat hippocampal neurons.[20] Using a whole-cell patch-clamp configuration on neurons, PHA-543613 evoked desensitizing inward currents that were concentration dependent and completely inhibited by selective α7 nAChR antagonist MLA (10 nM). These results suggested that PHA-543613 was modestly more active than PNU-282987 at native α7 nAChRs, which is consistent with the increased potency, demonstrated in both the binding and the FLIPR assays using the $\alpha7/5\text{-}HT_3$ chimera (Figure 15.7).

With significant interest in PHA-543613 as a clinical candidate, we sought to evaluate this compound fully using the battery of *in vivo* models we had established for the α7 nAChR program. PHA-543613 was thus tested in the validated rodent model of impaired sensory gating in which we had previously achieved proof of concept with PNU-282987.[23,38] Administration of the α7

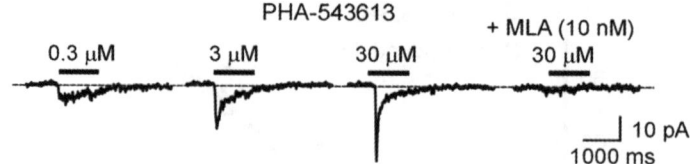

Figure 15.7   PHA-543613 evokes α7 nAChR-mediated currents from rat hippocampal neurons. Reproduced with permission of the American Chemical
Society from *J. Med. Chem.*, 2006, **49**, 4425.

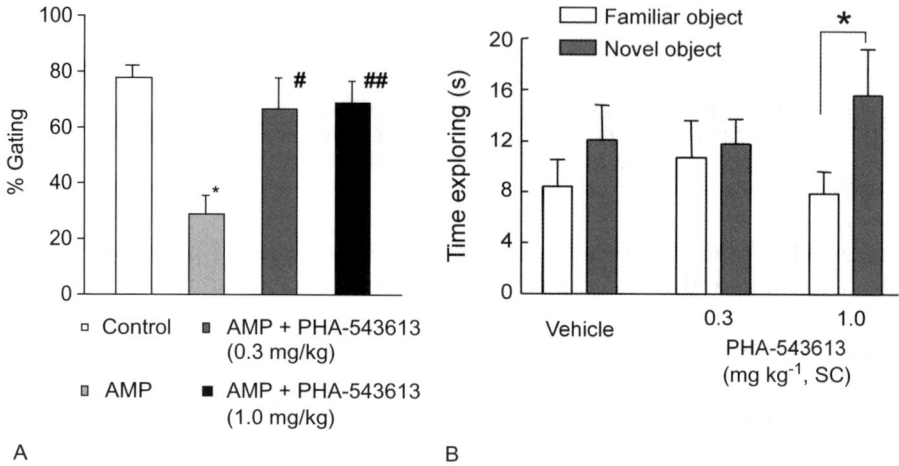

A   B

Figure 15.8   (A) Effect of PHA-543613 (0.3 and 1 mg kg^{-1} sc) on the auditory gating
deficit in amphetamine-treated rats: * $P < 0.001$ *vs.* control; # $P < 0.02$ *vs.*
d-amphetamine (AMP); ## $P < 0.005$ *vs.* AMP. (B) PHA-543613
improves performance in the rat novel object recognition task: * $P < 0.05$.
Adapted with permission of the American Chemical Society from
J. Med. Chem., 2006, **49**, 4425.

nAChR agonist PHA-543613 (0.3 or 1 mg kg^{-1} iv) significantly reverses the *d*-
amphetamine-induced gating deficit (Figure 15.8A). In contrast, application of
vehicle in control rats did not normalize the d-amphetamine-induced gating
deficit.

The potential for PHA-543613 to influence learning and memory was evaluated using the rat novel object recognition (NOR) task model.[39] This test is
based on the ethological observation that rats spend more time exploring
objects they have never seen (novel objects) than objects that are familiar.[40]
PHA-543613 was administered at 0.3 mg kg^{-1} or 1 mg kg^{-1} subcutaneously (sc)
to rats 30 min before each of three experimental sessions.[3] As illustrated in
Figure 15.8B, 1 mg kg^{-1} of PHA-543613 significantly improved the ability of
the test subjects to discriminate between the novel and familiar objects after a

24-h delay. The data from this and the auditory sensory gating *in vivo* models provided important evidence that in rats, PHA-543613 positively influenced sensory gating and memory, both of which are believed to be disrupted in patients with schizophrenia.

Of the 6,5-fused heterocyclic amides of 3-aminoquinuclidine evaluated, PHA-543613 proved to be a potent, high-affinity agonist of the α7 nAChR. The excellent *in vitro* PK profile of this compound was matched by rapid brain penetration, high oral bioavailability, and a favorable hERG profile. Furthermore, PHA-543613 demonstrated efficacy in two *in vivo* models of disease relevance. The data generated with this compound provide additional support for the hypothesis that α7 nAChR agonists represent a potential pharmacotherapy for the treatment of cognitive deficits in schizophrenia. The favorable *in vitro* and *in vivo* efficacy and safety profile of PHA-543613 made it an attractive candidate for clinical study. Regulatory toxicology studies before human studies uncovered no liabilities that would preclude further development of the compound. The main safety findings that required the development of a backup candidate were CNS signs in both rats and dogs and slight prolongation of the QTc interval in dogs at levels projected to be significantly above therapeutically relevant exposures. The CNS signs included seizures associated with maximal plasma concentrations $> 10\,\mu M$.

15.4 The Search for a Backup

15.4.1 Defining a Backup Strategy for Improving the Safety Profile

While the first-generation compound PHA-543613 was undergoing further profiling in preparation for clinical studies, the team felt it was critical to pursue a backup program for this important target. The program objective was to identify one or more novel, potent, and orally bioavailable α7 nAChR agonists with *in vitro* and *in vivo* profiles equal to or better than those of PHA-543613. The strategy also called for a backup compound that differentiated itself from PHA-543613, making it less susceptible to any potential failures seen in the PHA-543613 trials. Without human clinical data to guide the development of backup compounds, we focused on further improving the safety margins of PHA-543613 against two endpoints: QTc prolongation and convulsions in dog models. Although PHA-543613 showed adequate safety margins against these findings, QTc prolongation had the potential to be dose limiting in the clinic, and thus the team sought opportunities for further improvement. Because this medication would be administered chronically as an adjunctive therapy, it was believed that a compound with at least a threefold improvement in each of the safety areas mentioned above represented an opportunity for differentiation from PHA-543613.

The chemical strategy for identifying a backup compound focused on the quinuclidine aryl amide template (Figure 15.9). When PHA-543613 was

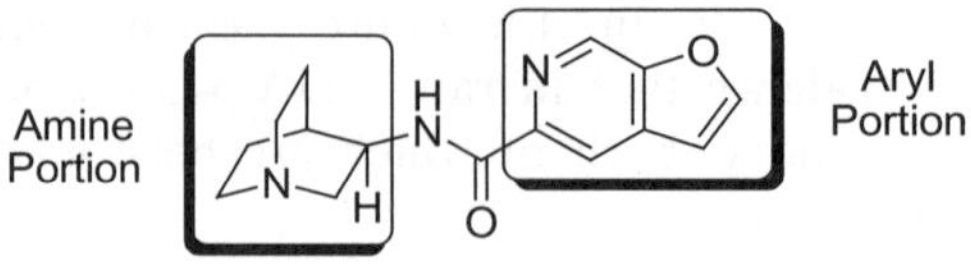

Figure 15.9   Chemical strategy for the identification of a PHA-543613 backup compound.

discovered, the team had begun to develop an extensive SAR around the aryl portion of the molecule, and these fused-ring systems offered improved potency and metabolic stability. Thus one tactic in the pursuit of a backup was to continue the exploration of fused aryl ring systems as well as the azabicyclic amine portion of the molecule.

15.4.2   Exploring Additional Novel Amine Templates

Based on previous studies, we found that aryl amides derived from a rigid diamine such as 3-aminoquinuclidine are potent α7 nAChR agonists, whereas aryl amides derived from a more flexible diamine are, at best, partial agonists.[17] The rigid quinuclidine locks the orientation of the bridgehead nitrogen lone electron pair orthogonal to one of the amide carbonyl oxygen lone pairs. This orientation is presumably important for optimal binding to the α7 nAChR.[41] Before the discovery of PHA-543613, limited efforts examined alternate azabicyclic amines as quinuclidine isosteres against this target. Given the ample literature reports demonstrating the effectiveness of other azabicyclics to function as quinuclidine isosteres,[42] we developed a second tactic focused on the preparation of alternate azabicyclic amine ring systems. More than 100 new quinuclidine aryl amide analogs were prepared;[20,27] the best quinuclidine analog candidates are shown in Figure 15.10.

In general, electron-rich aryl rings were more potent than electron-deficient rings. Also, 6,5-fused ring systems were more potent than similarly substituted 6,6-fused ring systems. With the exception of benzodioxane **18e**, which was four times less potent than PHA-543613, all new quinuclidine analogs with aryl

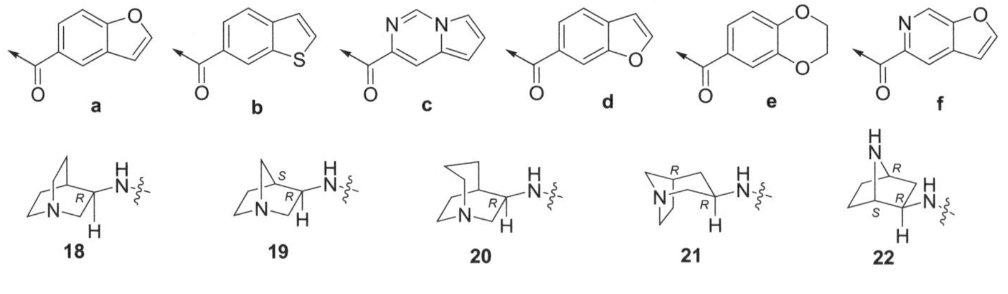

Figure 15.10   Most promising overall aryl fragments and azabicyclic amines used in the α7 nAChR backup program.

portion modifications (see Figure 15.10) showed α7 nAChR binding and functional activity similar to that of PHA-543613. Metabolic stabilities in RLM (expressed as a percent remaining) for analogs **18a–e** ranged from 10% to 77%, which met our initial metabolic stability criteria of ≥ 10%.[26] Early in the program, we had observed a rank-order correlation between RLM and *in vivo* rat clearance, affording a key attribute to drive our ADME (absorption, distribution, metabolism, and excretion) SAR based on *in vitro* properties. We had also observed, however, that RLM routinely underpredicted compound stability in HLM;[20,27] hence the low RLM threshold for compound advancement. The rationale for using RLM instead of HLM for routine screening was two-fold: (1) our primary *in vivo* efficacy and safety models were in rats (*vide infra*), and (2) high-throughput RLM data were readily available to us, whereas HLM data were not.

Focusing on the azabicyclic portion of PHA-543613, we prepared four new ring systems as quinuclidine isosteres (see Figure 15.10).[27,43,44] For each new system prepared (**19–22**), all possible stereoisomers were tested individually as pure stereoisomers. Once the most biologically relevant stereoisomer with regard to α7 nAChR agonist activity and selectivity was identified, we also identified its absolute configuration (see Figure 15.10). For some azabicyclic ring systems, α7 nAChR agonist activity resided in a single stereoisomer {*e.g.* 1-azabicyclo[3.2.1]octan-3-yl (**21**) and 7-azabicyclo[2.2.1]heptan-3-yl (**22**)}, whereas for others {*e.g.* 1-azabicyclo[2.2.1]heptan-3-yl (**19**)}, significant α7 nAChR activity was seen in more than one isomer. In the case of azabicyclic ring **19**, two stereoisomers with the same aryl fragment [*i.e.*, (3*R*,4*S*)- and (3*S*,4*S*)-isomers] typically showed significant α7 nAChR activity; however, the (3*S*,4*S*)-isomer also usually displayed significant 5-HT$_3$ activity. Each new azabicyclic amine **19–22** was coupled to the aryl fragments **a–f** shown in Figure 15.10.

Analogs derived from azabicyclic ring **19** were on average two to three times less potent than their corresponding quinuclidine analogs. The RLM stability of analogs derived from azabicyclic ring **19** was significantly enhanced with respect to the corresponding quinuclidine analogs, however, which may be due to their lower log *D* values.[27] Analogs with azabicyclic ring **20** were 8–20 times less potent than their corresponding quinuclidine analogs; however, analog **20a** still met our potency criteria, and it was stable in RLM. Aryl amides derived from azabicyclic ring **21** were equipotent to the quinuclidine analogs, and the RLM stability of type **21** analogs was also comparable to that of quinuclidine analogs. Compounds derived from azabicyclic ring **22** were in general two to three times less potent than the corresponding quinuclidine analogs. Surprisingly, even though type **22** analogs have an epibatidine-like 7-azabicyclo[2.2.1]heptan-3-yl skeleton, they did not exhibit significant off-target activity at other nicotinic receptors such as α4β2 and α3β4.[27] In fact, all of the analogs derived from azabicyclic rings **19–22** showed excellent selectivity against off-target nicotinic receptors such as α4β2, α3β4, and α1β1γδ.[27] Based on their excellent α7 nAChR agonist activity and acceptable metabolic stability in RLM, a number of amides derived from azabicyclic rings **19–22** were selected for further profiling[27] in a set of *in vitro* and *in vivo* assays that included 5-HT$_3$

binding, hERG functional assessment, CNS penetration in mice, *in vivo* rat clearance, and oral bioavailability.[20,26] Several compounds showed potent (<250 nM) binding to 5-HT$_3$, including **18b**, **18c**, **19f**, **20a**, and **21a**, but they were later determined to be antagonists of 5-HT$_3$, as was PNU-282987.

In vitro cardiovascular safety was evaluated in a patch-clamp hERG K$^+$ ion channel assay.[33] Given the efficacious free plasma concentration of PHA-543613 in the P50 auditory gating assay,[45] the goal for a backup was to identify compounds with $\leq 30\%$ inhibition of hERG at 20 µM. A number of analogs met this criterion, such as **18e**, **19a**, **19f**, **21a**, and **21f**; however, several analogs, including those with a benzothiophene (*i.e.*, **18b**, 65%) or a pyrrolopyrimidine (*i.e.*, **18c**, 71%), displayed significant hERG activity as expressed as a percent inhibition at 20 µM. In addition, analogs derived from azabicyclic ring **20** displayed higher hERG activity than that of the other amines. In general, these data suggested that increased hERG activity for type **18–22** compounds correlated with increases in lipophilicity (clog *P*). For instance, azabicyclo[2.2.1]-heptane **19a**, with a clog *P* of 2.1, showed lower hERG activity compared to that of azabicyclo[3.2.2]nonane **20a**, with a clog *P* of 2.6. Significant increases in hERG activity were seen in compounds with a polarizable atom, such as benzothiophene **18b**. A correlation with pK_a was also observed in that more basic compounds tended to show higher hERG activity (data not shown). These observations are consistent with descriptors that have now appeared in the literature to modulate hERG activity for a given chemical series.[46]

Compounds lacking a hERG signal were evaluated in a constant-infusion rat PK model, and the majority of azabicyclic aryl amides demonstrated moderate to low clearance and later showed good oral bioavailability.[20,27,44] Although there was no overall relationship to structure, the *in vivo* clearance and oral bioavailability (*F*) reasonably correlated with the *in vitro* RLM data, where **18e** and **22f** had the lowest clearance and **21f** had the highest (72 mL min^{-1} kg^{-1}). The high clearance of **18e** and **21f** in rats was not a concern, however, as these compounds were stable in an HLM assay (half-life >120 min).[20,27] As stated earlier, this trend was seen for many of the aryl bicyclic amides, including PHA-543613. Indeed, based on the human PK for PHA-543613, RLM appears to overestimate human systemic clearance for this class of compounds.[47]

CNS penetration was accessed for these analogs using a mouse brain uptake assay.[36] All compounds had robust uptake at either 5 or 60 min, and several (**18a**, **18d**, **19a**, **22a**) demonstrated a modest degree of CNS accumulation.[27,44] In particular, several analogs in the benzofuran series had significant accumulation, whereas the furanopyridines (PHA-543613, **18f**, **21a**, **22f**) maintained high CNS penetration without accumulation.

15.4.3   Understanding Reactive Metabolite Profiles

The benzofuran series had always been of some concern owing to the known metabolic liabilities of furans and benzofurans. The other positive properties of these compounds directed the team to fully understand any potential risk before deprioritizing them, however. It is well known that this ring system has the ability to undergo cytochrome P450 (cyp)-catalyzed bioactivation to provide electrophilic intermediates that, in turn, are capable of forming covalent links with proteins and enzymes (including cyp isoforms), often eliciting a toxic response.[48] This concern guided early synthetic efforts toward furanopyridines, which led to the identification of PHA-543613. Later in the program, cyp screening indicated that benzofurans **19a**, **21a**, and **22a** showed 50–60% inhibition of CYP2D6. Furanopyridine **22f**, containing a 7-azabicyclo[2.2.1]amine, also showed significant inhibition in this screen. Analogs derived from azabicyclic amine **22** were, in general, effective cyp inhibitors. Given these concerns, we evaluated PHA-543613, **18a**, **18b**, and **18d** for their ability to undergo oxidative bioactivation in human microsomes.[49] Glutathione ethyl ester (GSH-EE) was used to trap any activated electrophile formed during the incubations, of which only **18d** formed GSH-EE adducts. Through LC/MS/MS analysis of the liver microsomes containing **18d** and GSH-EE, we detected a single conjugate on the benzofuran ring. Presumably, PHA-543613 and **18a** were inactive in this assay as a result of conjugation of the furan oxygen to the amide carbonyl and the deactivation of the aryl system owing to the pyridine nitrogen in PHA-543613.[34,47] The inactivity of thiophene analog **18b** is surprising, given that thiophenes are particularly prone to cyp mediated bioactivation.[48]

15.4.4 Differentiating Potential Candidates

The combination of the above studies led us to focus on amides **18e** (PHA-568487), **19f**, and **21f** (PHA-709829). These compounds were run in the rodent screening model of impaired sensory gating, and both PHA-568487 and PHA-709829 reversed the *d*-amphetamine-induced auditory gating deficit ($>60\%$) at $1\,mg\,kg^{-1}$ iv.[27,43] Amide **19a** was less effective, reversing the gating deficit by approximately 40%.[27] To understand the pharmacokinetic/pharmacodynamic (PK/PD) relationship, a full study of dose–response relationships for analogs PHA-568487 and PHA-709829 was carried out. Each compound normalized the gating response, with PHA-568487 and PHA-709829 demonstrating activity across a wide dose range. For compound PHA-709829, the minimum effective dose was $0.1\,mg\,kg^{-1}$ (sc), which resulted in a 60 nM and 47 nM plasma and brain exposure, respectively, determined at 37 min post-dose (Table 15.1).[44] Plasma protein binding for PHA-709829 was 15–25%; thus, the minimum effective concentration of PHA-709829 in the rat auditory gating assay was comparable to its α7 nAChR rat K_i of 3.4 nM.[50] Surprisingly, the minimum effective dose for benzodioxane PHA-568487 was $0.02\,mg\,kg^{-1}$ (sc; $n = 9$), which resulted in plasma exposure of 16 ± 11 nM and brain exposure of 2.2 ± 0.9 nM, determined at 37 ± 3 min after drug administration. Based on the plasma protein binding of 15–25%, the minimum effective concentration for PHA-568487 was approximately 2.0 nM, which is considerably lower than its measured α7 nAChR rat K_i of 44 nM. Subsequent mechanistic investigations revealed that PHA-568487 activates and desensitizes

Table 15.1 Summary of biological data for key backup compounds compared to PHA-543613.

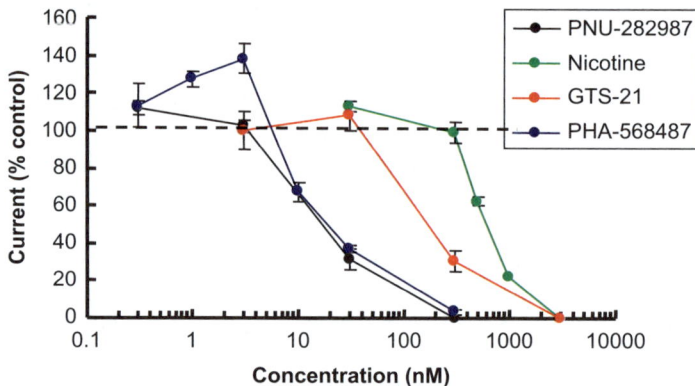

PHA-543613 (11)   **PHA-568487 (18e)**   **PHA-709829 (21f)**

	PHA-543613	*PHA-568487*	*PHA-709829*
$\alpha7/5\text{-HT}_3\text{EC}_{50}(\text{nM})$	65 ± 11	260	46
$\alpha7$ nAChR K_i (nM)	9.0	44	3.4
In vivo rat PK CL (mL/min/kg)	33 ± 5	58 ± 10	72 ± 25
Half-life (min)	36 ± 6	0.9 ± 0.1	48 ± 0
Vss (L/kg)/F(%)	1.8/74	2.1/18	2.4/40
In vivo RAG MED (mg/kg)/ plasma conc. (nM)	0.24/130	0.02/16	0.10/60
Acute Dog ECG QTc NOEL (μM)/TI	5.7/9 fold	$>16/>140$ fold	$>22/110$ fold
Dog seizure C_{max} (μM)/TI	2.6/17 fold	$>120/>560$ fold	$>51/>240$ fold

EC_{50}, half maximal effective concentration; K_i, binding affinity; RAG MED: rat auditory gating assay minimum efficacious dose; ECG, electrocardiography; QTc, Q wave, T wave cycle; NOEL, no adverse effects levels; N.D., not determined.

Figure 15.11 Desensitization concentration–response of PHA-568487 compared to reference compounds.

$\alpha7$ nAChRs in a concentration-dependent manner similar to that of other known $\alpha7$ nAChR agonists such as PNU-282987,[17] nicotine, and GTS-21 (Figure 15.11). In addition, at concentrations lower than those required to activate the receptor, PHA-568487 positively modulated $\alpha7$ nAChR function, whereas other known $\alpha7$ nAChR agonists lacked this modulatory effect (see Figure 15.11).[51] Thus, the broad range of *in vivo* activity of PHA-568487 appears to result from a unique combination of agonist and positive allosteric modulation.

With PHA-568487 and PHA-709829 identified as the key backups to PHA-543613, detailed safety studies were performed to compare these compounds to the lead (Table 15.1). To assess cardiovascular risk, both compounds were tested against the action potential duration in canine Purkinje cells and did not show significant effect. They were also evaluated in an *in vivo* cardiovascular safety study using an acute dog electrocardiographic model for their potential to cause QTc prolongation. In this study the NOELs (no observable effects levels) for PHA-568487 and PHA-709829 were $>15.8\,\mu M$ and $>22\,\mu M$, respectively. This value was compared to efficacy determined in the sensory gating model in which the minimum efficacious dose for PHA-568487 had a free maximally effective plasma concentration (C_{max}) of $\sim 0.11\,\mu M$ *versus* 0.075 μM for PHA-709829, leading to margins of >140-fold and >106-fold over any QTc signal, respectively. Both compounds demonstrated a greater than eight-fold improvement in therapeutic index over any signs of QTc effect compared to that of PHA-543613. Initial safety studies with PHA-543613 at a total C_{max} of $3.6\,\mu M$ (17-fold margin) produced seizures in dogs. Both PHA-568487 and PHA-709829 greatly attenuated this propensity, with >560-fold and >240-fold margins, respectively.

These high-quality candidates, PHA-568487 and PHA-709829, were confirmed as backups to PHA-543613. They differed chemically from PHA-543613 in that one compound contained an alternative bicyclic amine and the other had a different aryl core. Both compounds were potent α7 nAChR agonists, had robust oral bioavailability and PK, and displayed efficacy in an *in vivo* model of sensory gating. Interestingly, PHA-568487 was uniquely potent in the *in vivo* efficacy model, which was attributed to a unique combination of agonist and positive modulatory activity. Furthermore, both compounds addressed the known safety liabilities of PHA-543613.

15.5   Clinical Results and Challenges

Cardiovascular safety assessments were critical if agents in the α7 nAChR agonist class were to enter clinical studies. Nicotine, a nonselective nAChR agonist, is linked to cardiovascular changes,[52] but we believed that selective agents would lack cardiovascular safety issues. To date, a primary interest in our program has been ensuring that sufficient hERG selectivity is attained.[53] Many of the α7 nAChR compounds reported in the literature, including PNU-282987, displayed potent hERG blocking effects, whereas others, such as PHA-543613, PHA-568487, and PHA-709829, have shown adequate selectivity against this ion channel. Preclinical assessment of PHA-543613 and PHA-568487 in multispecies toxicology studies demonstrated not only sufficient hERG therapeutic indices but also overall safety profiles supporting their entry into Phase I clinical studies.

During Phase I single ascending dose (SAD) and multiple ascending dose (MAD) safety, tolerability, and PK studies, PHA-543613 demonstrated linear pharmacokinetics and a suitable half-life in human ($\sim$9–12 h).[47] In the Cog-State cognition testing battery,[54] patients showed a modest improvement in cognitive function at the low dose.[55] Similar Phase I SAD and MAD safety, tolerability, and PK studies were completed with PHA-568487.

During dosing of the first group of subjects in the PHA-568487 MAD study, two subjects experienced episodes of nonsustained ventricular tachycardia (NSVT) and one subject exhibited an increased frequency of premature ventricular contraction. During the PHA-543613 studies, one subject in the SAD study experienced an episode of NSVT, and one subject in the MAD study experienced NSVT. A low-frequency of cardiovascular arrhythmia was also observed during this study, and there was no evidence of QTc prolongation or other electrocardiographic interval changes associated with the episodes of arrhythmia. An analysis of the SAD and MAD studies ($N = 68$ for PHA-543613; $N = 25$ for placebo) found no evidence of a direct relationship between the drug concentration and cardiac arrhythmia.

Recently, however, the clinical development of both PHA-543613 and PHA-568487 has been discontinued owing to a low incidence (5%) of asymptomatic NSVT and premature ventricular contraction in healthy volunteers.[55] It is unclear whether this side effect is related to the target or the class of compounds, and we are pursuing additional research to elucidate its origin. A full understanding of the potential safety concerns should emerge with the data from clinical trials of partial and full agonists of the α7 nAChR.

15.6 Attempting to Overcome Clinical Toxicology Findings: Merging Novel Templates to Develop Additional Drug Candidates

Despite consultation with an external cardiology expert, we were unable to determine whether the cardiac side effects seen in Phase I studies of PHA-543613 and PHA-568487 were related to α7 nAChR agonism or to molecular structure. Because both compounds belong to the quinuclidine amide structural class, we decided that the best path forward would be to advance a compound from a novel, non-quinuclidine-containing chemotype. Fortunately the integration of two organizations, Pfizer and Pharmacia, afforded a unique opportunity to pursue a program focused on the 1,4-diazabicyclo[3.2.2]nonane–4-azabenzoxazole series of α7 nAChR agonists. This series had been discovered by the legacy Pfizer team, and could now be profiled in the deep *in vivo* biology assays developed by the Pharmacia team.

In previous work, we selected PHA-543613 as the target profile for a backup compound for several reasons: (1) PHA-543613 was characterized as a potent, subtype-selective, α7 nAChR full agonist with no α7 nAChR positive allosteric modulator activity; (2) the ADME profile of this compound in humans was ideal because it was metabolized slowly and had linear PK in the clinic. In addition, it demonstrated a modest improvement in the CogState test battery in healthy volunteers at the lowest dose tested (5 mg); and (3) it had fewer and less severe cardiovascular side effects than its counterpart, PHA-568487.[55]

Key design principles in this effort included favorable physicochemical properties, *in vitro* ADME assays, and a pharmacophore model for hERG to minimize this potential liability. Specifically, compounds with molecular weight <400 and clog $P<3.0$ were targeted since we were mindful of the basic amine within the molecule. We believed that these design criteria would assist in the rapid optimization of the PK profile as well as reduce known risks associated with lipophilic, basic compounds.[56] With respect to *in vitro* ADME assays, we targeted compounds with HLM clearance $<10\,\mathrm{mL\,min^{-1}\,kg^{-1}}$, high permeability [as shown in a Madin Darby canine kidney cell line (MDCK)] A → B $>10\times10^{-6}\,\mathrm{cm\,s^{-1}}$, and low potential for PgP efflux (MDR BA/AB ratio <2.0).

This lead series was derived from screening the Pfizer chemical file and from the primary and patent literature. Starting from known quinuclidine phenylurethane **23**, the first design cycle consisted of reversing the orientation of the carbamate and including the nitrogen atom of the carbamate within the bicyclic ring to give compound **24** (Figure 15.12). This design was greatly facilitated by our previous use of 1,4-diazabicyclo[3.2.2]nonane in a quinoline antibiotic program, giving us ample supply to establish the SAR.[57,58] This SAR revealed that one of the major liabilities with the phenyl carbamate series was maintaining the balance of good potency ($K_i<20\,\mathrm{nM}$) and good functional activity (*i.e.*, full agonist activity when compared to nicotine).[59] In an attempt to address this challenge, the team envisioned utilizing heterocyclic replacements of the carbamate functional group of **24** and developed the SAR of benzoxazole and azabenzoxazoles **25** as heterocyclic replacements of the phenyl carbamate.

The preparation of benzoxazole and azabenzoxazole analogs was straightforward.[60] The α7 nAChR and 5-HT$_3$ binding activities and α7 nAChR functional data for a number of key monosubstituted benzoxazole and unsubstituted azabenzoxazole analogs are shown in Figure 15.13. Unsubstituted benzoxazole analog **27** was characterized as a full agonist (*vs.* 50 μM nicotine) in a *Xenopus* oocyte assay, with robust α7 nAChR affinity and equivalent potency at the 5-HT$_3$ receptor. Replacing the carbamate functional

23, α7 K_i = 167 nM, 90% ag
WO 97 / 30998 Astra

24, α7 K_i = 36 nM, 158% ag
(SR-180711)

25: X=CH or N, Y=CH or N

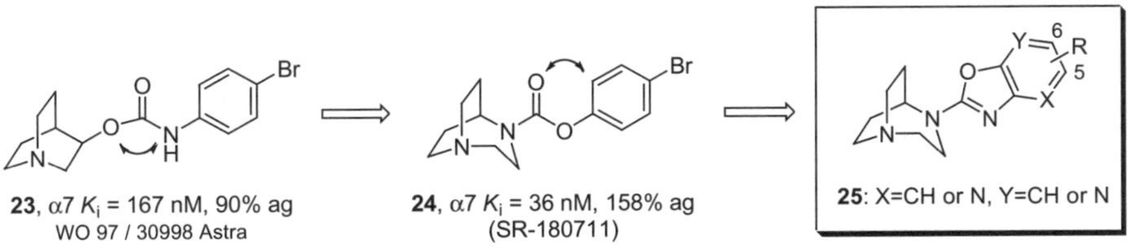

Figure 15.12 Rational design of the benzoxazole and azabenzoxazole series.

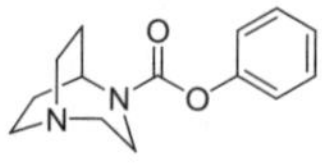

26 α7 K_i = 739 nM, 99% ag
5-HT$_3$ K_i = 3993 nM, antag

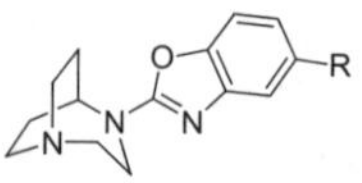

27 R=H; α7 K_i = 22.5 nM, 183% ag
5-HT$_3$ K_i = 13 nM, antag
28 R=Cl; α7 K_i = 1.92 nM, 122% ag
5-HT$_3$ K_i = 1.2 nM, antag.

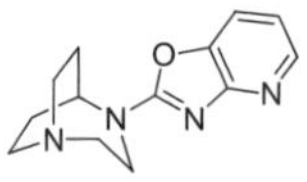

29 α7 K_i = 30.3 nM, 220% ag
5-HT$_3$ K_i = 579 nM, antag

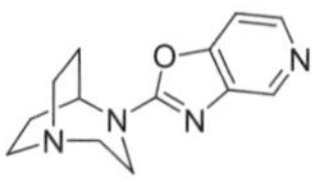

30 α7 K_i = 312 nM, 156% ag
5-HT$_3$ K_i = 2120 nM, antag

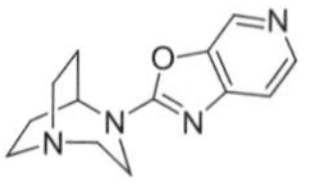

31 α7 K_i = 309 nM, 186% ag
5-HT$_3$ K_i = 162 nM, antag

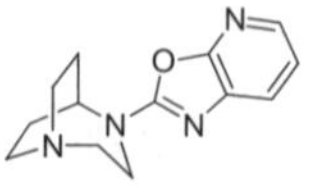

32 α7 K_i = 40.6 nM, 182% ag
5-HT$_3$ K_i = 92 nM, antag

Figure 15.13 Azabenzoxazole structure–activity relationships leading to the discovery of the 4-azabenzoxazole chemotype.

group with a heterocyclic benzoxazole proved a viable option, as **27** was ~32 times more potent than **26** at the α7 nAChR, confirming our design hypothesis. Substitution with a halogen or small alkyl group at the 5- or 6-position of the benzoxazole ring generally increased potency (exemplified by **28**) for the α7 nAChR, but these substitutions did not improve 5-HT$_3$ selectivity. Four monoaza-benzoxazole regioisomers (**29–32**) were also prepared in an attempt to broadly define the SAR for this core. Both 4-azabenzoxazole **29** and 7-azabenzoxazole **32** demonstrated potent α7 nAChR affinities, maintained full agonist activity in the functional assay, and demonstrated potential for improved selectivity over the 5-HT$_3$ receptor (20× and 2×, respectively), pointing to a promising direction for additional SAR studies.

Through extensive SAR development, we determined that analogs of the 4-azabenzoxazole template **29** generally displayed greater binding affinity at the α7 nAChR than their counterparts in the 7-azabenzoxazole template **32**. The SAR of the 4-azabenzoxazole series is summarized in Figure 15.14.[60] Substitution at the 7-position resulted in compounds with reduced affinity for the α7 nAChR, whereas 5- and 6-position substitutions resulted in compounds with robust potency (α7 nAChR, K_i < 20 nM) and full agonist activity relative to nicotine. Halogens, nitrile, and small alkyl groups (*e.g.* Me), phenyl groups, and aryl ethers were preferred at the 6-position, and phenyl, pyrrolidinyl, small alkyl groups (*e.g.* Me, Et), and aryl ethers were preferred at the 5-position. Additionally, bis-substitution at the 5- and 6-position was tolerated and in some cases synergistic when examining off-target pharmacology such as hERG affinity and *in vitro* micronucleus findings (*vide infra*).

A group of high-interest compounds (**33–36**) were further profiled in key selectivity assays (Table 15.2).[61] All of these compounds were potent at the α7 nAChR in binding assays and demonstrated functional agonism both in whole-cell patch-clamp assays, in which α7 nAChR was expressed in oocytes, as well as in an α7/5-HT$_3$ chimera FLIPR assay. These compounds also displayed selectivity over other nicotinic receptor subtypes (>90× *vs.* α4β2 and α3β4 nAChRs). This subset of compounds displayed modest to good selectivity

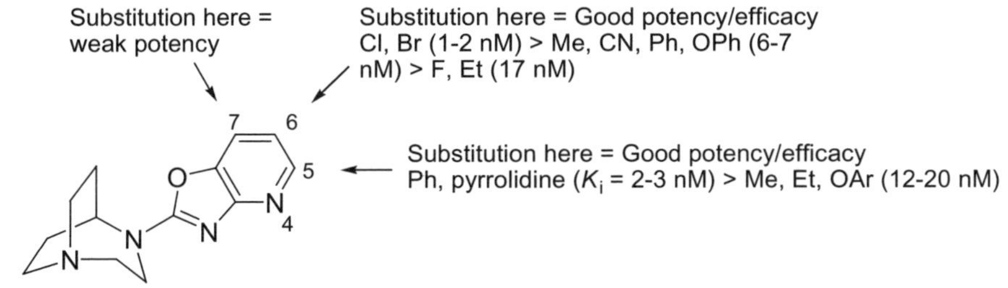

Substitution here = weak potency

Substitution here = Good potency/efficacy
Cl, Br (1-2 nM) > Me, CN, Ph, OPh (6-7 nM) > F, Et (17 nM)

Substitution here = Good potency/efficacy
Ph, pyrrolidine (K_i = 2-3 nM) > Me, Et, OAr (12-20 nM)

Figure 15.14 Structure–activity relationships of the 4-azabenzoxazole series.

Table 15.2 Pharmacologic profile of four high-interest 4-azabenzoxazole analogs.[61]

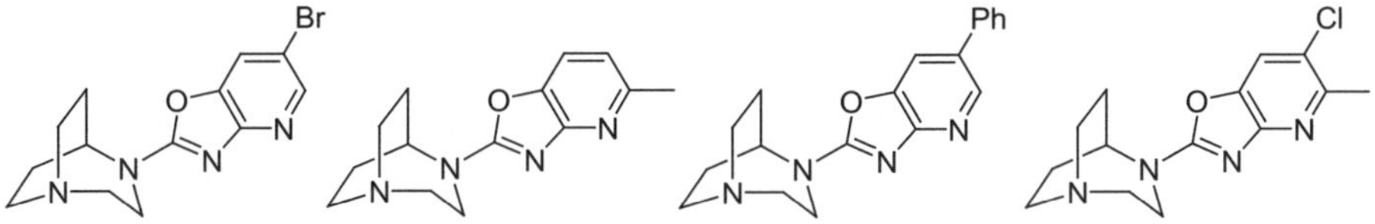

Compound ID	33	34	35	36
α7 K_i (nM)	2.1	13.3	7.0	3.2
% Agonist *vs.* nic	347%	195%	380%	n.d.
α7/5-HT$_3$ EC$_{50}$ (nM, %)	240(64%)	244(46%)	320(71%)	160(58%)
α4β2 IC$_{50}$ (nM)	771	1,350	>10,000	n.d.
α3β4 IC$_{50}$ (nM)	180	5,370	n.d.	n.d.
5-HT$_3$$K_i$ (nM)	926	248	2,117	244
5-HT$_3$IC$_{50}$ (nM)	779	5	n.d.	70
hERGIC$_{50}$ (nM)	2,300	40,000	1,700	16,200
In vitro micronucleus	Pos	Neg	Pos	Neg

EC$_{50}$, half maximal effective concentration; hERG, human ether-à-go-go; n.d., not determined; nic, nicotine; IC$_{50}$, the half-maximal inhibitory concentration; K_i, binding affinity.

versus 5-HT$_3$ (19–440×) and acted as antagonists in a functional assay for the 5-HT$_3$ receptor.

With the exception of **35**, the compounds in Table 15.2 displayed good selectivity *versus* the hERG channel (>1000× *vs.* α7 K_i). An interesting SAR trend emerged during *in vitro* micronucleus testing in that **34** and **36** were negative and **33** and **35** were positive, suggesting that substitution at the 5-position precludes *in vitro* genetic toxicity. The overall *in vitro* potency and selectivity profiles at α7 nAChR, α4β2 nAChR, α3β4 nAChR, and 5-HT$_3$, and in hERG, and *in vitro* micronucleus testing along with ease of synthesis (*vs.* **36**) prompted further evaluation of compound **34** in both *in vitro* and *in vivo* ADME studies and our *in vivo* efficacy models.

Table 15.3 Selected physicochemical properties, ADME, and pharmacokinetic data for **34**.[a]

hCL (mL/min/kg)	<5.3		
MDR BA/AB	0.8	MW 258	
MDCK (10–6 cm/s)	18.0	cLogP 1.57	
Oral bioavailability, F(%) (rat)	73	logD 0.36	
Plasma protein binding (fu) (rat/	0.78/	TPSA 45.4	
mouse/human)	0.91/	pKa 8.86	
	0.85	HBD 0	**34** (CP-810123)
Brain (ng/g)	3640		
Plasma (ng/ml)	1160		
Brain/Plasma	3.2		
CSF (nM)	3230		

[a]ADME, absorption, distribution, metabolism, and excretion; CSF, cerebrospinal fluid; HBD, hydrogen bond donor; MDCK, Madin Darby canine kidney; MDR, multidrug resistance; MW, molecular weight; TPSA, topological polar surface area.

The physicochemical properties, *in vitro* characteristics, ADME, and PK data for **34** are captured in Table 15.3.[60] These properties clearly aligned with our design goals of low molecular weight, low clog P and clog D, and topological polar surface area consistent with effective CNS drugs.[62] Compound **34** showed excellent solubility and ADME attributes and was metabolically stable in the presence of HLM ($hCl_h < 5.3$ mL min^{-1} kg^{-1}), which predicts a long half-life. The compound was also well absorbed (MDCK A → B ≥ 18×10^{-6} cm s^{-1}), with no evidence of PgP efflux (MDR1/MDCK B → A/A → B ≈ 1).[63] Subsequent *in vivo* studies validated these *in vitro* results: **34** demonstrated high oral bioavailability and favorable brain/plasma ratios in rats, with no evidence of impaired brain penetration using PgP knockout mice. Furthermore, the ratio of [cerebrospinal fluid (CSF)]/[total plasma] *versus* fraction unbound in PgP studies was 0.8, suggesting no disequilibrium and ready distribution to the brain.

The overall profile of **34** prompted further *in vivo* characterizations. *Ex vivo* binding studies evaluated the exposure of **34** at the target site after systemic administration. Receptor binding under these conditions is dependent on the access of the compound to the receptor and is therefore influenced by brain uptake, protein binding, and other partitioning that may occur in the brain. Receptor occupancy was determined in the hippocampal brain regions of rats 30 min after sc injections of **34** or vehicle (3 animals/dose group) using [^{125}I]-α-bungarotoxin. The dose- and *ex vivo*-derived receptor occupancy relationship yielded an ED$_{50}$ of 0.34 mg kg^{-1} for **34**, clearly establishing that **34** occupied its target of action.

Compound **34** was evaluated in two rodent models of cognition, *d*-amphetamine-induced gating and NOR. In the auditory gating experiment, **34** significantly improved auditory gating at 0.3 and 1.0 mg kg^{-1} sc ($p < 0.05$). These data were used to establish a minimum effective dose of **34** at 0.3 mg kg^{-1}, and the free plasma exposure achieved in rats associated with this dose was 326 nM approximately 40 min after dosing. Additionally in rats, quantitative electroencephalographic analysis showed that **34** significantly enhanced theta band power. These results demonstrate that α7 nAChRs agonists can synchronize ongoing hippocampal oscillations, leading to significantly augmented theta band power and suggesting that it could be a contributing mechanism to the procognitive actions of α7 nAChR agonists. Compound **34** was also evaluated in rat NOR, a model thought to be comparable to delayed-matching-to-sample tasks commonly used in nonhuman primates and humans to study working memory performance.[64] The compound demonstrated significant improvement of scopolamine-impaired performance at both 0.32 and 1.0 mg kg^{-1} sc. Importantly, the efficacious doses in the auditory gating assay and NOR aligned completely with the effective doses in the *ex vivo* binding experiment.

Compound **34** (CP-810123) exhibited an excellent balance of potency, selectivity, and high-affinity agonist activity at the α7 nAChR and represented a novel chemotype. Its *in vitro* profile, complemented by excellent brain penetration, high oral bioavailability in rats, and an acceptable preclinical cardiovascular safety profile with a favorable hERG and genetic toxicology profile, satisfied the desired characteristics of a distinct chemotype mimicking PHA-543613. Strong correlation among the *in vitro* data, *ex vivo* binding, and exposure required for *in vivo* efficacy demonstrate the solid *in vitro* to *in vivo* preclinical correlation of **34**. The structural diversity of this class of compounds, when compared to the prior clinical entries, combined with its preclinical evaluation package, supported the hypothesis that α7 nAChR agonists may have potential as a pharmacotherapy for the treatment of cognitive deficits in schizophrenia.

Phase I SD and MD studies evaluating safety, tolerability, and PK of CP-810123 demonstrated linear pharmacokinetics and a half-life in humans comparable to those of PHA-543613 and PHA-568487. CP-810123 also induced a very modest improvement in cognitive function at the low dose (5 mg) in the CogState cognition testing battery administrated during the SD portions of the studies. In the clinical setting, however, this compound also produced episodes of NSVT in volunteers and in light of the clinical data on PHA-543613 and PHA 568487, its development was also halted.

15.7 New Thoughts with Regard to Receptor Occupancy

The α7 nAChR is a homopentameric ligand-gated ion channel, with each subunit containing an orthosteric binding site for its endogenous ligand acetylcholine.[4] The transmembrane nature of this proteinaceous neuronal receptor means that its orthosteric binding sites are bathed by brain tissue extracellular fluid (ECF); hence ligand ECF concentration dictates ligand–orthosteric binding site interactions. Like other neuronal ion channels,[65,66] prolonged exposure of agonist at the α7 nAChR receptor can cause desensitization, a phenomenon by which continuous agonist-mediated receptor stimulation leads to receptor inactivation. From a medicinal chemistry perspective, this last concept should be considered carefully as too great of an exposure over a prolonged period could result in a loss of desired agonistic activity.

Based on these concepts, receptor occupancy (RO) calculations may be used to project the extent of *in vivo* ligand–receptor interaction, providing an idea of the percentage of total receptors occupied by a specific ligand concentration. For centrally acting compounds, RO–concentration relationships may be determined more definitively using a receptor-specific radioligand (*i.e.*, *in vivo* or *ex vivo* brain receptor binding)[67] or, more speculatively, incorporating ligand concentration at the site (or via a surrogate of that site, such as CSF) of ligand–receptor interaction at a specific time point (C_t) and ligand binding affinity (K_i) for the specific receptor.[68]

Because of the extracellular positioning of the α7 nAChR orthosteric binding sites, ligand ECF concentrations (C_{ECF}) within the brain would be most optimal for projecting RO in any species. Unfortunately, such values in humans receiving PHA-543613 are unknown. However, a clinical study with PHA-543613 [50 mg twice daily (*bid*) for 7 d; $N = 6$ active, 2 placebo] was conducted to determine the steady-state relationship between CSF concentration (C_{CSF}) and its unbound plasma concentration ($C_{p,u}$) in humans. This ratio was 1.2 for PHA-543613—a passively permeable compound not predicted by *in vitro* data to be an MDR 1 PgP substrate—consistent with a ratio of 0.80 determined in SD rat neuropharmacokinetic studies and suggesting intercompartmental equilibrium.[69,70] A combination of rat neuropharmacokinetic and micro-dialysis (medial prefrontal cortex) studies at identical doses of PHA-543613 determined a rat $C_{ECF}:C_{CSF}$ ratio of 0.42, suggesting a slight disequilibrium between these two compartments in favor of CSF.[70] Based on these inter-compartmental neuropharmacokinetic relationships in rats, the assumption of an equivalent rat-derived $C_{ECF}:C_{CSF}$ ratio in humans, and the clinically-determined steady-state human $C_{CSF}:C_{p,u}$ ratio, projections of human C_{ECF} (and hence RO) may be made from clinically determined total plasma concentrations (C_p) and human plasma free fraction.[47,68]

Using this approach, human α7 nAChR ROs were projected from total plasma concentrations after single (1, 5, 10, 20, and 60 mg) or multiple (6, 21, and 40 mg *bid* for 7 days) doses of PHA-543613 to healthy volunteers. Human ROs from SDs were calculated from mean total plasma concentration–time curves from 0.5 to 24 h post-dose (Figure 15.15), and ROs from MDs were estimated from day 1 and day 7 (steady-state) mean maximum ($C_{p,max}$) and minimum ($C_{p,min}$) total plasma concentrations (Figure 15.16).

When combined with the overall clinical datasets for PHA-543613, the projected human RO data for both SD and MD plasma concentrations may provide insight into (and prompt debate over) the most optimal clinical dose and dosing regimen for PHA-543613. From an efficacy perspective, the RO data suggest that a single 10-mg dose already attains a maximal RO of 88% and maintains an RO of 57% at 24 h post-dose, which would be expected for PHA-543613 owing to its long plasma half-life in humans ($\sim$10 h)[47] and assuming that humans demonstrate elimination rates from both ECF and plasma similar

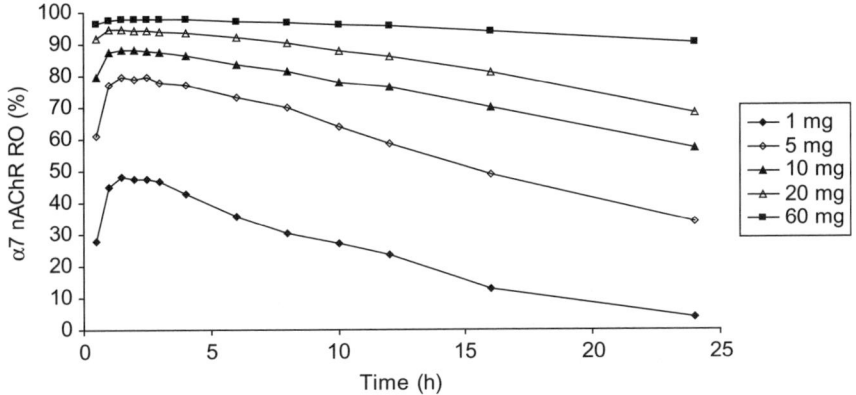

Figure 15.15 Mean projected α7 nAChR occupancies 0.5–24 h post-dose in humans after a single oral dose of PHA-543613.

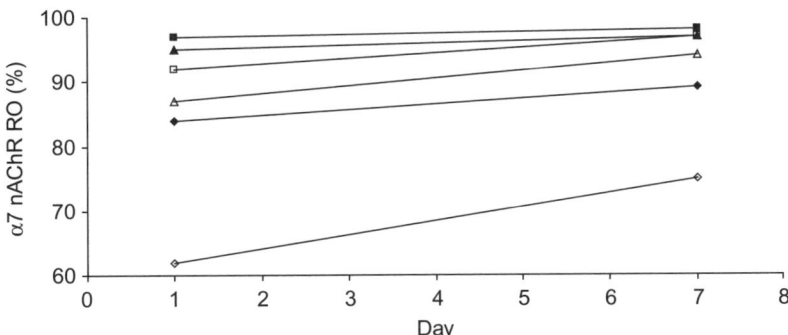

Figure 15.16 Projected α7 nAChR occupancies at maximum (C_{max}, filled symbols) and minimum plasma concentrations (C_{min}, unfilled symbols) of PHA-543613 on day 1 and day 7 (steady-state) in humans after a twice-daily oral dosing regimen of 6 (◇), 21 (▲), or 40 (■) mg.

to those in rats.[70] Furthermore, the SD data demonstrate a common observation for RO *versus* ligand concentration: the greatest changes in RO with the smallest changes in ligand concentration occur within the 30–70% RO range, whereas very small changes in RO occur with large changes in ligand concentration at RO <30% and >70%. In other words, increasing PHA-543613 $C_{p,max}$ 6-fold (from 35 to 224 ng mL^{-1} for 10 and 60 mg, respectively) would only increase maximal projected RO 1.1-fold (from 88% to 98%, respectively). This phenomenon is also seen in the MD data in which increasing the dose from 6 to 40 mg *bid* increases steady-state (*i.e.*, day 7) $C_{p,max}$ 7.5-fold and projected RO only 1.1-fold. Importantly, from the RO data derived from the three MDs tested, α7 nAChR occupancy is projected to have never fallen below 75% at steady-state total plasma trough ($C_{ss,min}$) for the low dose and never <94% at the middle or high dose (Figure 15.15).

These data have important implications from an efficacy perspective when considering the ability of neuronal ion channels to desensitize with excess activation. If high degrees of RO coupled with intrinsic agonist activity would lead to receptor desensitization, then efficacy, which in this case is agonist activity associated with improved cognition, could be lost at unnecessarily elevated ROs. This consideration is critical for cognition-enhancing targets that usually demonstrate hormetic dose–response curves.[71,72] Thus it is crucial to understand the RO range required for a desired effect from a ligand with a specific intrinsic agonism so that the optimal dose may be tested clinically. In the case of PHA-543613, the clinical doses evaluated in the MD study suggest from a RO standpoint that lower doses (*i.e.*, ≤6 mg *bid*) may have been most optimal for a cognition trial in patients with schizophrenia.

An additional influence for efficacy other than dose is dosing interval, which affects $C_{ss,max}$ and the steady-state total plasma trough ($C_{ss,min}$) concentrations and their temporal relationships. The implications for dosing interval are related to the amount of time required for the receptor to "reset" to its active form after a desensitizing exposure. If, for example, a desensitized receptor requires that agonist concentration fall below a specific threshold for a certain period of time to revert to its fully functional form, then the dosing interval of a chronic dosing regimen would be critical in ensuring optimized ligand–receptor interactions for testing agonist-mediated efficacy.

One last consideration building on these concepts is that if a specific RO <90% was required for efficacy and did not prompt receptor desensitization, then targeting such an RO would require lower ECF concentrations, which would directly translate to lower plasma concentrations. This detail is important in light of the deleterious cardiovascular events observed clinically with PHA-543613. Lower doses targeting lower ROs would require lower systemic concentrations, which might expand therapeutic indices for observed cardiovascular adverse events specifically and decrease total body load to compound generally. Taken together, these human RO estimations suggest that lower doses of PHA-543163 could have been evaluated clinically for optimizing efficacy (*i.e.*, not leading to receptor desensitization and hence overshooting efficacy) while expanding therapeutic indices for cardiovascular adverse events. Studies exploring these concepts, particularly dosing regimen, could be prepared more fully before further clinical evaluation of other α7 nAChR agonists to optimize clinical outcomes.

15.8   Conclusion

Our studies of α7 nAChR agonists clearly show the benefit of developing a deep and well-defined screening strategy. Starting with the right HTS assay to identify quality hits and pursuing only those that demonstrate desirable properties are keys for success. The application of parallel medicinal chemistry can rapidly enable an understanding of the SAR to drive decisions and directions for the program. Our willingness to tackle synthetically challenging azabicyclic amines and heterocycles was equally important in designing novel compounds with the right pharmacological and safety profiles. With the right pharmacologic tools in hand, we were able to identify efficacy and safety markers early in a program, which increased our understanding of the translational pharmacology into a clinical setting. Even with this success, however, surprises did occur with PHA-543613, and backup compounds were necessary to address those issues. Not every successful clinical entry results in the identification of a new medicine for the patients that our teams would like to help, but each attempt refines our techniques for the pursuit of safe molecules with which to test hypotheses.

References

1. *Neuronal Nicotinic Receptors: Pharmacology and Therapeutic Opportunities*, ed. S. P. Arneric and J. D. Brioni, Wiley-Liss, New York, 1999.
2. *Handbook of Experimental Pharmacology: Neuronal: Neuronal Nicotinic Receptors*, ed. F. Clementi, D. Fornasari and C. Gotti, Springer, Berlin, 2000.
3. For nAChR lead references, see: (a) A. Taly, P.-J. Corringer, D. Guedin, P. Lestage and J.-P. Changeux, *Nat. Rev. Drug Discovery*, 2009, **8**, 733; (b) M. R. Picciotto, B. J. Caldarone, S. L. King and V. Zachariou, *Neuropsychopharmacology*, 2000, **22**, 451; (c) J. A. Dani, *Biol. Psychiatry*, 2001, **49**, 166.
4. R. J. Lukas, J.-P. Changeux, N. Le Novère, E. X. Albuquerque, D. J. K. Balfour, D. K. Berg, D. Bertrand, V. A. Chiappinelli, P. B. S. Clarke, A. C. Collins, J. A. Dani, S. R. Grady, K. J. Kellar, J. M. Lindstrom, M. J. Marks, M. Quik, P. W. Taylor and S. Wonnacott, *Pharmacol. Rev.*, 1999, **51**, 397.
5. (a) J. A. Dani and D. Bertrand, *Annu. Rev. Pharmacol. Toxicol.*, 2007, **47**, 699; (b) C. Ulens, A. Akdemir, A. Jongejan, R. van Elk, S. Bertrand, A. Perrakis, R. Leurs, A. B. Smit, T. K. Sixma, D. Bertrand and I. J. P. de Esch, *J. Med. Chem.*, 2009, **52**, 2372.
6. S. Weiland, D. Bertrand and S. Leonard, *Behav. Brain Res.*, 2000, **113**, 43.
7. (a) P. Séguéla, J. Wadiche, K. Dineley-Miller, J. A. Dani and J. W. Patrick, *J. Neurosci.*, 1993, **13**, 596; (b) E. D. Levin and B. B. Simon, *Psychopharmacology*, 1998, **138**, 217.
8. L. E. Adler, A. Olincy, M. Waldo, J. G. Harris, J. Griffith, K. Stevens, K. Flach, H. Nagamoto, P. Bickford, S. Leonard and R. Freedman, *Schizophr. Bull.*, 1998, **24**, 189.
9. (a) L. E. Adler, L. D. Hoffer, A. Wiser and R. Freedman, *Am. J. Psychiatry*, 1993, **150**, 1856; (b) K. E. Stevens, W. R. Kem, V. M. Mahnir and R. Freedman, *Psychopharmacology*, 1998, **136**, 320.

10. A. Olincy, J. G. Harris, L. L. Johnson, V. Pender, S. Kongs, D. Allens-worth, J. Ellis, G. O. Zerbe, S. Leonard, K. E. Stevens, J. O. Stevens, L. Martin, L. E. Adler, F. Soti, W. R. Kem and R. Freedman, *Arch. Gen. Psychiatry*, 2006, **63**, 630.

11. H. T. Nagamoto, L. E. Adler, K. A. McRae, P. Huettl, E. Cawthra, G. Gerhardt, R. Hea and J. Griffith, *Neuropsychobiology*, 1999, **39**, 10.

12. For a recent review, see M. Hajos and B.N. Rogers, *Curr. Pharm. Des.*, 2010, **16**, 538.

13. For leading reviews, see: (a) A. P. Lightfoot, J. N. C. Kew and J. Skidmore, *Prog. Med. Chem.*, 2008, **46**, 131; (b) S. L. Cincotta, M. S. Yorek, T. M. Moschak, S. R. Lewis and J. S. Rodefer, *Curr. Opin. Investig. Drugs*, 2008, **9**, 47; (c) C. Chiamulera and G. Fumagalli, *CNS Agents Med. Chem.*, 2007, **7**, 269; (d) A. A. Jensen, B. Frolund, T. Liljefors and P. Krogsgaard-Larsen, *J. Med. Chem.*, 2005, **48**, 4705; (e) W. H. Bunnelle, M. J. Dart and M. R. Schrimpf, *Curr. Med. Chem.*, 2004, **4**, 299, and references therein.

14. (a) J. Eisele, S. Bertrand, J. Galzi, A. Devillers-Thiery, J. Changeux and D. Bertrand, *Nature*, 1993, **366**, 479; (b) V. E. Groppi, M. L. Wolfe and M. B. Berkenpas, *Pat. Appl.*, WO 2000/073431, 2000.

15. (a) D. Fancelli, C. Caccia, M. G. Fornaretto, R. McArthur, D. Severino, F. Vaghi and M. Varasi, *Bioorg. Med. Chem. Lett.*, 1996, **6**, 263; (b) for 5-HT3 reviews, see I. Israili and H. Zafar, Curr. Med. Chem.: CNS, 2001, 1, 171; (c) S. M. Walton, *Expert Opin. Pharmacother.*, 2000, **1**, 207.

16. J. E. Macor, D. Gurley, T. Lanthorn, J. Loch, R. A. Mack, G. Mullen, O. Tran, N. Wright and J. C. Gordon, *Bioorg. Med. Chem. Lett.*, 2001, **11**, 319.

17. A. L. Bodnar, L. A. Cortes-Burgos, K. K. Cook, D. M. Dinh, V. E. Groppi, M. Hajos, N. R. Higdon, W. E. Hoffmann, R. S. Hurst, J. K. Myers, B. N. Rogers, T. M. Wall, M. L. Wolfe and E. Wong, *J. Med. Chem.*, 2005, **48**, 905.

18. (a) J. M. Ward, V. B. Cockcroft, G. G. Lunt, F. S. Smillie and S. Won-nacott, *FEBS Lett.*, 1990, **270**, 45; (b) E. Palma, S. Bertrand, T. Binzoni and D. Bertrand, *J. Physiol.*, 1996, **15**, 151.

19. (a) J. P. Sullivan and A. W. Bannon, *CNS Drug Rev.*, 1996, **2**, 21; (b) J. D. Brioni, S. J. Morgan, A. B. O'Niell, T. M. Sykora, S. P. Postl, J. B. Pan, J. P. Sullivan and S. P. Arneric, *Med. Chem. Res.*, 1996, **487**.

20. D. G. Wishka, D. P. Walker, K. M. Yates, S. C. Reitz, S. Jia, J. K. Myers, K. L. Olson, E. J. Jacobsen, M. L. Wolfe, V. E. Groppi, A. J. Hanchar, B. A. Thornburgh, L. A. Cortes-Burgos, E. H. F. Wong, B. A. Staton, T. J. Raub, N. R. Higdon, T. M. Wall, R. S. Hurst, R. R. Walters, W. E. Hoffmann, M. Hajos, S. Franklin, G. Carey, L. H. Gold, K. K. Cook, S. B. Sands, S. X. Zhao, J. R. Soglia, A. S. Kalgutkar, S. P. Arneric and B. N. Rogers, *J. Med. Chem.*, 2006, **49**, 4425.

21. M. Hajós, *Trends Pharmacol. Sci.*, 2006, **27**, 391.

22. M. Krause, W. E. Hoffmann and M. Hajós, *Biol. Psychiatry*, 2003, **53**, 244.

23. M. Hajós, R. S. Hurst, W. E. Hoffmann, M. Krause, T. M. Wall, N. R. Higdon and V. E. Groppi, *J. Pharmacol. Exp. Ther.*, 2005, **312**, 1213.

24. A. Renaud, M. Langlois, R. J. Naylor and B. Naylor, *Pat. Appl.*, EP 311724, 1987.

25. W. J. Welstead, *US Pat.*, 4 605 652, 1986.

26. (a) D. G. Wishka, S. C. Reitz, D. W. Piotrowski and V. E. Groppi, *Pat. Appl.*, WO 2002/100857, 2002; (b) D. P. Walker, D. G. Wishka, J. W. Corbett, M. R. Rauckhorst, D. W. Piotrowski and V. E. Groppi, *Pat. Appl.*, WO 2002/100858; (c) D. P. Walker, D. W. Piotrowski, E. J. Jacobsen, B. A. Acker and V. E. Groppi, *Pat. Appl.*, WO 2003/070731; (d) B. N. Rogers, D. W. Piotrowski, D. P. Walker, E. J. Jacobsen, B. A. Acker, D. G. Wishka and V. E. Groppi, *Pat. Appl.*, WO 2003/070732, 2003.

27. D. P. Walker, D. G. Wishka, D. W. Piotrowski, S. Jia, S. C. Reitz, K. M. Yates, J. K. Myers, T. N. Vetman, B. J. Margolis, E. J. Jacobsen, B. A. Acker, V. E. Groppi, M. L. Wolfe, B. A. Thornburgh, P. M. Tinholt, L. A. Cortes-Burgos, R. R. Walters, M. R. Hester, E. P. Seest, L. A. Dolak, F. Han, B. A. Olson, L. Fitzgerald, B. A. Staton, T. J. Raub, M. Hajós, W. E. Hoffmann, K. S. Li, N. R. Higdon, T. M. Wall, R. S. Hurst, E. H. F. Wong and B. N. Rogers, *Bioorg. Med. Chem.*, 2006, **14**, 8219.

28. (a) N. M. Barnes and T. Sharp, *Neuropharmacology*, 1999, **38**, 1083; (b) A. V. Maricq, A. S. Peterson, A. J. Brake, R. M. Meyers and D. Julius, *Science*, 1991, **254**, 432; (c) D. A. Gurley and T. H. Lanthorn, *Neurosci. Lett.*, 1998, **247**, 107.

29. P. Sirota, T. Mosheva, H. Shabtay, N. Giladi and A. D. Korczyn, *Am. J. Psychiatry*, 2000, **157**, 287.

30. (a) L.-W. Fu and J. C. Longhurst, *J. Physiol.*, 2002, **544**, 897; (b) L. G. H. Bonagamba, C. Couche-Sévoz, A. N'Diaye, K. Louvet-Uygun, J.-C. Callera, B. H. Machado, M. Namon and R. Laguzzi, *Neuropharmacology*, 2000, **39**, 2336.

31. S. Viskin, *Lancet*, 1999, **354**, 1625.

32. J. S. Mitcheson, J. Chen, M. Lin, C. Culberson and M. C. Sanguinetti, *Proc. Natl. Acad. Sci. U. S. A.*, 2000, **97**, 12329.

33. Compounds screened in Chan Test, hERG Block Comparitor Screen.

34. C. L. Shaffer, M. Gunduz, B. A. Thornburgh and G. D. Fate, *Drug Metab. Dispos.*, 2006, **34**, 1615.

35. (a) B. J. Smith, A. C. Doran, S. McLean, F. D. Tingley, B. T. O'Neill and S. M. Kajiji, *J. Pharmacol. Exp. Ther.*, 2001, **298**, 1252; (b) E. Bakos, R. Evers, G. Szakacs, G. E. Tusnady, E. Welker, K. Szabo, M. De Haas, L. Van Deemter, P. Borst, A. Varadi and B. Sarkadi, *J. Biol. Chem.*, 1998, **273**, 32167.

36. P. Garberg, M. Ball, N. Borg, R. Cecchelli, L. Fenart, R. D. Hurst, T. Lindmark, A. Mabondzo, J. E. Nilsson, T. J. Raub, D. Stanimirovic, T. Terasaki, J.-O. Oeberg and T. Oesterberg, *Toxicol. In vitro*, 2005, **19**, 299.

37. Compound 11 was screened against a panel of 90 receptors, channels, and enzymes at Cerep and found to have no additional activities.

38. (a) K. E. Stevens and K. D. Wear, *Pharmacol. Biochem. Behav.*, 1997, **57**, 869; (b) K. E. Stevens, W. R. Kem, V. M. Mahnir and R. Freedman, *Psychopharmacology*, 1998, **136**, 320.

39. (a) A. Ennaceur and J. Delacour, *Behav. Brain Res.*, 1988, **31**, 47; (b) P. C. Moser, O. E. Bergis, S. Jegham, A. Lochead, E. Duconseille, J. Terranova, D. Caille, I. Berque-Bestel, F. Lezoualc'h, R. Fischmeister, A. Dumuis, J. Bockaert, P. George, P. Soubrié and P. Scatton, *J. Pharmacol. Exp. Ther.*, 2002, **302**, 731.

40. J. Prickaerts, A. Sik, F. J. van der Staay, J. de Vente and A. Blokland, *Psychopharmacology*, 2005, **177**, 381.

41. J. E. Spang, S. Bertrand, G. Westera, J. T. Patt, P. A. Schubiger and D. Bertrand, *Chem. Biol.*, 2000, **7**, 545.

42. (a) C. J. Swain, R. Baker, C. Kneen, J. Moseley, J. Saunders, E. M. Seward, G. Stevenson, M. Beer, J. Stanton and K. Watling, *J. Med. Chem.*, 1991, **34**, 140; (b) C. J. Swain, R. Baker, C. Kneen, R. Herbert, J. Moseley, J. Saunders, E. M. Seward, G. I. Stevenson, M. Beer, J. Stanton, K. Watling and R. G. Ball, *J. Med. Chem.*, 1992, **35**, 1019; (c) B. S. Orlek, F. E. Blaney, F. Brown, M. S. Clark, M. S. Hadley, J. Hatcher, G. J. Riley, H. E. Rosenberg, H. J. Wadsworth and P. Wyman, *J. Med. Chem.*, 1991, **34**, 2726.

43. D. P. Walker, B. A. Acker, E. J. Jacobsen and D. G. Wishka, *J. Heterocycl. Chem.*, 2008, **45**, 247.

44. B. A. Acker, E. J. Jacobsen, B. N. Rogers, D. G. Wishka, S. C. Reitz, D. W. Piotrowski, J. K. Myers, M. L. Wolfe, V. E. Groppi, B. A. Thornburgh, P. M. Tinholt, R. R. Walters, B. A. Olson, L. Fitzgerald, B. A. Staton, T. J. Raub, M. Krause, K. S. Li, W. E. Hoffman, M. Hojos, R. S. Hurst and D. P. Walker, *Bioorg. Med. Chem. Lett.*, 2008, **17**, 3611.

45. M. Hajós, M. Krause, W. E. Hoffmann, E. Hajós-Korcsok, J. H. Yu, D. D. Robinson, T. M. Wall, N. R. Higdon, K. A. Svensson, N. F. Nichols, R. R. Walters, V. E. Groppi, S. P. Arneric and R.S. Hurst, presented at the 37th Annual Meeting for the Society for Neuroscience, San Diego, 2007, abstr. 60.9/V17.2007.

46. C. Jamieson, E. M. Moir, Z. Rankovic and G. Wishart, *J. Med. Chem.*, 2006, **49**, 5029.

47. C. L. Shaffer, M. Gunduz, R. J. Scialis and A. F. Fang, *Drug Metab. Dispos.*, 2007, **35**, 1188.

48. A. S. Kalgutkar, I. Gardner, R. S. Obach, C. L. Shaffer, E. Callegari, K. R. Henne, A. E. Mutlib, D. K. Dalvie, J. S. Lee, Y. Nakai, J. P. O'Donnell, J. Boer and S. P. Harriman, *Curr. Drug Metab.*, 2005, **6**, 161.

49. J. R. Soglia, S. P. Harriman, S. Zhao, J. Barberia, M. J. Cole, J. G. Boyd and L. G. Contillo, *J. Pharm. Biomed. Anal.*, 2004, **36**, 105.

50. Assuming the brain protein binding is the same as the plasma protein binding.

51. R. S. Hurst, E. Hajós-Korcsok, R. Walters, D. Walker, B. N. Rogers, R. Gorczyca, T. M. Wall, N. R. Higdon, K. S. Lee, Z.-L. Mo, W. E. Hoffmann and M. Hajós, presented at the 37th Annual Meeting for the Society for Neuroscience, San Diego, 2007; Abstract 60.9.

52. N. L. Benowitz and S. G. Gourlay, *J. Am. Coll. Cardiol.*, 1997, **29**, 1422.

53. B. Fermini and A. A. Fossa, *Nat. Rev. Drug Discovery*, 2003, **2**, 439.

54. CogState Schizophrenia Battery, 195 Church St, New Haven, CT 06510, USA.

55. B. N. Rogers, *Emerging Frontiers in Basic Research and Clinical Sciences*, San Diego, 2007; Invited Lecture at Nicotinic Acetylcholine Receptors as Therapeutic Targets.

56. D. A. Price, J. Blagg, L. Jones, N. Greene and T. Wager, *Expert Opin. Drug Metab. Toxicol.*, 2009, **5**, 921.

57. P. R. McGuirk, M. R. Jefson, D. D. Mann, N. C. Elliott, P. Chang, E. P. Cisek, C. P. Cornell, T. D. Gootz, S. L. Haskell, M. S. Hindahl, L. J. LaFleur, M. J. Rosenfeld, T. R. Shryock, A. M. Silvia and F. H. Weber, *J. Med. Chem.*, 1992, **35**, 611.

58. Concurrently a group at Sanofi Aventis also discovered this series, as highlighted by SR-180711: (a) B. Biton, O. E. Bergis, F. Galli, A. Nedelec, A. W. Lochead, S. Jegham, D. Godet, C. Lanneau, R. Santamaria, F. Chesney, J. Léonardon, P. Granger, M. W. Debono, G. A. Bohme, F. Sgard, F. Besnard, D. Graham, A. Coste, A. Oblin, O. Curet, X. Vigé, C. Voltz, L. Rouquier, J. Souilhac, V. Santucci, C. Gueudet, D. Francon, R. Steinberg, G. Griebel, F. Oury-Donat, P. George, P. Avenet and B. Scatton, *Neuropsychopharmacology*, 2007, **32**, 1; (b) P. Pichat, O. E. Bergis, J.-P. Terranova, A. Urani, C. Duarte, V. Santucci, C. Gueudet, C. Volts, R. Steinberg, J. Stemmelin, F. Oury-Donat, P. Avenet, G. Griebel and B. Scatton, *Neuropsychopharmacology*, 2007, **32**, 17; (c) S. Barak, M. Arad, A. De Levie, M. D. Black, G. Griebel and I. Weiner, *Neuropsychopharmacology*, 2009, **34**, 1.

59. C. J. O'Donnell, L. Peng, B. T. O'Neill, E. P. Arnold, R. J. Mather, S. B. Sands, A. Shrikhande, L. A. Lebel, D. K. Spracklin and F. M. Nedza, *Bioorg. Med. Chem. Lett.*, 2009, **19**, 4747.

60. C. J. O'Donnell, B. N. Rogers, B. S. Bronk, D. K. Bryce, J. W. Coe, K. K. Cook, A. J. Duplantier, E. Evrard, M. Hajos, W. E. Hoffmann, R. S. Hurst, N. Maklad, R. J. Mather, S. McLean, F. M. Nedza, B. T. O'Neill, L. Peng, W. Qian, M. M. Rottas, S. B. Sands, A. W. Schmidt, A. V. Shrikhande, D. K. Spracklin, D. F. Wong, A. Zhang and L. Zhang, *J. Med. Chem.*, 2010, **53**, 1222.

61. See ref. 62 for details of the assays.

62. X. Hou, P. R. Verhoest, A. Villalobos and T. Wager, presented at the 238th National Meeting of the American Chemical Society, Washington, 2009, abstr. COMP-135.

63. B. Feng, J. B. Mills, R. E. Davidson, R. J. Mireles, J. S. Janiszewski, M. D. Troutman and S. M. de Morais, *Drug Metab. Dispos.*, 2008, **36**, 268.

64. A. Ennaceur and J. Delacour, *Behav. Brain Res.*, 1988, **31**, 47.

65. M. W. Quick and R. A. Lester, *J. Neurobiol.*, 2002, **53**, 457.

66. Y. Sun, R. Olson, M. Horning, N. Armstrong, M. Mayer and E. Gouaux, *Nature*, 2002, **417**, 245.

67. S. Grimwood and P. R. Hartig, *Pharmacol. Ther.*, 2009, **122**, 281.

68. C. L. Shaffer, *Annu. Rep. Med. Chem.*, 2010, **45**, in press.

69. M. F. Fromm, *Adv. Drug Deliv. Rev.*, 2003, **54**, 1295.
70. S. M. Osgood, S. L. Becker, R. Gorczyca, L. M. Buchholz, H. Rollema and C. L. Shaffer, presented at the 57th American Society for Mass Spectrometry Conference on Mass Spectrometry and Allied Topics, Philadelphia, 2009.
71. M. S. Lidow, G. V. Williams and P. S. Goldman-Rakic, *Trends Pharmacol. Sci.*, 1998, **19**, 136.
72. E. J. Calabrese, *Crit. Rev. Toxicol.*, 2008, **38**, 41.

Subject Index